Navier–Stokes Equations and Related Nonlinear Problems

Navier–Stokes Equations and Related Nonlinear Problems

Edited by

A. Sequeira
Technical University
Lisbon, Portugal

Plenum Press • New York and London

Library of Congress Cataloging-in-Publication Data

On file

Proceedings of the Third International Conference on Navier–Stokes Equations and Related Nonlinear Problems, held May 21–27, 1994, in Funchal, Madeira, Portugal

ISBN 0-306-45118-2

© 1995 Plenum Press, New York
A Division of Plenum Publishing Corporation
233 Spring Street, New York, N. Y. 10013

10 9 8 7 6 5 4 3 2 1

PREFACE

This volume contains the Proceedings of the *Third International Conference on Navier-Stokes Equations and Related Nonlinear Problems*. The conference was held in Funchal (Madeira, Portugal), on May 21-27, 1994. In addition to the editor, the organizers were Carlos Albuquerque (FC, University of Lisbon), Casimiro Silva (University of Madeira) and Juha Videman (IST, Technical University of Lisbon). This meeting, following two other successful events of similar type held in Thurnau (Germany) in 1992 and in Cento (Italy) in 1993, brought together, to the majestically beautiful island of Madeira, more than 60 specialists from all around the world, of which about two thirds were invited lecturers.

The main interest of the meeting was focused on the mathematical analysis of nonlinear phenomena in fluid mechanics. During the conference, we noticed that this area seems to provide, today more than ever, challenging and increasingly important problems motivating the research of both theoretical and numerical analysts.

This volume collects 32 articles selected from the invited lectures and contributed papers given during the conference. The main topics covered include: Flows in Unbounded Domains; Flows in Bounded Domains; Compressible Fluids; Free Boundary Problems; Non-Newtonian Fluids; Related Problems and Numerical Approximations. The contributions present original results or new surveys on recent developments, giving directions for future research. I express my gratitude to all the authors and I am glad to recognize the scientific level and the actual interest of the articles.

The success of the conference was largely due to the overwhelming generosity of several local institutions, which made the social program of the Conference unforgettable. On the other hand, the conference could not have been possible without the financial support of Universidade da Madeira, Secretaria Regional da Educação, Junta Nacional de Investigação Científica e Tecnológica, Fundação Luso-Americana para o Desenvolvimento, International Science Foundation and Human Capital and Mobility Program (Project ERB-CHRX-CT93-0407) of European Union. I am greatly indebted to all the sponsors and also to Fundação Calouste Gulbenkian for partially sponsoring this volume. Moreover, for the essential technical support I acknowledge Centro de Matemática e Aplicações Fundamentais (CMAF)/University of Lisbon.

I also wish to thank once more all the participants and my co-organizers for making this Conference a reality. I hope that the friendly and stimulating atmosphere experienced during the meeting, in the midst of the natural beauties of Madeira, will remain as a common memory to all of us.

My special thanks go to Professor Giovanni Paolo Galdi from the University of Ferrara (Italy) for his continuous cooperation and encouragement during the entire process of planning and organizing the conference and preparing this book.

Finally, I am deeply grateful to my dear collaborators, Carlos Albuquerque and Juha Videman, and also to Cristian Barbarosie from CMAF, for their invaluable assistance in turning this proceedings volume into its final form.

Adélia Sequeira
Instituto Superior Técnico
Departamento de Matemática
Av. Rovisco Pais, 1
1096 Lisboa Codex
Portugal

CONTENTS

Flows in Unbounded Domains

ON THE STABILITY OF A FLOW BETWEEN TWO VERTICAL PLANES CAUSED BY HOMOGENEOUS HEAT SOURCE AND PRESSURE GRADIENT

Francesco Mollica

Università di Ferrara, Istituto di Ingegneria
Via Scandiana, 21, 44100 Ferrara, Italy

ABSTRACT

The energy non-linear stability of a flow driven by homogeneous internal heat sources and pressure gradient is investigated. Critical Grashof numbers are plotted against Prandtl number for some values of the Reynolds number. It is found that stability decreases at large Prandtl numbers and at large Reynolds numbers. A comparison is made between the results of linearized stability (see Gershuni, Zhukhovitskii and Iakimov, 1973) and the ones of the present work in the case where Reynolds number is equal to zero. A more complete version of this work will appear elsewhere.

1. INTRODUCTION

As is known, fluid motions caused by internal heat generation play a central role in several questions of practical interest. In particular, the problem of their stability is of great relevance from both theoretical and experimental point of view (see *e.g.* Gershuni, Zhukhovitskii and Iakimov, 1970, 1973).

Recently, many significant contributions have been devoted to the study of stability of flow ($\mathcal{F}$, say) occurring between two parallel vertical planes in presence of a heat source uniformly distributed throughout the volume (cf. Gershuni, Zhukhovitskii and Iakimov, 1970, 1973). For this flow, the velocity and temperature are one-dimensional field polynomially related to the horizontal coordinate orthogonal to the planes, while the relevant non-dimensional parameters, individuating the stability region, are the Prandtl (Pr) and Grashof (Gr^2) numbers.

However, in all the previously mentioned works the authors restrict themselves to a linearized stability analysis which, of course, can only provide sufficient conditions for instability (see Chandrasekhar, 1961). Therefore, *a priori*, to our knowledge, there are not known conditions on Gr^2 and Pr under which $\mathcal{F}$ is stable. One way of pursuing this objective is to perform a nonlinear energy stability analysis of the problem. As is well-known (cf. Joseph, 1976), this method, due to Orr (1907) and, in its modern version, to Serrin (1959) and Joseph (1976), reveals itself very useful in the study of convective instability, see Joseph (1976); Galdi and Padula (1990).

The aim of this work is to give sufficient conditions for the nonlinear stability of the flow $\mathcal{F}'$ which is a generalization of $\mathcal{F}$ obtained by superimposing a constant pressure gradient in the vertical direction. This effect is measured via another non-dimensional parameter, the Reynolds number (Re). The conditions we find are expressed in terms of a relation between Gr, Pr and Re. In particular, the numerical analysis shows that, in the case of two-dimensional perturbations, the nonlinear stability bounds (for $Re = 0$) are not far from those obtained by the linearized theory in Gershuni, Zhukhovitskii and Iakimov (1973), thus restricting the region of possible subcritical instabilities.

The paper is organized as follows: in the next section we derive the basic flow, whose stability will be discussed in the rest of the paper; as costumary, the model which we use is the *Boussinesq Approximation* (see Joseph, 1976; Hills and Roberts, 1991) which is the most useful model for problems of this kind. We will assume, as prescribed quantities, the intensity of the internal heat source

$$Q = Q_0$$

and the flux of velocity

$$\int_{-h}^{h} v_0 dx = \varphi.$$

In the third section we prove the *energy equation* from the equations of perturbations. The energy equation is the starting point for energy stability method which furnishes *sufficient conditions of stability*. In the fourth section we obtain the best conditions of non-linear stability by solving a suitable variational problem. We will give also a sketch of the numerical scheme which has been used in the solution of the problem. Finally, in the fifth section we discuss and comment our results, and make a comparison with analogous results obtained from linearized stability in Gershuni, Zhukhovitskii and Iakimov (1970, 1973).

2. BASIC FLOW

In this section we state the problem and derive the *basic flow*, whose stability we shall investigate in subsequent sections. Assume we have a fluid between two parallel vertical planes whose distance is $2h$, perturbed by a homogeneous heat source whose intensity is Q and by a superimposed pressure gradient which makes the flux of velocity be equal to φ. Supposing that we are dealing with a newtonian incompressible fluid, we can use the Boussinesq approximation, whose equations are the following:

$$
\begin{cases}
\dfrac{\partial \mathbf{v}}{\partial t} + (\mathbf{v} \cdot \nabla)\mathbf{v} &= -\dfrac{1}{\rho}\nabla p + \nu \Delta \mathbf{v} + g\alpha T\mathbf{k} - g\mathbf{k} \\[2mm]
\dfrac{\partial T}{\partial t} + \mathbf{v} \cdot \nabla T &= \kappa \Delta T + \dfrac{Q}{\rho c_p} \\[2mm]
\nabla \cdot \mathbf{v} &= 0.
\end{cases}
\tag{2.1}
$$

The velocity and temperature vanish at the boundary:

$$\mathbf{v} = \mathbf{0} \qquad\qquad T = 0 \tag{2.2}$$

4

It is possible to prove that a solution for the above system is:

$$v_x = v_y = 0 \qquad v_z = v_0(x) \qquad p = p_0(z) \qquad T = T_0(x)$$

where

$$
\begin{cases}
v_0(x) &= \dfrac{\alpha g q h^4}{120\nu}\left[5\left(\dfrac{x}{h}\right)^4 - 6\left(\dfrac{x}{h}\right)^2 + 1\right] + \dfrac{3}{4}\dfrac{\varphi}{h}\left[1 - \left(\dfrac{x}{h}\right)^2\right] \\[3mm]
T_0(x) &= \dfrac{q h^2}{2}\left[1 - \left(\dfrac{x}{h}\right)^2\right] \\[3mm]
\dfrac{dp_0}{dz} &= \dfrac{2}{5}\rho g \alpha q h^2 - \dfrac{3}{2}\dfrac{\rho\nu\varphi}{h^3} - \rho g
\end{cases}
\tag{2.3}
$$

As we can see from the first equation of system (2.3), the velocity field is given by the sum of two terms: the first one is related to the heat source and has been found also in Gershuni, Zhukhovitskii and Iakimov (1970), while the other one is the consequence of the pressure gradient. Let us now adimensionalize the velocity and temperature field of system (2.3) by using the following units of length, modulus of velocity and temperature:

$$[L] = h \qquad\qquad [|\mathbf{v}|] = \frac{\nu}{h} \qquad\qquad [T] = \frac{q h^2}{2}$$

and the following adimensional parameters:

$$Gr^2 = \frac{\alpha g q h^5}{2\nu^2} \qquad\qquad\qquad Re = \frac{3\varphi}{4\nu}$$

where Gr^2 is the Grashof number and Re is the Reynolds number. The result is the following

$$
\begin{cases}
\mathbf{v_0} &= v_0(x)\mathbf{k} = Gr^2\,\mathbf{v_1}(x) + Re\,\mathbf{v_2}(x) \\[2mm]
T_0 &= T_0(x)
\end{cases}
\tag{2.4}
$$

where

$$\mathbf{v_1}(x) = \frac{1}{60}\left(5x^4 - 6x^2 + 1\right)\mathbf{k} \qquad\qquad \mathbf{v_2}(x) = (1 - x^2)\,\mathbf{k} \tag{2.5}$$

3. NON LINEAR STABILITY

In this section we obtain the *energy equation*. If $\mathbf{u}, \theta$ and π are the perturbations of velocity, temperature and pressure, to the basic flow (2.4), then they must satisfy the following system

$$
\begin{cases}
\dfrac{\partial \mathbf{u}}{\partial t} + (\mathbf{u}\cdot\nabla)\mathbf{u} + (\mathbf{u}\cdot\nabla)\mathbf{v_0} + (\mathbf{v_0}\cdot\nabla)\mathbf{u} &= -\dfrac{1}{\rho}\nabla\pi + \nu\Delta\mathbf{u} + \alpha g\theta\mathbf{k} \\[3mm]
\dfrac{\partial\theta}{\partial t} + \mathbf{u}\cdot\nabla\theta + \mathbf{u}\cdot\nabla T_0 + \mathbf{v_0}\cdot\nabla\theta &= \kappa\Delta\theta \\[3mm]
\nabla\cdot\mathbf{u} &= 0
\end{cases}
\tag{3.1}
$$

Let us adimensionalize the equations by using the units of the previous section and the following ones:

$$[t] = \frac{h^2}{\nu} \qquad\qquad [|\mathbf{u}|] = \sqrt{\frac{\alpha g q h^3}{2}} \qquad\qquad [p] = \rho\nu\sqrt{\frac{\alpha g q h}{2}}$$

5

respectively units of time, modulus of perturbation of velocity and of pressure:

$$\begin{cases} \dfrac{\partial \mathbf{u}}{\partial t} + Gr\,(\mathbf{u} \cdot \nabla)\mathbf{u} + (\mathbf{u} \cdot \nabla)\mathbf{v}_0 + (\mathbf{v}_0 \cdot \nabla)\mathbf{u} + \nabla \pi &= \Delta \mathbf{u} + Gr\,\theta \mathbf{k} \\[2mm] Pr\,\dfrac{\partial \theta}{\partial t} + Pr\,[Gr\,\mathbf{u} \cdot \nabla\theta + Gr\,\mathbf{u} \cdot \nabla T_0 + \mathbf{v}_0 \cdot \nabla\theta] &= \Delta\theta \\[2mm] \nabla \cdot \mathbf{u} &= 0 \end{cases} \qquad (3.2)$$

in which we have also used the usual Prandtl number: $Pr = \dfrac{\nu}{\kappa}$.

We introduce the following "energy" functional

$$E = \frac{1}{2}\int_\Omega \left[\mathbf{u}^2 + \lambda\,Pr\,\theta^2 \right] d\Omega \qquad\qquad \lambda \in (0, \infty) \qquad (3.3)$$

The energy equality can be easily obtained by multiplying scalarly by $\mathbf{u}$ the first equation of (3.2), multiplying by θ the second equation of (3.2) and integrating:

$$\begin{aligned} \frac{dE}{dt} &= -Gr^2 \int_\Omega \mathbf{u} \cdot \mathbf{D}_1 \cdot \mathbf{u}\,d\Omega - Re \int_\Omega \mathbf{u} \cdot \mathbf{D}_2 \cdot \mathbf{u}\,d\Omega - \int_\Omega \nabla\mathbf{u} : \nabla\mathbf{u}\,d\Omega \\[2mm] &\quad + Gr \int_\Omega \theta\mathbf{u} \cdot \mathbf{k}\,d\Omega - \lambda\,Gr\,Pr \int_\Omega \theta\mathbf{u} \cdot \nabla T_0\,d\Omega - \lambda \int_\Omega \nabla\theta \cdot \nabla\theta\,d\Omega \end{aligned} \qquad (3.4)$$

where

$$\mathbf{D}_i = \mathrm{sym}(\nabla\mathbf{v}_i) \quad i = 1, 2.$$

4. VARIATIONAL PROBLEM

As is known, the best conditions of non-linear stability can be derived by solving a suitable eigenvalue problem (cf. Joseph, 1976). To this end, we observe that the energy equality (3.4) can be written as follows:

$$\frac{dE}{dt} = -(\mathcal{I} + \mathcal{D}) \qquad (4.1)$$

if

$$\begin{cases} \mathcal{I} &= Gr^2 \int_\Omega \mathbf{u} \cdot \mathbf{D}_1 \cdot \mathbf{u}d\Omega + Re \int_\Omega \mathbf{u} \cdot \mathbf{D}_2 \cdot \mathbf{u}d\Omega \\[2mm] &\quad - Gr \int_\Omega \theta\mathbf{u} \cdot \mathbf{k}d\Omega + \lambda\,Gr\,Pr \int_\Omega \theta\mathbf{u} \cdot \nabla T_0 d\Omega \\[2mm] \mathcal{D} &= \int_\Omega \nabla\mathbf{u} : \nabla\mathbf{u}d\Omega + \lambda \int_\Omega \nabla\theta \cdot \nabla\theta d\Omega \end{cases}$$

Let us now consider the following functional

$$G(\mathbf{u}, \theta, \lambda) = \frac{\mathcal{I} + \mathcal{D}}{E}. \qquad (4.2)$$

It can be proved with the methods used in Galdi and Straughan (1985) that setting

$$\mathcal{S} = \{(\mathbf{u}, \theta) \ : \ \nabla \cdot \mathbf{u} = 0 \ ; \ \mathbf{u}|_{\partial\Omega} = \mathbf{0} \ ; \ \theta|_{\partial\Omega} = 0\},$$

there exists

$$m = \max_{\lambda \in [0, \infty)} \left[\min_{(\mathbf{u}, \theta) \in \mathcal{S}} G(\mathbf{u}, \theta, \lambda) \right]. \qquad (4.3)$$

From (4.1) and (4.2) and by using Gronwall's lemma we thus have

$$E \leq E_0 \, e^{-m\,t}.$$

Hence, if $m > 0$ the basic state (2.4) is asymptotically, inconditionally stable. Using standard methods (see Joseph, 1976), we show that m is the least eigenvalue μ of the following problem

$$
\begin{cases}
\nabla \pi + Gr^2 \, \mathbf{D}_1 \cdot \mathbf{u} + Re \, \mathbf{D}_2 \cdot \mathbf{u} + \\
\quad + \dfrac{\lambda}{2} Gr \, Pr \, \theta \nabla T_0 - \dfrac{1}{2} Gr \, \theta \mathbf{k} - \Delta \mathbf{u} = \mu \mathbf{u} \\
\dfrac{\lambda}{2} Gr \, Pr \, \mathbf{u} \cdot \nabla T_0 - \dfrac{1}{2} Gr \, \mathbf{u} \cdot \mathbf{k} - \lambda \Delta \theta = \mu \lambda \, Pr \, \theta \\
\nabla \cdot \mathbf{u} = 0 \\
\mathbf{u}(\mathbf{x})|_{x=\pm 1} = \mathbf{0} \\
\theta(\mathbf{x})|_{x=\pm 1} = 0
\end{cases}
\tag{4.4}
$$

We have considered for simplicity only two dimensional perturbations

$$
\mathbf{u} \cdot \mathbf{i} = -\frac{\partial \Psi}{\partial z} \qquad \mathbf{u} \cdot \mathbf{j} = 0 \qquad \mathbf{u} \cdot \mathbf{k} = \frac{\partial \Psi}{\partial x},
$$

where

$$\Psi = \Psi(x, z)$$

is the stream function. If we now suppose that all functions are periodic along the directions in which the domain is unbounded

$$
\begin{cases}
\Psi = \Phi(x)e^{ikz} \\
\theta = \Theta(x)e^{ikz} \\
\pi = q(x)e^{ikz},
\end{cases}
$$

system (4.4) becomes

$$
\begin{cases}
\Delta^2 \Phi + ik \left[Gr^2 \, v_1'(x) + Re \, v_2'(x) \right] \Phi' + \\
\quad + \dfrac{ik}{2} \left[Gr^2 \, v_1''(x) + Re \, v_2'' \right] \Phi + \\
\quad + \dfrac{1}{2} Gr \, \Theta' + \dfrac{\lambda}{2} ik \, Gr \, Pr \, T_0'(x)\Theta = -\mu \Delta \, \Phi \\
\lambda \Delta \Theta + \dfrac{1}{2} Gr \, \Phi' + \dfrac{\lambda}{2} ik \, Gr \, Pr \, T_0'(x)\Phi = -\mu \lambda \, Pr \, \Theta
\end{cases}
\tag{4.5}
$$

where

$$\Delta = \frac{d^2}{dx^2} - k^2$$

and with the following boundary conditions:

$$\Phi' = 0 \qquad \Phi = 0 \qquad \Theta = 0 \qquad \text{for} \quad x = \pm 1.$$

The system (4.5) has been discretized by using finite differences method which turns the differential eigenvalue problem into the following generalised algebrical eigenvalue problem

$$A\mathbf{w} = \mu B\mathbf{w}$$

which has been finally numerically solved by using an IMSL routine: the GVLCG.

5. RESULTS AND CONCLUSIONS

The results of the numerical analysis are shown in the following figures. Critical value of the Grashof number is chosen by finding the value which makes μ be close to zero and then by minimizing with respect to k and maximizing with respect to λ.

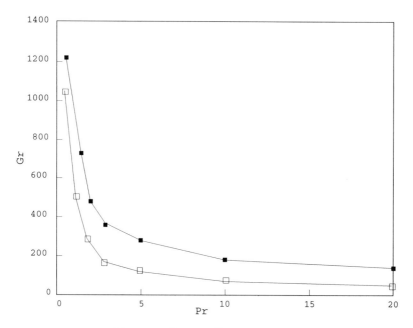

Figure 1. Curves $Gr^2 - Pr$ for $Re = 0$.

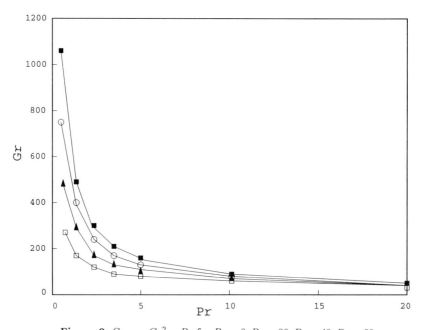

Figure 2. Curves $Gr^2 - Pr$ for $Re = 0, Re = 20, Re = 40, Re = 60$.

In Figure 1, the upper curve has been obtained in Gershuni, Zhukhovitskii and Iakimov (1973) in the case $Re = 0$ and by using linearized stability method; it implies that the region over this curve is a region of instability. The lower curve is the curve which we have obtained in the case $Re = 0$; this result implies that the region below this curve is a stability region. The region between the curves is unsure but, specially for low Prandtl numbers, is not large.

In Figure 2, it is shown the dependence of the critical Gr^2 on Pr for several Re. The upper curve is for $Re = 0$; it shows that the flow becomes more unstable by increasing the Reynolds number.

ACKNOWLEDGEMENTS

This work collects part of the results contained in the thesis of the author. He is grateful to his advisor Prof. Giovanni P. Galdi for his guidance and precious advice.

REFERENCES

CHANDRASEKHAR, S. (1961), *Hydrodynamic and Hydromagnetic Stability*, (Clarendon Press, Oxford).

GALDI, G.P. and M. PADULA (1990), A new approach to energy theory in the stability of fluid motions, *Arch. Rat. Mech. Anal.* **110**, 187.

GALDI, G.P. and B. STRAUGHAN (1985), Exchange of Stabilities, Simmetry and Non Linear Stability, *Arch. Rat. Mech. Anal.* **89**, 211.

GERSHUNI, G.Z., ZHUKHOVITSKII, E.M. and A.A. IAKIMOV (1970), On the stability of steady convective motion generated by internal heat sources, *Sov. J. Appl. Math. Mech.* **34**, 669.

GERSHUNI, G.Z., ZHUKHOVITSKII, E.M. and A.A. IAKIMOV (1973), Two kinds of instability of stationary convective motion generated by internal heat sources, *Sov. J. Appl. Math. Mech.* **37**, 564.

HILLS, R.N. and P.H. ROBERTS (1991), On the motion of a fluid that is incompressible in a generalized sense and its relationship to the Boussinesq approximation, *Stability Appl. Anal. Contin. Media* **3**, 205.

JOSEPH, D.D. (1976), *Stability of Fluid Motion* Vols. 1 and 2, Springer Tracts in Natural Philosophy (Springer, New York).

ORR, W. McF. (1907), The Stability or Instability of the Steady Motions of a Liquid – Part II : A Viscous Liquid, *Proc. Roy. Irish Acad.* (A) **27**, 69.

SERRIN, J. (1959), On the Stability of Viscous Fluid Motions, *Arch. Rational Mech. Anal.* **3**, 1.

STABILITY OF SOLUTIONS OF PARABOLIC EQUATIONS BY A COMBINATION OF THE SEMIGROUP THEORY AND THE ENERGY METHOD[1]

Jiří Neustupa

Czech Technical University,
Department of Technical Mathematics,
Karlovo nám. 13, 121 35 Praha 2, Czech Republic

ABSTRACT

We formulate sufficient conditions for asymptotic stability of the zero solution of a nonlinear parabolic differential equation in a Hilbert space by means of spectral properties and essential dissipativeness of certain linear operator. The spectrum of this operator may have a continuous part touching the imaginary axis.

1. INTRODUCTION

It is known that the question of stability of steady solutions of many types of parabolic equations can be transformed to the question of stability of the zero solution of the equation

$$\frac{du}{dt} + Lu = N(u) \tag{1.1}$$

where L is a linear operator and N is a nonlinear operator in a certain Hilbert space. There exist a lot of works containing sufficient conditions for stability of the zero solution of the equation of the type (1.1). Most of them use one of the following assumptions (with some exceptions, in explicit or implicit form): 1) There exists $\delta > 0$ so that $Re\,\lambda > \delta$ for all λ in the spectrum of L (see e.g. Kielhöfer, 1976; Prodi, 1962; Sattinger, 1970). 2) The operator $-L$ is dissipative or more generally, there exists an appropriate Lyapunov functional (see e.g. Henry, 1981; Maremonti, 1984; Galdi and Padula, 1990; Borchers and Miyakawa, 1992). (The works of Prodi, 1962; Sattinger, 1970; Maremonti, 1984 and Borchers and Miyakawa, 1992, deal with the special case of the equation (1.1)

---

[1]The research was supported by the National Committee for Research (CNR), Italy and the Grant Agency of Czech Republic (grant n. 201/93/2177).

which corresponds to the Navier-Stokes equations.) But in many cases none of these assumptions is satisfied – it is typical for example for parabolic equations in exterior domains when the operator L can have a continuous spectrum touching the imaginary axis and a symmetric part L_s of L can have some eigenvalues on the negative part of the real axis. Then the operator $-L$ is non–dissipative and under some conditions concerning the nonlinear operator N the zero solution of the equation

$$\frac{du}{dt} + L_s u = N(u)$$

is unstable. The stability of the zero solution of (1.1) can be guaranteed only by an appropriate influence of the skew–symmetric ($\equiv$ antisymmetric) part of L. This influence was studied by Galdi and Padula (1990) in the part dealing with so called weakly coupled systems. Our aim is to formulate sufficient conditions for stability which can be fulfilled also in the cases mentioned above and which do not involve the assumption that the equation (1.1) represents a weakly coupled system. The problem is solved on an abstract level at first and some examples are shown at the end of the paper.

2. FORMULATION OF THE PROBLEM AND BASIC ASSUMPTIONS

Let H be a real Hilbert space with a scalar product $(.,.)_0$ and an associated norm $\|.\|_0$. Assume that the operator L has the form $L = A + B_1 + B_2$, where A is a nonnegative selfadjoint operator in H. B_1, B_2 are linear operators with domains of definition $D(B_1)$, $D(B_2)$ containing $D(A)$. N is a nonlinear operator in H with the domain of definition $D(N)$ such that $D(A) \subset D(N)$. The operator N can generally depend on time, but in order not to complicate the notation, we prefer to write only N instead of $N(t)$.

Suppose that the space H can be expressed as a direct sum $H' \oplus H''$ so that if we denote by P', P'' the projections of H onto H', H'', associated with the decomposition $H = H' \oplus H''$, then the norms of P', P'' are bounded (let us denote by c_1 their upper bound), P', P'' map $D(A)$ into $D(A)$, $D(A)$ is dense in H' and in H'',

(i) H' and H'' are invariant under $A + B_1$,

(ii) $\exists \delta > 0$: $\operatorname{Re} \lambda > \delta$ for $\lambda \in \sigma((A + B_1 + P'B_2)|_{H'})$,

(iii) $((A + B_1)\phi'', \phi'')_0 \geq 0$ and $((A + B_1)\phi'', \phi'')_0 = 0 \iff \phi'' = 0$ for $\phi'' \in H''$.

If $\phi \in D(A)$ then we put

$$\|\phi\|_1 = \left[\|A^{1/2}P'\phi\|_0^2 + ((A + B_1)P''\phi, P''\phi)_0 \right]^{1/2},$$

$$\|\phi\|_2 = \left[\|AP'\phi\|_0^2 + \|(A + B_1)P''\phi\|_0^2 \right]^{1/2}.$$

Suppose further that

(iv) $\exists c_2 > 0$ $\exists r > 0$ $\forall \epsilon \in (0,1)$: $\|B_1\phi\|_0 \leq \epsilon\|\phi\|_2 + c_2\,\epsilon^{-r}\,\|\phi\|_0$ for $\phi \in D(A)$,

(v) $\exists c_3 > 0$ $\exists s > 0$ $\forall \epsilon \in (0,1)$: $\|B_2\phi\|_0 \leq \epsilon\|\phi\|_2 + c_3\,\epsilon^{-s}\,\|\phi\|_1$ for $\phi \in D(A)$.

It follows immediately from (iv) and (v) that

$$\|B_1\phi\|_0^2 \le 2\,\epsilon\,\|\phi\|_2^2 + 2\,c_2^2\,\epsilon^{-r}\,\|\phi\|_0^2, \tag{2.1}$$

$$\|B_2\phi\|_0^2 \le 2\,\epsilon\,\|\phi\|_2^2 + 2\,c_3^2\,\epsilon^{-s}\,\|\phi\|_1^2 \tag{2.2}$$

for $\phi \in D(A)$ and $\epsilon \in (0,1)$. It can be also derived from (iv) and (v) that if $\phi \in D(A)$ then

$$\|B_1\phi\|_0 \le c_4\,\|\phi\|_2^{\alpha_1}\,\|\phi\|_0^{1-\alpha_1}, \tag{2.3}$$

$$\|B_2\phi\|_0 \le c_5\,\|\phi\|_2^{\alpha_2}\,\|\phi\|_1^{1-\alpha_2}, \tag{2.4}$$

where $\alpha_1 = r/(r+1)$, $\alpha_2 = s/(s+1)$, $c_4 = (rc_2^2)^{1/(r+1)} + c_2^2\,(rc_2^2)^{-r/(r+1)}$, $c_5 = (sc_3^2)^{1/(s+1)} + c_3^2\,(sc_3^2)^{-s/(s+1)}$. If $\phi \in H'$ then (2.3) and (2.4) imply that

$$\|B_1\phi\|_0 \le c_4\,\|A\phi'\|_0^{\alpha_1}\,\|\phi'\|_0^{1-\alpha_1}, \tag{2.5}$$

$$\|B_2\phi\|_0 \le c_5\,\|A\phi'\|_0^{\alpha_2}\,\|A^{1/2}\phi'\|_0^{1-\alpha_2} \le c_5\,\|A\phi'\|_0^{(1+\alpha_2)/2}\,\|\phi'\|_0^{(1-\alpha_2)/2}. \tag{2.6}$$

It can be derived from the condition (iv) that the operator B_1 is A–bounded with an A–bound arbitrarily small. Hence if B_1 is symmetric then $A + B_1$ is selfadjoint (see Kato, 1966) and the decomposition $H = H' \oplus H''$ with the properties (i), (iii) is possible – we can choose a Borel set $M \subset [0, +\infty)$ and put $P'' = \int_M dE(\lambda)$, $P' = I - P''$, $H' = P'H$, $H'' = P''H$, where $E(\lambda)$ is the resolution of identity corresponding to the operator $A + B_1$.

Let us denote $L' = (A + B_1 + P'B_2)|_{H'}$. Since the operator A is selfadjoint and nonnegative in H, it is sectorial in H. Using (2.3) and (2.4), it can be shown (by means of the procedure analogous to that one used by Henry, 1981, p. 19) that the operator $A + B_1 + P'B_2$ is sectorial in H, too. Thus, the operator $L' \equiv (A + B_1 + P'B_2)|_{H'}$ is sectorial also in H' and $-L'$ is an infinitesimal generator of an analytic semigroup $e^{-L't}$ in H'. If $\kappa > 0$ then we can put

$$(L')^{-\kappa} = \frac{1}{\Gamma(\kappa)} \int_0^{+\infty} t^{\kappa-1}\,e^{-L't}\,dt$$

and by $(L')^\kappa$ we understand the inverse operator to $(L')^{-\kappa}$. $(L')^\kappa$ is a closed densely defined operator in H' which commutes with $e^{-L't}$ on $D((L')^\kappa)$ (see Henry, 1981, pp. 25, 26). If $\kappa \ge 0$ then there exists a constant $c_6(\kappa) > 0$ such that

$$\|(L')^\kappa\,e^{-L't}\phi\|_0 \le c_6(\kappa)\,t^{-\kappa}\,e^{-\delta t}\,\|\phi\|_0 \tag{2.7}$$

for $\phi \in H'$.

Let us define a bilinear form on $H' \times H'$:

$$h_\alpha[\phi, \psi] = \int_0^{+\infty} \left((L')^{\alpha/2}\,e^{-L't}\phi,\,(L')^{\alpha/2}\,e^{-L't}\psi\right)_0 dt.$$

It is clear that $h_\alpha[\phi, \phi] \ge 0$ for $\phi \in H'$. It follows from (2.7) that if $\alpha < 2$, $\phi \in H'$ and $\psi \in D((L')^{\alpha/2})$ then

$$h_\alpha[\phi, \psi] \le \int_0^{+\infty} \|(L')^{\alpha/2}\,e^{-L't}\phi\|_0 \cdot \|e^{-L't}(L')^{\alpha/2}\psi\|_0\,dt \le$$

$$\le \int_0^{+\infty} c_6(\alpha/2)\,c_6(0)\,t^{-\alpha/2}\,e^{-2\delta t}\,\|\phi\|_0 \cdot \|(L')^{\alpha/2}\psi\|_0\,dt =$$

$$= c_7(\alpha)\,\|\phi\|_0 \cdot \|(L')^{\alpha/2}\psi\|_0 \tag{2.8}$$

where $c_7(\alpha) = c_6(\alpha/2)\,c_6(0)\,\Gamma(1-\alpha/2)\,(2\delta)^{\alpha/2-1}$. Further, we have

$$h_\alpha[\,L'\phi,\phi\,] = \int_0^{+\infty} (L'\,\mathrm{e}^{-L't}\,(L')^{\alpha/2}\phi,\ \mathrm{e}^{-L't}\,(L')^{\alpha/2}\phi)_0\ \mathrm{d}t =$$

$$= -\tfrac{1}{2}\int_0^{+\infty}\frac{\mathrm{d}}{\mathrm{d}t}\left\|\mathrm{e}^{-L't}\,(L')^{\alpha/2}\phi\right\|_0^2\ \mathrm{d}t = \tfrac{1}{2}\left\|(L')^{\alpha/2}\phi\right\|_0^2 \qquad (2.9)$$

for $\phi \in D(L')$.

It can be derived from the estimates (2.5) and (2.6) that the operators $B_1 A^{-\kappa_1}$, $B_2 A^{-\kappa_2}$ are bounded in H' for $\kappa_1 \in (\alpha_1, 1\,]$ and $\kappa_2 \in ((1+\alpha_2)/2, 1\,]$ and this can be further used to show that the operator $A^{1/2}(L')^{-1/2}$ is bounded in H'. Namely, by standard estimates (see Henry, 1981, pp. 28, 29), it can be derived that if $\phi \in D(L')$ then

$$\|A^{1/2}\phi\|_0 \le c_8\,\|(L')^{1/2}\phi\|_0 \qquad (2.10)$$

where $c_8 = c_9 + c_{10}/\delta$ and c_9, c_{10} can be expressed by means of c_4, c_5, $c_6(0)$, $c_6(1)$, α_1 and α_2.

We shall use also the following assumption about the nonlinear operator N:

(vi) $\exists\,\beta \in [0,1]\ \exists\,\gamma \ge 2-\beta\ \exists\,c_{11} > 0\ :\ \|N(\phi)\|_0 \le c_{11}\,\|\phi\|_1^\gamma\,[\,\|\phi\|_2 + \|\phi\|_1\,]^\beta$
    for $\phi \in D(A)$.

(If N depends on time then the inequality in (vi) is supposed to be satisfied for a.a. $t \in [0,+\infty)$.)

Under solutions of the equation (1.1), we understand functions $u : [0,+\infty) \to H$ such that the norm $\|u\|_1$ is a right upper semicontinuous function of the variable t on $[0,+\infty)$ and u satisfies (1.1) for a.a. $t \in [0,+\infty)$. (It means that $u(t) \in D(A)$ and $\mathrm{d}u/\mathrm{d}t \in H$ for a.a. $t \in [0,+\infty)$.)

3. MAIN RESULT

Theorem 3.1. *If the conditions (i)–(vi) are satisfied then there exist $\xi > 0$, $\zeta > 0$ and $\eta > 0$ so that if*

(vii) $-(P''B_2\phi'',\phi')_0 - (P''B_2\phi',\phi'')_0 \le c_{12}\,\|\phi'\|_0^2 + c_{13}\,\|\phi'\|_2^2 + \tfrac{1}{2}\,\|\phi''\|_1^2 + c_{14}\,\|\phi''\|_2^2$
    for $\phi' \in H' \cap D(A)$, $\phi'' \in H'' \cap D(A)$, where $c_{12} < \tfrac{1}{8}\xi$, $c_{13} < \tfrac{1}{8}\eta$, $c_{14} < \tfrac{1}{8}\eta$

then the zero solution of the equation (1.1) is monotonically stable in the norm

$$\left[\xi\,h_0[\,P'u,P'u\,] + \zeta\,h_1[\,P'u,P'u\,] + \eta\,\|u\|_1^2 + \|P''u\|_0^2\right]^{1/2}$$

and asymptotically stable in the norm $\|u\|_1$. (The numbers ξ, ζ, η are expressed explicitly in (3.8).)

Proof: Each solution of (1.1) can be written in the form $u = u' + u''$ where $u' = P'u$, $u'' = P''u$. If we apply the projectors P' and P'' to (1.1) and use the mentioned form of u, we get

$$\frac{\mathrm{d}u'}{\mathrm{d}t} + Au' + B_1u' + P'B_2u' + P'B_2u'' = P'N(u'+u''), \qquad (3.1)$$

$$\frac{\mathrm{d}u''}{\mathrm{d}t} + Au'' + B_1u'' + P''B_2u'' + P''B_2u' = P''N(u'+u''). \qquad (3.2)$$

It follows from (2.2), (2.8), (2.9) and (3.1) that

$$\frac{1}{2}\frac{\mathrm{d}}{\mathrm{d}t}\,h_0[\,u',u'\,] = h_0[-L'u',u'] + h_0[-P'B_2u'',u'] + h_0[P'N(u'+u''),u'] \le$$

$$\le -\frac{1}{2}\|u'\|_0^2 + c_7(0)\,\|P'B_2u''\|_0\,\|u'\|_0 + c_7(0)\,\|P'N(u'+u'')\|_0\,\|u'\|_0 \le$$

$$\le -\frac{1}{4}\|u'\|_0^2 + 2\,c_1^2\,c_7^2(0)\,\|B_2u''\|_0^2 + 2\,c_1^2\,c_7^2(0)\,\|N(u'+u'')\|_0^2 \le$$

$$\le -\frac{1}{4}\|u'\|_0^2 + 4\,c_1^2\,c_7^2(0)\,\epsilon_1\,\|u''\|_2^2 + 4\,c_1^2\,c_3^2\,c_7^2(0)\,\epsilon_1^{-s}\,\|u''\|_1^2 +$$

$$+\, 2\,c_1^2\,c_7^2(0)\,\|N(u'+u'')\|_0^2\,,$$

$$\frac{\mathrm{d}}{\mathrm{d}t}\,h_0[\,u',u'\,] \le -\frac{1}{2}\|u'\|_0^2 + 8\,c_1^2\,c_7^2(0)\,\epsilon_1\,\|u''\|_2^2 + 8\,c_1^2\,c_3^2\,c_7^2(0)\,\epsilon_1^{-s}\,\|u''\|_1^2 +$$

$$+\, 4\,c_1^2\,c_7^2(0)\,\|N(u'+u'')\|_0^2\,. \tag{3.3}$$

Similarly, it follows from (2.2), (2.8), (2.9), (2.10) and (3.1) that

$$\frac{1}{2}\frac{\mathrm{d}}{\mathrm{d}t}\,h_1[\,u',u'\,] = h_1[-L'u',u'] + h_1[-P'B_2u'',u'] + h_1[P'N(u'+u''),u'] \le$$

$$\le -\frac{1}{2}\|(L')^{1/2}u'\|_0^2 + c_7(1)\,\|P'B_2u''\|_0\,\|(L')^{1/2}u'\|_0 +$$

$$+\, c_7(1)\,\|P'N(u'+u'')\|_0\,\|(L')^{1/2}u'\|_0 \le$$

$$\le -\frac{1}{4}\left(\|(L')^{1/2}u'\|_0^2 + 2\,c_1^2\,c_7^2(1)\,\|B_2u''\|_0^2 + 2\,c_1^2\,c_7^2(1)\,\|N(u'+u'')\|_0^2 \le$$

$$\le -\frac{1}{4}\left(\|(L')^{1/2}u'\|_0^2 + 4\,c_1^2\,c_7^2(1)\,\epsilon_2\,\|u''\|_2^2 + 4\,c_1^2\,c_3^2\,c_7^2(1)\,\epsilon_2^{-s}\,\|u''\|_1^2 +$$

$$+\, 2\,c_1^2\,c_7^2(1)\,\|N(u'+u'')\|_0^2\,,$$

$$\frac{\mathrm{d}}{\mathrm{d}t}\,h_1[\,u',u'\,] \le -\frac{1}{2\,c_8^2}\|u'\|_1^2 + 8\,c_1^2\,c_7^2(1)\,\epsilon_2\,\|u''\|_2^2 + 8\,c_1^2\,c_3^2\,c_7^2(1)\,\epsilon_2^{-s}\,\|u''\|_1^2 +$$

$$+\, 4\,c_1^2\,c_7^2(1)\,\|N(u'+u'')\|_0^2\,. \tag{3.4}$$

Multiplying the equation (3.1) by Au' and using the inequalities (2.1) and (2.2), we obtain

$$\left(\frac{\mathrm{d}u'}{\mathrm{d}t},Au'\right)_0 = \frac{1}{2}\frac{\mathrm{d}}{\mathrm{d}t}\,(u',Au')_0 = \frac{1}{2}\frac{\mathrm{d}}{\mathrm{d}t}\|u'\|_1^2 =$$

$$= -\|Au'\|_0^2 - (B_1u',Au')_0 - (P'B_2u',Au')_0 - (P'B_2u'',Au')_0 +$$

$$+\, (P'N(u'+u''),Au')_0 \le$$

$$\le -\frac{1}{2}\|u'\|_2^2 + 2\,\|B_1u'\|_0^2 + 2\,\|P'B_2u'\|_0^2 + 2\,\|P'B_2u''\|_0^2 + 2\,\|P'N(u'+u'')\|_0^2 \le$$

$$\le -\frac{1}{2}\|u'\|_2^2 + 4\,(\epsilon_3 + c_1^2\,\epsilon_4)\,\|u'\|_2^2 + 4\,c_2^2\,\epsilon_3^{-r}\,\|u'\|_0^2 + 4\,c_1^2\,c_3^2\,\epsilon_4^{-s}\,\|u'\|_1^2 +$$

$$+\, 4\,c_1^2\,\epsilon_5\,\|u''\|_2^2 + 4\,c_1^2\,c_3^2\,\epsilon_5^{-s}\,\|u''\|_1^2 + 2\,c_1^2\,\|N(u'+u'')\|_0^2\,,$$

$$\frac{\mathrm{d}}{\mathrm{d}t}\|u'\|_1^2 \le -\|u'\|_2^2 + 8\,(\epsilon_3 + c_1^2\,\epsilon_4)\,\|u'\|_2^2 + 8\,c_2^2\,\epsilon_3^{-r}\,\|u'\|_0^2 + 8\,c_1^2\,c_3^2\,\epsilon_4^{-s}\,\|u'\|_1^2 +$$

$$+\, 8\,c_1^2\,\epsilon_5\,\|u''\|_2^2 + 8\,c_1^2\,c_3^2\,\epsilon_5^{-s}\,\|u''\|_1^2 + 4\,c_1^2\,\|N(u'+u'')\|_0^2\,.$$

If we choose $\epsilon_3 = \frac{1}{32}$, $\epsilon_4 = 1/(32\,c_1^2)$, we have

$$\frac{\mathrm{d}}{\mathrm{d}t}\,\|u'\|_1^2 \;\leq\; -\tfrac{1}{2}\,\|u'\|_2^2 + 8\,c_1^2\,c_2^2\,(32)^r\,\|u'\|_0^2 + 8\,c_1^2\,c_3^2\,(32\,c_1^2)^s\,\|u'\|_1^2 +$$
$$+\; 8\,c_1^2\,\epsilon_5\,\|u''\|_2^2 + 8\,c_1^2\,c_3^2\,\epsilon_5^{-s}\,\|u''\|_1^2 + 4\,c_1^2\,\|N(u'+u'')\|_0^2\,. \qquad (3.5)$$

If we multiply the equation (3.2) by u'' and use the condition (vii), we obtain

$$\tfrac{1}{2}\frac{\mathrm{d}}{\mathrm{d}t}\,\|u''\|_0^2 \;=\; -\|u''\|_1^2 - (P''B_2 u'', u'')_0 - (P''B_2 u', u'')_0 + (P''N(u'+u''), u'')_0 \;\leq$$
$$\leq\; -\tfrac{1}{2}\,\|u''\|_1^2 + c_{12}\,\|u'\|_0^2 + c_{13}\,\|u'\|_2^2 + c_{14}\,\|u''\|_2^2 + \|P'N(u'+u'')\|_0\,\|u''\|_0\,,$$

$$\frac{\mathrm{d}}{\mathrm{d}t}\,\|u''\|_0^2 \;\leq\; -\|u''\|_1^2 + 2\,c_{12}\,\|u'\|_0^2 + 2\,c_{13}\,\|u'\|_2^2 + 2\,c_{14}\,\|u''\|_2^2 +$$
$$+\; 2\,c_1\,\|N(u'+u'')\|_0\,\|u''\|_0\,. \qquad (3.6)$$

Finally, multiplying the equation (3.2) by $(A+B_1)u''$ and using the estimate (2.3), we get

$$\left(\frac{\mathrm{d}u''}{\mathrm{d}t}, (A+B_1)u''\right)_0 \;=\; \tfrac{1}{2}\frac{\mathrm{d}}{\mathrm{d}t}\,(u'', (A+B_1)u'')_0 \;=\; \tfrac{1}{2}\frac{\mathrm{d}}{\mathrm{d}t}\,\|u''\|_1^2 \;=$$
$$=\; -\|u''\|_2^2 - (P''B_2 u'', (A+B_1)u'')_0 - (P''B_2 u', (A+B_1)u'')_0 +$$
$$+\; (P''N(u'+u''), (A+B_1)u'')_0 \;\leq$$
$$\leq\; -\tfrac{1}{2}\,\|u''\|_2^2 + \tfrac{3}{2}\,\|P''B_2 u''\|_0^2 + \tfrac{3}{2}\,\|P''B_2 u'\|_0^2 + \tfrac{3}{2}\,\|P''N(u'+u'')\|_0^2 \;\leq$$
$$\leq\; -\tfrac{1}{2}\,\|u''\|_2^2 + 3\,c_1^2\,\epsilon_6\,\|u''\|_2^2 + 3\,c_1^2\,c_3^2\,\epsilon_6^{-s}\,\|u''\|_1^2 + 3\,c_1^2\,\epsilon_7\,\|u'\|_2^2 +$$
$$+\; 3\,c_1^2\,c_2^2\,\epsilon_7^{-s}\,\|u'\|_1^2 + \tfrac{3}{2}\,c_1^2\,\|N(u'+u'')\|_0^2\,.$$

If we choose $\epsilon_6 = 1/(12\,c_1^2)$, we obtain

$$\frac{\mathrm{d}}{\mathrm{d}t}\,\|u''\|_1^2 \;\leq\; -\tfrac{1}{2}\,\|u''\|_2^2 + 6\,c_1^2\,c_3^2\,(12\,c_1^2)^s\,\|u''\|_1^2 + 6\,c_1^2\,\epsilon_7\,\|u'\|_2^2 +$$
$$+\; 6\,c_1^2\,c_2^2\,\epsilon_7^{-s}\,\|u'\|_1^2 + 3\,c_1^2\,\|N(u'+u'')\|_0^2\,. \qquad (3.7)$$

Let ξ, ζ and η be positive numbers. Put

$$E(u', u'') \;=\; \xi\,h_0[u', u'] + \zeta\,h_1[u', u'] + \eta\,\|u'\|_1^2 + \|u''\|_0^2 + \eta\,\|u''\|_1^2\,.$$

It follows from (3.3)–(3.7) that

$$\frac{\mathrm{d}}{\mathrm{d}t}\,E(u', u'') \;\leq\; [-\tfrac{1}{2}\xi + 8\,\eta\,c_1^2\,c_2^2\,(32)^r + 2\,c_{12}]\,\|u'\|_0^2 +$$
$$+\; [-\zeta/(2\,c_8^2) + 8\,\eta\,c_1^2\,c_3^2\,(32\,c_1^2)^s + 6\,\eta\,c_1^2\,c_2^2\,\epsilon_7^{-s}]\,\|u'\|_1^2 +$$
$$+\; [-\tfrac{1}{2}\eta + 2\,c_{13} + 6\,\eta\,c_1^2\,\epsilon_7]\,\|u'\|_2^2 +$$
$$+\; [8\,\xi\,c_1^2\,c_3^2\,c_7^2(0)\,\epsilon_1^{-s} + 8\,\zeta\,c_1^2\,c_3^2\,c_7^2(1)\,\epsilon_2^{-s} + 8\,\eta\,c_1^2\,c_3^2\,\epsilon_5^{-s} - 1 +$$
$$+\; 6\,\eta\,c_1^2\,c_3^2\,(12\,c_1^2)^s]\,\|u''\|_1^2 +$$
$$+\; [8\,\xi\,c_1^2\,c_7^2(0)\,\epsilon_1 + 8\,\zeta\,c_1^2\,c_7^2(1)\,\epsilon_2 + 8\,\eta\,c_1^2\epsilon_5 + 2\,c_{14} - \tfrac{1}{2}\eta]\,\|u''\|_2^2 +$$
$$+\; [4\,\xi\,c_1^2\,c_7^2(0) + 4\,\zeta\,c_1^2\,c_7^2(1) + 7\,\eta\,c_1^2]\,\|N(u'+u'')\|_0^2 + 2\,c_1\,\|N(u'+u'')\|_0\,\|u''\|_0\,.$$

Denote

$$c_{15} = 32\,c_1^2\,c_2^2\,(32)^r,$$

$$c_{16} = 4\,c_1^{2+2s}\,c_3^2\,c_8^2\,[\,8\,(32)^s + 6\,(24)^s\,].$$

Choose ϵ_1, ϵ_2, ϵ_5, ϵ_7 so that $8\,c_1^2\,c_7^2(0)\,c_{15}\,\epsilon_1 = \frac{1}{12}$, $8\,c_1^2\,c_7^2(1)\,c_{16}\,\epsilon_2 = \frac{1}{12}$, $8\,c_1^2\,\epsilon_5 = \frac{1}{12}$, $6\,c_1^2\,c_7 = \frac{1}{4}$. Put

$$c_{17} = 8\,c_1^2\,c_3^2\,[\,c_7^2(0)\,c_{15}\,\epsilon_1^{-s} + c_7^2(1)\,c_{16}\,\epsilon_2^{-s} + \epsilon_5^{-s} + \tfrac{3}{4}\,(12\,c_1^2)^s\,],$$

$$\eta = 1/(2\,c_{17})), \quad \xi = c_{15}\,\eta, \quad \zeta = c_{16}\,\eta \tag{3.8}$$

and $c_{18} = \frac{1}{4}\,\xi - 2\,c_{12}$, $c_{19} = \zeta/(4\,c_8^2)$, $c_{20} = \frac{1}{4}\,\eta - 2\,c_{13}$, $c_{21} = \frac{1}{4}\,\eta - 2\,c_{14}$,
$c_{22} = 4\,\xi\,c_1^2\,c_7^2(0) + 4\,\zeta\,c_1^2\,c_7^2(1) + 7\,\eta\,c_1^2$. Then we have

$$\frac{d}{dt}\,E(u',u'') \leq -c_{18}\,\|u'\|_0^2 - c_{19}\,\|u'\|_1^2 - c_{20}\,\|u'\|_2^2 - \tfrac{1}{2}\,\|u''\|_1^2 - c_{21}\,\|u''\|_2^2 +$$

$$+ c_{22}\,\|N(u'+u'')\|_0^2 + 2\,c_1\,\|N(u'+u'')\|_0\,\|u''\|_0. \tag{3.9}$$

It follows from the condition (vii) that $c_{18} > 0$, $c_{20} > 0$, $c_{21} > 0$. If we denote

$$\|(u',u'')\| = \left[\,c_{18}\,\|u'\|_0^2 + c_{19}\,\|u'\|_1^2 + c_{20}\,\|u'\|_2^2 + \tfrac{1}{2}\,\|u''\|_1^2 + c_{21}\,\|u''\|_2^2\,\right]^{1/2}$$

then we can obtain from the condition (vi) that there exist $c_{23} > 0$ and $c_{24} > 0$ so that

$$\|N(u'+u'')\|_0^2 \leq c_{23}\,E(u',u'')^{2(\beta+\gamma-1)}\,\|(u',u'')\|^2,$$

$$\|N(u'+u'')\|_0 \leq c_{24}\,E(u',u'')^{\beta+\gamma-2}\,\|(u',u'')\|^2.$$

Substituting this into (3.9) and taking into account that $\|u''\|_0 \leq E(u',u'')^{1/2}$, we obtain

$$\frac{d}{dt}\,E(u',u'') \leq \|(u',u'')\|^2\,\left[-1 + c_{22}\,c_{23}\,E(u',u'')^{2(\beta+\gamma-1)} + 2\,c_1\,c_{24}\,E(u',u'')^{\beta+\gamma-3/2}\right].$$

Thus, if

$$c_{22}\,c_{23}\,\left[\,E(u',u'')|_{t=0}\,\right]^{2(\beta+\gamma-1)} + 2\,c_1\,c_{24}\,\left[\,E(u',u'')|_{t=0}\,\right]^{\beta+\gamma-3/2} \leq 1$$

then

$$\frac{d}{dt}\,E(u',u'') \leq 0.$$

This implies the monotonic stability of the zero solution of the equation (1.1) with respect to the norm $E(u',u'')^{1/2}$.

If we have even

$$c_{22}\,c_{23}\,\left[\,E(u',u'')|_{t=0}\,\right]^{2(\beta+\gamma-1)} + 2\,c_1\,c_{24}\,\left[\,E(u',u'')|_{t=0}\,\right]^{\beta+\gamma-3/2} \leq \tfrac{1}{2}$$

then

$$\frac{d}{dt}\,E(u',u'') \leq -\tfrac{1}{2}\,\|(u',u'')\|^2,$$

$$\lim_{t\to+\infty} E(u',u'') + \tfrac{1}{2}\int_0^{+\infty} \|(u',u'')\|^2\,dt \leq E(u',u'')|_{t=0}.$$

Since $\|u\|_1^2 \leq const.\,\|\|(u',u'')\|\|^2$, $\|u\|_1^2$ is integrable on $(0,+\infty)$. We can show that $(d/dt)\|u\|_1^2$ is bounded from above a.e. on $(0,+\infty)$: Adding (3.5) and (3.7), we can obtain

$$\frac{d}{dt}\|u\|_1^2 \leq -\tfrac{1}{4}\|u\|_2^2 + c_{25}\,E(u',u'') + c_{26}\,E(u',u'')^\gamma\,[\,\|u\|_2 + E(u',u'')\,] \leq$$

$$\leq -\tfrac{1}{8}\|u\|_2^2 + c_{25}\,E(u',u'') + c_{26}\,E(u',u'')^{\gamma+1} + 2\,c_{26}^2\,E(u',u'')^{2\gamma}.$$

Taking into account also the right upper semicontinuity of $\|u(t)\|_1^2$ on $[0,+\infty)$, we have: $\lim_{t\to+\infty}\|u\|_1 = 0$. The theorem is proved.

4. SOME REMARKS AND EXAMPLES

Remark 4.1. Suppose that the operator B_1 is symmetric, the operator B_2 is skew–symmetric and the subspaces H' and H'' are orthogonal. (This is true for example if the projectors P' and P'' have the form $P'' = \int_M dE(\lambda)$ and $P' = I - P''$, where $E(\lambda)$ is the resolution of identity corresponding to the selfadjoint operator $A + B_1$ and M is a Borel set in $\mathbf{R}$.) Then the condition (vii) is satisfied if the operator B_2 is reduced in H' or in H'' or at least if it is "near enough" to such an operator.

Remark 4.2. In many cases the operator L has the form $L = A + B$ where A is selfadjoint and nonnegative and B is a linear perturbation of L with some convenient properties. The reason why we express B as the sum $B_1 + B_2$ and work with the operator L in the form $L \equiv A + B_1 + B_2$ is that we need to have the decomposition $H = H' \oplus H''$ such that except others as big part of L as possible is reduced in H' and in H''. Of course, since A is selfadjoint, H can be decomposed into a direct sum of subspaces which are invariant under A. But we can obtain better results if not only A, but "something more" is reduced in H' and in H''. That is why we assume that H', H'' can be chosen so that not A, but $A + B_1$ is reduced in H' and in H'' – as we have already mentioned, it is possible for example if B_1 is a symmetric part of B.

If there exists a decomposition $H = H' \oplus H''$ which is still more optimal in the sense that

a) the whole operator $L \equiv A + B$ is reduced in H' and in H'' and there exists $\delta > 0$ so that $Re\,\lambda > \delta$ for $\lambda \in \sigma(L|_{H'})$,

b) L is essentially dissipative in H'' (i.e. $(L\phi,\phi)_0 \geq 0$ for $\phi \in H''$ and 0 is not an eigenvalue of $L|_{H'}$)

then the stability of the zero solution of the equation (1.1) can be proved by means of some further conditions of the technical character (corresponding to the conditions (iv), (v), (vi)), but without the condition (vii)! (It is caused by the fact that we can put $B_1 = B$ and $B_2 = 0$.) It is a subject of a further investigation when such a decomposition of H is possible.

Remark 4.3. Suppose that the operator A does not have 0 as its eigenvalue and put $\|\phi\|_1^* = \|A^{1/2}\phi\|_0$, $\|\phi\|_2^* = \|A\phi\|_0$. Suppose that the condition (iii) is fulfilled in the form

(iii)* $\exists c_{27} > 0 :\ ((A + B_1)\phi'',\phi'')_0 \geq c_{27}\,\|\phi''\|_1^{*\,2}$ for $\phi'' \in H''$

and the conditions (iv)–(vi) are satisfied with the norms $\|\,.\,\|_1^*$ and $\|\,.\,\|_2^*$ instead of $\|\,.\,\|_1$ and $\|\,.\,\|_2$. Then the proof of Theorem 3.1 can be modified in such a way that we obtain the existence of positive numbers ξ, ζ and η such that if also the condition (vii) is satisfied with the norms $\|\,.\,\|_1^*$, $\|\,.\,\|_2^*$ instead of $\|\,.\,\|_1$, $\|\,.\,\|_2$ then the zero solution of the equation (1.1) is monotonically stable with respect to the norm

$$\left[\xi\,h_0[\,P'u,P'u\,] + \zeta\,h_1[\,P'u,P'u\,] + \eta\,\|u\|_1^{*\,2} + \|P''u\|_0^2\right]^{1/2}$$

and asymptotically stable with respect to the norm $\|u\|_1^*$.

Example 4.1. Suppose that $S_1,\ldots,S_m$ are bounded operators in H which commute with A in $D(A)$, $0 \le \alpha_1 < \ldots < \alpha_m < 1$, $T_1,\ldots,T_n$ are bounded operators in H which need not to commute with A in $D(A)$ and $\frac{1}{2} \le \beta_1 < \ldots < \beta_n < 1$. Put

$$B_1 = \sum_{i=1}^{m} S_i\,A^{\alpha_i}, \qquad B_2 = \sum_{i=1}^{n} T_i\,A^{\beta_i}.$$

Let M be a Borel set in $[0.+\infty)$ such that there exists $c_{28} > 0$ so that

$$\left([\,\lambda I + \sum_{i=1}^{m} S_i\,\lambda^{\alpha_i}\,]\phi,\,\phi\right)_0 \ge c_{28}\,\lambda\,\|\phi\|_0^2, \tag{4.1}$$

$$\left([\,\lambda I + \sum_{i=1}^{m} S_i\,\lambda^{\alpha_i}\,]\phi,\,[\,\lambda I + \sum_{i=1}^{m} S_i\,\lambda^{\alpha_i}\,]\phi\right)_0 \ge c_{29}\,\lambda^2\,\|\phi\|_0^2 \tag{4.2}$$

for $\lambda \in M$ and $\phi \in D(A)$. (The condition (4.2) follows from (4.1) with $c_{29} = c_{28}^2$ if $S_1,\ldots,S_m$ are symmetric.)

Let $E(\lambda)$ be the resolution of identity corresponding to the selfadjoint operator A. Put $P'' = \int_M \mathrm{d}E(\lambda)$, $P' = I - P''$, $H' = P'H$, $H'' = P''H$.

Since the operators $S_1,\ldots,S_m$ commute with A, they commute also with the resolution of identity corresponding to A and consequently, $A + B_1$ commutes with P' and P'' in $D(A)$. Hence the condition (i) is satisfied.

Using the inequality (4.1) and the expression of the operator $A + B_1$ in H'' by means of the resolution of identity $E(\lambda)$

$$(A + B_1)\phi'' = \int_M \left[\lambda I + \sum_{i=1}^{m} S_i\lambda^{\alpha_i}\right]\mathrm{d}E(\lambda)\phi'',$$

we can prove that

$$((A + B_1)\phi'',\phi'')_0 \ge c_{28}\,\|A^{1/2}\phi''\|_0^2.$$

Hence the condition (iii) is satisfied, too. Moreover, it can be derived from (4.2) that

$$((A + B_1)\phi'',(A + B_1)\phi'')_0 \ge c_{29}\,\|A\phi''\|_0^2 \tag{4.3}$$

for $\phi'' \in D(A) \cap H''$.

Further, we have

$$\|B_1\phi\|_0 \le \sum_{i=1}^{m} \left\|\int_M S_i A^{\alpha_i}\,\mathrm{d}E(\lambda)\phi\right\|_0 = \sum_{i=1}^{m} \left\|\int_M S_i\lambda^{\alpha_i}\,\mathrm{d}E(\lambda)\phi\right\|_0 \le$$

$$\le \sum_{i=1}^{m} \int_M c_{30}\,\lambda^{\alpha_i}\,\mathrm{d}\,\|E(\lambda)\phi\|_0 \le \sum_{i=1}^{m} \int_M c_{30}\,(\tau\lambda + C_i\,\tau^{-\alpha_i/(1-\alpha_i)})\,\mathrm{d}\,\|E(\lambda)\phi\|_0\,.$$

Let $\epsilon \in (0,1)$ be given. We can put $\tau = \epsilon/(c_{30}\, m)$ and $s = \alpha_m/(1 - \alpha_m)$. Then there exists $c_{31} > 0$ so that

$$\|B_1\phi\|_0 \ \leq \ \epsilon \int_M \lambda \, \mathrm{d}\, \|E(\lambda)\phi\|_0 + \sum_{i=1}^{m} c_{30}\, C_i \, (\epsilon/(c_{30}m))^{-\alpha_m/(1-\alpha_m)} \int_M \mathrm{d}\, \|E(\lambda)\phi\|_0 \ \leq$$

$$\leq \ \epsilon \|A\phi\|_0 + c_{31}\, \epsilon^{-s} \, \|\phi\|_0 \,.$$

If follows from this estimate and the inequality (4.3) that the condition (iv) is satisfied. Similarly, it can be shown that the condition (v) is satisfied, too.

Hence if N is a nonlinear operator satisfying the condition (vi) then the only conditions which remain and which must be assumed to apply Theorem 3.1 are the conditions (ii) and (vii). The condition (ii) is a natural one – the operator $[A + (B_1 + P'B_2)_s]|_{H'}$ (where $(B_1 + P'B_2)_s$ is the symmetric part of $B_1 + P'B_2$) may have some negative eigenvalues and the condition (ii) expresses the stabilizing influence of the skew–symmetric part of $B_1 + P'B_2$. The condition (vii) is satisfied if the operators T_i $(i = 1, \ldots, n)$ are skew–symmetric and commute with A or at least if they are "near enough" to such operators.

Example 4.2. Let Ω be an exterior of a bounded domain in $\mathbf{R}^2$ with a lipschitzian boundary and let $H = L^2(\Omega)^2$. Elements of H will be denoted by $[\phi_1, \phi_2]$. We study stability of the zero solution of the problem given by the system of equations

$$u_{i,t} - \Delta u_i + \sum_{j=1}^{2} a_{ij}(x)\, u_{i,j} + \sum_{j=1}^{2} b_{ij}(x)\, u_j \ = \ f_i(x, t, u_1, u_2, \nabla u_1, \nabla u_2) \quad (i = 1,2) \quad (4.4)$$

in Ω and the boundary conditions $u_1 = u_2 = 0$ on $\partial\Omega \times [0, +\infty)$. We suppose that $a_{ij} \in W^{1,\infty}(\Omega)$, $b_{ij} \in L^\infty(\Omega)$ for $i, j = 1, 2$ and

$$\int_\Omega |b_{12}(x)\,\phi|^2 \, \mathrm{d}x \ \leq \ c_{32} \int_\Omega |\nabla\phi|^2 \, \mathrm{d}x, \quad \int_\Omega |b_{21}(x)\,\phi|^2 \, \mathrm{d}x \ \leq \ c_{33} \int_\Omega |\nabla\phi|^2 \, \mathrm{d}x \quad (4.5)$$

for $\phi \in W_0^{1,2}(\Omega)$. The nonlinear functions f_1, f_2 are supposed to satisfy the estimates

$$\int_\Omega |f_i(x, t, \phi_1, \phi_2, \nabla\phi_1, \nabla\phi_2)|^2 \, \mathrm{d}x \ \leq$$

$$\leq \ c_{34} \Big[\int_\Omega \sum_{j=1}^{2} |\nabla\phi_j|^2 \, \mathrm{d}x \Big]^\gamma \cdot \Big[\int_\Omega \Big(\sum_{j=1}^{2} |\Delta\phi_j|^2 + \sum_{j=1}^{2} |\nabla\phi_j|^2 \Big) \, \mathrm{d}x \Big]^\beta \quad (4.6)$$

for some $\beta \in [0,1]$, $\gamma \geq 2 - \beta$, $c_{34} > 0$ and all $[\phi_1, \phi_2] \in W^{2,2}(\Omega)^2 \cap W_0^{1,2}(\Omega)^2$.

Let us further suppose that there exists $\delta > 0$ so that $Re\, \lambda > \delta$ for all λ from the spectrum of the operator

$$\mathcal{L}\phi \ \equiv \ -\Delta\phi + \sum_{j=1}^{2} a_{1j}(x)\, \phi_{,j} + b_{11}(x)\, \phi$$

in $L^2(\Omega)$ with the domain of definition $W^{2,2}(\Omega) \cap W_0^{1,2}(\Omega)$.

Finally, suppose that there exists $c_{35} > 0$ so that

$$(1 - c_{35}) \int_\Omega |\nabla\phi|^2 \, \mathrm{d}x + \int_\Omega \Big[-\tfrac{1}{2} \sum_{j=1}^{2} a_{2j,j}(x) + b_{22}(x) \Big] \phi^2 \, \mathrm{d}x \ \geq \ 0 \quad (4.7)$$

for $\phi \in W^{2,2}(\Omega) \cap W_0^{1,2}(\Omega)$.

Let us apply the general theory now. Put $D(A) = W^{2,2}(\Omega)^2 \cap W_0^{1,2}(\Omega)^2$,

$$A[\phi_1, \phi_2] = [-\triangle\phi_1, -\triangle\phi_2],$$

$$B_1[\phi_1, \phi_2] = \left[-\frac{1}{2}\sum_{j=1}^{2} a_{1j,j}(x)\,\phi_1 + b_{11}(x)\,\phi_1, \; -\frac{1}{2}\sum_{j=1}^{2} a_{2j,j}(x)\,\phi_2 + b_{22}(x)\phi_2 \right],$$

$$B_2[\phi_1, \phi_2] = \left[\frac{1}{2}\sum_{j=1}^{2}(a_{1j}(x)\,\phi_1)_{,j} + b_{12}(x)\,\phi_2, \; \frac{1}{2}\sum_{j=1}^{2}(a_{2j}(x)\,\phi_2)_{,j} + b_{21}(x)\,\phi_2 \right],$$

$$N[\phi_1, \phi_2] = [\,f_1(\,.\,,t,\phi_1,\phi_2,\nabla\phi_1,\nabla\phi_2), \; f_2(\,.\,,t,\phi_1,\phi_2,\nabla\phi_1,\nabla\phi_2)\,],$$

$$H' = \{\,[\phi_1,\phi_2] \in H; \; \phi_2 \equiv 0\,\}, \quad H'' = \{\,[\phi_1,\phi_2] \in H; \; \phi_1 \equiv 0\,\}.$$

It is seen that $A + B_1 : D(A) \cap H' \to H'$ and $A + B_1 : D(A) \cap H'' \to H''$ and so the condition (i) is satisfied. The assumption about the spectrum of the operator $\mathcal{L}$ implies that the condition (ii) is also satisfied. The validity of the condition (iii) follows from (4.7). The conditions (iv), (v) and (vi) follow from the regularity of the functions a_{ij}, b_{ij} $(i, j = 1, 2)$, (4.5) and (4.6). The term on the left hand side of the inequality in (vii) can be estimated:

$$-(P''B_2\phi'', \phi'')_0 - (P''B_2\phi_1, \phi_2)_0 = -\int_\Omega b_{21}(x)\,\phi_1\,\phi_2\,\mathrm{d}x \leq$$

$$\leq \epsilon \int_\Omega \phi_1^2\,\mathrm{d}x + \frac{1}{4\epsilon}\int_\Omega |b_{21}(x)\,\phi_2|^2\,\mathrm{d}x \leq \epsilon \int_\Omega \phi_1^2\,\mathrm{d}x + \frac{c_{33}}{4\epsilon}\int_\Omega |\nabla\phi_2|^2\,\mathrm{d}x. \quad (4.8)$$

(for $\phi' = [\phi_1, 0]$ and $\phi'' = [0, \phi_2]$). Applying Theorem 3.1, we can obtain the following result: There exist $\xi > 0$, $\zeta > 0$ and $\eta > 0$ so that if $c_{33} < \frac{1}{4}\xi$ then the zero solution of the system (4.4) is monotonically and asymptotically stable in the norms which are specified in Theorem 3.1. (We have used (4.8) with $\epsilon = c_{33}/2$.)

REFERENCES

BORCHERS, W. and MIYAKAWA, T. (1992), L^2–decay for Navier-Stokes flows in unbounded domains, with application to exterior stationary flows, *Arch. for Rat. Mech. and Anal.* **118**, 273–295.

GALDI, G.P. and PADULA, M. (1990), A new approach to energy theory in the stability of fluid motion, *Arch. for Rat. Mech. and Anal.* **110**, 187–286.

HENRY, D. (1981), Geometric theory of semilinear parabolic equations, *Lecture Notes in Mathematics* **840**, (Springer–Verlag, Berlin–Heidelberg–New York).

KATO, T. (1966), Perturbation theory for linear operators, (Springer–Verlag, Berlin–Heidelberg–New York).

KIELHÖFER, H. (1976), On the Lyapunov stability of stationary solutions of semilinear parabolic differential equations, *J. of Diff. Equations* **22**, 193–208.

MAREMONTI, P. (1984), Asymptotic stability theorems for viscous fluid motions in exterior domains, *Rend. Sem. Mat. Univ. Padova* **71**, 35–72.

PRODI, G. (1962), Teoremi di tipo locale per il sistema di Navier-Stokes e stabilità delle soluzioni stazionarie, *Rend. Sem. Mat. Univ. Padova* **32**, 374–397.

SATTINGER, D.H. (1970), The mathematical problem of hydrodynamic stability, *J. of Math. and Mech.* **19**, 797–817.

STEADY FLOW OF A VISCOUS INCOMPRESSIBLE FLUID IN AN UNBOUNDED "FUNNEL-SHAPED" DOMAIN

Arianna Passerini[1], M. Cristina Patria[2] and Gudrun Thäter[3]

[1] Università di Ferrara, Istituto di Ingegneria
Via Scandiana, 21, 44100 Ferrara, Italy

[2] Università di Ferrara, Dipartimento di Matematica
Via Machiavelli, 35, 44100 Ferrara, Italy

[3] University of Paderborn, Fachbereich Mathematik/Informatik
Warburger Straße 100, 33098 Paderborn, Germany

ABSTRACT

In this paper we consider a domain Ω which is tube-like at one exit to infinity and the halfspace at the other side. We prove the existence of steady motions for the Navier-Stokes problem and for the case in which the fluid is moving through a porous medium at rest filling Ω. In both cases the proof holds for arbitrary fluxes. We describe the asymptotic behaviour of the solutions in the halfspace for both the problems.

1. INTRODUCTION AND PROBLEM FORMULATION

In the following, the problem under consideration is to find a velocity-field $\mathbf{v}$ and a pressure p such that for $\mathbf{x}' \equiv (x_1, x_2), \quad \Sigma(l) \equiv \{(\mathbf{x}', l) \in I\!\!R^3 : |\mathbf{x}'| \leq R_0\}$

$$(PM) \quad \begin{cases} -\nu\Delta\mathbf{v} + \alpha^2\mathbf{v} + \mathbf{v}\cdot\nabla\mathbf{v} + \nabla p &= 0 \\ \nabla\cdot\mathbf{v} &= 0 \\ \mathbf{v}|_{\partial\Omega} &= 0 \\ \int_{\Sigma(0)} v_3 &= \Phi \end{cases}$$

in

$$\Omega \equiv \{(\mathbf{x}', x_3) \in I\!\!R^3 : |\mathbf{x}'| \leq R_0 \text{ or } x_3 \leq 0\}.$$

Hereby $\nu > 0$, α, R_0 and Φ are real constants denoting the kinematical viscosity, the

Navier-Stokes Equations and Related Nonlinear Problems
Edited by A. Sequeira, Plenum Press, New York, 1995

extra viscous force, the radius of the tube and the flux, $\partial\Omega$ is the boundary of the domain. We actually consider a regularization of $\partial\Omega$ in such a way that 'angles' are avoided. The equations (PM) describe the steady motion of a viscous incompressible fluid through a porous medium, occupying Ω, that is at rest. The term $\alpha^2 \mathbf{v}$ is an approximation for the viscous interaction between the fluid and the medium. For $\alpha = 0$ we get the usual steady Navier-Stokes equations and we denote this case with (NS).

In the background, there is the intention to use the results for considering the pollution transport with water through a sewer to the sea. The main question is, if the flux (as measurable parameter in the sewer) is given, what is predictable on the distribution of pollution in the sea, near the mouth, for different values of α and Φ. Therefore Ω consists of two main parts: the tube-like part for $x_3 > 0$ as a model for the sewer and the halfspace for $x_3 < 0$ as representative of the ocean. The mouth lies at $x_3 = 0$.

The problem (NS) in $I\!\!R_-^3$ was studied by Heywood (1980), Maremonti (1991), Chang (1992) and Coscia and Patria (1992). In particular in Coscia and Patria (1992) the optimal decay is proved, for zero boundary datum and external body force of compact support. The problem (PM) has already been studied in the halfspace $I\!\!R_-^3$ by Passerini (1995). In that paper existence and regularity of the solutions are proved and the asymptotic decay is estimated. The linear problem associated to (PM) is nothing else than the resolvent problem for the Stokes system studied in Fahrwig and Sohr (1993).

This paper is organized in the following way. In Section 2 we recall the properties of a flux carrier field $\mathbf{a}$, prescribed as in Galdi (1994), for the domain here considered. In Section 3 we show that the technique of Ladyzhenskaya and Solonnikov (1980), developed for infinite tubes, is useful for the tube-like part of Ω. All together enables us to prove existence of weak solutions to (PM) and (NS) in Ω for every flux Φ. On this subject, we recall that, for (NS), as well known, there exist no solutions with finite Dirichlet integral when Ω is the infinite cylinder.[1] Defining for any $l \geq 0$

$$\Omega_l \equiv \{\mathbf{x} \in \Omega : |x_3| \leq l\},$$

the estimate for the Dirichlet integral in the infinite cylinder gives for all $l \geq 0$

$$\int_{\Omega_l} |\nabla \mathbf{v}|^2 < c_1\, l + c_2,$$

where c_1, c_2 are positive constants, $c_1 = c(R_0, \Phi, \nu)$, $c_2 = c(R_0, \Phi, \nu)$.[2] This means, that the square of the Dirichlet norm grows linear with the x_3-direction. For details, see Ladyzhenskaya and Solonnikov (1980).

Section 4 is devoted to the decay of the solutions to both problems. Smallness assumptions on the flux are necessary for (NS). For (PM) the decay is as in Passerini (1995). For (NS) we follow the ideas of Coscia and Patria (1992).

In the present paper we omit most of the proofs, a detailed version will be published in a subsequent work.

---

[1] Dirichlet integral and Dirichlet norm are $(\int_\Omega |\nabla \mathbf{v}|^2)^{1/2} = \|\nabla \mathbf{v}\|_{2,\Omega}$.

[2] Here and in the sequel by c (c_1, c_2) we denote a generic constant whose possible dependence on parameters will be specified when necessary. If its numerical value is inessential for our aims, then it may have several different values.

2. DEFINITIONS AND PRELIMINARY RESULTS

The functions which appear can be scalar, vector or tensor valued functions according to the context. The following notation for function spaces will be used

$$\mathcal{D}(\Omega) \equiv \{\varphi \in C_0^\infty(\Omega) : \nabla \cdot \varphi = 0\},$$

$$H^1(\Omega) \equiv \overline{\mathcal{D}(\Omega)}^{\|\cdot\|_{1,2,\Omega}},$$

$$\mathcal{D}_0^{1,2}(\Omega) \equiv \overline{\mathcal{D}(\Omega)}^{\|\nabla\cdot\|_{2,\Omega}}.$$

The space $\mathcal{D}_0^{1,2}(\Omega)$ is useful for (NS) and $H^1(\Omega)$ for (PM). Here and in what follows $\|.\|_{q,\Omega}$ is the L_q-norm and $(.,.)_\Omega$ the L_2-scalar product in Ω. All functions φ that together with their first derivatives belong to $L^2(\Omega)$, belong to the space $W^{1,2}(\Omega)$ and $\|\varphi\|_{1,2,\Omega}^2 \equiv \int_\Omega(\varphi^2 + |\nabla\varphi|^2)$. The subspace of functions with zero traces on $\partial\Omega$ is denoted by $W_0^{1,2}(\Omega)$. When unnecessary we omit the subscript for the domain in the symbols of the norm and the scalar product.

We shall prove the existence of solutions to problems (PM) and (NS) of the kind $\mathbf{v}=\mathbf{u}+\mathbf{a}$, where $\mathbf{u}$ is searched as solution of

$$\begin{cases} -\nu\Delta\mathbf{u} + \alpha^2\mathbf{u} + \mathbf{u}\cdot\nabla\mathbf{u} + \mathbf{u}\cdot\nabla\mathbf{a} + \mathbf{a}\cdot\nabla\mathbf{u} + \nabla p &= \nu\Delta\mathbf{a} - \alpha^2\mathbf{a} - \mathbf{a}\cdot\nabla\mathbf{a}, \\[2mm] \nabla\cdot\mathbf{u} &= 0, \qquad \text{in } \Omega \\[2mm] \mathbf{u}|_{\partial\Omega} &= 0, \\[2mm] \int_{\Sigma(0)} u_3 &= 0 \end{cases} \tag{2.1}$$

and where $\mathbf{a} \in C^\infty(\Omega)$ is a prescribed flux carrier function. In particular, $\mathbf{a} = \Phi\,\nabla\times(\zeta\,\mathbf{h})$, $\Phi \in \mathbb{R}$, where

$$\mathbf{h} \equiv \frac{1}{2\pi|\mathbf{x}'|^2}(-x_2, x_1, 0)$$

and ζ can be defined as in Galdi (1994, IX.7, Lemma 7.1) and Ladyzhenskaya and Solonnikov (1980). As a consequence of its definition, the vector field $\mathbf{a}$ is such that as $|\mathbf{x}| \to \infty$, for $x_3 < 0$

$$|\mathbf{a}(\mathbf{x})| = O(|\mathbf{x}|^{-2}), \ |\nabla\mathbf{a}(\mathbf{x})| = O(|\mathbf{x}|^{-3}), \ |\Delta\mathbf{a}(\mathbf{x})| = O(|\mathbf{x}|^{-4}). \tag{2.2}$$

Moreover, for $x_3 > 0$ $\mathbf{a}$ is bounded and does not depend on x_3. Finally, the following properties hold

(i) $\nabla\cdot\mathbf{a} = 0$, $\quad \mathbf{a}|_{\partial\Omega} = 0$, $\quad \int_{\Sigma(0)} a_3 = \Phi$,

(ii) $\int_\Omega \mathbf{v}^2\mathbf{a}^2 < \varepsilon \cdot c \int_\Omega |\nabla\mathbf{v}|^2$, $\quad \forall\,\mathbf{v} \in \mathcal{D}(\Omega)$, $\quad c = c(R_0)$ and arbitrary small ε.

The proof of these properties is completely analogous to that given in Galdi (1994) for the halfspace and the aperture domain. In particular, in order to obtain **(ii)** in the pipe it is sufficient to apply the Poincaré inequality. Note that **(ii)** is true also when the domain of integration is Ω_l, $\forall l \geq 0$.

At this stage we are able to prove an existence result for the weak formulation of the problem (2.1) in Ω_l.

Lemma 2.1. *For fixed $l \in [0, +\infty)$ and $\mathbf{a}$ from Galdi (1994) there exists a solution $\mathbf{u}^l$ to the problem*

$$\nu(\nabla\mathbf{u}, \nabla\varphi)_{\Omega_l} + (\mathbf{u}\cdot\nabla\mathbf{u}, \varphi)_{\Omega_l} + (\mathbf{a}\cdot\nabla\mathbf{u}, \varphi)_{\Omega_l} + (\mathbf{u}\cdot\nabla\mathbf{a}, \varphi)_{\Omega_l} + \alpha^2(\mathbf{u}, \varphi)_{\Omega_l} =$$
$$= (-\nu\nabla\mathbf{a} + \mathbf{a}\otimes\mathbf{a}, \nabla\varphi)_{\Omega_l} - \alpha^2(\mathbf{a}, \varphi)_{\Omega_l}, \quad \forall\varphi \in \mathcal{D}(\Omega_l),$$
$$(\mathbf{u}, \nabla\Psi)_{\Omega_l} = 0 \quad \forall\Psi \in C_0^\infty(\Omega_l), \tag{2.3}$$

such that $\mathbf{u}^l \in \mathcal{D}_0^{1,2}(\Omega_l)$ if $\alpha = 0$, $\mathbf{u}^l \in H^1(\Omega_l)$ if $\alpha \neq 0$. Moreover

$$\|\nabla\mathbf{u}^l\|_{2,\Omega_l}^2 \leq c_1 l + c_2 \quad \text{if } \alpha = 0, \quad \|\mathbf{u}^l\|_{1,2,\Omega_l}^2 \leq c_1 l + c_2 \quad \text{if } \alpha \neq 0, \tag{2.4}$$

with $c_1 = c_1(R_0, \Phi, \nu)$, $c_2 = c_2(\Phi, \nu)$, independent of l.

Proof: In order to show the existence of $\mathbf{u}^l$ we apply the well-known Galerkin method (see Galdi 1994), taking into account that

$$\int_{\Omega_l}(|\nabla\mathbf{a}|^2 + \mathbf{a}^4 + \mathbf{a}^2) \leq d_1 l + d_2 < \infty, \tag{2.5}$$

where d_1, d_2 are constants depending on R_0 and Φ. Notice that, after estimating with the aid of (ii), one has $c_1 > d_1/\nu$ and $c_2 > d_2/\nu$.

Remark 2.1. By the Helmholtz-Weyl decomposition we get also the existence of a pressure field $p^l \in L^2(\Omega_l)$, $\forall l \geq 0$ (see Galdi 1994).

Remark 2.2. For any α, $\mathbf{u}^l \in L^6(\Omega_l)$ ($l \geq 0$). In fact $\mathbf{u}^l \in L^6(\mathbb{R}_-^3)$ (see Coscia and Patria, Lemma 5) and $\mathbf{u}^l \in L^6(\Omega_{0,l})$, where $\Omega_{k,l} \equiv \Omega_l\backslash\overline{\Omega_k}$ ($k \leq l$).

Remark 2.3. Since $\partial\Omega$ is regular, $\mathbf{u}^l$ and p^l are $C^\infty(K)$ for any K such that $\overline{K} \subset \Omega$, $\overline{K}$ compact.

Remark 2.4. If we look at (2.5) and recall that $\mathbf{v} = \mathbf{u} + \mathbf{a}$ it is clear that we cannot get, with the same procedure, a solution in the whole domain Ω having finite Dirichlet integral but at most

$$\int_{\Omega_l}|\nabla\mathbf{v}|^2 \leq c_1 l + c_2 \quad \forall l \geq 0.$$

Now, in order to prove the existence of a weak solution $\mathbf{v}$ in Ω we want to show that

$$\|\nabla\mathbf{u}^l\|_{2,\Omega_k}^2 \leq c_1 k + c_2 \quad \forall k \in [0, l], \quad \alpha = 0, \tag{2.6}$$
$$\|\mathbf{u}^l\|_{1,2,\Omega_k}^2 \leq c_1 k + c_2 \quad \forall k \in [0, l], \quad \alpha \neq 0.$$

By standard techniques we have for (2.1), $\alpha = 0$, the following 'energy' inequality.

Lemma 2.2. *Let $\mathbf{u}^l \in \mathcal{D}_0^{1,2}(\Omega_l)$ be a solution to (2.3) for $\alpha = 0$. Then $\forall k \in [0, l]$*

$$\nu\|\nabla\mathbf{u}^l\|_{2,\Omega_k}^2 \leq (\mathbf{u}^l\cdot\nabla\mathbf{u}^l, \mathbf{a})_{\Omega_k} + (\nabla\mathbf{u}^l, \mathbf{a}\otimes\mathbf{a} - \nu\nabla\mathbf{a})_{\Omega_k} + \tag{2.7}$$
$$+ \int_{\Sigma(k)}\left(\frac{\nu\partial}{\partial x_3}(\mathbf{u}^l + \mathbf{a})\cdot\mathbf{u}^l - p^l u_3^l - \frac{1}{2}(\mathbf{u}^l)^2(a_3 + u_3^l) - a_3(\mathbf{a}\cdot\mathbf{u}^l) - u_3^l(\mathbf{a}\cdot\mathbf{u}^l)\right).$$

Remark 2.5. We note that, for $k = l$, the surface integral in (2.7) vanishes.

Remark 2.6. If $\alpha \neq 0$ (i.e. problem (PM)), (2.7) must be substituted by

$$\nu \|\nabla \mathbf{u}^l\|_{2,\Omega_k}^2 + \alpha^2 \|\mathbf{u}^l\|_{2,\Omega_k} = (\mathbf{u}^l \cdot \nabla \mathbf{u}^l, \mathbf{a})_{\Omega_k} + (\mathbf{a} \otimes \mathbf{a} - \nu \nabla \mathbf{a}, \nabla \mathbf{u}^l)_{\Omega_k} - \alpha^2(\mathbf{a}, \mathbf{u}^l)_{\Omega_k} +$$

$$+ \int_{\Sigma(k)} \left(\frac{\nu \partial}{\partial x_3}(\mathbf{u}^l + \mathbf{a}) \cdot \mathbf{u}^l - p^l u_3^l - \frac{1}{2}(\mathbf{u}^l)^2(a_3 + u_3^l) - a_3(\mathbf{a} \cdot \mathbf{u}^l) - u_3^l(\mathbf{a} \cdot \mathbf{u}^l) \right). \quad (2.8)$$

At this stage, recall the following result, consequence of Lemma 2.3 in Ladyzhen-skaya and Solonnikov (1980)

Lemma 2.3. *Assume that the nonnegative smooth function $z(t)$, not identically equal to zero, satisfies on the segment $[1, T]$ the inequalities*

$$z(t) \leq A(z'(t) + z'(t)^{3/2}) + (Bt + C)(1 - \delta), \quad \delta \in (0, 1),$$
$$C \geq \frac{1}{\delta} A(B + B^{3/2}),$$

where A, B, C, δ are positive constants. In this case, if

$$z(T) \leq B \cdot T + C, \quad then$$
$$z(t) \leq B \cdot t + C \quad \forall t \in [1, T].$$

Finally we are able to prove the inequalities (2.6) and hence the existence theorem. To this we devote the next section.

3. EXISTENCE OF A WEAK SOLUTION

Let us begin by proving inequalities (2.6).

Lemma 3.1. *Let $\mathbf{u}^l$ be the solution to (2.3) in Ω_l, $\mathbf{u}^l \in \mathcal{D}_0^{1,2}(\Omega_l)$ if $\alpha = 0$, $\mathbf{u}^l \in H^1(\Omega_l)$ if $\alpha \neq 0$. Then, for all $k \in [0, l]$ we have*

$$\|\nabla \mathbf{u}^l\|_{2,\Omega_k}^2 \leq c_1 k + c_2' \quad if \quad \alpha = 0, \quad (3.1)$$
$$\|\mathbf{u}^l\|_{1,2,\Omega_k}^2 \leq c_1 k + c_2' \quad if \quad \alpha \neq 0,$$

where c_1, c_2' are constants independent of k and l.

Proof: The inequalities follow from (2.7), (2.8) and Lemma 2.3. For details, see the forthcoming paper.

Remark 3.1. Inequalities (3.1) still hold if $k > l$ by extending $\mathbf{u}^l$ with zero for $x_3 > l$.

Now we have all the tools to prove the main result of this section.

Theorem 3.1. *Problems (PM) and (NS) for Ω admit a weak solution $\mathbf{v}$ such that, for any $l \geq 0$*

$$\|\nabla \mathbf{v}\|_{2,\Omega_l}^2 \leq c_1 l + c_2' \quad if \quad \alpha = 0, \quad (3.2)$$
$$\|\mathbf{v}\|_{1,2,\Omega_l}^2 \leq c_1 l + c_2' \quad if \quad \alpha \neq 0,$$

where $c_1 = c_1(R_0, \nu, \Phi)$, $c_2' = c_2'(R_0, \nu, \Phi)$.

Proof: We take $\mathbf{v} = \mathbf{u} + \mathbf{a}$, with $\mathbf{a}$ as above. In order to construct $\mathbf{u}$ in whole Ω we consider a family of subdomains

$$\Omega_0 \subset \Omega_1 \subset \Omega_2 \subset \dots \quad \text{such that} \quad \bigcup_{i=0}^{\infty} \Omega_i = \Omega.$$

Let $k > 0$ and consider for all l the solution $\mathbf{u}^l$ to (2.3) in Ω_l. Lemma 3.1 and Remark 3.1 ensure that

$$\|\nabla \mathbf{u}^l\|_{2,\Omega_k}^2 \le c_1 k + c_2' \quad \text{if } \alpha = 0, \quad \|\mathbf{u}^l\|_{1,2,\Omega_k}^2 \le c_1 k + c_2' \quad \text{if } \alpha \ne 0.$$

So $\{\mathbf{u}^l\}_{l \in \mathbb{N}}$ is a sequence of vector fields whose gradient is uniformly bounded in $L^2(\Omega_k)$. Therefore there exists a subsequence $\{\mathbf{u}_k^{l_j}\}_{j \in \mathbb{N}}$ converging weakly in the norm of the gradient.

Then consider Ω_{k+1}. It holds $\|\nabla \mathbf{u}_k^{l_j}\|_{2,\Omega_{k+1}}^2 \le c_1(k+1) + c_2'$. We can extract a new subsequence $\{\mathbf{u}_{k+1}^{l_j}\}_{j \in \mathbb{N}}$ from the previous one such that it converges in Ω_{k+1}. Since $L^2(\Omega_k) \subset L^2(\Omega_{k+1})$, for the uniqueness of the limit, the limits of $\{\mathbf{u}_k^{l_j}\}_{j \in \mathbb{N}}$ and $\{\mathbf{u}_{k+1}^{l_j}\}_{j \in \mathbb{N}}$ coincide in Ω_k. And so on.

A classical procedure of diagonalization furnishes the final sequence whose limit $\mathbf{u}$ satisfies (2.1) and is such that (3.2) holds.

4. ASYMPTOTIC DECAY IN THE HALFSPACE

In this section we establish the asymptotic behaviour of the solution $\mathbf{v}$ as $|\mathbf{x}| \to +\infty$ for $x_3 < 0$. To this end, we will deal with restrictions to $\mathbb{R}_-^3$ of the solutions we found. The restrictions will be indicated with the same symbol as the solutions in Ω. We don't give any detailed proof, but anyway in both cases (NS) and (PM) representations are used.

4.1. Case $\alpha = 0$

As far as the Navier-Stokes problem is concerned, we must suitably modify the results of Coscia and Patria (1992), which are obtained in the halfspace, with zero boundary data and compact support body force. In fact, we will bring back our problem to that studied in Coscia and Patria (1992), but with different hypotheses on the force.

Let us consider the solution $\mathbf{u}(\mathbf{x})$ of problem (2.1), found in Theorem 3.1. Since $\int_{\Sigma(0)} u_3 = 0$ holds true, we can write

$$\mathbf{u} = \mathbf{u}_0 + \mathbf{W},$$

where the trace $\mathbf{u}_0^*$ of $\mathbf{u}_0$ on $\Sigma(0)$ is zero and the solenoidal vector field $\mathbf{W}(\mathbf{x})$ has compact support and zero flux through $\Sigma(0)$. $\mathbf{W}$ is actually the extension of the boundary data of $\mathbf{u}$ on $\Sigma(0)$. Let us see how to construct it.

Let $\mathbf{u}^*$ be the trace of $\mathbf{u}$ on $\Sigma(0)$ and $\Omega' \subset \mathbb{R}_-^3$ a bounded regular domain, such that $\Sigma(0) \subset \partial\Omega' \cap \partial\mathbb{R}_-^3$. Then, we can choose $\mathbf{W}$ as the solution of the auxiliary Stokes problem

$$\begin{cases} -\nu\Delta\mathbf{W} + \nabla\tau = 0, \quad \nabla\cdot\mathbf{W} = 0 \quad \text{in } \Omega', \\ \mathbf{W}|_{\Sigma(0)} = \mathbf{u}^*, \\ \mathbf{W}|_{\partial\Omega'\backslash\Sigma(0)} = 0. \end{cases}$$

The solution $(\mathbf{W}, \tau)$ verifies, as well known

$$\|\nabla \mathbf{W}\|_{2,\Omega'} + \|\tau\|_{2,\Omega'} \leq c\|\mathbf{u}^*\|_{1/2,2,\partial\Omega'}, \tag{4.1}$$

where the norm on the right hand side is the usual one for trace spaces (see Galdi 1994, Chapter 2, Section 3). Moreover, it is well-known that, since the data on the boundary are $C^\infty(\partial\Omega')$, the solution $(\mathbf{W}, \tau)$ belongs to $C^\infty(\Omega')$. Clearly, $\mathbf{W}$ can be extended to be zero out of Ω' in $\mathbb{R}^3$. Again, the function in $C_0^\infty(\overline{\mathbb{R}^3_-})$ will be indicated with $\mathbf{W}$.

Recalling that $\mathbf{v} = \mathbf{u} + \mathbf{a}$, we get

$$\mathbf{v} = \mathbf{u}_0 + \mathbf{b}$$

with

$$\mathbf{b} = \mathbf{a} + \mathbf{W}.$$

If $x_3 < 0$, and for $|\mathbf{x}|$ sufficiently large, $\mathbf{b}$ behaves as $|\mathbf{x}|^{-2}$, because this is the decay of $\mathbf{a}$ and $\mathbf{W}$ has compact support. Thus, we have to check the asymptotic behaviour of $\mathbf{u}_0$ only. Setting

$$G = \nu\nabla\mathbf{b} - \mathbf{b} \otimes \mathbf{b} - \mathbf{u}_0 \otimes \mathbf{b} - \mathbf{b} \otimes \mathbf{u}_0, \tag{4.2}$$

$$F = \nabla \cdot G,$$

we see that $\mathbf{u}_0 \in \mathcal{D}_0^{1,2}(\mathbb{R}^3_-)$ verifies

$$-\nu\Delta\mathbf{u}_0 + \mathbf{u}_0 \cdot \nabla\mathbf{u}_0 + \nabla p_0 = F, \quad \nabla \cdot \mathbf{u}_0 = 0 \quad \text{in } \mathbb{R}^3_-, \tag{4.3}$$

$$\mathbf{u}_0|_{\partial\mathbb{R}^3_-} = 0.$$

In order to extend the results of Coscia and Patria (1992), we deduce from (4.2) the properties of F. Let us start by defining for $q \in (1, \infty)$

$$D_0^{1,q}(\mathbb{R}^3_-) \equiv \overline{C_0^\infty(\mathbb{R}^3_-)}^{\|\nabla \cdot\|_{q,\mathbb{R}^3_-}}.$$

Denoting the Hölder conjugate[3] of q with q', $D_0^{-1,q'}(\mathbb{R}^3_-)$ is defined as the dual space of $D_0^{1,q}(\mathbb{R}^3_-)$. As known, the class of functionals on $D_0^{1,q}(\mathbb{R}^3_-)$ of the form

$$\mathcal{F}(\varphi) = \int_{\mathbb{R}^3_-} f\varphi, \quad f \in C_0^\infty(\mathbb{R}^3_-)$$

is dense in $D_0^{-1,q'}(\mathbb{R}^3_-)$ with respect to the norm

$$|f|_{-1,q',\mathbb{R}^3_-} \equiv \sup_{\|\nabla\varphi\|_{q,\mathbb{R}^3_-}=1} \left| \int_{\mathbb{R}^3_-} f\varphi \right|.$$

Actually, the completion of this class of functionals in the negative norm is isomorphic and isometric to $D_0^{-1,q'}(\mathbb{R}^3_-)$ (see Galdi 1994). Once these definitions are given, since $\mathbf{W}$ belongs to $C_0^\infty(\overline{\mathbb{R}^3_-})$, by recalling (2.2) and Remark 2.2 and 2.3, it is easy to see that for all $q \in (6/5, 6)$ and $r \in [1, 2]$

$$G \in L^q(\mathbb{R}^3_-),$$

$$F \in D_0^{-1,q}(\mathbb{R}^3_-) \cap L^r(\mathbb{R}^3_-). \tag{4.4}$$

Indeed, F and G are smooth in any compact domain $K \subset \mathbb{R}^3_-$.

---

[3]The Hölder conjugate is defined as $\frac{1}{q} + \frac{1}{q'} = 1$.

Remark 4.1. When the flux Φ is zero, we have $\mathbf{b} = \mathbf{W}$ in (4.2). Thus, G and F have compact support and we can immediately infer from Coscia and Patria (1992) the expected decay $O(|\mathbf{x}|^{-2})$.

The most general result on the decay of solutions of problem (4.3) should be obtained by taking F in weighted spaces (see Borchers and Pileckas 1992). On the other hand, our force F is dependent on the solution $\mathbf{u}_0$. Thus, instead of giving general statements, we deal with F as in definition (4.2).

We call *D-solution* any solution having finite Dirichlet integral; since $\nabla \mathbf{u}_0 \in L^2(I\!\!R^3_-)$, $\mathbf{u}_0$ is a D-solution. Extending the results of Coscia and Patria (1992) we find that, if F verifies certain smallness assumptions, the D-solution is unique and satisfies

$$p_0, \ \nabla \mathbf{u}_0 \ \in \ L^q(I\!\!R^3_-) \quad \forall \, q \in (6/5, 3],$$

$$\mathbf{u}_0 \ \in \ L^q(I\!\!R^3_-) \quad \forall \, q \in (2, \infty).$$

All the results on the decay are summarized in the following.

Theorem 4.1. *Let F be defined as in (4.2). There exists a positive constant c, such that if*

$$|F|_{-1,3/2} \le c, \tag{4.5}$$

then the D-solution $\mathbf{u}_0$ of the problem (4.3) verifies, if $\Phi \ne 0$, for any $\varepsilon > 0$,

$$|\mathbf{u}_0(\mathbf{x})| = O(|\mathbf{x}|^{-2+\varepsilon}) \quad \text{as } |\mathbf{x}| \to +\infty.$$

If $\Phi = 0$ then it holds

$$|\mathbf{u}_0(\mathbf{x})| = O(|\mathbf{x}|^{-2}) \quad \text{as } |\mathbf{x}| \to +\infty.$$

Remark 4.2. The smallness assumption (4.5) to F can be reformulated in terms of the flux Φ as follows

$$|\Phi| < c.$$

4.2. Case $\alpha \ne 0$

Let us consider directly problem (PM) in $I\!\!R^3$, with compact support data on the boundary and zero body force. The uniqueness of D-solutions for this problem is trivial, because of the summability properties (given in Passerini, 1995)

$$\mathbf{v} \in \ W^{2,q}(I\!\!R^3_-) \quad \forall \, q \in (3/2, +\infty), \quad \text{if } \Phi \ne 0,$$

$$\mathbf{v} \in \ W^{2,q}(I\!\!R^3_-) \quad \forall \, q \in (1, +\infty), \quad \text{if } \Phi = 0.$$

Thus, we have only to summarize the results of Passerini (1995), which hold true without any restriction on $|\Phi|$.

Theorem 4.2. *Let $\mathbf{v}(\mathbf{x})$ be the D-solution of problem (PM). Then everywhere in $I\!\!R^3_-$, with exception of the exterior of a cone of arbitrary aperture, whose axis is x_3 and whose vertex is the origin,*

$$|\mathbf{v}(\mathbf{x})| = O(|\mathbf{x}|^{-1+\varepsilon}) \quad \forall \, \varepsilon > 0, \quad \text{if } \Phi \ne 0,$$

$$|\mathbf{v}(\mathbf{x})| = O(|\mathbf{x}|^{-3+\varepsilon}) \quad \forall \, \varepsilon > 0, \quad \text{if } \Phi = 0,$$

as $|\mathbf{x}| \to +\infty$.

ACKNOWLEDGEMENTS

The authors thank Professor G.P. Galdi and Professor M.R. Padula for helpful discussions. This work was supported by the Italian M.U.R.S.T. (60% contract at the University of Ferrara) and the German DFG (50% contract at the University of Paderborn).

REFERENCES

BORCHERS, W. and K. PILECKAS (1992), Existence, uniqueness and asymptotics of steady jets, *Arch. Rational Mech. Anal.* **120**, 1-49.

CHANG, H. (1992), Speedy jets into a halfspace governed by the Navier-Stokes equations, in: *Proc. IUTAM Symp. 1991*, Lecture Notes in Mathematics **1530**, 85-96 (Springer, Berlin).

COSCIA, V., PATRIA, M.C. (1992), Existence, uniqueness and asymptotic decay of steady incompressible flows in a halfspace, *Stud. Appl. Anal. Cont. Media* **2**, 111-131.

FARWIG, R. and H. SOHR (1993), An approach to resolvent estimates for the Stokes equations in L^q-spaces, Preprint Universität Bonn.

GALDI, G. P. (1994), *An Introduction to the Mathematical Theory of the Navier-Stokes equations* Vols. I-II, Springer Tracts in Natural Philosophy (Springer, Heidelberg).

HEYWOOD, J.G. (1980), The Navier-Stokes equations: on the existence, regularity and decay of solutions, *Indiana Univ. Math. J.* **29**, 639-681.

LADYZHENSKAYA, O. A. and V.A. SOLONNIKOV (1980), Determination of the solutions of boundary value problems for stationary Stokes and Navier-Stokes equations having an unbounded dirichlet integral, translated from *Zapiski nauchnykh Seminarov Leningradskogo Otdeleniya Matematicheskogo Instituta im. V. A. Steklova Akad. Nauk SSSR* **96**, 117-160 (in Russian).

MAREMONTI, P. (1991), Some theorems of existence for solutions of the Navier-Stokes equations with slip conditions in half-space, *Ricerche Mat.*, Vol. **XL**, 81-135.

PASSERINI, A. (1995), Steady flows of a viscous incompressible fluid in a porous half-space, M^3AS (to appear).

WEIGHTED L^q-THEORY AND POINTWISE ESTIMATES FOR STEADY STOKES AND NAVIER-STOKES EQUATIONS IN DOMAINS WITH EXIT TO INFINITY

Konstantin Pileckas

Universität GH Paderborn, Fachbereich Mathematik-Informatik,
Warburger Str. 100, 33098 Paderborn, Germany.

ABSTRACT

In the paper we consider the stationary Stokes and Navier-Stokes equations of a viscous incompressible fluid in domains Ω with $m > 1$ exits to infinity, which have in some coordinate system the following form

$$\Omega_i = \{x : \ |x'| < g(x_n), \ x_n > 0\},$$

where g_i are functions satisfying the global Lipschitz condition and $g_i'(x_n) \to 0$ as $x_n \to \infty$. We present the theory concerning the solvability of the Stokes and Navier-Stokes systems with prescribed fluxes in weighted Sobolev and Hölder spaces and we show the pointwise decay of the solutions.

INTRODUCTION

The solvability of the boundary and initial–boundary value problems for Stokes and Navier-Stokes equations is one of the most important questions in the mathematical hydrodynamics. It has been studied in many papers and monographs (e.g. Ladyzhenskaya, 1969; Temam, 1977; Galdi, 1994). The existence theory which is developed there concerns mainly the domains with compact boundaries (bounded or exterior). Although some of these results do not depend on the shape of the boundary, many problems of scientific interest concerning the flow of a viscous incompressible fluid in domains with noncompact boundaries were unsolved. Therefore, it is not surprising that during the last 17 years the special attention was given to problems in such domains.

J. Heywood (1976) has shown that in domains with noncompact boundaries the motion of a viscous fluid is not always uniquely determined by the applied external forces and by the usual initial and boundary conditions. Moreover, certain physically important quantities (as fluxes of the velocity field or limiting values of the pressure at infinity) should be prescribed additionally.

For instance, J. Heywood considered the aperture domain

$$\Omega = \{x = (x_1, x_2, x_3) \in I\!\!R^3 : x_3 \neq 0 \text{ or } x_3 = 0, \ x' = (x_1, x_2) \in S\}, \qquad (0.1)$$

where S is a bounded region in the plane $I\!\!R^2$. For such domain the Stokes problem

$$\begin{cases} -\nu\Delta\vec{u} + \nabla p &= \vec{f} \quad \text{in} \ \Omega, \\ \operatorname{div}\vec{u} &= 0 \quad \text{in} \ \Omega, \\ \vec{u} &= 0 \quad \text{on} \ \partial\Omega \end{cases} \qquad (0.2)$$

admits infinitely many solutions with a finite Dirichlet integral. The unique solution of (0.2) can be specified by prescribing either the total flux F of the fluid through the aperture S, i.e.

$$\int_S u_3(x', x_3)dx' = F, \qquad (0.3)$$

or the "pressure drop"

$$p_* = p_+ - p_- = \lim_{\substack{|x|\to\infty \\ x_3 > 0}} p(x) - \lim_{\substack{|x|\to\infty \\ x_3 < 0}} p(x) \qquad (0.4)$$

The necessity to prescribe additional conditions (0.3), (0.4) is related to the fact that the spaces of divergence free vector fields $\widehat{H}(\Omega)$ and $H(\Omega)^1$ are not identical. For the domain (0.1) one has

$$\dim \widehat{H}(\Omega)/H(\Omega) = 1 \qquad (0.5)$$

(see Heywood, 1976). Moreover, in Heywood (1976) the existence of weak solutions to (0.2), (0.3) (or (0.2), (0.4)) has been proved for arbitrary data $\vec{f}, F$ and p_*. The analogous problems for the nonlinear stationary Navier-Stokes system was solved in Heywood (1976) for small data.

In the subsequent papers of Ladyzhenskaya and Solonnikov (1976, 1977), Solonnikov and Pileckas (1977), Solonnikov (1983), Kapitanskii (1981), Kapitanskii and Pileckas (1983) it was found that the spaces $\widehat{H}(\Omega)$ and $H(\Omega)$ are different for a wide class of domains $\Omega \subset I\!\!R^n$, $n = 2, 3$, having more than one exit to infinity, i.e. the set $\{x \in \Omega : |x| > R_0\}$ is a union of $m > 1$ unbounded disjoint domains $\Omega_1, ..., \Omega_m$ which are called "exits to infinity". The number of additional conditions which should be prescribed for the correct formulation of the Stokes problem is equal to $\dim \widehat{H}(\Omega)/H(\Omega)$. In turn, the dimension of $\widehat{H}(\Omega)/H(\Omega)$ is less than m and depends on the number of exits Ω_i blowing up at infinity "not to slowly". The solution to problem (0.2), having the finite Dirichlet integral, is completely determined by prescribing exactly $\dim \widehat{H}(\Omega)/H(\Omega)$ additional conditions in "wide" exits to infinity. The weak solvability of stationary Stokes and Navier-Stokes problems with prescribed additional side conditions in "wide" exits to infinity was proved for arbitrary data in Ladyzhenskaya and Solonnikov (1976, 1977), Solonnikov and Pileckas (1977), Solonnikov (1981, 1983), Kapitanskii and Pileckas (1983). Notice that the domain (0.1) has two "wide" exits to infinity and, therefore, in Ladyzhenskaya and Solonnnikov (1977), Solonnikov and Pileckas (1977), Solonnikov (1981), Kapitanskii and Pileckas (1983), the results of Heywood (1976), concerning the nonlinear problem, are extended to arbitrary large data.

---

[1] $H(\Omega)$ is the completion in the Dirichlet norm of all divergence free vector funtions with compact supports in Ω and $\widehat{H}(\Omega)$ is the space of all divergence free vector functions having finite Dirichlet norm and zero traces on $\partial\Omega$.

The next step in the mathematical theory of viscous incompressible fluids in domains with noncompact boundaries is to consider the physically natural problems with prescribed additional conditions also in "narrow" exits to infinity (for example, in pipes). The dimension of $\widehat{H}(\Omega)/H(\Omega)$ depends only on the number of "wide" exits to infinity and any vector field with a finite Dirichlet integral has zero fluxes through sections of "narrow" exits. Therefore, such problems can not be solved in a class of divergence free vector fields with the finite Dirichlet integral.

In such formulation the Stokes and Navier–Stokes problems were studied in a row of papers: Ladyzhenskaya and Solonnikov (1980, 1983), Pileckas (1981b, 1984), Solonnikov (1982, 1983). In the basic paper of Ladyzhenskaya and Solonnikov (1980) the weak solvability of the Navier–Stokes equations with prescribed fluxes through all exits to infinity was proved for arbitrary data in a class of solutions with an infinite Dirichlet integral. In these papers the case of the domain Ω having $m > 1$ exits to infinity Ω_i of the form

$$\Omega_i = \{x \in I\!\!R^n : |x'| < g_i(x_n), \ x_n > 1\}. \tag{0.6}$$

is considered. The estimates of the growth of the Dirichlet integral are given in terms of the functions g_i by mean of differential inequalities techniques (so called "techniques of the Saint-Venant's principle"). In particular, the solvability of the Navier–Stokes equations is proved for arbitrary data in domains with cylindrical exits to infinity and it is shown that for small data this solution is unique and tends as $|x| \to \infty$ to the Poiseuille flow. Notice that for domains with two cylindrical exits to infinity the problem with prescribed flux has been considered also by Amick (1977, 1978), where the solution approaching exponentially the Poiseuille flow (and, therefore having an infinite Dirichlet integral) is constructed.

The similar results for domains with layer–like exits to infinity are obtained by Pileckas (1981b).

Solonnikov (1982, 1983, 1991) developed the existence theory for stationary and nonstationary Navier–Stokes equations in a very general class of domains with exits to infinity. He avoided to make the assumptions on the shape of exits to infinity and only impose certain general restrictions. The function spaces in which the problem is solved are also very general. Roughly speaking, in Solonnikov (1982, 1983, 1991) the axiomatic approach is developed. The methods used in Solonnikov (1982, 1983, 1991) are closed to that of Ladyzhenskaya and Solonnikov (1980). However, in Solonnikov (1982, 1983, 1991) the differential inequalities for energy integrals (due to the techniques of Saint–Venant's principle) are replaced by more convenient (in our opinion) difference inequalities.

The theory of regular solutions to Navier–Stokes equations in noncompact domains is not so extended and very little is known about the decay properties of solutions as $|x| \to \infty$. In this connection we mention the papers of Pileckas (1981a, 1981c, 1988a, 1988b, 1989), Pileckas and Specovius–Neugebauer (1989), Nazarov and Pileckas (1983), Solonnikov (1989), Pileckas and Solonnikov (1989), Nazarov and Pileckas (1993), Abergel and Bona (1992), where certain existence theorems for regular solutions are proved in domains with cylindrical and periodical exits to infinity. For more general two-dimensional domains certain results concerning the pointwise decay and the asymptotics of the solutions are obtained by Amick and Fraenkel (1980). Some of mentioned above papers are related with free boundary problems in noncompact domains and there arise additional difficulties due to the presence of an unknown boundary and more complicated boundary conditions.

We also mention the papers of Borchers and Pileckas (1992), Borchers, Galdi and Pileckas (1993), Chang (1992), Coscia and Patria (1992), where the decay estimates are proved for the aperture problem (0.1)–(0.3).

The advanced theory of Stokes and Navier–Stokes equations in exterior domains is presented in the books of Galdi (1994). There one can find also certain results concerning the domains with noncompact boundaries and exhaustive list of references for mathematical hydrodynamics.

In this work we present the existence results for stationary Stokes and Navier–Stokes equations in domains with $m > 1$ exits to infinity of form (0.6). As we have mentioned above, remaining in the class of solutions with a bounded Dirichlet integral, we can prescribe the fluxes only in "wide" exits to infinity. For exits of form (0.6) this means that

$$\int_0^\infty g_i^{-n-1}(t)\, dt < \infty. \tag{0.7}$$

On the other hand, in Pileckas (1980, 1983), (see also Maslennikova and Bogovskii, 1981) it is shown that solenoidal vector fields belonging to the space $\widehat{H}^q(\Omega)$ (i.e. having the finite L^q–Dirichlet integral) can have nonzero fluxes even if condition (0.7) is violated. In this case (0.7) is changed to

$$\int_0^\infty g_i^{-(n-1)(q-1)-q}(t)\, dt < \infty \tag{0.8}$$

and the dimension of the quotient space $\widehat{H}^q(\Omega)/H^q(\Omega)$ depends on the number of exits Ω_i satisfying (0.8). Therefore, it is natural to suppose that, if the functions $g_i(t)$ grow as $t \to \infty$, we can find the numbers $q_i > 1$ such that the Stokes and Navier–Stokes systems have solutions $\vec{u}$ with prescribed fluxes and $\nabla \vec{u} \in L^{q_i}(\Omega)$.

This article has a survey character and is based on the author's papers Pileckas (1994a, 1994b, 1994c). It is organized as follows. In Section 1 we present the main notations and definitions of function spaces as well as certain auxiliary results, concerning weighted imbedding theorems and the existens of special solenoidal vector functions.

Sections 2 and 3 contain the results on solvability of the linear Stokes problem in weighted Sobolev and Hölder spaces. Namely, Section 2 is related to the Stokes problem with zero fluxes and Section 3 to the same problem with nonzero fluxes.

Finally, in Section 4 we present the analogous results for the nonlinear Navier–Stokes problem.

In this paper only the main ideas of the proofs are given. All missing details and computations can be found in Pileckas (1994a, 1994b, 1994c).

1. MAIN NOTATIONS AND FUNCTION SPACES

We indicate by $C_0^\infty(\Omega)$ the set of all infinitely differentiable real vector functions with compact supports in Ω and by $J_0^\infty(\Omega)$ the subset of all solenoidal (i.e., satisfying the condition $\operatorname{div} \vec{u}(x) = \sum_{k=1}^n \partial u_k(x)/\partial x_k = 0$) vector functions from $C_0^\infty(\Omega)$; $W^{l,q}(\Omega)$ is the usual Sobolev space with the norm

$$\|\vec{u};\, W^{l,q}(\Omega)\| = \left(\sum_{\alpha=0}^l \int_\Omega |D^\alpha u|^q\, dx \right)^{1/q},$$

where $D_x^\alpha = \partial^{|\alpha|}/\partial x_1^{\alpha_1}...\partial x_n^{\alpha_n}$, $|\alpha| = \alpha_1 + ...\alpha_n$; $C^{l,\delta}(\Omega)$, l being an integer, $0 < \delta < 1$, is a Hölder space of continuous in Ω functions u which have continuous derivatives

$D^\alpha u = \partial^{|\alpha|} u / \partial x_1^{\alpha_1} ... x_n^{\alpha_n}$, $|\alpha| = \alpha_1 + ... + \alpha_n$, up to the order l and the finite norm

$$\|u; \, C^{l,\delta}(\Omega)\| = \sum_{|\alpha| \le l} \sup\{|D^\alpha u(x)|\} + \sum_{|\alpha| = l} \sup\{[D^\alpha u]_\delta(x)\},$$

where the supremum is taken over $x \in \Omega$ and

$$[u]_\delta(x) = \sup_{\substack{0 < |x-y| < |x|/2 \\ y \in \Omega}} \frac{|u(x) - u(y)|}{|x-y|^\delta}.$$

$W_{loc}^{l,q}(\Omega)$ and $C_{loc}^{l,\delta}(\Omega)$ are the spaces of functions which belong $W^{l,q}(\Omega')$ and $C^{l,\delta}(\Omega')$ for every strictly interior subdomain Ω' of Ω.

Denote by $D_0^q(\Omega)$ and $H^q(\Omega)$ the completions of $C_0^\infty(\Omega)$ and $J_0^\infty(\Omega)$ in the norm

$$\|\nabla \vec{u}\|_{q,\Omega} = \left(\int_\Omega \sum_{i=1}^n \left| \frac{\partial \vec{u}}{\partial x_i} \right|^q dx \right)^{1/q}$$

and by $\widehat{H}^q(\Omega)$ the subspaces of all solenoidal vector functions from $D_0^q(\Omega)$. It is obvious that

$$\widehat{H}^q(\Omega) \supset H^q(\Omega).$$

For simplicity in notations we put $L^q(\Omega) = W^{0,q}(\Omega)$, $\|\cdot; L^q(\Omega)\| = \|\cdot\|_{q,\Omega}$, $D_0(\Omega) = D_0^2(\Omega)$, $H(\Omega) = H^2(\Omega)$, $\widehat{H}(\Omega) = \widehat{H}^2(\Omega)$.

Let $D^{-1}(\Omega)$ be the dual space to $D_0(\Omega)$ with the norm

$$\|\vec{f}\|_{-1,2,\Omega} = \sup_{\vec{\eta} \in D_0(\Omega)} \frac{|\int_\Omega \vec{f} \cdot \vec{\eta} \, dx|}{\|\nabla \vec{\eta}\|_{2,\Omega}}.$$

Let us consider now a domain $\Omega \subset \mathbb{R}^n$, $n = 2, 3$, having m exits to infinity. We suppose that outside the sphere $|x| = R_0$ the domain Ω splits into m connected components Ω_i (exits to infinity) which in some coordinate systems $x^{(i)}$ are given by the relations

$$\Omega_i = \{x^{(i)}: \, |x^{(i)'}| < g_i(x_n^{(i)}), x_n^{(i)} > 0\}, \tag{1.1}$$

where $|x^{(i)'}| \equiv |x_1^{(i)}|$ if $n = 2$ and $|x^{(i)'}| \equiv \sqrt{x_1^{(i)2} + x_2^{(i)2}}$ if $n = 3$, and $g_i(t)$ are functions satisfying the conditions

$$|g_i(t) - g_i(t')| \le M_i |t - t'|, \quad \forall t, t' > 0; \qquad g_i(t) \ge g_0 > 0, \tag{1.2}$$

$$\lim_{t \to \infty} g_i'(t) = 0, \qquad |g_i'(t)| \le M_i, \quad i = 1, \dots, m. \tag{1.3}$$

Below we omit the index i in the notations for local coordinates. In what follows we use the notations:

$$\sigma_i(t) = \{x \in \Omega_i : \, x_n = t = const.\};$$

$$R_{i0} = 0, \qquad R_{ik+1} = R_{ik} + (2M_i)^{-1} g_{ik}, \qquad g_{ik} = g_i(R_{ik}), \quad i = 1, \dots, m,$$

$$\Omega_{ik} = \{x \in \Omega_i : \, x_n < R_{ik}\}, \qquad \omega_{ik} = \Omega_{ik+1} \setminus \Omega_{ik}, \qquad \Omega_{(k)} = \Omega_0 \bigcup \left(\bigcup_{i=1}^m \Omega_{ik} \right);$$

$$N_i(x_n) = g_i(x_n) g_{ik_0}^{-1}, \quad i = 1, \ldots, m,$$

$$N(x)^{\vec{\kappa}} = \begin{cases} N_i(x)^{\kappa_i}, & x = \Omega_i \setminus \Omega_{(k_0)}, \ i=1,\ldots,m, \\ 1, & x \in \Omega_{(k_0)}, \end{cases}$$

$$\vec{\kappa} = (\kappa_1, \ldots, \kappa_m);$$

$$\alpha_i(t) = \int_0^t g_i(\tau)^{-1} \, d\tau, \quad \alpha_{ik} = \alpha_i(R_{ik}), \quad i = 1, \ldots, m.$$

Lemma 1.1. (Pileckas, 1994a) *There hold the following relations*

$$\frac{1}{2} g_{ik} \le g_i(t) \le \frac{3}{2} g_{ik}, \quad t \in [R_{ik}, R_{ik+1}], \tag{1.4}$$

$$\int_0^\infty g_i^{-1}(\tau) \, d\tau = \infty, \tag{1.5}$$

$$\mu_* |l - k| \le |\alpha_{il} - \alpha_{ik}| \le \mu^* |l - k|, \quad i = 1, \ldots, m, \tag{1.6}$$

where μ_* and μ^* are positive constants independent of k and l.

Let us introduce the weighted function spaces in the domain Ω with $m > 1$ exits to infinity. $L_{(\vec{\kappa}, \vec{\beta})}^{\vec{q}}(\Omega)$, $\vec{q} = (q_0, q_1, \ldots, q_m)$, $\vec{\beta} = (\beta_1, \ldots, \beta_m)$, is the space of functions with the finite norm

$$\|\vec{f}; L_{(\vec{\kappa}, \vec{\beta})}^{\vec{q}}(\Omega)\| = \left(\int_{\Omega_{(k_0)}} |\vec{f}|^{q_0} dx \right)^{1/q_0} + \sum_{i=1}^m \left(\int_{\Omega_i \setminus \Omega_{(k_0)}} N_i^{q_i \kappa_i} \exp\left(q_i \beta_i \alpha_i(x_n)\right) |\vec{f}|^{q_i} dx \right)^{1/q_i}$$

and $\widetilde{L}_{(\vec{\kappa}, \vec{\beta})}^{\vec{q}}(\Omega)$ is the space of functions with the norm

$$\|\vec{f}; \widetilde{L}_{(\vec{\kappa}, \vec{\beta})}^{\vec{q}}(\Omega)\| = \left(\int_{\Omega_{(k_0+1)}} |\vec{f}|^{q_0} dx \right)^{1/q_0} + \sum_{i=1}^m \left(\int_{\Omega_i \setminus \Omega_{(k_0-1)}} N_i^{q_i \kappa_i} \exp\left(q_i \beta_i \alpha_i(x_n)\right) |\vec{f}|^{q_i} dx \right)^{1/q_i}.$$

$\widehat{H}_{\vec{\kappa}}^{\vec{q}}(\Omega)$ is the space of all solenoidal vector functions, having zero traces on $\partial\Omega$ and the finite norm $\|\nabla \vec{u}; L_{(\vec{\kappa}, \vec{0})}^{\vec{q}}(\Omega)\|$ and $H_{\vec{\kappa}}^{\vec{q}}(\Omega)$ is the closure of $J_0^\infty(\Omega)$ in the norm $\| \cdot ; \widehat{H}_{\vec{\kappa}}^{\vec{q}}(\Omega)\|$. We put $H_{\vec{0}}^{\vec{q}}(\Omega) = H^{\vec{q}}(\Omega)$, $\widehat{H}_{\vec{0}}^{\vec{q}}(\Omega) = \widehat{H}^{\vec{q}}(\Omega)$. Notice that for $q_0 = q_1 = \ldots = q_m = q$ we have $H^{\vec{q}}(\Omega) = H^q(\Omega)$, $\widehat{H}^{\vec{q}}(\Omega) = \widehat{H}^q(\Omega)$.

Theorem 1.1. *Assume that $\Omega \subset \mathbb{R}^n$, $n = 2, 3$, is the domain with m exits to infinity Ω_i, possessing the properties described above, and let for all $k > k_0$ the domain $\Omega_{(k)}$ has the Lipschitz boundary. If among the integrals*

$$\int_0^\infty g_i^{-(n-1)(q_i-1)-q_i+q_i\kappa_i}(t) \, dt, \quad i = 1, \ldots, m,$$

there are exactly r which converge, then

$$\dim \widehat{H}_{\vec{\kappa}}^{\vec{q}}(\Omega) / H_{\vec{\kappa}}^{\vec{q}}(\Omega) = r - 1.$$

The space $H_{\vec{\kappa}}^{\vec{q}}(\Omega)$ consists of those and only those $\vec{u} \in \widehat{H}_{\vec{\kappa}}^{\vec{q}}(\Omega)$ which satisfy the condition

$$\int_{\sigma_i} \vec{u} \cdot \vec{n} \, ds = 0, \quad i = 1, \ldots, m.$$

The proof of this theorem is completely analogous to that for the case $\vec{\kappa} = \vec{0}$, $q_0 = q_1 = ... = q_m = q$ (see Pileckas, 1980, 1983) and is based on the following lemma.

Lemma 1.2. (Solonnikov and Pileckas, 1977) *Let Ω be a domain with $m > 1$ exits to infinity Ω_i of the form (1.1). Then there exists a vector field $\vec{A}$ satisfying the following conditions*

$$\begin{cases} \operatorname{div}\vec{A} & = & 0 & in & \Omega, \\ \vec{A} & = & 0 & on & \partial\Omega, \\ \displaystyle\int_{\sigma_i(t)} \vec{A} \cdot \vec{n}\, dS & = & F_i, & i = 1, ..., m, \end{cases}$$

where $\vec{n}$ is a unit vector of the normal to σ_i and F_i are given numbers with

$$\sum_{i=1}^{m} F_i = 0.$$

Moreover, there holds the estimate

$$|D_x^\alpha \vec{A}(x)| \leq C(|\vec{F}|)\, g_i^{-n+1-|\alpha|}(x_n), \tag{1.7}$$

$i = 1, ..., m$, $x \in \Omega_i$, $|\alpha| \geq 0$. *In (1.7) $|\vec{F}| = (\sum_{i=1}^{m} F_i^2)^{1/2}$.*

Let $V_{(\vec{\kappa},\vec{\beta})}^{l,\vec{q}}(\Omega)$, $l \geq 1$, be the completion of $C_0^\infty(\Omega)$ in the norm

$$\|\vec{f}; V_{(\vec{\kappa},\vec{\beta})}^{l,\vec{q}}(\Omega)\| = \sum_{|\alpha|=0}^{l} \|D^\alpha \vec{f};\, L_{(\vec{\kappa}+|\alpha|-l,\vec{\beta})}^{\vec{q}}(\Omega)\|.$$

and $V_{(\vec{\kappa},\vec{\beta})}^{-1,\vec{q}}(\Omega)$ be the space of functions $\vec{f}$ which can be represented in the form

$$\vec{f} = \vec{f}^{(0)} + (\operatorname{div} \vec{f}^{(1)}, ..., \operatorname{div} \vec{f}^{(n)}) \tag{1.8}$$

with $\vec{f}^{(0)} \in L_{(\vec{\kappa}+1,\vec{\beta})}^{\vec{q}}(\Omega)$, $\vec{f}^{(j)} \in L_{(\vec{\kappa},\vec{\beta})}^{\vec{q}}(\Omega)$, $j = 1, ..., n$, and

$$\|\vec{f};\, V_{(\vec{\kappa},\vec{\beta})}^{-1,\vec{q}}(\Omega)\| = \|\vec{f}^{(0)};\, L_{(\vec{\kappa}+1,\vec{\beta})}^{\vec{q}}(\Omega)\| + \sum_{j=1}^{n} \|\vec{f}^{(j)};\, L_{(\vec{\kappa},\vec{\beta})}^{\vec{q}}(\Omega)\|.$$

Here $\vec{\kappa} + b = (\kappa_1 + b, ..., \kappa_m + b)$.

The spaces $\tilde{V}_{(\vec{\kappa},\vec{\beta})}^{l,\vec{q}}(\Omega)$, $l \geq -1$, are defined by the same formulas with the only difference that $L_{(\cdot,\vec{\beta})}^{\vec{q}}(\Omega)$ in the definitions of the norms are replaced by $\tilde{L}_{(\cdot,\vec{\beta})}^{\vec{q}}(\Omega)$.

The weighted Hölder space $C_{(\vec{\kappa},\vec{\beta})}^{l,\delta}(\Omega), l \geq 0, 1 > \delta > 0$, consists of functions u, continuously differentiable up to the order l in Ω, for which the norm

$$\|u; C_{(\vec{\kappa},\vec{\beta})}^{l,\delta}(\Omega)\| = \|u; C^{l,\delta}(\Omega_{(k_0)})\| + \sum_{i=1}^{m} \sum_{|\alpha|\leq l} \sup_{x \in \Omega_i} \{g_i^{\kappa_i - l - \delta + |\alpha|}(x_n) \exp(\beta_i \alpha_i(x_n)) |D^\alpha u(x)|\}$$

$$+ \sum_{i=1}^{m} \sum_{|\alpha|=l} \sup_{x \in \Omega_i} \{g_i^{\kappa_i}(x_n) \exp(\beta_i \alpha_i(x_n)) [D^\alpha u]_\delta(x)\}$$

is finite.

The weighted function spaces in subdomains Ω' of Ω are defined by the same formulas with only difference that either integrals or supremum are taken over Ω' instead of Ω. For example,

$$\|f;\ L^{q_i}_{(\kappa_i,\beta_i)}(\omega_{ik})\| = \left(\int_{\omega_{ik}} g_i^{q_i\kappa_i}(x_n) \exp\left(q_i\beta_i\alpha_i(x_n)\right)|f|^{q_i}dx \right)^{1/q_i}, \quad i = 1,\dots,m.$$

The following weighted imbedding theorems hold true.

Theorem 1.2. (Pileckas, 1994b) Let $u \in V^{l,q_i}_{(\kappa_i,0)}(\omega_{ik})$.
(i) If

$$q_i l \leq n, \quad q_i \leq s_i \leq n\,q_i/(n-q_i l), \tag{1.9}$$

then $u \in L^{s_i}_{(\kappa_i-l+n/q_i-n/s_i,0)}(\omega_{ik})$ and

$$\|u;\ L^s_{(\kappa_i-l+n/q_i-n/s,0)}(\omega_{ik})\| \leq c\|u;\ V^{l,q_i}_{(\kappa_i,0)}(\omega_{ik})\|. \tag{1.10}$$

(ii) If

$$q_i l > n, \quad m+\delta \leq (q_i l - n)/q_i, \quad \delta \in (0,\,1), \tag{1.11}$$

then $u \in C^{m,\delta}_{(m+\delta-l+\kappa_i+n/q_i,0)}(\omega_{ik})$ and

$$\|u;\ C^{m,\delta}_{(m+\delta-l+\kappa_i+n/q_i,0)}(\omega_{ik})\| \leq c\|u;\ V^{l,q_i}_{(\kappa_i,0)}(\omega_{ik})\|. \tag{1.12}$$

The constants in (1.10), (1.2) are independent of k and u.

Theorem 1.3. (Pileckas, 1994b) Let $u \in V^{l,q_i}_{(\kappa_i,\beta_i)}(\Omega_i), \beta \in \mathbb{R}^1$.
(i) If s_i satisfies conditions (1.9), then $u \in L^{s_i}_{(\kappa_i-l+n/q_i-n/s_i,\beta_i)}(\Omega_i)$ and

$$\|u;\ L^{s_i}_{(\kappa_i-l+n/q_i-n/s_i,\beta_i)}(\Omega_i)\| \leq c\|u;\ V^{l,q_i}_{(\kappa_i,\beta_i)}(\Omega_i)\|. \tag{1.13}$$

(ii) If conditions (1.11) are fulfilled, then $u \in C^{m,\delta}_{(m+\delta-l+\kappa_i+n/q_i,\beta_i)}(\Omega_i)$ and

$$\|u;\ C^{m,\delta}_{(m+\delta-l+\kappa_i+n/q_i,\beta_i)}(\Omega_i)\| \leq c\|u;\ V^{l,q_i}_{(\kappa_i,\beta_i)}(\Omega_i)\|. \tag{1.14}$$

Proofs of Theorems 1.1, 1.2 are based on classical imbedding results (e.g. Adams 1975) and on simple scaling arguments applied in the domains ω_{ik}.

2. SOLVABILITY OF THE STOKES PROBLEM WITH ZERO FLUXES

Let us consider in the domain $\Omega \subset \mathbb{R}^n, n = 2,3$, with $m > 1$ exits to infinity the Stokes problem

$$-\nu\Delta\vec{u} + \nabla p = \vec{f} \quad \text{in } \Omega, \tag{2.1}$$

$$\text{div}\,\vec{u} = 0 \quad \text{in } \Omega, \tag{2.2}$$

$$\vec{u} = 0 \quad \text{on } \partial\Omega, \tag{2.3}$$

supplemented by the additional flux conditions

$$\int_{\sigma_i} \vec{u}\cdot\vec{n}\,ds = F_i, \quad i = 1,\dots,m, \quad \sum_{i=1}^m F_i = 0. \tag{2.4}$$

In what follows we denote the problem (2.1)–(2.4) with zero fluxes, i.e. $F_i = 0$, $i = 1, \ldots, m$, by (2.1)-(2.4)$_0$.

By a weak solution of problem (2.1)–(2.4) we understand the vector function $\vec{u} \in W^{1,2}_{\text{loc}}(\Omega)$ such that equations (2.2)–(2.4) hold and

$$\nu \int_\Omega \nabla \vec{u} \cdot \nabla \vec{\eta} \, dx = \int_\Omega \vec{f} \cdot \vec{\eta} \, dx \tag{2.5}$$

for every test function $\vec{\eta} \in J_0^\infty(\Omega)$.

By using the well known local estimates for elliptic problems (see Agmon, Douglis and Nirenberg, 1964, Solonnikov, 1964) and the scaling argument in domains ω_{il}, one can prove the following weighted local estimates.

Theorem 2.1. (i) (Pileckas, 1994a) *Let the function $\vec{f}$ has the representation* (1.8) *and* $\vec{f}^{(0)}$, $\vec{f}^{(j)} \in L^{q_i}_{\text{loc}}(\Omega_i)$, $j = 1, 2, 3$, $q_i \geq 2$, $i = 0, 1, \ldots, m$. *Then the weak solution $\vec{u}$ of the Stokes problem* $(2.1) - (2.4)$ *satisfies the local estimate*

$$\|\nabla \vec{u}; \, L^{q_i}_{(\kappa_i, \beta_i)}(\omega_{is})\| \leq c \Big(\|\vec{f}^{(0)}; \, L^{q_i}_{(\kappa_i + 1, \beta_i)}(\omega^*_{is})\| +$$

$$+ \sum_{j=1}^{3} \|\vec{f}^{(j)}; \, L^{q_i}_{(\kappa_i, \beta_i)}(\omega^*_{is})\| + \|\nabla \vec{u}; \, L^2_{(\kappa_i + \hat{\kappa}_i, \beta_i)}(\omega^*_{is})\| \Big), \tag{2.6}$$

where $\hat{\kappa}_i = n(2 - q_i)/2q_i$, $\omega^*_{is} = \omega_{is-1} \bigcup \omega_{is} \bigcup \omega_{is+1}$.

(ii) (Pileckas, 1994a) *Assume that* $\partial\Omega \in C^{l+2}$, $\vec{f} \in W^{l,q_i}_{\text{loc}}(\Omega_i)$, $q_i > 1$, $l \geq 0$. *Then the solution $(\vec{u}, p)$ of problem* $(2.1) - (2.4)$ *satisfies the local estimates*

$$\|\vec{u}; \, V^{l+2,q_i}_{(\kappa_i, \beta_i)}(\omega_{is})\| + \|\nabla p; \, V^{l,q_i}_{(\kappa_i, \beta_i)}(\omega_{is})\| \leq$$

$$\leq c \Big(\|\vec{f}; \, V^{l,q_i}_{(\kappa_i, \beta_i)}(\omega^*_{is})\| + \|\nabla \vec{u}; \, L^{q_i}_{(\kappa_i - l - 1, \beta_i)}(\omega^*_{is})\| \Big), \tag{2.7}$$

$$\|\vec{u}; \, V^{l+2,q_i}_{(\kappa_i, \beta_i)}(\omega_{is})\| + \|\nabla p; \, V^{l,q_i}_{(\kappa_i, \beta_i)}(\omega_{is})\| \leq$$

$$\leq c \Big(\|\vec{f}; \, V^{l,q_i}_{(\kappa_i, \beta_i)}(\omega^*_{is})\| + \|\nabla \vec{u}; \, L^{2_i}_{(\kappa_i - l - n/2 - 1 + n/q_i, \beta_i)}(\omega^*_{is})\| \Big). \tag{2.8}$$

(iii) (Pileckas, 1994b) *Assume that* $\partial\Omega \in C^{l+2,\delta}$ *and let* $\vec{f} \in C^{l,\delta}_{\text{loc}}(\Omega_i)$, $l \geq 0$, $0 < \delta < 1$. *Then the solution $(\vec{u}, p)$ of* $(2.1) - (2.4)$ *satisfies the local estimate*

$$\|\vec{u}; \, C^{l+2,\delta}_{(\kappa_i, \beta_i)}(\omega_{is})\| + \|\nabla p; \, C^{l,\delta}_{(\kappa_i, \beta_i)}(\omega_{is})\| \leq$$

$$\leq c \Big(\|\vec{f}; \, C^{l,\delta}_{(\kappa_i, \beta_i)}(\omega^*_{is})\| + \|\nabla \vec{u}; \, L^2_{(\kappa_i - l - 1 - n/2 - \delta, \beta_i)}(\omega^*_{is})\| \Big). \tag{2.9}$$

The constants in (2.6)–(2.9) *are independent of s and $\vec{f}$.*

Let us formulate the main results of the section.

Theorem 2.2. (i) (Pileckas, 1994a) *Assume that* $\partial\Omega \in C^{l+2}$, $\vec{f} \in \widetilde{V}^{l,\vec{q}}_{(\vec{\kappa}, \vec{\beta})}(\Omega)$, [2] $l \geq -1$, $q_j > 1$, $|\beta_j| < \beta_*$, $\vec{\kappa}$ *is arbitrary. Then there exists a unique solution $(\vec{u}, p)$[3] of problem* $(2.1) - (2.4)_0$ *with* $\vec{u} \in V^{l+2,\vec{q}}_{(\vec{\kappa}, \vec{\beta})}(\Omega)$ *and*

$$\|\vec{u}; \, V^{l+2,\vec{q}}_{(\vec{\kappa}, \vec{\beta})}(\Omega)\| \leq c \|\vec{f}; \, \widetilde{V}^{l,\vec{q}}_{(\vec{\kappa}, \vec{\beta})}(\Omega)\|. \tag{2.10}$$

---

[2] If $l = -1$, it is enough to assume that $\partial\Omega$ satisfies only the Lipschitz condition.
[3] Speaking about the uniqueness of the solution $(\vec{u}, p)$, we have in mind that the pressure p is unique up to an additive constant.

Moreover, if $l \geq 0$, *we have* $\nabla p \in V_{(\vec{\kappa}, \vec{\beta})}^{l, \vec{q}}(\Omega)$ *and there holds the estimate*

$$\|\vec{u}; \ V_{(\vec{\kappa}, \vec{\beta})}^{l+2, \vec{q}}(\Omega)\| + \|\nabla p; \ V_{(\vec{\kappa}, \vec{\beta})}^{l, \vec{q}}(\Omega)\| \leq c \|\vec{f}; \ \widetilde{V}_{(\vec{\kappa}, \vec{\beta})}^{l, \vec{q}}(\Omega)\|. \tag{2.11}$$

(ii) (Pileckas, 1994b) *Assume that* $\partial \Omega \in C^{l+2, \delta}$ *and* $\vec{f} \in C_{(\vec{\kappa}, \vec{\beta})}^{l, \delta}(\Omega)$, *where* $l \geq 0$, $\delta \in (0, 1)$, $|\beta_i| < \beta_*$ *and* $\vec{\kappa}$ *is arbitrary. Then problem* $(2.1)-(2.4)_0$ *has a unique solution* $(\vec{u}, p)$ *with* $\vec{u} \in C_{(\vec{\kappa}, \vec{\beta})}^{l+2, \delta}(\Omega)$, $\nabla p \in C_{(\vec{\kappa}, \vec{\beta})}^{l, \delta}(\Omega)$. *Moreover, there holds the estimate*

$$\|\vec{u}; \ C_{(\vec{\kappa}, \vec{\beta})}^{l+2, \delta}(\Omega)\| + \|\nabla p; \ C_{(\vec{\kappa}, \vec{\beta})}^{l, \delta}(\Omega)\| \leq c \|\vec{f}; \ C_{(\vec{\kappa}, \vec{\beta})}^{l, \delta}(\Omega)\|. \tag{2.12}$$

Remark 2.1. *In particular, if* $\vec{f} \in C_{(\vec{\kappa}, \vec{\beta})}^{l, \delta}(\Omega)$ *with* $\beta_i > 0$ *(for example,* $\vec{f}$ *has a compact support), then from Theorem 2.2 follows the exponential decay estimates for the solution* $(\vec{u}, \ p)$ *of problem* $(2.1) - (2.4)_0$, *i.e. there hold the estimates*

$$|D^\alpha \vec{u}(x)| \leq c \exp \left(- \beta_i \int_0^{x_n^{(i)}} g_i^{-1}(t) \, dt \right) g_i^{l+2+\delta - |\alpha| - \kappa_i}(x_n^{(i)}), \quad x \in \Omega_i, \ 0 \leq |\alpha| \leq l+2,$$

$$|D^\alpha \nabla p(x)| \leq c \exp \left(- \beta_i \int_0^{x_n^{(i)}} g_i^{-1}(t) \, dt \right) g_i^{l+\delta - |\alpha| - \kappa_i}(x_n^{(i)}), \quad x \in \Omega_i, \ 0 \leq |\alpha| \leq l.$$

We discuss shortly the most important steps of the proof. The detailed calculations can be found in Pileckas (1994a, 1994b).

Step 1. First of all, we consider the case, when $q_i \geq 2$ for all $i \in \{0, 1, ..., m\}$. We introduce in the domain Ω the partition of unity $\{\varphi_{k_0}, \varphi_{ik}\}$, $i = 1, \ldots, m$, $k = k_0+1, \ldots$, subordinated to the covering of Ω by the domains $\{\Omega_{(k_0+1)}, \omega_{ik}^*, \}$, $i = 1, \ldots, m$, $k = k_0 + 1, \ldots$, i.e. supp $\varphi_{k_0} \subset \bar{\Omega}_{(k_0+1)}$, supp $\varphi_{ik} \subset \bar{\omega}_{ik}^*$, $\varphi_{k_0}, \varphi_{ik} \in C_0^\infty(\mathbb{R}^n)$,

$$\Omega = \Omega_{(k_0+1)} \cup \left(\bigcup_{i=1}^m \left(\bigcup_{k=k_0+1}^\infty \omega_{ik}^* \right) \right)$$

and

$$\sum_{i=1}^m \sum_{k=k_0+1}^\infty \varphi_{ik}(x) + \varphi_{k_0}(x) = 1 \quad \text{in} \ \Omega.$$

Let

$$\vec{f}_{k_0}^{(j)} = \varphi_{k_0} \vec{f}^{(j)}, \qquad \vec{f}_{ik}^{(j)} = \varphi_{ik} \vec{f}^{(j)}, \quad j = 0, 1, 2, 3,$$

$$\vec{f}_{k_0} = \vec{f}_{k_0}^{(0)} + (\text{div} \ \vec{f}_{k_0}^{(1)}, \ \text{div} \ \vec{f}_{k_0}^{(2)}, \ \text{div} \ \vec{f}_{k_0}^{(3)}), \quad .$$

$$\vec{f}_{ik} = \vec{f}_{ik}^{(0)} + (\text{div} \ \vec{f}_{ik}^{(1)}, \ \text{div} \ \vec{f}_{ik}^{(2)}, \ \text{div} \ \vec{f}_{ik}^{(3)}),$$

$$\vec{f}_i^{[N]} = \sum_{k=k_0+1}^N \vec{f}_{ik}, \ \vec{f}^{[N]} = \vec{f}_{k_0} + \sum_{i=1}^m \vec{f}_i^{[N]}.$$

Obviously, the functions $\vec{f}_{k_0}$, $\vec{f}_{ik}$ have compact supports:

$$\text{supp} \ \vec{f}_{k_0} \subset \bar{\Omega}_{k_0}, \ \text{supp} \ \vec{f}_{ik} \subset \bar{\omega}_{ik}^*.$$

Then to each $\vec{f}_{k_0}$, $\vec{f}_{ik}$ corresponds a unique weak solution $\vec{u}_{k_0}$, $\vec{u}_{ik}$ of problem $(2.1) - (2.4)_0$ with the finite Dirichlet integral (see Ladyzhenskaya, 1969).

Step 2. Applying the difference inequality techniques (techniques of the Saint–Venant principle) we prove that each $\vec{u}_{ik}$ admits the estimate

$$\|\nabla \vec{u}_{ik}; \ L_{(\kappa_i, \beta_i)}^{q_i}(\omega_{il}^*)\| \leq c \exp \left(- \varepsilon_0 c_0 |k - l| \right) \|\vec{f}_{ik}; \ V_{(\kappa_i + \hat{\kappa}_i, \beta_i)}^{-1, 2}(\omega_{il}^*)\| \leq$$

$$c \exp\left(-\varepsilon_0 c_0 |k - l|\right) \|\vec{f}_{ik}; \ V^{-1,q_i}_{(\kappa_i,\beta_i)}(\omega^*_{il})\|, \tag{2.13}$$

where $\widehat{\kappa}_i = n(2 - q_i)/2q_i$ and $|\beta_i| < \beta_*$.

Step 3. Let

$$\vec{u}^{[N]} = \sum_{i=1}^{m} \sum_{k=k_0+1}^{N} \vec{u}_{ik} + \vec{u}_{k_0}.$$

The repeated application of (2.13), Hölder and triangle inequalities allows us to prove that $\vec{u}^{[N]} \in V^{1,\vec{q}}_{(\vec{\kappa},\vec{\beta})}(\Omega)$ and

$$\|\vec{u}^{[N]}; \ V^{1,\vec{q}}_{(\vec{\kappa},\vec{\beta})}(\Omega)\| \leq c \|\vec{f}^{[N]}; \ \widetilde{V}^{-1,\vec{q}}_{(\vec{\kappa},\vec{\beta})}(\Omega)\| \leq c \|\vec{f}; \ \widetilde{V}^{-1,\vec{q}}_{(\vec{\kappa},\vec{\beta})}(\Omega)\|$$

Therefore, the sequence $\{\vec{u}^{[N]}\}$ converges in the norm of the space $V^{1,\vec{q}}_{(\vec{\kappa},\vec{\beta})}(\Omega)$ to the solution $\vec{u} \in V^{1,\vec{q}}_{(\vec{\kappa},\vec{\beta})}(\Omega)$ of problem (2.1)–(2.4)$_0$.

Step 4. Applying to $\{\vec{u}^{[N]}\}$ the weighted local estimates (2.6), (2.7), we conclude that $\vec{u} \in V^{l+2,\vec{q}}_{(\vec{\kappa},\vec{\beta})}(\Omega)$, $\nabla p \in V^{l+2,\vec{q}}_{(\vec{\kappa},\vec{\beta})}(\Omega)$ and obtain the estimates (2.10), (2.11) for $\vec{u}$ and ∇p.

Step 5. The proof for the case $q_i < 2$ for certain $i \in \{0, 1, ..., m\}$ follows now by the duality arguments.

Step 6. The weighted Hölder solvability of problem (2.1)–(2.4)$_0$ is proved by the same scheme with only difference that the local estimate (2.9) is applied in place of (2.6) or (2.7).

3. SOLVABILITY OF THE STOKES PROBLEM WITH NONZERO FLUXES

In this section we consider the Stokes problem (2.1)–(2.4) with arbitrary nonzero fluxes F_i, $i = 1, ..., m$. Let us assume that for each $i \in \{1, ..., m\}$ there exists a number q_i^* such that

$$\int_0^\infty g_i^{-(n-1)(q_i^*-1)-q_i^*}(t)\, dt < \infty. \tag{3.1}$$

We look for the solution $\vec{u}$ in the form

$$\vec{u} = \vec{A} + \vec{v},$$

where $\vec{A}$ is the solenoidal vector function from Lemma 1.1, satisfying the flux condition (2.4) and estimates (1.7), and $\vec{v}$ is the solution of the Stokes problem with zero fluxes (2.1)–(2.4)$_0$, corresponding to the right–hand side $\vec{f} + \nu \Delta \vec{A}$. In virtue of (1.7) it is easy to verify that there hold the estimates

$$\|\vec{A}; \ \widehat{H}^{\vec{q}}(\Omega)\| \leq c \sum_{i=1}^{m} |F_i| \left(\int_0^\infty g_i^{-(n-1)(q_i^*-1)-q_i^*}(t)dt\right)^{1/q_i^*} \leq c \sum_{i=1}^{m} |F_i| \tag{3.2}$$

$$\|\Delta \vec{A}; \ \widetilde{V}^{l,\vec{q}}_{(\vec{\kappa}^*,\vec{0})}(\Omega)\| \leq c \sum_{i=1}^{m} |F_i|. \tag{3.3}$$

$$\|\vec{A}; \ C^{l+2,\delta}_{(\vec{\kappa},\vec{0})}(\Omega)\| \leq c \sum_{i=1}^{m} |F_i|, \tag{3.4}$$

where

$$\vec{\kappa}^* = (\kappa_1^*, ..., \kappa_m^*), \qquad \kappa_i^* = n + 1 + l - nq_i^*/q_i, \quad i = 1, ..., m, \tag{3.5}$$

$$\vec{\kappa} = (\kappa_1, ..., \kappa_m), \qquad \kappa_i = n + 1 + l + \delta, \qquad i = 1, ..., m. \tag{3.6}$$

Using these estimates and applying to the solution $(\vec{v}, p)$ of problem (2.1)–(2.4)$_0$ the results of Section 2, we derive the following theorem.

43

Theorem 3.1. (i) (Pileckas, 1994a) *Let condition (3.1) holds. Then for arbitrary* $\vec{f} \in \widetilde{V}_{(\vec{0},\vec{0})}^{-1,\vec{q}}(\Omega)$ *and* $F_i \in \mathbb{R}^1$, $i = 1, \ldots, m$, *problem (2.1)–(2.4) has a unique solution* $\vec{u} \in \widehat{H}^{\vec{q}}(\Omega)$ *satisfying the estimate*

$$\|\vec{u}; \ \widehat{H}^{\vec{q}}(\Omega)\| \leq c\Big(\|\vec{f}; \ \widetilde{V}_{(\vec{0},\vec{0})}^{-1,\vec{q}}(\Omega)\| + \sum_{i=1}^{m} |F_i|\Big). \tag{3.7}$$

(ii) (Pileckas, 1994a) *Let* $\partial\Omega \in C^{l+2}$, $F_i \in \mathbb{R}^1$, $i = 1, \ldots, m$, $\vec{f} \in \widetilde{V}_{(\vec{\kappa}^*,\vec{0})}^{l,\vec{q}}(\Omega)$, *where* $l \geq -1$, $q_i > 1$ *and* κ_i^* *are defined by the formula (3.5). Then there exists a unique solution* $(\vec{u}, p)$ *of problem (2.1)–(2.4) with* $\vec{u} \in V_{(\vec{\kappa}^*,\vec{0})}^{l+2,\vec{q}}(\Omega)$, $\nabla p \in V_{(\vec{\kappa}^*,\vec{0})}^{l,\vec{q}}(\Omega)$ *and*

$$\|\vec{u}; \ V_{(\vec{\kappa}^*,\vec{0})}^{l+2,\vec{q}}(\Omega)\| + \|\nabla p; \ V_{(\vec{\kappa}^*,\vec{0})}^{l,\vec{q}}(\Omega)\| \leq c\Big(\sum_{i=1}^{m} |F_i| + \|\vec{f}; \ \widetilde{V}_{(\vec{\kappa}^*,\vec{0})}^{l,\vec{q}}(\Omega)\|\Big). \tag{3.8}$$

In particular, if $\vec{f} \in V_{(\vec{\kappa}^*,\vec{0})}^{-1,\vec{q}}(\Omega)$, *then* $\vec{u} \in \widehat{H}_{\vec{\kappa}^*}^{1,\vec{q}}(\Omega)$.

(iii) (Pileckas, 1994b) *Let* $\partial\Omega \in C^{l+2,\delta}$, $\vec{f} \in C_{(\vec{\kappa},\vec{0})}^{l,\delta}(\Omega)$, $l \geq 0$, $\delta \in (0,1)$, *where* κ *is defined by (3.6). Then there exists a unique solution* $(\vec{u}, p)$ *of problem (2.1)–(2.4) such that* $\vec{u} \in C_{(\vec{\kappa},\vec{0})}^{l+2,\delta}(\Omega)$, $\nabla p \in C_{(\vec{\kappa},\vec{0})}^{l,\delta}(\Omega)$ *and there holds the estimate*

$$\|\vec{u}; \ C_{(\vec{\kappa},\vec{0})}^{l+2,\delta}(\Omega)\| + \|\nabla p; \ C_{(\vec{\kappa},\vec{0})}^{l,\delta}(\Omega)\| \leq c\Big(\sum_{i=1}^{m} |F_i| + \|\vec{f}; \ C_{(\vec{\kappa},\vec{0})}^{l,\delta}(\Omega)\|\Big), \tag{3.9}$$

In particular, from (3.9) follows that

$$|D^{\alpha}\vec{u}(x)| \leq c\Big(\sum_{i=1}^{m} |F_i| + \|\vec{f}; \ C_{(\vec{\kappa},\vec{0})}^{l,\delta}(\Omega)\|\Big) g_i^{-n+1-|\alpha|}(x_n), \ x \in \Omega_i, \ |\alpha| \geq 0, \tag{3.10}$$

$$|D^{\alpha}\nabla p(x)| \leq c\Big(\sum_{i=1}^{m} |F_i| + \|\vec{f}; \ C_{(\vec{\kappa},\vec{0})}^{l,\delta}(\Omega)\|\Big) g_i^{-n-1-|\alpha|}(x_n), \ x \in \Omega_i,$$

$$|\alpha| \geq 0, \tag{3.11}$$

Moreover,

$$|p(x)| \leq c\Big(\sum_{i=1}^{m} |F_i| + \|\vec{f}; \ C_{(\vec{\kappa},\vec{0})}^{l,\delta}(\Omega)\|\Big)\Big(\int_0^{x_n} g_i^{-n-1}(t)dt\Big) + c_1, \ x \in \Omega_i. \tag{3.12}$$

Concerning the proof of the theorem, we should mention additionally, that in oder to prove (3.12), we represent the function $p(x)$ in the form

$$p(x) = p(x_0) + \int_{x_0}^{x} \frac{\partial p}{\partial\gamma} \, d\gamma,$$

where $\gamma \subset \Omega_i$ is a smooth counter, connecting some fixed point $x_0 \in \Omega_i$ with arbitrary point $x \in \Omega_i$, such that γ is given by the equations $x_j = \gamma_j(x_n)$, $j = 1, \ldots, n-1$, and $(1 + \gamma_1'(x_n)^2 + \ldots + \gamma_{n-1}'(x_n)^2)^{1/2} \leq$ const. Then,

$$|p(x)| \leq |p(x)| + \Big|\int_{x_0}^{x} \frac{\partial p}{\partial\gamma} \, d\gamma\Big| \leq |p(x_0)|$$

$$+ \sup_{x \in \Omega_i} |\nabla p(x) g_i^{n+1}(x_n)| \int_{x_{0n}}^{x_n} g_i^{-n-1}(t)\sqrt{1 + \gamma_1'(t)^2 + \ldots + \gamma_{n-1}'(t)^2}dt$$

$$\leq |p(x_0)| + c\|\nabla p; \ C_{(\vec{\kappa},\vec{0})}^{l,\delta}(\Omega)\| \int_0^{x_n} g_i^{-n-1}(t) \, dt.$$

Remark 3.1. *The estimates for the pressure $p(x)$ in domains with exits to infinity have been obtained by* Solonnikov (1981). *It is proved in* Solonnikov (1981) *that for each exit to infinity Ω_i with $\int_0^\infty g_i^{-n-1}(t)\,dt < \infty$ the pressure $p(x)$ tends to a constant p_i as $|x| \to \infty$, $x \in \Omega_i$. Estimate (3.12) gives in this case $|p(x)| \le$ const., which agrees with results from* Solonnikov (1981).

4. SOLVABILITY OF THE NONLINEAR NAVIER–STOKES PROBLEM

Let us consider in the domain $\Omega \subset I\!R^n$ with $m > 1$ exits to infinity the nonlinear Navier–Stokes system

$$-\nu \Delta \vec{u} + (\vec{u} \cdot \nabla)\vec{u} + \nabla p = \vec{f} \quad \text{in} \quad \Omega, \tag{4.1}$$

$$\operatorname{div} \vec{u} = 0 \quad \text{in} \quad \Omega, \tag{4.2}$$

$$\vec{u} = 0 \quad \text{on} \quad \partial\Omega, \tag{4.3}$$

supplemented by the flux conditions

$$\int_{\sigma_i} \vec{u} \cdot \vec{n}\, ds = F_i, \quad i = 1,\dots,m, \quad \sum_{i=1}^{m} F_i = 0, \tag{4.4}$$

The problem (4.1)–(4.4) with zero fluxes, i.e. $F_i = 0$, $i = 1,\dots,m$, is denoted by (4.1)–(4.4)$_0$.

4.1. Solvability of problem (4.1)–(4.4)$_0$

The Navier–Stokes problem with zero fluxes is studied in weighted Sobolev and Hölder spaces $V_{(\vec{\kappa},\vec{\beta})}^{l,\vec{q}}(\Omega)$ and $C_{(\vec{\kappa},\vec{\beta})}^{l,\delta}(\Omega)$, $\beta_i > 0$, $i = 1,...,m$, of exponentially vanishing at infinity functions. The solvability of it is proved for small data.

Theorem 4.1. (Pileckas, 1994c) (i) *Let $\partial\Omega \in C^{l+2}$ and $\vec{f} \in \widetilde{V}_{(\vec{\kappa},\vec{\beta})}^{l,\vec{q}}(\Omega)$ with*

$$l \ge 0, \quad q_i > 1, \quad \beta_* > \beta_i > 0, \quad \kappa_i \quad is \quad arbitrary. \tag{4.5}$$

Then for sufficiently small $\|\vec{f};\ \widetilde{V}_{(\vec{\kappa},\vec{\beta})}^{l,\vec{q}}(\Omega)\|$ problem (4.1)–(4.4)$_0$ has a unique solution $(\vec{u},\ p)$ with $\vec{u} \in V_{(\vec{\kappa},\vec{\beta})}^{l+2,\vec{q}}(\Omega)$, $\nabla p \in V_{(\vec{\kappa},\vec{\beta})}^{l,\vec{q}}(\Omega)$ and the following estimate holds true

$$\|\vec{u};\ V_{(\vec{\kappa},\vec{\beta})}^{l+2,\vec{q}}(\Omega)\| + \|\nabla p;\ V_{(\vec{\kappa},\vec{\beta})}^{l,\vec{q}}(\Omega)\| \le c\|\vec{f};\ \widetilde{V}_{(\vec{\kappa},\vec{\beta})}^{l,\vec{q}}(\Omega)\|. \tag{4.6}$$

(ii) *Let $\partial\Omega \in C^{l+2,\delta}$ and $\vec{f} \in C_{(\vec{\kappa},\vec{\beta})}^{l,\delta}(\Omega)$,*

$$l \ge 0, \quad 1 > \delta > 0, \quad \beta_* > \beta_i > 0, \quad \kappa_i \quad is \quad arbitrary. \tag{4.7}$$

If the norm $\|\vec{f};\ C_{(\vec{\kappa},\vec{\beta})}^{l,\delta}(\Omega)\|$ is sufficiently small, then (4.1)–(4.4)$_0$ has a unique solution $(\vec{u},\ p)$ with $\vec{u} \in C_{(\vec{\kappa},\vec{\beta})}^{l+2,\delta}(\Omega)$, $\nabla p \in C_{(\vec{\kappa},\vec{\beta})}^{l,\delta}(\Omega)$ and

$$\|\vec{u};\ C_{(\vec{\kappa},\vec{\beta})}^{l+2,\delta}(\Omega)\| + \|\nabla p;\ C_{(\vec{\kappa},\vec{\beta})}^{l,\delta}(\Omega)\| \le c\|\vec{f};\ C_{(\vec{\kappa},\vec{\beta})}^{l,\delta}(\Omega)\|. \tag{4.8}$$

To prove the statements of the theorem we put the nonlinear term $(\vec{u} \cdot \nabla)\vec{u}$ to the right and we consider the nonlinear problem (4.1)–(4.4)$_0$ as the linear one with the right–hand side equal to $\vec{f} - (\vec{u} \cdot \nabla)\vec{u}$. Applying the results of Section 2, we reduce now the problem to the operator equation either in the space $V^{l+2,\vec{q}}_{(\vec{\kappa},\vec{\beta})}(\Omega)$ or in the space $C^{l+2,\delta}_{(\vec{\kappa},\vec{\beta})}(\Omega)$:

$$\vec{u} = \mathcal{A}\vec{u}.$$

By using the weighted imbedding Theorem 1.3, we prove that for sufficiently small data the nonlinear terms define a contraction operator in a small ball of the spaces $V^{l+2,\vec{q}}_{(\vec{\kappa},\vec{\beta})}(\Omega)$ and $C^{l+2,\delta}_{(\vec{\kappa},\vec{\beta})}(\Omega)$. Therefore, the theorem follows from the Banach contraction principle.

4.2. Solvability of problem (4.1)–(4.4) with nonzero fluxes

The main result for the problem (4.1)–(4.4) with nonzero fluxes F_i, $i = 1, ..., m$, reads as follows

Theorem 4.2. (Pileckas, 1994c) *Let $\Omega \subset \mathbb{R}^3$ be a domain with $m > 1$ exits to infinity. Assume that, in addition to (1.2), (1.3) the functions g_i satisfy the conditions*

$$\int_0^\infty g_i^{-4/3}(t)\,dt = \infty, \quad i = 1, ..., m, \tag{4.9}$$

$$|g_i'(t)g_i^{1/3}(t)| \leq \gamma << 1 \quad for \quad t > k_0, \quad i = 1, ..., m. \tag{4.10}$$

Let $\partial\Omega \in C^{l+2,\delta}$, $l \geq 0$, $0 < \delta < 1$, $\vec{f} = 0$ and suppose that for each $i \in \{1, \ldots, m\}$ there exists a number $q_i^* \geq 3/2$ such that

$$\int_0^\infty g_i^{-3q_i^*+2}(t)\,dt < \infty. \tag{4.11}$$

(i) *Then for arbitrary fluxes $F_i \in \mathbb{R}^1, i = 1, ..., m$, there exists a solution $\vec{u}$ of problem (4.1)–(4.4), belonging to the space $V^{l+2,\vec{q}}_{(\vec{\kappa}^*,\vec{0})}(\Omega)$ with*

$$\kappa_i^* = 4 + l - 3q_i^*/q_i, \quad i = 1, \ldots, m. \tag{4.12}$$

Moreover, there exists the pressure function $\nabla p \in V^{l,\vec{q}}_{(\vec{\kappa}^*,\vec{0})}(\Omega)$ and the following estimate holds true

$$\|\vec{u};\ V^{l+2,\vec{q}}_{(\vec{\kappa}^*,\vec{0})}(\Omega)\| + \|\nabla p;\ V^{l,\vec{q}}_{(\vec{\kappa}^*,\vec{0})}(\Omega)\| \leq C(|\vec{F}|). \tag{4.13}$$

In particular, $\vec{u} \in \widehat{H}^{\vec{q}}_{\vec{\kappa}^*}(\Omega)$ $(\vec{u} \in \widehat{H}^{\vec{q}}(\Omega))$.

(ii) *The solution $(\vec{u}, p)$ of problem (1.1)–(1.4) admits the pointwise estimates*

$$|D^\alpha \vec{u}(x)| \leq C(|\vec{F}|)g_i^{-2-|\alpha|}(x_3), \quad x \in \Omega_i, \quad 0 \leq |\alpha| \leq l, \tag{4.14}$$

$$|D^\alpha \nabla p(x)| \leq C(|\vec{F}|)g_i^{-3-|\alpha|}(x_3), \quad x \in \Omega_i, \quad 0 < |\alpha| \leq l, \tag{4.15}$$

$$|p(x)| \leq C(|\vec{F}|)\int_0^{x_3} g_i^{-4}(t)\,dt + c_1, \quad x \in \Omega_i. \tag{4.16}$$

The proof of this theorem is based on the estimate obtained for the weak solution of the problem with the unbounded Dirichlet integral by Ladyzhenskaya and Solonnikov (1980). Namely, there holds

Theorem 4.3. (Ladyzhenskaya and Solonnikov, 1980) *Let $\vec{f} = 0$. Suppose that the functions g_i satisfy conditions* (4.9), (4.10). *Then for arbitrary fluxes $F_i \in I\!\!R^1$, $i = 1, ..., m$, there exists at least one weak solution $\vec{u} \in W^{1,2}_{loc}(\Omega)$ to* (4.1)–(4.4) *with an infinite Dirichlet integral. This solution admits the representation as a sum*

$$\vec{u} = \vec{A} + \vec{v}, \tag{4.17}$$

where $\vec{A}$ is a divergence free vector field satisfying flux condition (4.4) *and estimates* (1.7) *(see Lemma 1.2). The following estimates hold true*

$$\int_{\Omega_{(k)}} |\nabla \vec{v}|^2 \, dx \leq c(|\vec{F}|) \sum_{k=1}^{m} \int_0^{R_{k+1}} g_i^{-4}(t) \, dt, \tag{4.18}$$

$$\int_{\omega_{ik}} |\nabla \vec{v}|^2 \, dx \leq c(|\vec{F}|) g_{ik}^{-8/3}, \, k > k_0, i = 1, ..., m. \tag{4.19}$$

Moreover, if the fluxes $F_i, i = 1, ..., m$, are sufficiently small, then any other (different from $\vec{u}$) solution $\vec{u}'$ of problem (4.1)–(4.4) *satisfies the relation*

$$\liminf_{k \to \infty} k^{-3} \int_{\Omega_{(k)}} |\nabla(\vec{u}' - \vec{u})|^2 \, dx > 0. \tag{4.20}$$

On the heuristical level one can see that each function $\vec{v}$ satisfying estimates (4.19) and admitting the decay estimates

$$|D^\alpha \vec{v}(x)| \leq c \, g_i^{\gamma - |\alpha|}(x_3), \quad x \in \Omega_i, \, |\alpha| \geq 0,$$

should have, minimally, the decay rate $\vec{v} \sim g_i^{-11/6 - |\alpha|}(x_3)$, i.e. $\gamma = -11/6$. Then

$$\Delta \vec{v} \sim g_i^{-23/6}(x_3), \quad (\vec{v} \cdot \nabla)\vec{v} \sim g_i^{-28/6}(x_3).$$

Thus, the nonlinear term $(\vec{v} \cdot \nabla)\vec{v}$ decays at infinity faster than the linear one $\Delta \vec{v}$ and by bootstrap argument we can improve the estimates for $\vec{v}$ and derive (4.14). Having this simple idea as background, Theorem 4.2 is proved by repeated application of weighted imbedding Theorem 1.2, the local weighted estimates and results on the linear Stokes problem (2.1)–(2.4)$_0$ (see Theorem 2.1, 2.2).

For two dimensional domain Ω with exits to infinity the analogous result can not be proved by the same method. This is related to the fact that in this case the solenoidal vector $\vec{A}$ satisfying the flux conditons (4.4), has decay rate as $g_i^{-1}(x_2)$ and hence,

$$\Delta \vec{A} \sim g_i^{-3}(x_2), \quad (\vec{A} \cdot \nabla)\vec{A} \sim g_i^{-3}(x_2) \quad \text{as } |x| \to \infty, \, x \in \Omega_i.$$

Thus, the linear and nonlinear terms have the same power at infinity. For domains $\Omega \subset I\!\!R^2$ with $m > 1$ we have the results only for small data. The proof of the following theorem is based on the Banach contraction principle.

Theorem 4.4. (Pileckas, 1994c) *Let $\Omega \subset I\!\!R^2$ be a domain with $m > 1$ exits to infinity. For sufficiently small $|\vec{F}|$ problem* (4.1)–(4.4) *has a unique solution $(\vec{u}, p)$ satisfying representation* (4.17) *and the estimates*

$$|D^\alpha \vec{u}(x)| \leq C(|\vec{F}|) g_i^{-1 - |\alpha|}(x_2), \quad x \in \Omega_i, \quad 0 \leq |\alpha| \leq l,$$

$$|D^\alpha \nabla p(x)| \leq C(|\vec{F}|) g_i^{-2 - |\alpha|}(x_2), \quad x \in \Omega_i, \quad 0 < |\alpha| \leq l,$$

$$|p(x)| \leq C(|\vec{F}|) \int_0^{x_2} g_i^{-3}(t) \, dt + c_1, \quad x \in \Omega_i.$$

Remark 4.1. *Theorems 4.2– 4.4 are valid also for nonzero right–hand sides $\vec{f}$ having an appropriate decay at infinity.*

REFERENCES

ABERGEL, F., BONA, J.L. (1992), A mathematical theory for viscous free–surface flows over a perturbed plane, *Arch. Rational Mech. Anal.*, **118**, 71–93.

AGMON, S., DOUGLIS, A., NIRENBERG, L. (1964), Estimates near the boundary for solutions of elliptic partial differential equations satisfying general boundary conditions II. *Commun. Pure Appl. Math.*, **17**, 35–92.

ADAMS, R.A. (1975), *Sobolev Spaces*, (Academic Press, New York, San Francisco, London).

AMICK, C.J. (1977), Steady solutions of the Navier–Stokes equations in unbounded channels and pipes, *Ann. Scuola Norm. Sup. Pisa*, **4**, 473–513.

AMICK, C.J. (1978), Properties of steady Navier–Stokes solutions for certain unbounded channels and pipes, *Nonlinear analisys, Theory, Meth., Appl.*, **2**, 689–720.

AMICK, C.J., FRAENKEL, L.E. (1980), Steady solutions of the Navier –Stokes equations representing plane flow in channels of various types, *Acta Math.*, **144**, 81–152.

BORCHERS, W., PILECKAS, K. (1992), Existence, uniqueness and asymptotics of steady jets, *Arch. Rational Mech. Anal.*, **120**, 1-49.

BORCHERS, W., GALDI, G.P. and K. PILECKAS (1993), On the uniqueness of Leray-Hopf solutions for the flow through an aperture, *Arch. Rational Mech. Anal.*, **122**, 19–33.

CHANG, H. (1992), The steady Navier–Stokes problem for low Reynolds number viscous jets, *Proc. of the conference "The Navier-Stokes equations: Theory and Numerical Methods," Lecture Notes in Math.*, **1530**, 85–96 (Springer, Berlin).

COSCIA, V., PATRIA, M.C. (1992), Existence, uniqueness and asymptotic decay of steady flow of an incompressible fluid in a half space, *Stability Appl. Anal. Cont. Media*, **2**, 101–127..

GALDI, G. P. (1994), *An Introduction to the Mathematical Theory of the Navier-Stokes Equations* Vols. I-II, Springer Tracts in Natural Philosophy (Springer, Heidelberg).

HEYWOOD, J.G. (1976), On Uniqueness questions in the theory of viscous flow, *Acta. Math.*, **136**, 61–102.

KAPITANSKII, L.V. (1981), Coincidence of the spaces $\overset{o}{J}{}^{1}_{2}(\Omega)$ and $\hat{J}^{1}_{2}(\Omega)$ for plane domains Ω having exits at infinity, *Zapiski Nauch. Semin. LOMI*, **110**, 74–81. English Transl. (1984): *J. Sov. Math.*, **25**, 850–855.

KAPITANSKII, L.V., PILECKAS, K. (1983), On spaces of solenoidal vector fields and boundary value problems for the Navier-Stokes equations in domains with non-compact boundaries, *Trudy Mat. Inst. Steklov*, **159**, 5–36. English Transl. (1984): *Proc. Math. Inst. Steklov*, **159**, issue 2, 3–34.

LADYZHENSKAYA, O.A. (1969), *The Mathematical Theory of Viscous Incompressible Flow*, (Gordon and Breach, New York, London, Paris).

LADYZHENSKAYA, O.A., SOLONNNIKOV, V.A. (1976), On some problems of vector analysis and generalized formulations of boundary value problems for the Navier-Stokes equations, *Zapiski Nauchn. Sem. LOMI*, **59**, 81–116. English Transl. (1978): *J. Sov. Math.*, **10**, No. 2, 257–285.

LADYZHENSKAYA, O.A., SOLONNNIKOV, V.A. (1977), On the solvability of boundary value problems for the Navier-Stokes equations in regions with noncompact boundaries, *Vestnik Leningrad. Univ.*, **13** (*Ser. Mat. Mekh. Astr.*, vyp. 3), 39–47. English Transl. (1982): *Vestnik Leningrad Univ. Math.*, **10**, 271–280.

LADYZHENSKAYA, O.A., SOLONNNIKOV, V.A. (1980), Determination of the solutions of boundary value problems for stationary Stokes and Navier-Stokes equations having an unbounded Dirichlet integral, *Zapiski Nauchn. Sem. LOMI*, **96**, 117–160. English Transl. (1983): *J. Sov. Math.*, **21**, No.5, 728–761.

LADYZHENSKAYA, O.A., SOLONNNIKOV, V.A. (1983), On initial–boundary value problem for the linearized Navier–Stokes system in domains with noncompact boundaries., *Trudy Mat. Inst. Steklov*, **159**. English Transl. (1984): *Proc. Math. Inst. Steklov*, **159**, issue 2, 35–40.

MASLENNIKOVA, V.N., BOGOVSKII, M.E. (1981), Sobolev spaces of solenoidal vector fields, *Sibirsk. Mat. Zh.*, **22**, 91–118. English Transl. (1982): *Siberian Math. J.*, **22**, 399–420.

NAZAROV, S.A., PILECKAS, K. (1983), On the behaviour of solutions of the Stokes and Navier–Stokes systems in domains with periodically varying section, *Trudy Mat. Inst. Steklov*, **159**, 137–149. English Transl. (1984): *Proc. Math. Inst. Steklov*, **159**, issue 2, 141–154.

NAZAROV, S.A., PILECKAS, K. (1993), On noncompact free boundary problems for the plane stationary Navier–Stokes equations, *J. für reine und angewandte Math.*, **438**, 103–141.

PILECKAS, K. (1980), Three–dimensional solenoidal vectors, *Zapiski Nauchn. Sem. LOMI*, **96**, 237–239. English Transl. (1983): *J. Sov. Math.*, **21**, 821–823.

PILECKAS, K. (1981a), On the solvability of certain problem of a plane motion of viscous incompressible liquid with a noncompact free boundary, *Zapiski Nauchn. Sem. LOMI*, **110**, 174–179. English Transl. (1984): *J. Sov. Math.*, **25**, No.1, 927–931.

PILECKAS, K. (1981b), Existence of solutions for the Navier–Stokes equations having an infinite dissipation of energy, in a class of domains with noncompact boundaries, *Zapiski Nauchn. Sem. LOMI*, **110**, 180–202. English Transl. (1984): *J. Sov. Math.*, **25**, No.1, 932–947.

PILECKAS, K. (1981c), Solvability of a problem of a plane motion of viscous fluid with noncompact boundary, *Differential Equations and their Application*, Vilnius, **30**, 57–96 (in Russian).

PILECKAS, K. (1983), On spaces of solenoidal vectors, *Trudy Mat. Inst. Steklov*, **159**, 137–149. English Transl. (1984): *Proc. Math. Inst. Steklov*, **159**, issue 2, 141–154.

PILECKAS, K. (1984), Existence of axisymmetric solutions of the stationary system of Navier-Stokes equations in a class of domains with noncompact boundary, *Litovskii Mat. Sb.*, **24**, 145–154. English Transl. (1984): *Lithuanian Math. J.*, **24**, 53-59.

PILECKAS, K. (1988a), On the problem of motion of heavy viscous incompressible fluid with noncompact free boundary, *Litovskii Mat. Sb.*, **28**, 315-333. English Transl. (1988): *Lithuanian Math. J.*, **28**, No. 2.

PILECKAS, K. (1988b), The example of nonuniqueness of the solutions to a noncompact free boundary problem for the Navier-Stokes system, *Differential Equations and their Application*, Vilnius, **42**, 59-65 (in Russian).

PILECKAS, K. (1989), On plane motion of a viscous incompressible cappilary liquid with a noncompact free boundary, *Arch. Mech.*, **41**, No. 2-3, 329-342.

PILECKAS, K. (1994a), Weighted L^q–solvability for the steady Stokes system in domains with noncompact boundaries, to appear.

PILECKAS, K. (1994b), Classical solvability and uniform estimates for the steady Stokes system in domains with noncompact boundaries, to appear.

PILECKAS, K. (1994c), Strong solutions of the steady nonlinear Navier–Stokes system in domains with exits to infinity, to appear.

PILECKAS, K., SOLONNIKOV, V.A. (1989), On stationary Stokes and Navier-Stokes systems in an open infinite channel. I., *Litovskii Mat. Sb.*, **29**, No.1, 90-108. English Transl. (1989): *Lithuanian Math. J.*, **29**, No.1; II., *Litovskii Mat. Sb.*, **29**, No.2, 347-367. English Transl. (1989): *Lithuanian Math. J.*, **29**, No.2.

PILECKAS, K., SPECOVIUS–NEUGEBAUER, M. (1989), Solvability of a noncompact free boundary problem for the stationary Navier-Stokes system. I., *Litovskii Mat. Sb.*, **29**, No.3, 532-547. English Transl. (1989): *Lithuanian Math. J.*, **29**, No.3; II., *Litovskii Mat. Sb.*, **29**, No.4, 773-784 (in Russian).

SOLONNIKOV, V.A. (1964), On the boundary value problems for systems elliptic in the sence of A. Douglis, L. Nirenberg, I. *Izv. Akad. Nauk SSSR, Ser. Mat.*, **28**, 665–706; II. *Trudy Mat. Inst. Steklov*, **92**, 233–297. English Transl. (1966): I. *Amer. Math. Soc. Transl.*, **56** (2), 192–232, II. *Proc. Steklov Inst. Math.*, **92**, 269–333.

SOLONNIKOV, V.A. (1981), On the solvability of boundary and initial–boundary value problems for the Navier–Stokes system in domains with noncompact boundaries, *Pacific J. Math.*, **93**, No.2, 443–458.

SOLONNIKOV, V.A. (1982), On solutions of stationary Navier–Stokes equations with an infinite Dirichlet integral, *Zapiski Nauchn. Sem. LOMI*, **115**, 257–263. English Transl. (1985): *J. Sov. Math.*, **28**, No.5, 792–799.

SOLONNIKOV, V.A. (1983), Stokes and Navier–Stokes equations in domains with noncompact boundaries, *College de France Seminar*, **4**, 240-349.

SOLONNIKOV, V.A. (1989), Solvability of the problem of effluence of a viscous incompressible fluid into an open basin, *Trudy Mat. Inst. Steklov*, **179**, 193–225. English Transl. (1989): *Proc. Math. Inst. Steklov*, **179**, issue 2, 193–225.

SOLONNIKOV, V.A. (1991), Boundary and initial–boundary value problems for the Navier–Stokes equations in domains with noncompact boundaries, *Math. Topics in Fluid Mechanics, Pitman Research Notes in Mathematics Series*, **274**, 117–162.

SOLONNIKOV, V.A., PILECKAS, K. (1977), Certain spaces of solenoidal vectors and the solvability of the boundary value problem for the Navier-Stokes system of equations in domains with noncompact boundaries, *Zapiski Nauchn. Sem. LOMI*, **73**, 136–151. English Transl. (1986): *J. Sov. Math.*, **34**, No.6 , 2101–2111.

TEMAM, R. (1977), *Navier–Stokes equations*, (Northolland Pub. Co., Amsterdam, N.Y., Tokyo).

GENERALIZED SOLUTIONS FOR THE STOKES EQUATION IN EXTERIOR DOMAINS

Roberta Pulidori[1] and Maria Specovius-Neugebauer[2]

[1] Università di Ferrara, Dipartimento di Matematica
Via Machiavelli, 21, 44100 Ferrara, Italy

[2] University of Paderborn, Fachbereich 17
Warburger Straße 100, D–33098 Paderborn, Germany

ABSTRACT

In this paper the idea of q-generalized solutions, i.e. $\nabla u \in L^q$, for the exterior Stokes problem is extended to weighted Sobolev spaces with weights proportional to a power $|x|^\beta$. Existence and uniqueness results are proved for suitable $\beta \in \mathbb{R}$.

1. INTRODUCTION

We study the Stokes system in an exterior domain $\Omega \subset \mathbb{R}^n$ with $0 \notin \overline{\Omega}$, $n \geq 3$. Let f be a given force. We look for a velocity field v and a scalar pressure p which fulfill

$$\Delta v - \nabla p = f, \qquad \operatorname{div} v = 0 \quad \text{in } \Omega, \tag{1.1}$$

$$v|_{\partial\Omega} = 0. \tag{1.2}$$

During the last years several authors (see, e.g. Galdi and Simader, 1990;, Kozono and Sohr, 1992) proved existence and uniqueness results for q-generalized solutions, i.e. $\nabla u \in L^q(\Omega)$. It turned out, that the problem (1.1), (1.2) has a unique q-generalized solution if and only if $n/(n-1) < q < n$. In this paper we extend the idea of q-generalized solutions to the case of homogeneous weighted Sobolev spaces , i.e. $|x|^\beta \nabla u \in L^q(\Omega)$. It is possible then to remove the restrictions on q and to derive explicit decay formulas for the solutions.

The paper is organized as follows: In Section 2 we list the definitions and some auxiliary results on weighted Sobolev spaces. In Section 3 the main theorems are presented, and the proofs are given in Section 4.

Navier-Stokes Equations and Related Nonlinear Problems
Edited by A. Sequeira, Plenum Press, New York, 1995

2. NOTATIONS AND PRELIMINARY RESULTS

Let $\Omega \subset I\!R^n$, $n \geq 3$ be an exterior domain, i.e. an open unbounded connected subset with compact boundary $\partial\Omega \in C^2$. For simplicity we assume that $0 \notin \overline{\Omega}$.

We recall the notation of some function spaces: For $1 < q < \infty$, $m \in I\!N$, $L^q(\Omega)$ denotes the usual Lebesgue space, $W^{m,q}(\Omega)$ the Sobolev space of all functions in $L^q(\Omega)$ with $D^\alpha f \in L^q(\Omega)$[1] for all $|\alpha| \leq m$ and $W_0^{m,q}(\Omega)$ the closure of $C_0^\infty(\Omega)$[2] in $W^{m,q}(\Omega)$. Since Ω is unbounded, we also use local spaces. By $W_{loc}^{m,q}(\overline{\Omega})$ we mean the space of all functions f with $f \in W^{m,q}(\Omega \cap K)$ for any bounded domain K with $K \cap \Omega \neq \emptyset$. For $1 < q < \infty$, $\beta \in I\!R$, we consider weighted L^q-spaces

$$L_\beta^q(\Omega) = \{f \text{ measurable }, \|f\|_{q,\beta,\Omega} = \||x|^\beta f\|_{L^q(\Omega)} < \infty\}.$$

We define weighted homogeneous Sobolev spaces

$$D_\beta^{1,q}(\Omega) = \{u \in L_{loc}^q(\overline{\Omega}), |x|^\beta \nabla u \in L^q(\Omega)\}.$$

$D_\beta^{1,q}(\Omega)$ is complete with respect to the seminorm

$$|u|_{1,q,\Omega}^{(\beta)} = \left(\int_\Omega (|x|^\beta |\nabla u|)^q\right)^{1/q}.$$

On the subspace $C_0^\infty(\Omega)$ $|\cdot|_{1,q,\Omega}^{(\beta)}$ does not only define a seminorm, but a norm. We denote the closure of $C_0^\infty(\Omega)$ with respect to this norm by $D_{0,\beta}^{1,q}(\Omega)$[3]. Some essential properties are characterized in the following lemma.

Lemma 2.1. *Let $\Omega \subset I\!R^n$, $n \geq 3$ be an exterior domain with $\partial\Omega \in C^1$ and $0 \notin \overline{\Omega}$. For $\beta \neq n/q + 1$ the norm*

$$\|u\|_{1,q,\beta-1,\Omega} := \|u\|_{q,\beta-1,\Omega} + |u|_{1,q,\Omega}^{(\beta)}$$

is equivalent to $|u|_{1,q,\Omega}^{(\beta)}$.

For $u \in L_{loc}^q(\overline{\Omega})$ with $\nabla u \in L_\beta^q(\Omega)$ it holds : $u \in D_{0,\beta}^{1,q}(\Omega)$ if and only if $\gamma u = 0$ and $u \in L_{\beta-1}^q(\Omega)$, where $\gamma u \in W^{1-1/q,q}(\partial\Omega)$ is the trace of u on the boundary.

Proof: The equivalence of the norms follows from Hardy's inequality (see Kufner, 1980, p. 34; , Benci and Fortunato, 1979, p. 321). For $u \in C_0^\infty(\Omega)$ it holds:

$$\|u\|_{q,\beta-1,\Omega} \leq C|u|_{1,q,\Omega}^{(\beta)} \tag{2.1}$$

hence $D_{0,\beta}^{1,q}(\Omega) \subset L_{\beta-1}^q(\Omega)$. Let $K \subset I\!R^n$ be an open ball which contains $\partial\Omega$. The trace theorem for bounded domains (see Adams, 1975, p. 216) leads to

$$\|\gamma u\|_{W^{1-1/q,q}(\partial\Omega)} \leq C\|u\|_{W^{1,q}(\Omega \cap K)} \leq C|u|_{1,q,\Omega}^{(\beta)}.$$

Hence $\gamma u = 0$ for $u \in D_{0,\beta}^{1,q}(\Omega)$. Conversely, let $u \in L_{loc}^q(\Omega)$ with $\nabla u \in L_\beta^q(\Omega)$, $u \in L_{\beta-1}^q(\Omega)$ with $\gamma u = 0$. Let $\psi_R \in C_0^\infty(\Omega)$, $R \geq 0$ be a system of cut off functions with $\psi_R = 1$ for $|x| \leq R$, $\psi_R = 0$ for $|x| \geq 2R$ and $|\nabla \psi_R(x)| \leq CR^{-1}$. Elementary

---

[1]We use the multiindex notation: For $\alpha \in I\!N_0^n$ we set $D^\alpha = \frac{\partial}{\partial x_1} \cdots \frac{\partial}{\partial x_n}$, $|\alpha| = \alpha_1 + \ldots + \alpha_n$.

[2]$C_0^\infty(\Omega)$ denotes the space of all infinitely differentiable functions with compact support in Ω.

[3]We use the same notations for the spaces of scalar functions and of vector fields, when no confusion arises.

calculations show that $|u - (\psi_R u)|^{(\beta)}_{1,q,\Omega} \to 0$ as $R \to \infty$. So for $\varepsilon > 0$ we find R_0 with $|u - \psi_{R_0} u|^{(\beta)}_{1,q,\Omega} < \varepsilon/2$. Since $\psi_{R_0} u \in W^{1,q}_0(\Omega_{2R_0})$, we find $\bar{u} \in C^\infty_0(\Omega_{2R_0})$ with

$$|\psi_{R_0} u - \bar{u}|^{(\beta)}_{1,q,\Omega} \leq C(R_0)\|\psi_{R_0} u - \bar{u}\|_{W^{1,q}(\Omega_{2R_0})} < \frac{\varepsilon}{2},$$

which proves $u \in D^{1,q}_{0,\beta}(\Omega)$. $\qquad\square$

Moreover, we have the following lemma on the decay properties in $D^{1,q}_{0,\beta}(\Omega)$. Let $S_n = \{x \in I\!\!R^n : |x| \leq 1\}$.

Lemma 2.2. *Let $\Omega \subset I\!\!R^n$, $n \geq 3$ be an exterior domain with $0 \notin \bar{\Omega}$. Let $1 < q < \infty$ and $\beta \in I\!\!R$ with $\beta > 1 - n/q$. Then for any $u \in D^{1,q}_{0,\beta}(\Omega)$ there exists $u^* \in L^q(S_n)$ such that*

$$\lim_{|x|\to\infty} \int_{S_n} |u - u^*|^q \mathrm{d}\vartheta = 0.$$

Moreover, if we set $u_0 = (n\omega_n)^{-1}\int_{S_n} u^*\mathrm{d}\vartheta$, $w = u - u_o$, we have

$$\int_{S_n} |w(r,\vartheta)|^q \, d\vartheta \leq \gamma_o r^{q-n-\beta q}\int_{\Omega_r} |x|^{\beta q}|\nabla u|^q \mathrm{d}x,$$

where $\Omega_r = \Omega \cap \{x : |x| < r\}$ and $\gamma_0 = \left[\frac{q-1}{q-n-\beta}\right]^{q-1}$.

This lemma is a generalization of Lemma 5.2 in Galdi (1994), p. 60. It can be proved in quite an analogous way by using the projection of $|x|^\beta \nabla u$ onto the sphere S_n and Wirtinger's inequality.

We define homogeneous weighted Sobolev spaces of order -1

$$D^{-1,q}_\beta(\Omega) = (D^{1,q'}_{0,-\beta}(\Omega))^*,$$

where $(\ )^*$ indicates the dual space normed by

$$\|f\|_{D^{-1,q}_\beta(\Omega)} = |f|^{(\beta)}_{-1,q,\Omega} = \sup\{|\langle f, u\rangle|; |u|^{(\beta)}_{1,q',\Omega} \leq 1\}.$$

From the definition it follows that $D^\alpha : D^{1,q}_{0,\beta}(\Omega) \to D^{-1,q}_\beta(\Omega)$ is a continuous linear operator for any $\alpha \in I\!\!N^n_0$ with $|\alpha| \leq 2$. Moreover, for $\beta \neq -n/q' - 1$ we get $L^q_{\beta+1}(\Omega) \subset D^{-1,q}_\beta(\Omega)$ by setting $\langle f, u\rangle = \int_\Omega fu\mathrm{d}x$ for any $f \in L^q_{\beta+1}(\Omega)$ and $u \in D^{1,q'}_{0,-\beta}(\Omega)$. Then it follows by Hölder's and Hardy's inequalities:

$$\left|\int_\Omega fu\mathrm{d}x\right| \leq \|f\|_{q,\beta+1,\Omega}\|u\|_{q',-\beta-1,\Omega} \leq C\|f\|_{q,\beta+1,\Omega}|u|^{(-\beta)}_{1,q',\Omega},$$

hence $|f|^{(\beta)}_{-1,q,\Omega} \leq C\|f\|_{q,\beta+1,\Omega}$.

We give now the definition of a weak solution for the problem (1.1), (1.2).

Definition 2.1. *Suppose $f \in D^{-1,q}_\beta(\Omega)$. A vector field $v \in D^{1,q}_{0,\beta}(\Omega)$ is called a q-β-generalized solution of (1.1), if and only if $\mathrm{div}\, v = 0$ and*

$$(\nabla v, \nabla\phi) = \int_\Omega \nabla v \nabla\phi\mathrm{d}x = -\langle f, \phi\rangle$$

for all $\phi \in C^\infty_0(\Omega)$.

3. MAIN THEOREMS

With the notations of the previous section we can prove the following theorems.

Theorem 3.1. *Let $\Omega \subseteq \mathbb{R}^n$, $n \geq 3$ be an exterior domain with $0 \notin \overline{\Omega}$ and $\partial\Omega \in C^2$. Let $1 < q < \infty$, $\beta \in \mathbb{R}$ with $-n/q + 1 < \beta < n/q' - 1$. Then for any $f \in D_\beta^{-1,q}(\Omega)$ there exists a unique q-β-generalized solution $u \in D_{0,\beta}^{1,q}(\Omega)$ of the system (1.1), (1.2).*

Moreover, there exists a unique pressure $p \in L_\beta^q(\Omega)$ such that the system (1.1) is fulfilled in the distributional sense. The following estimate holds:

$$|u|_{1,q,\Omega}^{(\beta)} + \|p\|_{q,\beta,\Omega} \leq C|f|_{-1,q,\Omega}^{(\beta)}$$

If $\beta < -n/q + 1$, the functions $u \in D_{0,\beta}^{1,q}(\Omega)$ may grow polynomially as $|x|$ tends to infinity. So there exist nontrivial solutions $(u, p) \in D_{0,\beta}^{1,q}(\Omega) \times L_\beta^q(\Omega)$ of the homogeneous system, i.e. $f = 0$ in (1.1). Let us denote by $S_\beta^q(\Omega)$ the subspace of $D_{0,\beta}^{1,q}(\Omega)$ generated by the solutions of the homogeneous problem. Now we recall Green's formula

$$\int_\Omega (\Delta v - \nabla p)u - (\Delta u - \nabla \pi)v \, dx = \int_\Omega p \operatorname{div} u - \pi \operatorname{div} v \, dx, \qquad (3.1)$$

which can be extended from the smooth functions with compact support to any $(v, p) \in D_{0,\beta}^{1,q}(\Omega) \times L_\beta^q(\Omega)$, $(u, \pi) \in D_{0,-\beta}^{1,q'}(\Omega) \times L_{-\beta}^{q'}(\Omega)$. (3.1) leads to the necessary condition

$$\langle f, u \rangle = 0$$

for any $u \in S_{-\beta}^{q'}(\Omega)$, here the brackets $\langle ., . \rangle$ indicate the duality $\langle D_\beta^{-1,q}(\Omega), D_{0,-\beta}^{1,q'}(\Omega) \rangle$. We characterize the generalized solutions in $D_{0,\beta}^{1,q}(\Omega)$ for $\beta \in (-n/q, -n/q + 1)$ and $\beta \in (n/q' - 1, n/q')$ in the following theorem:

Theorem 3.2. *For $\beta \in (-n/q, -n/q+1)$ the space $S_\beta^q(\Omega) = S_0$ is an n-dimensional subspace of $D_{0,\beta}^{1,q}(\Omega)$ not depending on β. For any $f \in D_\beta^{-1,q}(\Omega)$ there exists a unique generalized solution $u \in D_{0,\beta}^{1,q}(\Omega)/S_0$ of the system (1.1) and (1.2).*

If $\beta \in (n/q' - 1, n/q')$, then there exists a unique generalized solution if and only if $\langle f, v \rangle = 0$ for any $v \in S_0$.

4. PROOFS

For the results in the exterior domains we need some results for the inhomogeneous Stokes system

$$\Delta u - \nabla p = f, \quad \operatorname{div} u = g \quad \text{in } \mathbb{R}^n, \qquad (4.1)$$

where f and g are prescribed. Since the weight $|x|^\beta$ assigns an exceptional role to the point zero, we modify the weight function in the case $\Omega = \mathbb{R}^n$. Moreover, in the whole space Hardy's inequality is valid only for $\beta > -n/q + 1$. Thus we set $\omega(x) = (1 + |x|^2)^{1/2}$ and define

$$L_\beta^q = \{p \in L_{loc}^q(\mathbb{R}^n), \omega^\beta p \in L^q(\mathbb{R}^n)\},$$

normed by $\|u\|_{L_\beta^q} = \|u\|_{q,\beta} = \|u \cdot \omega^\beta\|_{L^q(\mathbb{R}^n)}$. Furthermore, we set

$$W_\beta^{1,q} = \{u \in W_{loc}^{1,q}(\mathbb{R}^n), \omega^{\beta-1}u \in L^q(\mathbb{R}^n), \nabla u \cdot \omega^\beta \in L^q(\mathbb{R}^n)\}.$$

$W_\beta^{1,q}$ is a Banach space provided with the norm $\|u\|_{W_\beta^{1,q}} = \|u \cdot \omega^{\beta-1}\|_{q,\mathbb{R}^n} + \|\omega^\beta \nabla u\|_{q,\mathbb{R}^n}$, and contains $C_0^\infty(\Omega)$ as a dense subspace. Let $W_\beta^{-1,q} = \left(W_{-\beta}^{1,q'}\right)^*$ denote the dual space of $W_{-\beta}^{1,q'}$. We summarize the results for the problem (4.1) in the following lemma (see Specovius-Neugebauer, 1995).

Lemma 4.3. *Let* $1 < q < \infty$, $f \in W_\beta^{-1,q}$, $g \in L_\beta^q$. *Then it holds*
(i) *For* $\beta \in (-n/q+1, n/q'-1)$ *there exists a unique solution* $u \in W_\beta^{1,q}$, $p \in L_\beta^q$, *which fulfils the system (4.1) in the distributional sense.*
(ii) *For* $\beta \in (n/q'-1, n/q')$ *there exists a unique solution* $u \in W_\beta^{1,q}$, $p \in L_\beta^q$ *if and only if* $\langle f, 1 \rangle = 0$.

Moreover, for $\beta > -n/q+1$, *and smooth data* f, g *the solution has the representation*

$$(u, p) = E * (f, g), \qquad (4.2)$$

where $E = (E_{jk})_{j,k=1}^n$ *is the fundamental solution matrix of the Stokes system,*

$$E_{jk}(x) = d_n \left\{ \frac{x_j x_k}{|x|^n} + \delta_{jk} \frac{|x|^{2-n}}{n-2} \right\},$$

$$E_{n+1,k}(x) = E_{k,n+1}(x) = \bar{d}_n \frac{x_k}{|x|^n},$$

$$E_{n+1,n+1}(x) = \delta(x).$$

$d_n, \bar{d}_n$ *are constants depending only on the dimension* n, *and* $\delta(x)$ *means Dirac's delta distribution, see Galdi and Simader (1990).*

(iii) *For* $\beta \in (-n/q, -n/q+1)$ *there exists a unique solution* $u \in W_\beta^{1,q}/\mathbb{R}$, $p \in L_\beta^q$ *of (4.1).*

Proof of Theorem 1:
Step 1. (Uniqueness for $\beta > -n/q+1$).
Let $(u, p) \in D_{0,\beta}^{1,q}(\Omega) \times L_\beta^q(\Omega)$ be a solution of the homogeneous problem (1.1), i.e. $f = 0$. Since $\partial\Omega$ is smooth, Cattabriga's theorem implies $u \in C^\infty(\Omega) \cap W_{loc}^{2,q}(\overline{\Omega})$, $p \in C^\infty(\Omega) \cap W_{loc}^{1,q}(\overline{\Omega})$. Let $\varphi \in C_0^\infty(\mathbb{R}^n)$ be a cut off function with $\varphi \equiv 1$ for $|x| \leq R$; $R > \max\{|y|, y \in \partial\Omega\}$ and $\varphi \equiv 0$ for $|x| > 2R$. We decompose $(u, p) = \varphi(u, p) + (1-\varphi)(u, p)$. Since $\varphi(u, p)$ has a compact support in $\overline{\Omega}$, we may use the representation formula for bounded domains (see, e.g. Galdi, 1994, p.230) in any region $\Omega_{\overline{R}} = \Omega \cap \{x, |x| < \overline{R}\}$ with $\overline{R} > 2R$ and obtain for $x \in \Omega$:

$$\varphi(u(x), p(x)) = E *_\Omega (\Delta(\varphi u) - \nabla(\varphi p), \operatorname{div}(\varphi u)) + V(T(u,p) \cdot \nu|_{\partial\Omega}). \qquad (4.3)$$

Here the index Ω indicates the convolution on Ω, V denotes the single layer potential with momentum $T(u,p) \cdot \nu$, and $T(u,p)$ the stress tensor. Now we extend $(1-\varphi)(u,p)$ by zero to the whole $\mathbb{R}^n$. From (4.2) we get

$$\begin{aligned}(1-\varphi)(u,p) &= E * (\Delta((1-\varphi)u) - \nabla((1-\varphi)p), \operatorname{div}((1-\varphi)u)) \\ &= E *_\Omega (\Delta((1-\varphi)u) - \nabla((1-\varphi)p), \operatorname{div}((1-\varphi)u)) \qquad (4.4)\end{aligned}$$

since $\operatorname{supp}(1-\varphi) \subset \Omega$. Adding (4.3) and (4.4) leads to $(u(x), p(x)) = V(T(u,p) \cdot \nu|_{\partial\Omega})$ which implies $|u(x)| = O(|x|^{2-n})$, $|p(x)| = O(|x|^{1-n})$ as $|x| \to \infty$. Hence $u = 0$, $p = 0$ by the classical uniqueness result (see Ladyženskaja, 1966, p.59)

Step 2. (A priori estimate for $\beta > -n/q + 1$).
We fix $u \in (C_0^\infty(\Omega))^n$, $p \in C_0^\infty(\Omega)$ and set $f = \Delta u - \nabla p$. We show:

$$|u|_{1,q,\Omega}^{(\beta)} + \|p\|_{q,\beta,\Omega} \leq C\left(|f|_{-1,q,\Omega}^{(\beta)} + \|\mathrm{div}\; u\|_{q,\beta,\Omega}\right). \tag{4.5}$$

Let $\varphi \in C_0^\infty(I\!\!R^n)$ be chosen as in Step 1. Then $(u_1, p_1) = \varphi(u, p) \in W^{1,q}(\Omega_{2R})$. Furthermore, we calculate

$$\begin{aligned}
f_1 &= \Delta u_1 - \nabla p_1 = \varphi f + 2\nabla u \cdot \nabla \varphi + (\Delta \varphi)u - (\nabla \varphi)p, &&(4.6)\\
\mathrm{div}\; u_1 &= \varphi \mathrm{div}\; u + (\nabla \varphi)u. &&(4.7)
\end{aligned}$$

From Cattabriga's inequality (Cattabriga, 1964), we deduce:

$$\begin{aligned}
\|u_1\|_{W^{1,q}(\Omega_{2R})} &+ \|p_1\|_{L^q(\Omega_{2R})}\\
&\leq C\left(\|f_1\|_{W^{-1,q}(\Omega_{2R})} + \|p_1\|_{W^{-1,q}(\Omega_{2R})} + \|\mathrm{div}\; u_1\|_{L^q(\Omega_{2R})}\right).
\end{aligned} \tag{4.8}$$

To estimate the right hand side of (4.8) we use the definition of $\|.\|_{W^{-1,q}(\Omega_{2R})}$

$$\|f_1\|_{W^{-1,q}(\Omega_{2R})} = \sup |\int_{\Omega_{2R}} f_1\, v dx|,$$

where the supremum is taken over all $v \in (C_0^\infty(\Omega_{2R}))^n$ with $\|v\|_{W^{1,q'}(\Omega_{2R})} \leq 1$. So we fix such a v. Using (4.6), Hölder's inequality and Hardy's inequality (2.1) we obtain

$$\begin{aligned}
\left|\int_{\Omega_{2R}} f_1 v\, dx\right| &= \left|\int_\Omega f_1 v\, dx\right|\\
&\leq \left|\int_{\Omega_{2R}} \varphi f \cdot v\, dx\right| + \left|\int_{T_R}(2\nabla u\nabla\varphi + (\Delta\varphi)u - p\nabla\varphi)v\, dx\right|\\
&\leq |f|_{-1,q,\Omega}^{(\beta)}|\varphi v|_{1,q',\Omega}^{(-\beta)} + \|u\|_{q,T_R}\|(\nabla\varphi + \nabla^2\varphi)v\|_{q',T_R}\\
&\quad + \|p\|_{W^{-1,q}(T_R)}\|\nabla\varphi v\|_{W^{1,q'}(T_R)}\\
&\leq C(\varphi,R,\beta)\left(|f|_{-1,q,\Omega}^{(\beta)} + \|u\|_{q,T_R} + \|p\|_{W^{-1,q}(T_R)}\right)\|v\|_{W^{1,q'}(\Omega_{2R})},
\end{aligned} \tag{4.9}$$

where $T_R = \{x : R < |x| < 2R\}$. To estimate the term $(1 - \varphi)(u, p)$ we extend $(1-\varphi)(u,p)$ by zero to $u_2 \in W_\beta^{1,q}$, $p_2 \in L_\beta^q$. From Lemma 3 we get with $f_2 = \Delta u_2 - \nabla p_2$:

$$\|u_2\|_{W_\beta^{1,q}} + \|p_2\|_{L_\beta^q} \leq \left(\|f_2\|_{W_\beta^{-1,q}} + \|\mathrm{div}\; u_2\|_{L_\beta^q}\right). \tag{4.10}$$

Similar as in (4.9) we use the definition:

$$\|f_2\|_{W_\beta^{-1,q}} = \sup\left\{\left|\int_{I\!\!R^n} f_2 v dx\right| ; v \in (C_0^\infty)^n, \|v\|_{W_{-\beta}^{1,q'}} \leq 1\right\}.$$

For $v \in (C_0^\infty(I\!\!R^n))^n$ with $\|v\|_{W_{-\beta}^{1,q'}} \leq 1$ we obtain:

$$\left|\int_{I\!\!R^n} f_2 v\, dx\right| \leq \left|\int_{I\!\!R^n}(1 - \varphi)f \cdot v\, dx\right| + \left|\int_{T_R}(2\nabla u \cdot \nabla\varphi + (\Delta\varphi)u - p(\nabla\varphi))v\, dx\right| \tag{4.11}$$

To treat the first term on the right hand side we use again Hölder's inequality and Hardy's inequality in Ω:

$$\begin{aligned}
\left|\int_{I\!\!R^n}(1 - \varphi)fv\, dx\right| &= \left|\int_\Omega f(1-\varphi)v\, dx\right| \leq |f|_{-1,q,\Omega}^{(\beta)}|(1 - \varphi)v|_{1,q',\Omega}^{(\beta)} \leq\\
&\leq |f|_{-1,q,\Omega}^{(\beta)}\left(\|(1 - \varphi)v\|_{q',\beta-1,\Omega} + |(1 - \varphi)v|_{1,q',\Omega}^{(\beta)}\right)\\
&\leq c(\varphi)|f|_{-1,q,\Omega}^{(\beta)}\|v\|_{W_\beta^{1,q'}}.
\end{aligned} \tag{4.12}$$

The last terms in (4.11) can be handled similar as in (4.9). Summing up (4.8) and (4.10) and using (4.7), (4.9) and (4.11) leads to

$$
\begin{aligned}
|u|_{1,q,\Omega}^{(\beta)} + \|p\|_{q,\beta,\Omega} &\leq C\left(|f|_{-1,q,\Omega}^{(\beta)} + \|\text{div } u\|_{q,\beta,\Omega} + \|u\|_{q,T_R} + \|p\|_{W^{-1,q}(T_R)}\right) \\
&\leq C\left(|f|_{-1,q,\Omega}^{(\beta)} + \|\text{div } u\|_{q,\beta,\Omega} + \|u\|_{r,\beta',\Omega} + |p|_{-1,q,\Omega}^{(\beta')}\right), \quad (4.13)
\end{aligned}
$$

where $\beta' < \beta - 1$ can be chosen arbitrarily. For the last inequality we used the equivalence of all norms $\|.\|_{q,\beta}$ on the bounded domain T_R. By density arguments (4.13) can be extended to all $u \in D_{0,\beta}^{1,q}(\Omega)$, $p \in L_\beta^q(\Omega)$. Since the embeddings $D_{0,\beta}^{1,q}(\Omega) \hookrightarrow L_\beta^q(\Omega)$, $L_\beta^q \hookrightarrow D_\beta^{-1,q}(\Omega)$ are compact, we may cancel the last two terms in (4.13) using the uniqueness and standard arguments (see, e.g. Choquet-Bruhat and Christodoulou, 1981, p.144). Hence the a priori inequality (4.5) is proved.

Step 3. (Existence of solutions).

From (4.5) we obtain

$$
|u|_{1,q,\Omega}^{(\beta)} + \|p\|_{q,\beta,\Omega} \leq c|f|_{-1,q,\Omega}^{(\beta)}, \quad (4.14)
$$

for any $u \in D_{0,\beta}^{1,q}(\Omega)$ with div $u = 0$, $p \in L_\beta^q(\Omega)$.

Hence it is sufficient to prove the existence of q-β-generalized solutions for $f \in C_0^\infty(\Omega)$. For a smooth f there exists a weak solution (u, p) with finite Dirichlet integral, which fulfills (see Solonnikov and Scadilov, 1973)

$$
\|\nabla u\|_{2,\Omega} + \|p\|_{2,\Omega} \leq c\|f\|_{W^{-1,2}(\Omega)}.
$$

We have $(u, p) \in C^\infty(\Omega)$ due to Cattabriga's inequality. It remains to investigate the behavior of $u(x), p(x)$ for $|x| \to \infty$. Lemma (4.1) in Galdi and Simader (1990) leads to the asymptotic expansion:

$$
u(x) = \tau U(x) + \sigma(x), \quad p(x) = \tau P(x) + \eta(x) \quad (4.15)
$$

with

$$
U(x) = (E_{ij}(x))_{i,j=1}^n, \quad P(x) = (E_{j,n+1}(x))_{j=1}^n, \quad \tau(x) = \int_{\Omega_R} f\, dx - \int_{\partial\Omega_R} T(v, p) \cdot \nu\, do,
$$

and $\sigma(x), \eta(x)$ are smooth functions with

$$
|D^\alpha \sigma(x)| = O(|x|^{1-n-|\alpha|}), \quad |D^\alpha \eta(x)| = O(|x|^{-n-|\alpha|}) \text{ as } |x| \to \infty.
$$

Using the decay properties of the fundamental solution $E(x)$ we arrive at $u \in D_{0,\beta}^{1,q}(\Omega)$, $p \in L_\beta^q(\Omega)$ for any $\beta < n/q' - 1$, and Theorem 1 is proved.

Proof of Theorem 2.

Step 1. The case $\beta < -n/q + 1$.

First we consider the space S_0. To each $v \in S_0$ we can assign a $\pi \in L_\beta^q(\Omega)$ such that (1.1) is fulfilled in the distributional sense, we set

$$
\overline{S}_0 = \{(v, \pi), v \in D_{0,\beta}^{1,q}(\Omega), \pi \in L_\beta^q(\Omega), \Delta v = \nabla\pi, \text{ div } v = 0\}.
$$

We recall a classical result for the exterior Stokes problem: For any $h \in C^1(\partial\Omega)$ there exists a unique solution $v \in C^\infty(\overline{\Omega})$, $\pi \in C^\infty(\Omega)$ with

$$
\Delta v = \nabla\pi, \quad \text{div } v = 0 \quad \text{in } \Omega, \quad v|_{\partial\Omega} = h, \quad (4.16)
$$

and

$$|v(x)| = O(|x|^{2-n}), \quad |p(x)| = O(|x|^{1-n}) \quad \text{as } |x| \to \infty. \tag{4.17}$$

We claim that

$$\overline{S}_0 = \{(v, \pi) = (v_0, 0) - (v_1, \pi_1);$$
$$v_0 \in I\!\!R^n, (v_1, \pi_1) \text{ solution of (4.16), (4.17) with } h = v_0\}. \tag{4.18}$$

Then $\dim \overline{S}_0 = \dim S_0 = n$ is obvious.

For the proof we assume first that $(v, \pi) \in \overline{S}_0$. We extend (v, π) to $(\bar{v}, \bar{\pi}) \in W_{loc}^{2,q}(I\!\!R^n) \times W_{loc}^{1,q}(\Omega)$, which is possible, since $(v, \pi) \in W_{loc}^{2,q}(\overline{\Omega}) \times W_{loc}^{1,q}(\Omega)$ (due to Cattabriga's inequality). $(\Delta \bar{v} - \nabla \bar{\pi}, \text{div } \bar{v})$ has a compact support which is contained in $I\!\!R^n \backslash \Omega$. Thus the convolution $E * (-(\Delta \bar{v} - \nabla \bar{\pi}, \text{div } \bar{v})) = (v_1, \pi_1)$ is well defined and has the decay properties (4.17). Moreover, using Lemma 1 for v we get $(\bar{v}, \bar{\pi}) + (v_1, \pi_1) \in W_\beta^{1,q} \times L_\beta^q$ and satisfies the homogeneous Stokes system in $I\!\!R^n$. Since $\bar{v}|_{\partial \Omega} = 0$, we have $v_1|_{\partial \Omega} = v_0$ and so $(\bar{v}, \bar{\pi})$ has the form asserted in (4.17). The other inclusion is obvious.

To prove the existence of q-β-generalized solutions for any $f \in D_\beta^{-1,q}(\Omega)$, we use similar arguments as in the proof of Theorem 1. We choose $u \in (C_0^\infty(\Omega))^n$, $p \in C_0^\infty(\Omega)$ and $(v, \pi) \in \overline{S}_0$, (u_1, p_1), (u_2, p_2) in an analogous way. For (u_1, p_1) we have again (4.8), (4.9). Part (iii) of Lemma 3 leads to

$$
\begin{aligned}
|u_2|_{1,q,\Omega}^{(\beta)} + \|p_2\|_{q,\beta,\Omega} &\leq C \left(\|\nabla u\|_{L_\beta^q} + \|p_2\|_{L_\beta^q} \right) \\
&\leq C \left(\|f_2\|_{W_\beta^{-1,q}} + \|\text{div } u_2\|_{L_\beta^q} \right).
\end{aligned}
$$

Now we proceed like in the estimate (4.11) and obtain (4.12). The same arguments are still valid if we replace (u, p) by $(u - v, p - \pi)$ so we arrive at

$$|u - v|_{1,q,\Omega}^{(\beta)} + \|p - \pi\|_{q,\beta,\Omega} \leq c \left(|f|_{-1,q,\Omega}^{(\beta)} + \|\text{div } u\|_{q,\beta,\Omega} + \|u - v\|_{q,\beta',\Omega} + |p - \pi|_{-1,q,\Omega}^{(\beta)} \right).$$

Taking the infimum with respect to $(v, p) \in \overline{S}_0$ and using again the compactness arguments lead to

$$\|(u, p)\|_{D_{0,\beta}^{1,q}(\Omega) \times L_\beta^q(\Omega)/\overline{S}_0} \leq C \left(|f|_{-1,q,\Omega}^{(\beta)} + \|\text{div } u\|_{q,\beta,\Omega} \right). \tag{4.19}$$

Since we already know the existence of a q-β-generalized solution $u \in D_{0,\beta}^{1,q}(\Omega)$, $p \in L_\beta^q(\Omega)$ for $f \in C_0^\infty(\Omega)$ by Step 3 of Theorem 1, the existence of solutions for arbitrary $f \in D_\beta^{-1,q}(\Omega)$ follows from (4.19).

Step 2. The case $\beta > n/q' - 1$.
The uniqueness of the solutions is already proved in Step 1 of Theorem 1. Moreover, we have the a priori inequality (4.5). As already mentioned, a necessary condition for the existence of a q-β-generalized solution for $\beta > n/q' - 1$ is $f \in S_0^\perp = \{f \in D_\beta^{-1,q}(\Omega), \langle f, v \rangle = 0, v \in S_0\}$. It remains to show the existence of solutions for a dense subset in $S_0^\perp$. For this purpose we construct a projection $P : D_\beta^{-1,q}(\Omega) \to S_0^\perp$ in the following way: Let $v_1, \ldots, v_n$ be a basis of S_0. Since $v_i \in L_{-\beta-1}^{q'}$ for all i, there exist $f_1, \ldots, f_n \in L_{\beta+1}^q \subset D_\beta^{-1,q}(\Omega)$ with

$$\langle f_j, v_i \rangle = \int_\Omega f_j v_i = \delta_{ji}.$$

We define for $f \in D_\beta^{-1,q}(\Omega)$

$$Pf = f - \sum_{i=1}^{n} \langle f, v_i \rangle f_i.$$

Obviously $Pf = f$ for all $f \in S_0^\perp$, hence P is a continuous projection.

Now we fix $f \in S_0^\perp$ arbitrary. Let $(f_n)_{n \in \mathbb{N}} \subset C_0^\infty(\Omega)$ be a sequence which converges to f in $D_\beta^{-1,q}(\Omega)$. Then $Pf_n \in L_{\beta+1}^q$ and $\lim_{n \to \infty} Pf_n = Pf = f$ in $D_\beta^{-1,q}(\Omega)$. Let ω denote the weight function of Lemma 1, i.e. $\omega(x) = (1 + |x|^2)^{1/2}$. By Theorem B in Specovius-Neugebauer (1986), we obtain a solution (u_n, p_n) with $\omega^{\beta-1}u \in L^q(\Omega)$, $\omega^\beta \nabla u \in L^q(\Omega)$ and $\omega^\beta p \in L^q(\Omega)$ which fulfills

$$\Delta u_n - \nabla p_n = Pf_n, \quad \text{div } u_n = 0 \quad \text{in } \Omega, \qquad \gamma u = 0.$$

Since $0 \notin \overline{\Omega}$, we have $u_n \in D_{0,\beta}^{1,q}(\Omega)$ by Lemma 1, and $p_n \in L_\beta^q$. Due to (4.5) u_n converges in $D_{0,\beta}^{1,q}(\Omega)$ to the $q - \beta$–generalized solution u belonging to f and p_n converges in L_β^q to the corresponding pressure.

AKNOWLEDGEMENTS

This research was supported by the DFG research group "Equations of Hydrodynamics", Universities of Bayreuth and Paderborn. The authors are very grateful to Prof. G.P. Galdi, Prof. H. Sohr and Mrs. G. Thäter for helpful discussions and advices.

REFERENCES

ADAMS, R.A. (1975), *Sobolev spaces*, (Academic press, New York).

BENCI, V. and D. FORTUNATO (1979), Weighted Sobolev spaces and the nonlinear Dirichlet problem, *Ann. Mat. Pur. Appl.* **121**, p. 319-336.

CANTOR, M. (1975), Spaces of functions with asymptotic conditions, *Indiana Univ. Math. J.* **24**, p. 897-902.

CATTABRIGA, L. (1964), Su un problema al contorno relativo al sistema di equazioni di Stokes, *Sem. Mat. Univ. Padova* **31**, 308-340.

CHOQUET-BRUHAT, Y. and D. CHRISTODOULOU (1981), Elliptic systems in $H_{s,\delta}$-spaces on manifolds which are Euclidic at infinity, *Acta math.* **146**, p. 129-150.

GALDI, G.P. (1994), An Introduction to the Mathematical Theory of the Navier-Stokes Equations, Vol. I, *Linearized problems*, Springer tracts in natural philosophy, (New York).

GALDI, G.P. and C.G. SIMADER (1990), Existence, uniqueness and L^q-estimates for the Stokes problem in an exterior domain, *Arch. Rat. Mech. Anal.* **112**, p. 291-318.

GIRAULT, V. and A. SEQUEIRA (1991), A well-posed problem for the exterior Stokes equation in two and three dimensions, *Arch. Rat. Mech. Anal.* **114**, p. 313-333.

KOZONO, H. and H. SOHR, H. (1992), On a new class of generalized solutions for the Stokes equations in exterior domains, *Sc. Norm. Sup. Pisa* **19**, p. 155-181.

KUFNER, A. (1980), *Weighted Sobolev Spaces*, (Teubner, Leipzig).

LADYŽENSKAJA, O.A. (1966), *The mathematical theory of viscous incompressible flow*, (Gordon & Breach, New York).

NEČAS, J. (1967), *Les méthodes directes en théorie des équations elliptiques*, (Masson, Paris).

NIRENBERG, L. and H. WALKER (1973), The null spaces of elliptic partial differential operators in $I\!R^n$, *J. Math. Anal. App.* **42**, p. 271-301.

NERI, U. (1977), *Singular integrals*, (Springer, New York).

SOLONNIKOV, V.A. and V.E. SCADILOV (1973), On a boundary value problem for a stationary system of Navier-Stokes equations, *Proc. Steklov Inst. Math.* **125**, p. 186-199.

SPECOVIUS-NEUGEBAUER, M. (1986), Exterior Stokes problems and decay at infinity, *Math. Meth. in the Appl. Sci.* **8**, p. 351-167.

SPECOVIUS-NEUGEBAUER, M. (1995), Weak solutions of the Stokes problem in weighted Sobolev spaces, *Acta Applicandae Mathematicae*, (to appear).

YOSHIDA, K. (1965), *Functional Analysis*, (Springer, New York).

ON THE EXISTENCE OF PERIODIC WEAK SOLUTIONS ON THE NAVIER–STOKES EQUATIONS IN EXTERIOR REGIONS WITH PERIODICALLY MOVING BOUNDARIES

Rodolfo Salvi

Dipartimento di Matematica, Politecnico di Milano
Via L. da Vinci, 32, 20133 Milano, Italy

1. INTRODUCTION

In this paper we consider the existence of periodic weak solutions for the Navier–Stokes equations in an exterior domain with periodically moving boundaries. To be more precise we consider a domain

$$\Omega_T = \bigcup_{0 \le t \le T} \Omega(t) \times \{t\}$$

where each $\Omega(t)$ is the exterior of a bounded connected domain $\Omega^c(t)$ in R^3, T is a finite number, and $\Omega(0) = \Omega(T)$.

The existence of periodic solutions of the exterior problem for the Navier-Stokes equations consists of finding in the region Ω_T exterior to a closed bounded surface, the velocity v and the pressure p which together solve the system (1), given below, and are such that the velocity assumes a given value on the surface, for $|x| \to \infty$, and $v(0) = v(T)$, $p(0) = p(T)$. When $\Omega(t)$ is a bounded domain, this problem has been investigated by several authors, and in particular by H. Morimoto (1971), by T. Miyakawa and Y. Teramoto (1982), and by R. Salvi (1994). Morimoto (1971) proved the existence by means of a kind of penalty method (this method was introduced by H. Fujita and N. Sauer, 1970), and Miyakawa and Teramoto (1982) reduced the problem to one in a cylindrical domain. In any case the proofs in Miyakawa and Teramoto (1982) and in Morimoto (1971) require that the domains move smoothly. In Salvi (1994), the author proves the existence of periodic solutions under very general conditions on $\Omega(t)$ making use of the method developed in Salvi (1985a, 1985b, 1988, 1990a, 1990b).

Concerning the unbounded domains (Ω depending or not on t), to our knowledge, it seems that, in literature, there exist only the results of P. Maremonti (1991) in R^3, for strong solutions. Probably, the reason is that the usual method to prove the

Navier-Stokes Equations and Related Nonlinear Problems
Edited by A. Sequeira, Plenum Press, New York, 1995

existence of periodic solutions makes use of a fixed point argument with the aid of $\| \nabla v \|_{L^2} \leq c \| v \|_{L^2}$. And this relation holds for bounded domains. To our knowledge, the problem is open even for a linearized version of the equations.

In this paper we prove the existence of periodic weak and strong solutions by use of the elliptic regularization as in Salvi (1994). In §2 we describe the problem, we define functional spaces, and we give the definition of periodic weak solutions. In §3 we give an existence proof for weak solutions. In §4 we prove the existence of strong solutions.

We remark that, in the case $\Omega(t) = \Omega(0)$ (bounded domain) for every t, the problem of periodic solutions has been studied extensively (see Prodi 1960; Prouse, 1963; Takeshita, 1970; for the Navier–Stokes equations, and Salvi, 1974; Salvi, 1983 for the Navier–Stokes type equations).

2. STATEMENT OF THE PROBLEM AND NOTATIONS

Let $\Omega(t)$ be an open bounded set in R^3 dependent on $t \in [0,T]$, T is a finite positive number. As t increases over $[0,T]$, $\Omega(t)$ generates a (x,t)- domain Ω_T and the boundary $\Gamma(t)$ of $\Omega(t)$ generates a (x,t)–hypersurface Γ_T. The motion in $\Omega(t)$ of an incompressible fluid with viscosity and density 1 , subjected to the external force $f = (f_1(x,t), f_2(x,t), f_3(x,t))$, is governed by the equations

$$\begin{cases} \partial_t v - \Delta v + v \cdot \nabla v + \nabla p &= f \\ \nabla \cdot v &= 0 \end{cases} \qquad (1)$$

where $v = v(t) = v(x,t) = (v_1(x,t), v_2(x,t), v_3(x,t))$ denotes the velocity, $p = p(x,t)$ the pressure, $\partial_t = \frac{\partial}{\partial t}$ and $v \cdot \nabla v = \sum_{i=1}^3 v_i \frac{\partial}{\partial x_i} v$.

We shall consider the boundary conditions

$$v(x,t) = 0 \qquad \text{on} \quad \Gamma(t), \quad \text{and for} \quad |x| \to \infty, \quad \text{and} \quad t \in [0,T]. \qquad (2)$$

We shall study the existence of a solution of the problem (1), (2) such that

$$v(0) = v(T) \qquad \text{(periodic condition)}. \qquad (3)$$

Let us begin by giving some definitions and basic notations.

$$(u,v)_{\Omega(t)} = \sum_{i=1}^3 \int_{\Omega(t)} u_i v_i dx; \qquad |u|^2_{\Omega(t)} = (u,u)_{\Omega(t)};$$

$$((u,v))_{\Omega(t)} = \sum_{i=1}^3 \int_{\Omega(t)} \nabla u_i \nabla v_i dx; \qquad \| u \|^2_{\Omega(t)} = ((u,u));$$

$$|u|^2 = \sum_{i=1}^3 u_i u_i.$$

For functions defined in Ω_T, we define $\gamma(u)$ by

$$\gamma^2(u) = \int_0^T \| \nabla u \|^2_{\Omega(t)} dt$$

whenever the integral makes sense.
We assume $\Omega(0) = \Omega(T)$.

Then we introduce

$$D(\Omega_T) = \{\phi | \phi \in C^\infty([0,T];(C_0^\infty(\Omega(t)))^3), \nabla \cdot \phi = 0, \phi(0) = \phi(T)\};$$
$$H(\Omega_T) = \text{the completion of } D(\Omega_T) \text{ under the } L^2(\Omega_T) - \text{norm};$$
$$V(\Omega_T) = \text{the completion of } D(\Omega_T) \text{ under the norm } \gamma(\phi);$$
$$H^{-1}(\Omega(t)) = \text{the dual of } V(\Omega(t));$$
$$\bar{H}^1(\Omega_T) = \text{the completion of } D(\Omega_T) \text{ under the } H^1(\Omega_T) - \text{norm}$$
$$H^1(\Omega_T) = \text{Sobolev space of order } 1 \text{ on } L^2(\Omega_T).$$

We set the definition of weak solution of (1), (2), (3).

Definition. v will be a weak solution of (1), (2), (3) if

$$i) \quad v \in V(\Omega_T)$$
$$ii) \quad \int_0^T \{(v, \partial_t \phi)_{\Omega(t)} + (v, v \cdot \nabla \phi)_{\Omega(t)} - ((v, \phi))_{\Omega(t)} + (f, \phi)_{\Omega(t)}\} dt = 0$$

holds for every $\phi \in D(\Omega_T)$.

The following theorems hold

Theorem 2.1. *Let $\Gamma_T \in C^1$. and $f \in L^2(0, T; H^{-1}(\Omega(t)) \cap L^2(\Omega_T)$. Then there exists a weak solution v of (1), (2), (3) such that $(v(t), \phi(t))_{\Omega(t)}$ is a continuous function in $[0, T]$ and T–periodic.*

Theorem 2.2. *Let $\Gamma_T \in C^3$ and $f \in L^2(0, T; H^{-1}(\Omega(t)) \cap L^2(\Omega_T)$, sufficiently small. Then there exists (v, p) that satisfies (1), (2) and (3) a.e. in Ω_T and $v \in V(\Omega_T)$ and $\partial_t v, D^2 v \in L^2(\Omega_T)$.*

Theorems 2.1 and 2.2 give the existence of weak and strong periodic solutions of the system (1), (2), (3).

3. PROOF OF THEOREM 2.1

Let us begin by considering the following approximating problem.

3.1. Auxiliary Problem

We look for $v^m \in \bar{H}^1(\Omega_T)$ such that for every $\phi \in \bar{H}^1(\Omega_T)$

$$\int_0^T \{\frac{1}{m}(\partial_t v^m, \partial_t \phi)_{\Omega(t)} + \frac{1}{m}(v^m, \phi)_{\Omega(t)} + ((v^m, \phi))_{\Omega(t)} + (v^{m-1} \cdot \nabla v^m, \phi)_{\Omega(t)} -$$
$$-(v^m, \partial_t \phi)_{\Omega(t)} - (f, \phi)_{\Omega(t)}\} dt = 0, \quad \text{with } \phi \in \bar{H}^1(\Omega_T) \text{ and } v^0 = 0.$$

We set

$$a_{\Omega_T}(v^m, \phi) = \int_0^T \{\frac{1}{m}(\partial_t v^m, \partial_t \phi)_{\Omega(t)} - (v^m, \partial_t \phi)_{\Omega(t)} +$$
$$+ ((v^m, \phi))_{\Omega(t)} + (v^{m-1} \cdot \nabla v^m, \phi)_{\Omega(t)} + \frac{1}{m}(v^m, \phi)_{\Omega(t)}\} dt;$$
$$L(\phi) = \int_0^T (f, \phi)_{\Omega(t)} dt.$$

65

By the following well-known theorem (see for example Girault and Raviart, 1979, page 106 or Leray and Lions, 1965), one obtains the existence of a solution in $\bar{H}^1(\Omega_T)$ of the equation

$$a_{\Omega_T}(v^m, \phi) = L(\phi). \tag{4}$$

(in the following we denote different constants by the same symbol c).

Theorem 3.1. *If*

i) there exists a constant $c_m > 0$ such that

$$a_{\Omega_T}(v^m, v^m) \geq c_m \parallel v^m \parallel_{\bar{H}^1(\Omega_T)}^2;$$

ii) the form

$$v^m \to a_{\Omega_T}(v^m, \phi)$$

is weakly continuous in $\bar{H}^1(\Omega_T)$ i.e.

$$v_n^m \to v^m \quad weakly \quad in \quad \bar{H}^1(\Omega_T) \quad implies$$
$$lim_{n\to\infty} a_{\Omega_T}(v_n^m, \phi) = a_{\Omega_T}(v^m, \phi).$$

Then there exists a solution in $\bar{H}^1(\Omega_T)$ of (4).

Proof: The condition i) can be easily proved; in fact, bearing in mind $v^m(0) = v^m(T)$, $\nabla \cdot v^m = 0$, we have

$$a_{\Omega_T}(v^m, v^m) = \int_0^T \{\frac{1}{m}(|\partial_t v^m|_{\Omega(t)}^2 + |v^m|_{\Omega(t)}^2) + \parallel v^m \parallel_{\Omega(t)}^2\}dt \geq$$
$$\geq c_m \parallel v^m \parallel_{\bar{H}^1(\Omega_T)}^2 .$$

To prove the condition ii), we note that

$$v_n^m \to v^m \quad weakly \quad in \quad \bar{H}^1(\Omega_T) \quad implies \quad that$$

$$\lim_{n\to\infty} \int_0^T (v^{m-1} \cdot \nabla v_n^m, \phi)_{\Omega(t)} dt = \int_0^T (v^{m-1} \cdot \nabla v^m, \phi)_{\Omega(t)} dt$$

for every $\phi \in \bar{H}^1(\Omega_T)$. Besides that, it is clear that all terms in $a_{\Omega_T}(v_n^m, \phi)$ converge. Therefore

$$\lim_{n\to\infty} a_{\Omega_T}(v_n^m, \phi) = a_{\Omega_T}(v^m, \phi)$$

for every $\phi \in \bar{H}^1(\Omega_T)$.

Now to pass to the limit $m \to \infty$ in (4), we shall need a priori estimates of the approximations v^m.

3.2. Standard a priori estimates

Replacing in (4) ϕ by v^m, it gives

$$\int_0^T \{\frac{1}{m}(|\partial_t v^m|_{\Omega(t)}^2 + |v^m|_{\Omega(t)}^2) + \parallel v^m \parallel_{\Omega(t)}^2 - (v^m, \partial_t v^m)_{\Omega(t)} +$$
$$+ (v^{m-1} \cdot \nabla v^m, v^m)_{\Omega(t)} - (f, v^m)_{\Omega(t)}\}dt = 0. \tag{5}$$

Bearing in mind $v^m(0) = v^m(T)$, $\nabla \cdot v^m = 0$, one has

$$\int_0^T \frac{1}{m}(|\partial_t v^m|^2_{\Omega(t)} + |v^m|^2_{\Omega(t)})dt \leq c; \int_0^T \| v^m \|^2_{\Omega(t)} dt \leq c \qquad (6)$$

(c is independent of m).

It follows

$$v^m \rightarrow v \qquad \text{weakly} \quad \text{in} \qquad V(\Omega_T).$$

To pass to the limit in the non linear term in (4) we need the convergence of $\{v^m\}$ in a suitable strong topology, for example in $L^2_{loc}(\Omega_T)$. To do this we will prove appropriate estimates.

3.3. Time Difference Quotient Estimates

We denote by $\bar{v}^m(x,t)$ the extension by zero to R^3 of $v^m(x,t)$ for every $t \in [0,T]$ and the periodic extension of v^m for $t < 0$ and $t > T$. We let

$$v_h^m = \frac{1}{h} \int_{t-h}^t \bar{v}^m(x,s)ds.$$

We have

$$v_h^m(0) = \frac{1}{h} \int_{-h}^0 \bar{v}^m(x,s)ds = \frac{1}{h} \int_{T-h}^T \bar{v}^m(x,s)ds = v_h^m(T).$$

Let $w_h^m(x,t)$ be the solution of the Stokes problem

$$\begin{cases} -\Delta v_h^m + \lambda w_h^m + \nabla p &=& 0 & \text{in} \quad \Omega(t) \\ \nabla \cdot w_h^m &=& 0 & \text{in} \quad \Omega(t) \\ w_h^m &=& v_h^m & \text{on} \quad \Gamma(t) \\ w_h^m &=& 0 & \text{as} \quad |x| \rightarrow \infty. \end{cases}$$

From standard results, we have

$$\| w_h^m \|_{\Omega(t)} \leq c \| v_h^m \|_{\Omega(t)}$$

(by the regularity of Ω_T we can consider c independent of t).
For the estimate of $\partial_t w_h^m$, we, formally, differentiate the above system with respect to t and consider $\partial_t w_h^m$ as a generalized solution of the problem

$$\begin{cases} -\Delta \partial_t w_h^m + \lambda \partial_t w_h^m + \nabla \partial_t p &=& 0 & \text{in} \quad \Omega(t), \\ \nabla \cdot \partial_t w_h^m &=& 0 & \text{in} \quad \Omega(t), \\ \partial_t w_h^m &=& \partial_t v_h^m & \text{on} \quad \Gamma(t) \\ \partial_t w_h^m &\rightarrow& 0 & \text{as} \quad |x| \rightarrow \infty \end{cases}$$

From standard results (see Fujita and Sauer, 1970 or Salvi, 1988), and bearing in mind that $\bar{v}^m(t)$ and $\bar{v}^m(t-h)$ are zero on $\Gamma(t)$ and on $\Gamma(t-h)$ respectively, we have

$$|\partial_t w_h^m|_{\Omega(t)} \leq c(\Omega(t)) \| \partial_t w_h^m \|_{H^{1/2}(\Gamma(t))} \leq$$

$$\leq \frac{c}{h} \| \bar{v}^m(t-h) \|_{H^{1/2}(\Gamma(t))} \leq \frac{c}{h}(\text{measure} \Delta\Omega(t))^{1/2} \| \bar{v}^m \|_{R^3}$$

Here $\Delta\Omega(t) = (\Omega(t) - \Omega(t-h)) \cup (\Omega(t-h) - \Omega(t))$, and $c = sup_t c(\Omega(t))$.

Now we have that in (4) all the noleading terms belong to $L^2(\Omega_T)$ for every m. It is routine matter that there exists a function $p^m(x,t)$ such that

$$-\frac{1}{m}\partial_{t^2}^2 v^m - \Delta v^m + \partial_t v^m + \nabla p^m = -v^{m-1} \cdot \nabla v^m + f \qquad (7)$$

67

is satisfied a.e. in Ω_T.

Furthermore p^m satisfies in $\Omega(t)$

$$\Delta p^m = -\nabla \cdot (v^{m-1} \cdot \nabla v^m + f)$$

and

$$\| p^m \|_{L^q_{loc}(\Omega(t))} \leq c \| v^{m-1}v^m \|_{L^q_{loc}(\Omega(t))}$$

Let $\theta \in C^\infty(R^3)$ a cut-off function such that $\theta = 1$ for $|x| < d$ and $= 0$ for $|x| \geq 2d$ (d is a positive number big enough).

We set

$$\psi = \theta^2(v_h^m - w_h^m).$$

Now, multiplying (7) by ψ and integrating over Ω_T, we obtain

$$\int_0^T \frac{1}{m}(\partial_t v^m, \theta^2((\bar{v}^m(t) - \bar{v}^m(t-h))/h - \partial_t w_h^m)_{\Omega(t)}dt -$$
$$-\frac{1}{h}\int_0^T (v^m(t), \theta^2(\bar{v}^m(t) - \bar{v}^m(t-h)))_{\Omega(t)}dt + \int_0^T ((v^m, \theta^2(v_h^m - w_h^m)))_{\Omega(t)}dt -$$
$$+\int_0^T (v^{m-1} \cdot \nabla v^m, \theta^2(v_h^m - w_h^m))_{\Omega(t)}dt - \int_0^T (f - \nabla p^m, \theta^2(v_h^m - w_h^m))_{\Omega(t)}dt = 0. \qquad (8)$$

By virtue of (6), Jensen's inequality and the smoothness of Γ_T, one has

$$\frac{1}{m}\int_0^T \{(|\partial_t \theta v^m|_{\Omega(t)}|\theta(\bar{v}^m(t) - \bar{v}^m(t-h))/h|_{\Omega(t)} + |\partial_t v^m|_{\Omega(t)}|\partial_t w_h^m|_{\Omega(t)}\}dt \leq$$
$$\leq \frac{c}{m}\int_0^T (|\partial_t \theta v^m|^2_{\Omega(t)} + \frac{1}{\sqrt{h}} \| \bar{v}^m \|_{R^3} |\partial_t v^m|_{\Omega(t)})dt \leq \frac{c}{\sqrt{h}}; \qquad (9)$$

$$|\int_0^T ((\theta v^m, \theta(v_h^m - w_h^m))_{\Omega(t)}dt| \leq \int_0^T \| \bar{v}^m \|_{\Omega(t)} \frac{1}{h}\int_{t-h}^t \bar{v}^m(s)ds \|_{R^3} dt \leq$$
$$\leq \frac{c}{\sqrt{h}}\int_0^T \| \bar{v}^m \|_{\Omega(t)} (\int_{t-h}^t \| \bar{v}^m \|_{R^3} ds)^{1/2}dt \leq \frac{c}{\sqrt{h}}. \qquad (10)$$

Analogously one obtains

$$|\int_0^T (\theta^2 v^{m-1} \cdot \nabla v^m, v_h^m - w_h^m)_{\Omega(t)}dt| \leq \frac{c}{\sqrt{h}};$$
$$|\int_0^T (\theta^2 f, v_h^m - w_h^m)_{\Omega(t)}dt| \leq \frac{c}{\sqrt{h}};$$
$$|\int_0^T (\theta^2 \nabla p^m, v_h^m - w_h^m)dt| \leq \frac{c}{\sqrt{h}}. \qquad (11)$$

Finally we estimate

$$-\frac{1}{h}\int_0^T (\theta v^m(t), \theta(\bar{v}(t) - \bar{v}^m(t-h))_{\Omega(t)}dt.$$

Bearing in mind the smoothness of Γ_T , one has

$$-\frac{1}{h}\int_0^T (\theta v^m(t), \theta(\bar{v}(t) - \bar{v}^m(t-h)))_{\Omega(t)}dt =$$

$$= -\frac{1}{h}\int_0^T |\theta\bar{v}^m(t)|^2_{\Omega(t)}dt + \frac{1}{2h}\int_0^T |\theta\bar{v}^m(t)|^2_{\Omega(t)}dt +$$

$$+\frac{1}{2h}\int_0^T |\theta\bar{v}^m(t-h)|^2_{\Omega(t)} - \frac{1}{2h}\int_0^T |\theta(\bar{v}^m(t) - \bar{v}^m(t-h))|^2_{\Omega(t)}dt \le$$

$$\le \frac{1}{2h}\int_h^T |\theta(\bar{v}^m(t) - \bar{v}^m(t-h))|^2_{\Omega(t)}dt + \frac{1}{2h}\int_h^T |\theta\bar{v}^m(t-h)|^2_{\Omega(t)-\Omega(t-h)}dt \le$$

$$\le \frac{c}{\sqrt{h}} - \frac{1}{2h}\int_h^T |\bar{v}^m(t) - \bar{v}^m(t-h)|^2_{\Omega(t)}dt. \tag{12}$$

From (10), (11), (12) we can conclude

$$\int_h^T |\theta(\bar{v}^m(t) - \bar{v}^m(t-h))|^2_{\Omega(t)}dt \le \frac{ch}{m}\int_0^T (|\partial_t v^m|^2_{\Omega(t)} +$$

$$+c\sqrt{h}\parallel \bar{v}^m \parallel_{R^3} |\partial_t v^m|_{\Omega(t)})dt + h|\int_0^T ((v^m, \theta^2(v_h^m - w_h^m)))_{\Omega(t)}dt|$$

$$h\int_0^T (f, \theta^2(v_h^m - w_h^m))_{\Omega(t)}dt| + c\sqrt{h} \le c\sqrt{h}. \tag{13}$$

By the classical characterization of M. Riesz and A. Kolmogorov of compact sets in $L^2(B \times (0,T))$, B a bounded domain, (see Nečas, 1965), we can prove that the set $\{v^m\}$ of v^m satisfying (6), (13) is relatively compact in $L^2_{loc}(\Omega_T)$.

From (6) and the relatively compactness of $\{v^m\}$ in $L^2_{loc}(\Omega_T)$, we can choose a subsequence, again denoted by $\{v^m\}$, such that for every $\phi \in D(\Omega_T)$

$$\lim_{m \to \infty}\int_0^T (v^{m-1} \cdot \nabla v^m, \phi)_{\Omega(t)}dt = \int_0^T (v \cdot \nabla v, \phi)_{\Omega(t)}dt.$$

Now passing to the limit $m \to \infty$ in (4) we obtain (ii) of the definition of weak solutions. Now we prove the continuity and the T–periodicity of $(v(t), \phi(t))_{\Omega(t)}$. From (7) we have

$$\frac{\partial^2 v^m}{\partial t^2} \in L^2(\Omega_T) \quad \forall m.$$

Consequently, from (4) and (6)

$$\{\frac{1}{m}(\partial_t v^m(t), \phi(t))_{\Omega(t)} + (v^m(t), \phi(t))_{\Omega(t)}\} \tag{14}$$

is uniformly bounded and equicontinuous on [0,T] $\forall\phi$ in $D(\Omega(t))$. By Ascoli–Arzelá theorem we can select a subsequence of (14) which converges to J(t) uniformly in $C[0, T]$. Thanks to the estimates (6) and (13) we have

$$\frac{1}{m}(\partial_t v^m(t), \phi(t))_{\Omega(t)} \to 0, \quad (v^m(t), \phi(t))_{\Omega(t)} \to (v(t), \phi(t))_{\Omega(t)} \quad a.e. \quad \text{in} \quad (0, T).$$

Then $J(t) = (v(t), \phi(t))_{\Omega(t)}$ is continuous in $[0, T]$ and T-periodic. Theorem 2.1 is completely proved.

4. PROOF OF THEOREM 2.2

Now we prove the existence of a strong solution of the problem (1),(2),(3) using the method developed in Salvi (1988) and Salvi (1990b).

Let

$$\mathcal{F} = \{\phi | \phi \in L^2(0,T;H^2(\Omega(t)) \cap V(\Omega(t))) \quad \text{with the natural norm}\};$$

$$\mathcal{P} = \{\phi | \phi \in L^2(0,T;H^2(\Omega(t)) \cap V(\Omega(t))), \quad \phi(0) = \phi(T),$$

$$\partial_t \phi \in L^2(0,T;H^2(\Omega(t)) \cap V(\Omega(t))) \quad \text{with the } \mathcal{F} \text{ norm}\}.$$

We consider the following problem.

Find a $v^m \in \mathcal{F}$ such that for all $\phi \in \mathcal{P}$

$$\int_0^T \{-(v^m, \lambda I + A)\partial_t \phi)_{\Omega(t)} + (Av^m, (\lambda I + A)\phi)_{\Omega(t)} +$$

$$+ \frac{1}{m}(v^m, (\lambda I + A)\phi)_{\Omega(t)}\}dt = \int_0^T (-u \cdot \nabla u +$$

$$+ f, (\lambda I + A)\phi)_{\Omega(t)})dt \qquad (15)$$

where λ is a suitable positive number, I is the unit operator, $u \in \mathcal{P}$ is a given function, and $A = -P\Delta$ (the Stokes operator) where P is the projector operator from $L^2(\Omega(t))$ onto $H(\Omega(t))$. We let

$$E(v^m, \phi) = \int_0^T \{-(v^m, (\lambda I + A)\partial_t \phi)_{\Omega(t)} + (Av^m, (\lambda I + A)\phi)_{\Omega(t)} +$$

$$+ \frac{1}{m}(v^m, (\lambda I + A)\phi_{\Omega(t)}\}dt;$$

$$L(\phi) = \int_0^T -(u \cdot \nabla u + f, (\lambda I + A)\phi)_{\Omega(t)})dt.$$

First $L(\phi)$ is a continuous form on $\mathcal{F}$ with respect to the norm $\| \phi \|_{\mathcal{P}}$.

Now bearing in mind that

$$\| \nabla \phi \|_{L^2(\Gamma(t))} \leq c(|D^2 \phi|_{\Omega(t)}^{1/2} \| \phi \|_{\Omega(t)}^{1/2} + \| \phi \|_{\Omega(t)}) \leq$$

$$\leq c(|A\phi|_{\Omega(t)}^{1/2} \| \phi \|_{\Omega(t)}^{1/2} + \| \phi \|_{\Omega(t)})$$

one has

$$E(\phi, \phi) = \int_0^T \{-(\phi, (\lambda I + A)\partial_t \phi)_{\Omega(t)} + (\lambda + \frac{1}{m}) \| \phi \|_{\Omega(t)}^2 +$$

$$+ \frac{\lambda}{m}|\phi|_{\Omega(t)}^2 + |A\phi|_{\Omega(t)}^2\}dt \geq \frac{1}{2}\int_0^T |A\phi|_{\Omega(t)}^2 dt +$$

$$\frac{\lambda}{m}\int_0^T |\phi|_{\Omega(t)}^2 dt + \int_0^T ((\lambda + \frac{1}{m}) \| \phi \|_{\Omega(t)}^2 - c \| \phi \|_{\Omega(t)}^2)dt \geq c \| \phi \|_{\mathcal{P}}^2 .$$

for suitable λ. Then there exists a $v^m \in \mathcal{F}$ such that (15) is satisfied for every $\phi \in \mathcal{P}$ (see Salvi, 1988). Now $(\lambda I + A)$ is one one and onto from $H^2(\Omega(t)) \cap V(\Omega(t))$ to $H(\Omega(t))$, and if $h(t) = (\lambda I + A)\phi(t)$, then, by density, we have that $v \in \mathcal{F}$ satisfies (15) with $h(t), \partial_t h \in H(\Omega_T)$. Furthermore, if $h(t) \in C_0^\infty(0,T;H(\Omega(t)))$, one obtains

$$| \int_0^T (v^m(t), \partial_t h(t))_{\Omega(t)}dt \leq | \int_0^T \{(Av^m(t), h(t))_{\Omega(t)} + \frac{1}{m}(v^m(t), h(t))_{\Omega(t)} +$$

$$+ (u \cdot \nabla u - f, h(t))_{\Omega(t)}\}dt| \leq c(\int_0^T |h|_{\Omega(t)}^2 dt)^{1/2}.$$

By results in Salvi (1990b) we have $\partial_t v^m \in L^2(\Omega_T)$, and by standard arguments,

$$P(\partial_t v^m - \Delta v^m + \frac{1}{m}v^m + (u \cdot \nabla u - f) = 0 \quad a.e. \quad \text{in} \quad \Omega_T. \qquad (16)$$

situation. As in the case of bounded domains ($q = 2$ see Agmon, 1965; $1 < q < \infty$ see Simader, 1972) higher differentiability properties of the data are transmitted to higher differentiability properties of the weak solutions (Theorem 1.4) if the boundary ∂G is sufficiently smooth. This is crucial for the further procedure.

At this point we emphasize a further difference between the case of bounded and that of exterior domains. If $1 < q < \infty$ and $q' = \frac{q}{q-1}$ denotes the dual exponent to q then as an easy consequence of the Hahn-Banach-theorem and the $L^q(G)^n$- $L^{q'}(G)^n$-duality any continuous linear functional $F^* \in (\hat{H}^{1,q'}_\bullet(G))^*$ can be norm-preservingly represented with a uniquely determined $\mathbf{f} \in L^q(G)^n$ in the form

$$F^*(\phi) = < \mathbf{f}, \nabla\phi > \quad \text{for } \phi \in \hat{H}^{1,q'}_\bullet(G). \tag{0.5}$$

Conversely the right hand side defines a continuous linear functional. Then by the representation theorem (Theorem 1.2) there exists a unique $u \in \hat{H}^{1,q}_\bullet(G)$ such that

$$< \nabla u, \nabla\phi > \quad = \quad < \mathbf{f}, \nabla\phi > \quad = \quad F^*(\phi) \quad \text{for } \phi \in \hat{H}^{1,q}_\bullet(G). \tag{0.6}$$

Therefore we construct weak solutions of the boundary value problem "$-\Delta u = -\operatorname{div} \mathbf{f}$, $u|_{\partial G} = 0$". But if $f \in L^q(G)$ is given, by

$$F^*(\phi) := < f, \phi > \quad \text{for } \phi \in C^\infty_0(G) \tag{0.7}$$

a linear functional on $C^\infty_0(G)$ is defined which in general *needs not* to be continuous with respect to the $\|\nabla.\|_{q'}$-norm. In addition for $\phi \in \hat{H}^{1,q'}_\bullet(G)$ in general it is not even defined. Therefore boundary value problems of the type "$-\Delta u = f$, $u|_{\partial G} = 0$" in general admit *no* weak L^q-solutions in exterior domains. But if G is a *bounded domain* with boundary $\partial G \in C^{2+k}$ and if $f \in H^{k,q}(G)$, where $H^{k,q}(G)$ denotes the usual Sobolev space ($k = 0, 1, \ldots$, $H^{0,q}(G) = L^q(G)$) then there exists a unique solution $u \in H^{2+k,q}(G) \cap H^{1,q}_0(G)$ of the equation $-\Delta u = f$ vanishing in the trace sense at the boundary. For $k = 0$ these solutions are called strong solutions and for $k \in \mathbb{N}$ we want to call them $(2+k)$-solutions. For an *exterior domain* $G \in \mathbb{R}^n$ we construct such strong solutions in section 2. What we may expect we can again see from the example G_1 (0.2). Let for $x \in G_1$

$$h_i(x) := \quad x_i - \frac{x_i}{|x|^n} \quad \text{for } n \geq 2, \, i = 1, \ldots, n \tag{0.8}$$

Then $\Delta h_i = 0$ in G_1, $h_i|_{\partial G_1} = 0$. Furthermore $h_i \notin L^s(G)$, $\nabla h_i \notin L^s(G)^n$ for $1 \leq s < \infty$ and $i = 1, \ldots, n$. But for $|\alpha| \geq 2$ we see $D^\alpha h_i \in L^q(G)$ for $1 < q < \infty$. Clearly a reasonable theory of strong solutions has to cover this example, too. Therefore we have to look for strong solutions in such generalized Sobolev spaces having second order derivatives in $L^q(G)$ whereas the function itself and the lower order derivatives have only local L^q-properties. These spaces are studied in section 2. Our procedure is as follows. If $f \in L^q(G)$ is given then we extend it by zero to the whole space $\mathbb{R}^n$ and try to solve the equation $-\Delta v = f$ with a function $v \in L^q_{loc}(\mathbb{R}^n)$ satisfying $\nabla v \in L^q_{loc}(\mathbb{R}^n)^n$ and

$\partial_i \partial_k v \in L^q(\mathbb{R}^n)$ for $i, k = 1, \ldots, n$. This is done by means of certain approximations. For getting a-priori-estimates we have to exclude the first order polynomials belonging to the null space of Δ in this Sobolev space. This is done by a normalization. Then we can solve weakly the boundary value problem "$\Delta h = 0$, $h|_{\partial G} = -v|_{\partial G}$" and put $u := v + h$ to get the desired strong solutions (see section 2.4). In a similar way we can determine the null space of the strong solutions of "$-\Delta h = 0$ in G and $h|_{\partial G} = 0$". Again it turns out that in the case G_1 the functions defined by (0.3), (0.8) reflect typically the general situation.

As we already indicated above in the case of an exterior domain there are data admitting strong solutions but no weak solutions and vice versa. In this context naturally the question for $(2 + k)$-solution arises from two points of view. For $\gamma > 0$ and for an exterior domain G let

$$f(x) := |x|^\gamma \tag{0.9}$$

Then $f \notin L^s(G)$ for $1 \leq s < \infty$. If $k \in \mathbb{N}$ satisfies $k - 1 \leq \gamma < k$ then for $|\alpha| = k$ and $q > \frac{n}{k-\gamma}$ we see $D^\alpha f \in L^q(G)$ whereas the lower order derivatives have only local L^q-properties. The question arises if the boundary value problem "$-\Delta u = f$, $u|_{\partial G} = 0$" admits solutions with $D^\alpha u \in L^q(G)$ for $|\alpha| = 2 + k$ whereas the lower order derivatives have only local integrability properties.

Further we consider again (0.2), (0.3). If $|\alpha| \geq 1$ then we easily see that $f_\alpha(x) := |x|^{n-2+2|\alpha|} D^\alpha h_0(x)$ is a homogeneous harmonic polynomial of degree $|\alpha|$. Further if we put for $x \in G_1$ and $|x| \geq 1$

$$h_\alpha(x) := D^\alpha h_0(x) - f_\alpha(x) \tag{0.10}$$

then h_α solves "$\Delta h_\alpha = 0$ in G_1, $h_\alpha|_{\partial G_1} = 0$". Further $D^\beta h_\alpha \in L^q(G_1)$ for $|\beta| = k$, whereas $D^\sigma h_\alpha \notin L^s(G)$ for $1 \leq s < \infty$ and $|\sigma| \leq k - 1$.
Such a theory of $(2 + k)$-solutions in $L^q(G)$ for exterior domains we sketch in section 3.

In a similar way based on the results of Simader and Sohr (1992) the Neumann problem could be considered.

The results for the Dirichlet problem can easily be translated to get $(2+k)$-solutions of Stokes' system. This is done in section 5. The construction is similar to that one for the Dirichlet problem for Δ. As a basis we use the existence results from Galdi and Simader (1990) (resp. its extension to the spaces $\hat{H}^{1,q}_\bullet(G)$) for weak solutions in $L^q(G)$ of Stokes' system in an exterior domain).

1. WEAK SOLUTIONS

1.1. The appropriate spaces

For an exterior domain $G \subset \mathbb{R}^n$ and $R > 0$ we put $G_R := G \cap B_R$ ($B_R := \{x \in \mathbb{R}^n : |x| < R\}$) and we define

$$\hat{H}^{1,q}_\bullet(G) := \{ \ u : G \to \mathbb{R} \text{ measurable }, u \in L^q(G_R) \tag{1.1}$$

for each $R > 0, \nabla u \in L^q(G)^n$ and

$$\eta u \in H_0^{1,q}(G) \text{ for all } \eta \in C_0^\infty(\mathbb{R}^n)\}$$

Here $H_0^{1,q}(G)$ denotes the usual Sobolev space (the closure of $C_0^\infty(G)$ with respect to the norm $\|v\|_{1,q} := (\|v\|_q^q + \|\nabla v\|_q^q)^{\frac{1}{q}}$). It turns out that $\|\nabla.\|_q$ is a norm on $\hat{H}_\bullet^{1,q}(G)$ and equipped with this norm $(\hat{H}_\bullet^{1,q}(G), \|\nabla.\|_q)$ is a reflexive Banach space ($q = 2$: Hilbert space). By (1.1) clearly holds $C_0^\infty(G) \subset \hat{H}_\bullet^{1,q}(G)$ and therefore we may define

$$\hat{H}_0^{1,q}(G) := \overline{C_0^\infty(G)}^{\|\nabla.\|_q} \tag{1.2}$$

Then in case of an exterior domain we have $H_0^{1,q}(G) \subsetneq \hat{H}_0^{1,q}(G) \subseteq \hat{H}_\bullet^{1,q}(G)$. (Let us remark that in the case of a *bounded* domain all the three spaces coincide). We choose now an arbitrary $r > 0$ such that $K = \mathbb{R}^n \setminus \overline{G} \subset B_r$ and

$$\varphi_r \in C^\infty(\mathbb{R}^n), \quad 0 \leq \varphi_r \leq 1 \quad \text{such that} \tag{1.3}$$

$$\varphi_r(x) = 0 \quad, \text{ for } |x| \leq r \quad \text{and} \quad \varphi_r(x) = 1 \quad, \text{ for } |x| \geq 2r.$$

We see immediately by (1.1) that $\varphi_r \in \hat{H}_\bullet^{1,q}(G)$ for $1 < q < \infty$. But in case $1 < q < n$ holds $\varphi_r \notin \hat{H}_0^{1,q}(G)$ (because otherwise by Sobolev's embedding theorem we would conclude $\varphi_r \in L^{q^*}(G)$ where $q^* = \frac{nq}{n-q}$, but $\varphi_r \notin L^s(G)$ for $1 < s < \infty$). More precisely we get (for the proof of this and the subsequent theorems see Simader and Sohr, 1995)

Theorem 1.1. *Let $G \subset \mathbb{R}^n$ satisfy (GE) and let $1 < q < \infty$. Then*

i) in case $q \geq n$ holds $\hat{H}_\bullet^{1,q}(G) = \hat{H}_0^{1,q}(G)$

ii) in case $1 < q < n$ holds in the sense of a direct sum

$$\hat{H}_\bullet^{1,q}(G) = \hat{H}_0^{1,q}(G) \oplus \{\alpha\varphi_r : \alpha \in \mathbb{R}\} \tag{1.4}$$

and with a constant $C = C(q, G, \varphi_r) > 0$ holds for $f = f_0 + \alpha\varphi_r$, $f_0 \in \hat{H}_0^{1,q}(G)$ and $\alpha \in \mathbb{R}$

$$C(|\alpha| + \|\nabla f_0\|_q) \leq \|\nabla f_0 + \alpha\nabla\varphi_r\|_q \leq \|\nabla f_0\|_q + |\alpha|\|\nabla\varphi_r\|_q. \tag{1.5}$$

1.2. Existence

The main result reads as follows:

Theorem 1.2. *Let $G \subset \mathbb{R}^n$ satisfy (GE) and let $1 < q < \infty$.*

a) There exists a constant $C_q = C(q, G) > 0$ such that

$$\|\nabla u\|_q \leq C_q \sup_{0 \neq \phi \in \hat{H}_\bullet^{1,q'}(G)} \frac{<\nabla u, \nabla\phi>}{\|\nabla\phi\|_{q'}} \tag{1.6}$$

holds for all $u \in \hat{H}_\bullet^{1,q}(G)$ where $q' := \frac{q}{q-1}$.

b) *Given* $F^* \in \left(\hat{H}_\bullet^{1,q'}(G) \right)^*$ *then there exists a unique* $u \in \hat{H}_\bullet^{1,q}(G)$ *such that*

$$< \nabla u, \nabla \phi >= F^*(\phi), \text{ for all } \phi \in \hat{H}_\bullet^{1,q'}(G) \tag{1.7}$$

and with $C_q > 0$ *by a) holds*

$$\|\nabla u\|_q \le C_q \|F^*\|_{q'}^* \tag{1.8}$$

(with $\|F^*\|_{q'}^* := \sup\{F^*(\phi) : \phi \in \hat{H}_\bullet^{1,q'}(G), \|\nabla \phi\|_{q'} = 1\}$)

Remark 1.1. We would like to emphasize that an estimate analogous to (1.6) in general fails if we replace "$u \in \hat{H}_\bullet^{1,q}(G)$" by "$u \in \hat{H}_0^{1,q}(G)$" and "$\phi \in \hat{H}_\bullet^{1,q'}(G)$" by "$\phi \in \hat{H}_0^{1,q'}(G)$". Such an estimate holds true in case $n \ge 3$ if and only if $1 < q < n$ and in case $n = 2$ if $1 < q \le 2$ (see Simader, 1990, Theorem 4.2 and Simader and Sohr, 1995). This phenomenon we easily understand if we consider the example (0.3) in the case $G_1 := \mathbb{R}^n \setminus \overline{B_1}$. The function h_0 satisfies $h_0 \in \hat{H}_\bullet^{1,q}(G_1)$ if $q > \frac{n}{n-1}$ for $n \ge 2$. But $h_0 \in \hat{H}_0^{1,q}(G_1)$ if and only if $q \ge n$ in case $n \ge 3$ whereas in case $n = 2$ we have to suppose $q > 2 = n$. Assume now that q satisfies the last conditions and let $\phi \in \hat{H}_0^{1,q'}(G_1)$. By (1.2) there exists a sequence $(\phi_i) \subset C_0^\infty(G_1)$ such that $\|\nabla \phi - \nabla \phi_i\|_{q'} \to 0$ and since $\Delta h_0 = 0$

$$< \nabla h_0, \nabla \phi >= \lim_{i \to \infty} < \nabla h_0, \nabla \phi_i >= \lim_{i \to \infty} < -\Delta h_0, \phi_i >= 0$$

Therefore $\sup_{0 \ne \phi \in \hat{H}_0^{1,q'}(G_1)} \frac{<\nabla h_0, \nabla \phi>}{\|\nabla \phi\|_{q'}} = 0$ and we *cannot* estimate $\|\nabla h_0\|_q$ by this supremum. Roughly speaking the reason that (1.6) holds for the space $\hat{H}_\bullet^{1,q}(G)$ with variational class $\hat{H}_\bullet^{1,q'}(G)$ in the sup at the right hand side but *not* with these spaces replaced by the $\hat{H}_0^{1,s}(G)$-spaces ($s = q$ resp. q') is the one-dimensional complementary space $\{\alpha \varphi_r : \alpha \in \mathbb{R}\}$ in the decomposition (1.4). For details we refer to Simader and Sohr (1995).

Remark 1.2. It is an easy functional analytical exercise to show that part a) of Theorem 1.2 holds as well for q as for q' (in interchanged roles, $q'' = q$) then part b) holds as well for q and q' (in interchanged roles) and vice versa. In this sense the statements a) and b) of Theorem 1.2 are equivalent.

Remark 1.3. Given $\mathbf{f} = (f_1, \ldots, f_n) \in L^q(G)^n$ and define

$$F^*(\phi) :=< \mathbf{f}, \nabla \phi >= \int_G \sum_{i=1}^n f_i \partial_i \phi \quad, \text{ for } \phi \in \hat{H}_\bullet^{1,q'}(G). \tag{1.9}$$

Then $F^* \in (\hat{H}_\bullet^{1,q'}(G))^*$ and $\|F^*\|_{q'}^* \le \|\mathbf{f}\|_q$. In fact any $F^* \in (\hat{H}_\bullet^{1,q'}(G))^*$ admits with a (unique) $\mathbf{f} \in L^q(G)^n$ satisfying $\|\mathbf{f}\|_q = \|F^*\|_{q'}^*$ the representation (1.9): We have only to regard $\{\nabla \phi : \phi \in \hat{H}_\bullet^{1,q'}(G)\}$ as a closed subspace of $L^{q'}(G)^n$ and F^* as a continuous linear functional on this subspace. If we extend last functional norm-preserving according to the Hahn-Banach extension theorem to $L^{q'}(G)^n$ and finally use the $L^q(G)^n - L^{q'}(G)^n$ duality we get this representation.

Therefore we may rewrite (1.7) in the form

$$< \nabla u, \nabla \phi > = < \mathbf{f}, \nabla \phi > \quad \text{for } \phi \in \hat{H}_\bullet^{1,q'}(G). \tag{1.10}$$

with $u \in \hat{H}_\bullet^{1,q}(G)$. If we suppose for simplicity u and $\mathbf{f}$ to be sufficiently regular and that in latter case $u \in \hat{H}_\bullet^{1,q}(G)$ implies $u|_{\partial G} = 0$ we immediately conclude that u solves the boundary value problem (BVP)

$$(\text{BVP}) \begin{cases} -\Delta u = -\text{div } \mathbf{f} \text{ in } G \\ \qquad u|_{\partial G} = 0 \end{cases} \tag{1.11}$$

But let us regard our example (0.3) in G_1 for simplicity in case $n \geq 3$ and $q \geq n$. Then we may add to u in (BVP) (1.11) any multiple of h_0 and we again get a solution whereas – according to Theorem 1.2 – the solution of (1.10) is unique. Does this fact form a contradiction? By no means. It only shows that the problems (1.11) and (1.10) are *not* equivalent and the functional equation (1.10) contains more informations than the BVP (1.11).

Roughly speaking we come from (1.10) to (1.11) by testing with $C_0^\infty(G)$-functions ϕ. But then we loose informations if we only use this test space. In fact (1.10) contains "hidden additional boundary values" not regarded in (1.11) (see Simader and Sohr, 1995).

1.3. The null space of the Dirichlet problem

Theorem 1.3. *Let $1 < q < \infty$ and put*

$$N^q(G) := \{h \in \hat{H}_\bullet^{1,q}(G) : \Delta h = 0 \text{ weakly in } G\} \tag{1.12}$$

Then

i) $N^q(G) = \{0\}$ *for* $1 < q \leq \frac{n}{n-1}$

ii) *There exists* $0 \neq h_0 \in \bigcap_{\{\frac{n}{n-1} < t < \infty\}} \hat{H}_\bullet^{1,t}(G)$ *such that* $h_0 \in C^\infty(G)$, $\Delta h_0 = 0$ *and*

$$N^q(G) = \{\alpha h_0 : a \in \mathbb{R}\} \quad \text{for} \quad \frac{n}{n-1} < q < \infty. \tag{1.13}$$

iii) *For* $\frac{n}{n-1} < q < n$ *(therefore $n \geq 3$) holds*

$$\hat{H}_\bullet^{1,q}(G) = \hat{H}_0^{1,q}(G) \oplus \{\alpha h_0 : \alpha \in \mathbb{R}\} \tag{1.14}$$

in the sense of a direct decomposition and for $q = 2$ as orthogonal decomposition. Further with constants $C_i = C_i(q, G, h_0) > 0$ $(i = 1, 2)$ holds for $f_0 \in \hat{H}_0^{1,q}(G)$ and $\alpha \in \mathbb{R}$

$$C_1(\|\nabla f_0\|_q + |\alpha|) \leq \|\nabla f_0 + \alpha \nabla h_0\|_q \leq C_2(\|\nabla f_0\|_q + |\alpha|) \tag{1.15}$$

This theorem shows that our explicit example (0.3) perfectly reflects the general situation.

1.4. Higher regularity of weak solutions

As is well known for the solutions of elliptic boundary value problems local and, if the boundary is sufficiently smooth, global differentiability properties of the right hand side of the equation are transmitted to the solution. This can be proved in an analogous way to Simader (1972) (compare with Simader and Sohr, 1995). For the construction of strong and $(2 + k)$-solutions we only need the following result.

For $j \in \mathbb{N}$ we consider the usual Sobolev space $H^{j,q}(G) = \{g \in L^q(G) : D^\alpha g \in L^q(G)$ for $|\alpha| \leq j\}$ equipped with the norm $\|g\|_{j,q} := \left(\sum_{|\alpha| \leq j} \|D^\alpha g\|_q^q \right)^{\frac{1}{q}}$.

Theorem 1.4. *Let $j \in \mathbb{N}$ and assume that $\partial G \in C^j$. Let $1 < q < \infty$ and let $R > R_0(G)$. Suppose that $g \in H^{j,q}(G)$ such that $g(x) = 0$ for $x \in G \setminus B_R$. Therefore $g \in H^{j,t}(G)$ for $1 < t \leq q$. Then there exists a unique $p \in \hat{H}^{1,q}_\bullet(G)$ with the property $p \in \bigcap_{1 < t \leq q} \hat{H}^{1,t}_\bullet(G)$ and*

$$< \nabla p, \nabla \phi >=< \nabla g, \nabla \phi > , \text{ for } \phi \in \hat{H}^{1,t'}_\bullet(G), \quad q' \leq t' < \infty \qquad (1.16)$$

Further for $1 < t \leq q$ there are constants $C_{t,q} = C(t, q, C_t, |G_R|) > 0$ (with $C_t > 0$ by (1.6)) such that

$$\|\nabla p\|_t \leq C_t \|\nabla g\|_{t,G_R} \leq C_{t,q} \|\nabla g\|_{q,G_R}, \text{ for } 1 < t \leq q. \qquad (1.17)$$

If furthermore $j \geq 2$ then for $1 \leq |\alpha| \leq j$ holds $D^\alpha p \in L^t(G)$ $(1 < t \leq q)$ and there are constants $C_{t,j} = C(t, j, G) > 0$, $C_{t,q,j} = C(C_{t,j}, q, |G_R|) > 0$ such that for $1 < t \leq q$ holds

$$\sum_{1 \leq |\alpha| \leq j} \|D^\alpha p\|_t \leq C_{t,j} \|g\|_{j,t;G_R} \leq C_{t,q,j} \|g\|_{j,q;G_R}. \qquad (1.18)$$

2. STRONG SOLUTIONS

2.1. The appropriate spaces

Throughout this section we assume $\partial G \in C^2$. Like as in the case of weak solutions we first have to consider appropriate spaces for possible strong solutions of the Dirichlet problem. First we define for $1 < q < \infty$

$$\hat{H}^{2,q}(G) := \{ \ u : G \to \mathbb{R} \text{ measurable}, u \in L^q(G_R), \qquad (2.1)$$
$$\partial_i u \in L^q(G_R) \quad (i = 1, \ldots, n) \text{ for all } R > 0$$
$$\text{and } \partial_i \partial_k u \in L^q(G) \quad (i, k = 1, \ldots, n) \ \}$$

For the Dirichlet problem with vanishing boundary data we consider

$$\hat{H}^{2,q}_D(G) := \{ \ u \in \hat{H}^{2,q}(G) : (\eta u) \in H^{1,q}_0(G) \qquad (2.2)$$
$$\text{for each } \eta \in C^\infty_0(\mathbb{R}^n) \ \}$$

For $u \in \hat{H}_D^{2,q}(G)$ we consider the semi-norm

$$|u|_{2,q} := \left(\sum_{i,k=1}^{n} \|\partial_i \partial_k u\|_q^q \right)^{\frac{1}{q}} \tag{2.3}$$

Then it turns out that $|.|_{2,q}$ is even a norm on $\hat{H}_D^{2,q}(G)$ such that $\left(\hat{H}_D^{2,q}(G), |.|_{2,q} \right)$ is a reflexive Banach space (and for $q = 2$ a Hilbert space). In addition let us define

$$C_{D,C}^2(G) := \{ \; u \in C^2(\overline{G}) : u|_{\partial G} = 0 \text{ and there exists an} \tag{2.4}$$
$$R = R(u) > 0 \quad \text{such that } u(x) = 0 \text{ for } |x| > R \; \}$$

Clearly $C_{D,C}^2(G) \subset \hat{H}_D^{2,q}(G)$ and we may define

$$\hat{H}_{D,C}^{2,q}(G) := \overline{C_{D,C}^2(G)}^{|.|_{2,q}}. \tag{2.5}$$

From the viewpoint of uniqueness of strong solutions e.g. in case $n \geq 3$ and $1 < q < \frac{n}{2}$ we will see that the space $\hat{H}_{D,C}^{2,q}(G)$ plays an important role. Again via the Sobolev embedding theorem we get

Theorem 2.1. *Let $1 < q < \infty$.*

i) If $n \leq q < \infty$ then $\hat{H}_D^{2,q}(G) = \hat{H}_{D,C}^{2,q}(G)$.

Let $r > R_0(G)$ be fixed and consider φ_r defined by (1.3).

ii) If $\frac{n}{2} \leq q < n$ $(1 < q < 2$ if $n = 2)$ then for $u \in \hat{H}_D^{2,q}(G)$ holds

$$u \in \hat{H}_{D,C}^{2,q}(G) \quad \Longleftrightarrow \quad \left\{ u \in \hat{H}_\bullet^{1,q^*}(G), \quad q^* = \frac{nq}{n-q} \right\} \tag{2.6}$$

Therefore $\hat{H}_{D,C}^{2,q}(G) = \hat{H}_D^{2,q}(G) \cap \hat{H}_\bullet^{1,q^*}(G)$. In the sense of a direct decomposition holds

$$\hat{H}_D^{2,q}(G) = \hat{H}_{D,C}^{2,q} \oplus \left\{ \left(\sum_{i=1}^{n} d_i x_i \right) \varphi_r(x) : d_i \in \mathbb{R}, i = 1, \ldots, n \right\} \tag{2.7}$$

iii) If $1 < q < \frac{n}{2}$ (that means $n \geq 3$) then for $u \in \hat{H}_D^{2,q}(G)$ holds

$$u \in \hat{H}_{D,C}^{2,q}(G) \quad \Longleftrightarrow \quad \{ \nabla u \in L^{q^*}(G)^n \text{ and } u \in L^{q^{**}}(G) \tag{2.8}$$
$$\text{where } q^* = \frac{nq}{n-q} \text{ and } q^{**} = \frac{nq}{n-2q} \}$$

Therefore $\hat{H}_{D,C}^{2,q}(G) = \hat{H}_D^{2,q}(G) \cap \hat{H}_0^{1,q^*}(G)$. In the sense of a direct decomposition holds

$$\hat{H}_D^{2,q}(G) = \hat{H}_{D,C}^{2,q}(G) \oplus \left\{ \left(c + \sum_{i=1}^{n} d_i x_i \right) \varphi_r(x) : c \in \mathbb{R}, d_i \in \mathbb{R}, i = 1, \ldots, n \right\} \tag{2.9}$$

Later we will give more "natural" decompositions than (2.7) and (2.9).

2.2. Strong solutions in the whole space

For given $f \in L^q(\mathbb{R}^n)$ we will seek for solutions of the equation $-\Delta u = f$ in appropriate spaces. For $1 < q < \infty$ we define

$$
\hat{H}^{2,q}(\mathbb{R}^n) := \{ \; u : \mathbb{R}^n \to \mathbb{R} : u, \partial_i u \in L^q_{loc}(\mathbb{R}^n) \tag{2.10}
$$
$$
(i = 1, \ldots, n), \partial_i \partial_k u \in L^q(\mathbb{R}^n), i, k = 1, \ldots, n \; \}
$$

Clearly $|.|_{2,q}$ is no longer a norm on $\hat{H}^{2,q}(\mathbb{R}^n)$. But if $u \in \hat{H}^{2,q}(\mathbb{R}^n)$ satisfies $|u|_{2,q} = 0$ then u is a polynomial of degree at most one. We denote the space of polynomials of degree at most one in n variables by $P(n; 1)$ and abbreviate $P(1) := P(n; 1)$. We could consider the abstract quotient space $\hat{H}^{2,q}(\mathbb{R}^n)/P(1)$ and we could equip this space with its natural norm. But we prefer to consider a "realization" of this quotient space depending on a parameter set G_0. This procedure seems to us more transparent and we are able to generalize it easily to the case of $(2 + k)$-solutions. Let us choose an arbitrary

$$
\emptyset \neq G_0 \subset\subset \mathbb{R}^n. \tag{2.11}
$$

Then we define

$$
\hat{H}^{2,q}(\mathbb{R}^n; G_0) := \left\{ u \in \hat{H}^{2,q}(\mathbb{R}^n) : \int_{G_0} u dx = \int_{G_0} \partial_i u dx = 0 \text{ for } i = 1, \ldots, n \right\} \tag{2.12}
$$

From the preceding remarks it is obvious that $|.|_{2,q}$ is now a norm on $\hat{H}^{2,q}(\mathbb{R}^n; G_0)$ and equipped with this norm $\left(\hat{H}^{2,q}(\mathbb{R}^n; G_0), |.|_{2,q} \right)$ is a reflexive Banach space ($q = 2$: Hilbert space) topologically equivalent to the quotient space $\hat{H}^{2,q}(\mathbb{R}^n)/P(1)$ equipped with the quotient-space-norm. If $u \in \hat{H}^{2,q}(\mathbb{R}^n)$ and $\Delta u = 0$ then also $\Delta(\partial_i \partial_k u) = 0$ $(i, k = 1, \ldots, n)$. Since $\partial_i \partial_k u \in L^q(\mathbb{R}^n)$ we get by the Liouville theorem for harmonic functions in $L^q(\mathbb{R}^n)$ (in fact being an elementary application of the mean value theorem for harmonic functions (see e.g. Simader, 1992)) that $\partial_i \partial_k u = 0$ and therefore $u \in P(1)$. We observe that in case of order 1 the space $P(1)$ coincides with the space of harmonic polynomials of degree less or equal to 1 (later in case of $(2 + k)$-solutions for $k \in \mathbb{N}$, $k \geq 1$, we have to be more careful!). If we assume in addition $u \in \hat{H}^{2,q}(\mathbb{R}^n; G_0)$ and $\Delta u = 0$ we conclude $u = 0$. Therefore $-\Delta : \hat{H}^{2,q}(\mathbb{R}^n; G_0) \to L^q(\mathbb{R}^n)$ is a one-to-one map in $L^q(\mathbb{R}^n)$. But we easily can prove more.

Theorem 2.2. *Let $1 < q < \infty$. Then*

$$
-\Delta : \hat{H}^{2,q}(\mathbb{R}^n; G_0) \to L^q(\mathbb{R}^n) \tag{2.13}
$$

is bijective and there exists a constant $C(q; G_0) > 0$ such that

$$
|u|_{2,q} \leq C(q; G_0) \| - \Delta u \|_q \text{ for all } u \in \hat{H}^{2,q}(\mathbb{R}^n; G_0). \tag{2.14}
$$

Remark 2.1. There are different approaches to the proof of Theorem 2.2. One could consist in using the fact that $\overline{C_0^\infty(\mathbb{R}^n)}^{\|.\|_q} = L^q(\mathbb{R}^n)$. Therefore given $f \in L^q(\mathbb{R}^n)$

there exists a sequence $(f_i) \subset C_0^\infty(\mathbb{R}^n)$ such that $\|f - f_i\|_q \to 0$. Then we may regard $\tilde{u}_i := S_n * f_i$ where S_n denotes the fundamental solution for the Laplacian. Via the Calderon-Zygmund-inequality we get for $g \in C_0^\infty(\mathbb{R}^n)$ and $v := S_n * g$

$$\left(\sum_{r,s=1}^n \|\partial_r \partial_s v\|_q^q \right)^{\frac{1}{q}} \leq C(q,n)\|g\|_q \qquad (2.15)$$

with a constant $C(q,n)$ depending solely on q and n. If we choose $p_i \in P(1)$ such that with $u_i := \tilde{u}_i - p_i$ holds $u_i \in \hat{H}^{2,q}(\mathbb{R}^n; G_0)$ then we readily see by means of completeness of $\hat{H}^{2,q}(\mathbb{R}^n; G_0)$ and (2.3) that there is a unique $u \in \hat{H}^{2,q}(\mathbb{R}^n; G_0)$ such that $|u - u_i|_{2,q} \to 0$ and $-\Delta u = f$. We keep this procedure in mind with respect to the concept of $(2 + k)$-solutions. The essential fact was that the given datum f (in a suitable space) can be approximated (in suitable norms) by functions f_i with *compact support*. The last property (or at least a certain decay of f_i at infinity) is necessary even to define the convolution $S_n * f_i$ and to verify its properties.

Remark 2.2. In fact in the contrary to the sketch of the proof given above we proceed via an elementary observation (for the proof see e.g. Simader, 1990, Theorem 3.2):

$$L^q(\mathbb{R}^n) = \overline{\{\Delta\phi : \phi \in C_0^\infty(\mathbb{R}^n)\}}^{\|\cdot\|_q}. \qquad (2.16)$$

The next considerations are based on an approximation property we first need only in case $k = 2$.

Lemma 2.1. *Let $1 < q < \infty$ and let $k \in \mathbb{N}$, $k \geq 2$. Given $u \in \hat{H}^{k,q}(\mathbb{R}^n)$ then there exists a sequence $(u_j) \subset C_0^\infty(\mathbb{R}^n)$ such that $|u - u_j|_{k,q} \to 0$.*

We have to observe that the sequence (u_j) needs *not* to converge even in local L^q-norms of the functions or its lower order derivatives to the given function u. The elementary proof of Lemma 2.1 uses the same ideas as that of Theorem 2.4 from Simader (1990). On this basis we get

Theorem 2.3.   *a) Let $1 < q < n$ and $q^* = \frac{nq}{n-q}$. We put*

$$\hat{H}_{0,1}^{2,q}(\mathbb{R}^n; G_0) := \{ \ u \in \hat{H}^{2,q}(\mathbb{R}^n) : u \in L_{loc}^{q^*}(\mathbb{R}^n), \qquad (2.17)$$
$$\nabla u \in L^{q^*}(\mathbb{R}^n)^n \ and \int_{G_0} u \, dx = 0\}$$

Then $\|\nabla u\|_{q^*} + |u|_{2,q}$ is a norm on $\hat{H}_{0,1}^{2,q}(\mathbb{R}^n; G_0)$ and equipped with this norm $\hat{H}_{0,1}^{2,q}(\mathbb{R}^n; G_0)$ is a reflexive Banach space. Further

$$- \Delta : \hat{H}_{0,1}^{2,q}(\mathbb{R}^n; G_0) \to L^q(\mathbb{R}^n) \qquad (2.18)$$

is bijective and there exists a constant $C_1 = C(q, n, G_0)$ such that

$$\|\nabla u\|_{q^*} + |u|_{2,q} \leq C_1 \|\Delta u\|_q \ for \ u \in \hat{H}_{0,1}^{2,q}(\mathbb{R}^n; G_0). \qquad (2.19)$$

b) Let $n \geq 3$ and $1 < q < \frac{n}{2}$. We put

$$\hat{H}_0^{2,q}(\mathbb{R}^n) := \{u \in \hat{H}^{2,q}(\mathbb{R}^n) : u \in L^{q^{**}}(\mathbb{R}^n), \nabla u \in L^{q^*}(\mathbb{R}^n)^n\}. \tag{2.20}$$

Then $|.|_{2,q}$ is a norm on this space such that $\left(\hat{H}_0^{2,q}(\mathbb{R}^n), |.|_{2,q}\right)$ is a reflexive Banach space.

$$\hat{H}_0^{2,q}(\mathbb{R}^n) = \overline{C_0^\infty(\mathbb{R}^n)}^{|.|_{2,q}} \tag{2.21}$$

and

$$-\Delta : \hat{H}_0^{2,q}(\mathbb{R}^n) \to L^q(\mathbb{R}^n) \tag{2.22}$$

is bijective and there exists a constant $C_2 = C_2(q,n) > 0$ such that

$$\|u\|_{q^{**}} + \|\nabla u\|_{q^*} + |u|_{2,q} \leq C_2 \|\Delta u\|_q \text{ for } u \in \hat{H}_0^{2,q}(\mathbb{R}^n). \tag{2.23}$$

Theorem 2.3 can be derived easily from Theorem 2.2 using Lemma 2.1 and Sobolev's inequality.

2.3. The null space of the strong Dirichlet problem in G

As we have seen in the preceding section in $\hat{H}^{2,q}(\mathbb{R}^n)$ the null space of Δ is spanned by $\tilde{h}_0(x) := 1$ and $\tilde{h}_i(x) := x_i$ $(i = 1, \ldots, n)$ for $x \in \mathbb{R}^n$.

$$\left. \begin{array}{l} \text{Let } \eta \in C_0^\infty(\mathbb{R}^n), \text{ with } 0 \leq \eta(x) \leq 1 \text{ and } \eta(x) = 1 \text{ for } |x| \leq 1, \\ \eta(x) = 0 \text{ for } |x| \geq 2. \text{ For } r > 0 \text{ we put } \eta_r(x) := \eta(r^{-1}x). \end{array} \right\} \tag{2.24}$$

We choose a fixed $R > 2R_0(G)$ and put $\tilde{\eta} := \eta_{\frac{R}{2}}$. Then $\tilde{\eta}(x) = 0$ for $|x| \geq R$ and $\tilde{\eta} \equiv 1$ in a neighborhood of ∂G. Given $1 < q < \infty$ then according to Theorem 1.4 there exists for $i = 0, 1, \ldots, n$ a unique $p_i \in \bigcap_{1 < t \leq q} \hat{H}_\bullet^{1,t}(G)$ with

$$< \nabla p_i, \nabla \phi > = < \nabla(\tilde{\eta}\tilde{h}_i), \nabla\phi > \text{ for } \phi \in \hat{H}_\bullet^{1,t'}(G), \quad q' \leq t < \infty. \tag{2.25}$$

Furthermore $\partial_j \partial_k p_i \in L^t(G)$ $(1 < t \leq q)$ for $j, k = 1, \ldots, n$. If we put $h_i := p_i + (1-\tilde{\eta})\tilde{h}_i$ then $\Delta h_i = 0$ in G and for $\varphi \in C_0^\infty(\mathbb{R}^n)$ holds $\varphi h_i \in H_0^{1,q}(G)$. Therefore $h_i \in \hat{H}_D^{2,q}(G)$. It is easy to see that the function h_0 constructed by this procedure can be identified with the function h_0 by Theorem 1.3. A careful analysis leads to

Theorem 2.4. *For $1 < q < \infty$ let*

$$N^{2,q}(G) := \{h \in \hat{H}_D^{2,q}(G) : \Delta h = 0\}$$

Then there exist linearly independent

$$h_i \in \bigcap_{1 < t < \infty} \hat{H}_D^{2,t}(G), \quad i = 0, 1, \ldots, n \tag{2.26}$$

with the additional property $h_0 \in \hat{H}_\bullet^{1,t}(G)$ for $\frac{n}{n-1} < t < \infty$ but $h_i \notin \hat{H}_\bullet^{1,t}(G)$ for all $1 < t < \infty$ and $i = 1, \ldots, n$. Further

$$N^{2,q}(G) = \left\{ \sum_{i=0}^n \lambda_i h_i : \lambda_i \in \mathbb{R}, \quad i = 0, 1, \ldots, n \right\}. \tag{2.27}$$

Therefore $\dim N^{2,q}(G) = n + 1$.

2.4. Existence and uniqueness problems for strong solutions

We proceed similarly to the beginning of Section 2.3. For $R > 0$ we consider the annullus

$$A_R := \{x \in \mathbb{R}^n : R < |x| < 2R\}. \tag{2.28}$$

Given $f \in L^q(G)$ we continue it by zero to the whole of $\mathbb{R}^n$ and we denote the continued function again by f. If we fix an $R > R_0(G)$ (see (0.1)) and put $G_0 := A_R$ then due to Theorem 2.2 there exists a unique $v \in \hat{H}^{2,q}(\mathbb{R}^n; A_R)$ with

$$-\Delta v = f \quad \text{and} \quad |v|_{2,q} \leq C(q; A_R)\|f\|_q \tag{2.29}$$

(see (2.14)). Since A_1 is a C^1-domain, it supports the Poincaré inequality. By an easy scaling argument (compare e.g. with Simader, 1990, part i) of proof of Theorem 2.4, p. 182/183) holds with a constant $C_q = C(q, A_1) > 0$:

$$\|\phi\|_{q,A_R} \leq C_q R \|\nabla \phi\|_{q,A_R}, \quad \text{for } \phi \in H^{1,q}(A_R) \tag{2.30}$$
$$\text{satisfying } \int_{A_R} \phi \, dy = 0.$$

Iterated application of (2.30) gives with a purely numerical constant $D_q = D(n,q) > 0$

$$\|\phi\|_{q,A_R} + \|\nabla \phi\|_{q,A_R} \leq D_q C_q R (1 + R C_q) |\phi|_{2,q;A_R}, \tag{2.31}$$
$$\text{for } \phi \in \hat{H}^{2,q}(\mathbb{R}^n) \text{ with } \int_{A_R} \phi \, dy = \int_{A_R} \partial_i \phi \, dy = 0, \quad i = 1, \ldots, n$$

where $|\phi|_{2,q;A_R} := \left(\sum_{i,k=1}^n \|\partial_i \partial_k \phi\|_{q,A_R}^q \right)^{\frac{1}{q}}$. Since the solution by (2.29) satisfies the conditions necessary for (2.31) to hold we may conclude via the product formula with η_R defined by (2.24) via (2.14):

$$\|\nabla(\eta_R v)\|_q + |\eta_R v|_{2,q} \leq C(q, n, R) |v|_{2,q} \tag{2.32}$$
$$\leq C(q, n, R) \, C(q; A_R) \|f\|_q$$

Similar to (2.25) we now solve the equation

$$< \nabla p, \nabla \phi > = < \nabla(\eta_R v), \nabla \phi >, \text{ for } \phi \in \hat{H}_\bullet^{1,t'}(G), \, q' \leq t' < \infty \tag{2.33}$$

with a unique $p \in \bigcap_{1 < t \leq q} \hat{H}_\bullet^{1,t}(G)$ and $\partial_i \partial_j p \in L^t(G)$ for $1 < t \leq q$. By (1.17), (1.18) and (2.32) we get

$$\|\nabla p\|_t + |p|_{2,t} \leq C(t, q, R, n) \|f\|_q \quad , \text{ for } 1 < t \leq q. \tag{2.34}$$

If we put $h := p - \eta_R v$ by means of Weyl's lemma for the Laplacian (see e.g. Simader, 1992) we immediately conclude $\Delta h = 0$. Finally we put

$$u := v + h = (1 - \eta_R)v + p.$$

By the properties of v and p we readily see $u \in \hat{H}_D^{2,q}(G)$ and $-\Delta u = -\Delta v = f$. Further combining (2.29), (2.32) and (2.34) we get with a constant $C_q = C(q, R, G) > 0$ for the solution u constructed in this way

$$|u|_{2,q} \leq C_q \|f\|_q. \tag{2.35}$$

We emphasize that our *construction* of u produced a unique solution. But with respect to Theorem 2.4 in general the solution is by no means unique. Clearly we may consider the quotient space $\hat{H}_D^{2,q}(G)/N^{2,q}(G)$ equipped with the quotient space norm $\|[u]\|_{2,q} := \inf \{|u + h|_{2,q} : h \in N^{2,q}(G)\}$ for $[u] \in \hat{H}_D^{2,q}(G)/N^{2,q}(G)$. Then $-\Delta : \hat{H}_D^{2,q}(G)/N^{2,q}(G) \to L^q(G)$ is bijective and by (2.35) we get $\|[u]\|_{2,q} \leq C_q\|f\|_q$ if $-\Delta[u] = f$. We may gain a realization of this procedure if we observe that by (2.26) there exists a basis $h_i \in \hat{H}_D^{2,t}(G)$ $(i = 0, 1, \ldots, n)$ for *all* $1 < t < \infty$. For $u, v \in \hat{H}_D^{2,2}(G)$ we put

$$< u, v >_2 := \sum_{i,k=1}^{n} \int_G \partial_i \partial_k u \, \partial_i \partial_k v \, dx. \tag{2.36}$$

We orthonormalize the h_i with respect to this inner product (beginning with h_0) and get $\hat{h}_i$ $(i = 0, 1, \ldots, n)$. We observe that for $u \in \hat{H}_D^{2,q}$ and $v \in \hat{H}_D^{2,q'}(G)$ $(q' = \frac{q}{q-1})$ the dual pair (2.36) is well defined. For $f \in L^q(G)$ we consider the solution $u \in \hat{H}_D^{2,q}(G)$ of $-\Delta u = f$ constructed above and we put $u_0 := u - \sum_{i=0}^{n} \lambda_i \hat{h}_i \in \hat{H}_D^{2,q}(G)$ with $\lambda_i \in \mathbb{R}$. Since $< \hat{h}_i, \hat{h}_j >_2 = \delta_{ij}$ $(i, j = 0, 1, \ldots, n)$ we get $< u_0, \hat{h}_i >_2 = 0$ for $i = 0, 1, \ldots, n$ if and only if

$$\lambda_i = < u, \hat{h}_i >_2 . \tag{2.37}$$

By (2.35) we see with

$$C(N^{2,q}(G)) := \sum_{i=0}^{n} |\hat{h}_i|_{2,q} |\hat{h}_i|_{2,q'}$$

$$|u_0|_{2,q} \leq C_q(1 + C(N^{2,q}(G)))\|f\|_q. \tag{2.38}$$

Therefore we proved

Theorem 2.5. *For $1 < q < \infty$ we put*

$$\hat{H}_{D,N\perp}^{2,q}(G) := \left\{ u_0 \in \hat{H}_D^{2,q}(G) : < u_0, \hat{h}_i >_2 = 0 \quad , \text{ for } i = 0, 1, \ldots, n \right\}$$

Then

$$\hat{H}_D^{2,q}(G) = \hat{H}_{D,N\perp}^{2,q}(G) \oplus N^{2,q}(G) \tag{2.39}$$

in the sense of a direct $(q = 2$: orthogonal$)$ decomposition.
If $u \in \hat{H}_D^{2,q}(G)$ then $u = u_0 + h$ with $h = \sum_{i=0}^{n} < u, \hat{h}_i > \hat{h}_i \in N^{2,q}(G)$ and $u_0 = (u - h) \in \hat{H}_{D,N\perp}^{2,q}(G)$. Further with (2.37)

$$
\begin{aligned}
|h|_{2,q} &\leq C(N^{2,q})|u|_{2,q}, \tag{2.40} \\
|u_0|_{2,q} &\leq (1 + C(N^{2,q}))|u|_{2,q} \\
|u_0|_{2,q} + |h|_{2,q} &\leq (1 + 2C(N^{2,q}))|u|_{2,q}
\end{aligned}
$$

In addition $-\Delta : \hat{H}^{2,q}_{D,N^\perp}(G) \to L^q(G)$ is bijective and (2.38) holds with $f = -\Delta u_0$ for $u_0 \in \hat{H}^{2,q}_{D,N^\perp}(G)$.

In the cases $\frac{n}{2} \le q < n$ and $1 < q < \frac{n}{2}$ (if $n \ge 3$) due to the construction of the solution we performed above we derive via Theorems 1.4, 2.1 (cases ii) iii)), 2.3, 2.4 the following sharper results.

Theorem 2.6. *i) Let $\frac{n}{2} \le q < n$ ($1 < q < 2$ if $n = 2$). Let $h_0 \in N^{2,q}(G)$ denote that element with $h_0 \in \hat{H}^{1,t}_\bullet(G)$ for $\frac{n}{n-1} < t < \infty$. Then $h_0 \in \hat{H}^{2,q}_{D,C}(G)$. We put*

$$\hat{H}^{2,q}_{D,C,h_0^\perp}(G) := \left\{ u \in \hat{H}^{2,q}_{D,C}(G) : \ < u, h_0 >_2 = 0 \right\} \tag{2.41}$$

Then in the sense of a direct decomposition holds

$$\hat{H}^{2,q}_{D,C}(G) = \hat{H}^{2,q}_{D,C,h_0^\perp}(G) \oplus \{\alpha h_0 : \alpha \in \mathbb{R}\}. \tag{2.42}$$

If $u \in \hat{H}^{2,q}_{D,C}(G)$ then $u_0 := (u - < u, h_0 >_2 h_0) \in \hat{H}^{2,q}_{D,C,h_0^\perp}(G)$ and $u = u_0 + < u, h_0 >_2 h_0$. Further

$$\hat{H}^{2,q}_D(G) = \hat{H}^{2,q}_{D,C,h_0^\perp}(G) \oplus N^{2,q}(G)$$

and $-\Delta : \hat{H}^{2,q}_{D,C,h_0^\perp}(G) \to L^q(G)$ is bijective. With a constant $C_q = C\left(q, G, C\left(N^{2,q}(G)\right)\right) > 0$ holds ($q^* = \frac{nq}{n-q}$)

$$\|\nabla u\|_{q^*} + |u|_{2,q} \le C_q \| - \Delta u\|_q \quad , \text{for } u \in \hat{H}^{2,q}_{D,C,h_0^\perp}(G). \tag{2.43}$$

ii) Let $n \ge 3$ and $1 < q < \frac{n}{2}$. Then in the sense of a direct decomposition holds

$$\hat{H}^{2,q}_D(G) = \hat{H}^{2,q}_{D,C}(G) \oplus N^{2,q}(G)$$

and $-\Delta : \hat{H}^{2,q}_{D,C}(G) \to L^q(G)$ is bijective. With a constant $C'_q = C'(q,G) > 0$ holds

$$\|u\|_{q^{**}} + \|\nabla u\|_{q^*} + |u|_{2,q} \le C'_q \| - \Delta u\|_q \text{ for } u \in \hat{H}^{2,q}_{D,C}(G). \tag{2.44}$$

3. $(2 + k)$-SOLUTIONS OF THE DIRICHLET PROBLEM

Throughout this section let $k \in \mathbb{N}_0 := \mathbb{N} \cup \{0\}$. The case $k = 0$ coincides with that of strong solutions considered in Section 2. In case $k \ge 1$ we indicate the major steps.

3.1. The appropriate spaces for the Dirichlet problem

We always assume $\partial G \in C^k$.

Let us define for $m \in \mathbb{N}$ and $1 < q < \infty$ for Ω either denoting G or $\mathbb{R}^n$ (with $\Omega_R := \Omega \cap B_R$)

$$\hat{H}^{m,q}(\Omega) := \{ \ g : \Omega \to \mathbb{R} \text{ measurable}, \ D^\beta g \in L^q(\Omega_R) \tag{3.1}$$
$$\text{for each } R > 0 \text{ and } |\beta| \le m - 1,$$
$$D^\alpha g \in L^q(\Omega) \text{ for } |\alpha| = m \ \}.$$

We consider on $\hat{H}^{m,q}(\Omega)$ the semi-norm

$$|g|_{m,q} := \left(\sum_{|\alpha|=m} \|D^\alpha g\|_{q,\Omega}^q \right)^{\frac{1}{q}}. \tag{3.2}$$

Let now $\emptyset \neq G_0 \subset\subset \Omega$ and define for $g \in \hat{H}^{m,q}(\Omega)$

$$|g|_{m-1,q;G_0} := \left(\sum_{|\beta| \leq m-1} \|D^\beta g\|_{q,G_0}^q \right)^{\frac{1}{q}} \tag{3.3}$$

Then by

$$\||g\||_{m,q;G_0} := \left(|g|_{m,q}^q + |g|_{m-1,q;G_0}^q \right)^{\frac{1}{q}} \tag{3.4}$$

a norm is defined on $\hat{H}^{m,q}(\Omega)$ such that $\left(\hat{H}^{m,q}(\Omega), \||.\||_{m,q;G_0} \right)$ is a reflexive Banach space ($q=2$: Hilbert space). Sometimes it turns out to be very convenient to consider for $\emptyset \neq G_0 \subset\subset \Omega$ the space

$$\hat{H}^{m,q}(\Omega; G_0) := \left\{ g \in \hat{H}^{m,q}(\Omega) : \int_{G_0} D^\sigma g \, dx = 0 \quad , \text{ for } |\sigma| \leq m-1 \right\}. \tag{3.5}$$

Let G_1 denote a domain with boundary $\partial G_1 \in C^1$ sucht that $\emptyset \neq G_0 \subset\subset G_1 \subset\subset \Omega$. Then with a constant $C_p = C_p(G_1;q) > 0$ the Poincaré inequality

$$\|v\|_{q,G_1} \leq C_p \|\nabla v\|_{q,G_1} \tag{3.6}$$

holds for all $v \in H^{1,q}(G_1)$ satisfying $\int_{G_1} v \, dx = 0$. As is readily seen there exists a constant $C_p' = C_p'(C_p, |G_0|, |G_1|) > 0$ such that (3.6) holds with C_p replaced by C_p' and all $v \in H^{1,q}(G_1)$ satisfying $\int_{G_0} v \, dx = 0$. Iterated application of the last mentioned estimate yields with $C_p'' = C_p''(C_p', m) > 0$

$$|g|_{m-1,q;G_0} \leq |g|_{m-1,q;G_1} \leq C_p'' |g|_{m,q;G_1} \leq C_p'' |g|_{m,q} \tag{3.7}$$

for $g \in \hat{H}^{m,q}(\Omega; G_0)$. This proves

Lemma 3.1. *Let $1 < q < \infty$ and let $\Omega \subset \mathbb{R}^n (n \geq 2)$ denote either $\Omega := G$ an exterior domain or $\Omega := \mathbb{R}^n$. Let $m \in \mathbb{N}$ and let $\emptyset \neq G_0 \subset\subset \Omega$. Then by (3.2) a norm is defined on $\hat{H}^{m,q}(\Omega; G_0)$ such that $\left(\hat{H}^{m,q}(\Omega; G_0), |.|_{m,q} \right)$ is a reflexive Banach space ($q=2$: Hilbert space). Further there is a constant $C_1 = C_1(G_0, m, q)$ such that*

$$|g|_{m,q} \leq \||g\||_{m,q;G_0} \leq C_1 |g|_{m,q} \tag{3.8}$$

holds for $g \in \hat{H}^{m,q}(\Omega; G_0)$.

Let $P(n; m-1)$ denote the linear space of all polynomials of degree less or equal to $m-1$. Then in the sense of a direct decomposition holds

$$\hat{H}^{m,q}(\Omega) = \hat{H}^{m,q}(\Omega; G_0) \oplus P(n; m-1). \tag{3.9}$$

Further there are constants $C_i = C_i(m, q; G_0) > 0$ $(i = 1, 2)$ *such that*

$$C_1 \left(|g_1|_{m,q}^q + |g_2|_{m-1,q;G_0}^q \right)^{\frac{1}{q}} \leq |||g_1 + g_2|||_{m,q;G_0} \tag{3.10}$$

$$\leq C_2 \left(|g_1|_{m,q}^q + |g_2|_{m-1,q;G_0}^q \right)^{\frac{1}{q}}$$

for all $g_1 \in \hat{H}^{m,q}(\Omega; G_0)$ *and all* $g_2 \in P(n; m-1)$.

If we regard an exterior domain $G \subset \mathbb{R}^n$ and $k \in \mathbb{N}$ we will consider the (DP) $-\Delta u = f$ in G, $u|_{\partial G} = 0$ with data $f \in \hat{H}^{k,q}(G)$ given. We will seek solutions u in the space

$$\hat{H}_D^{2+k,q}(G) := \left\{ u \in \hat{H}^{2+k,q}(G) : \quad (\eta u) \in H_0^{1,q}(G) \text{ for } \eta \in C_0^\infty(\mathbb{R}^n) \right\} \tag{3.11}$$

In the contrary to the case $k = 0$ if $k \geq 1$ by $|.|_{2+k,q}$ no longer a norm needs to be defined on $\hat{H}_D^{2+k,q}(G)$. Consider again e.g. $G_1 := \mathbb{R}^n \setminus B_1$ and $k = 1$. Then $u(x) := |x|^2 - 1$ satisfies $u \in \hat{H}_D^{3,q}(G)$ for $1 < q < \infty$ and $|u|_{3,q} = 0$. Therefore we choose a set $\emptyset \neq G_0 \subset\subset G$ and the $|||.|||_{2+k,q;G_0}$-norm constructed according to (3.4). Then $\left(\hat{H}_D^{2+k,q}(G), |||.|||_{2+k,q;G_0} \right)$ is a reflexive Banach space ($q = 2$: Hilbert space). Clearly $-\Delta : \hat{H}_D^{2+k,q}(G) \to \hat{H}^{k,q}(G)$ is continuous. Conversely to find for given $f \in \hat{H}^{k,q}(G)$ a solution $u \in \hat{H}_D^{2+k,q}(G)$ of the equation $-\Delta u = f$ we proceed similarly to the case of strong solutions (Section 2). For this purpose we observe that there exists an extension operator $E_{k,q} : \hat{H}^{k,q}(G) \to \hat{H}^{k,q}(\mathbb{R}^n)$ such that with a constant $D_q = D(q, k, G) > 0$ holds

$$|E_{k,q}f|_{k,q} \leq D_q |||f|||_{k,q;G_0} \text{ for } f \in \hat{H}^{k,q}(G). \tag{3.12}$$

and $E_{k,q}f|_G = f$. We abbreviate $\tilde{f} := E_{k,q}f$ and look for solutions $v \in \hat{H}^{2+k,q}(\mathbb{R}^n)$ of $-\Delta v = \tilde{f}$ and possibly for estimates of v by $\tilde{f}$. Then via Theorem 1.4 we solve $\Delta h = 0$ in G and $h|_{\partial G} = -v|_{\partial G}$ and we put finally $u := v + h$ to get the desired solutions.

3.2. The equation $-\Delta u = f$ for $u \in \hat{H}^{2+k,q}(\mathbb{R}^n)$

For our approach we need some auxiliary tools.

3.2.1. Harmonic polynomials

Let $k \in \mathbb{N}_0$ and let $u \in \hat{H}^{2+k,q}(\mathbb{R}^n)$ satisfy $\Delta u = 0$. Then by Weyl's lemma $u \in C^\infty(\mathbb{R}^n)$ and for $|\alpha| = 2 + k$ we get $\Delta D^\alpha u = 0$. Since $D^\alpha u \in L^q(\mathbb{R}^n)$ by Liouville's theorem in $L^q(\mathbb{R}^n)$ we see $D^\alpha u = 0$ for $|\alpha| = 2 + k$. Therefore u is a polynomial of degree less or equal to $1 + k$. Since in addition $\Delta u = 0$ we see that u is a harmonic polynomial. Therefore the null space $N^{2+k,q}(\mathbb{R}^n)$ of Δ in $\hat{H}^{2+k,q}(\mathbb{R}^n)$ consists of all harmonic polynomials of degree $\leq 1 + k$. We list now some classical facts (compare e.g. Müller, 1966) on harmonic polynomials.

Definition and Lemma 3.2. *Let $n \geq 2$ and $j \in \mathbb{N}_0$.*

i) Let

$$A(n;j) := \{ \ p : \mathbb{R}^n \to \mathbb{R} : p \ \text{is a homogeneous polynomial} \tag{3.13}$$
$$\text{of degree } j \ \text{in } n \ \text{variables} \ \}$$
$$= \{ \ p(x) = \sum_{|\alpha|=j} a_\alpha x^\alpha : a_\alpha \in \mathbb{R} \ \}$$

Then

$$M(n;j) = \dim A(n;j) = \binom{n-1+j}{j} = \binom{n-1+j}{n-1} \tag{3.14}$$

ii) Let

$$H(n;j) := \{p \in A(n;j) : \Delta p = 0\} \tag{3.15}$$

Then

$$N(n;j) := \begin{cases} \dim H(n;j) = \binom{n-3+j}{n-2} \frac{(n-2+2j)}{j} & \text{if } j \geq 1 \\ 1 & \text{if } j = 0 \end{cases} \tag{3.16}$$

iii) $A(n;j) = H(n;j)$ for $j = 0, 1$. Further for $j \geq 2$

$$M(n;j) - N(n;j) = \binom{n-3+j}{n-1} = M(n;j-2) \tag{3.17}$$

iv) Let $j \in \mathbb{N}$ and let $S_n := \{\xi \in \mathbb{R}^n : |\xi| = 1\}$ denote the unit sphere in $\mathbb{R}^n$. For $p, q \in C^0(S_n)$ define

$$< p, q >_{S_n} := \int_{S_n} p(\xi)\, q(\xi)\, d\omega \tag{3.18}$$

the (L^2)-inner product on the unit sphere. By ii) and by the E. Schmidt orthogonalization process we may assume that for $0 \leq r \leq j$ ($r \in \mathbb{N}_0$) there are $\{h_i^{(r)} \in H(n;r) : i = 1, \ldots, N(n;r)\}$ forming an orthonormal basis of $H(n;r)$ with respect to $< ., . >_{S_n}$. If $0 \leq r, s \leq j$ with $r, s \in \mathbb{N}_0$ and $r \neq s$ then automatically $< h_i^{(r)}, h_l^{(s)} >_{S_n} = 0$ for $i = 1, \ldots, N(n;r)$ and $l = 1, \ldots, N(n;s)$. Let $\left[\frac{i}{2}\right]$ denote as usual the biggest integer less or equal to $\frac{i}{2}$. Given $p \in A(n;j)$ we define

$$a_{r,m} := < p, h_m^{(j-2r)} >_{S_n} \ \text{for } r = 0, \ldots, \left[\frac{j}{2}\right] \ \text{and} \ m = 1, \ldots, N(n;j-2r) \tag{3.19}$$

Then p has the unique representation ($N(j - 2r) := N(n;j - 2r)$)

$$p(x) = \sum_{r=0}^{\left[\frac{j}{2}\right]} \sum_{m=1}^{N(j-2r)} a_{r,m} |x|^{2r} h_m^{(j-2r)}(x) \ \text{for } x \in \mathbb{R}^n \tag{3.20}$$

v) Let for $j \in \mathbb{N}_0$

$$Q(n;j) := A(n;j) \ominus H(n;j) \tag{3.21}$$

(e.g. in the sense of an orthogonal difference with the respect to $< ., . >_{S_n}$ or regarded as a direct decomposition). Then $Q(n;j) = \{0\}$ for $j = 0, 1$.

vi) For $k \in \mathbb{N}_0$ we put

$$P(n;k) := \left\{ p(x) := \sum_{j=0}^{k} p_j(x) : p_j \in A(n;j), j = 0 \ldots k \right\} = \bigoplus_{j=0}^{k} A(n;j) \quad (3.22)$$

$$\tilde{H}(n;k) \; := \; \left\{ h(x) := \sum_{j=0}^{k} h_j(x) : h_j \in H(n;j), j = 0, \ldots, k \right\} =$$

$$= \bigoplus_{j=0}^{k} H(n;j) \qquad (3.23)$$

and for $k \geq 2$

$$\tilde{Q}(n;k) = \left\{ q(x) := \sum_{j=2}^{k} q_j(x) : q_j \in Q(n;j), j = 2, \ldots, k \right\} = \bigoplus_{j=2}^{k} Q(n;j) \quad (3.24)$$

where $\oplus$ has to be understood in the sense of a direct sum.
Then

$$P(n;k) = \tilde{H}(n;k) \oplus \tilde{Q}(n;k) \qquad (3.25)$$

and

$$\dim P(n;k) = \sum_{j=0}^{k} M(n;j) = \binom{n+k}{k} = \binom{n+k}{n}.$$

$$\dim \tilde{H}(n;k) = \left\{ \begin{array}{ll} \sum\limits_{j=0}^{k} N(n;j) = \binom{n+k-2}{n-1} \frac{n-1+2k}{k} = N(n+1;k) & \text{if } k \geq 1 \\ 1 & \text{if } k = 0 \end{array} \right.$$

$$\dim \tilde{Q}(n;k) = \left\{ \begin{array}{ll} \sum\limits_{j=2}^{k} M(n;j-2) = \binom{n+k-2}{k-2} = \binom{n+k-2}{n} & \text{if } k \geq 2 \\ 0 & \text{if } k = 0,1 \end{array} \right.$$

As we indicated above the null space of $-\Delta$ in $\hat{H}^{2+k,q}(\mathbb{R}^n)$ is exactly $\tilde{H}(n;1+k)$. Furthermore we easily get from (3.20)

Lemma 3.3. *Let $j \in \mathbb{N}_0$ and let $P \in A(n;j)$ be given,*

$$P(x) = \sum_{i=0}^{[\frac{j}{2}]} \sum_{m=1}^{N(j-2i)} b_{i,m} |x|^{2i} h_m^{(j-2i)}(x) \qquad (3.26)$$

with $b_{i,m} \in \mathbb{R}$ and let $p \in Q(n;2+j)$ be defined by

$$p(x) = \sum_{i=1}^{[\frac{j+2}{2}]} \sum_{m=1}^{N(j+2-2i)} a_{i,m} |x|^{2i} h_m^{(2+j-2i)}(x) \qquad (3.27)$$

with

$$a_{i,m} := \frac{b_{i-1,m}}{2i[2(j-i+2)+n]} \quad, \; for \quad m = 1, \ldots, N(j-2i+2), \qquad (3.28)$$

$$i = 1, \ldots, [\frac{j}{2}]+1$$

Then $p \in Q(n; 2 + j)$ is the unique solution of the equation $-\Delta p = P$ and there are constants $C_i(n; j) > 0$ $(i = 1, 2)$ such that

$$C_1 \|p\|_{L^2(S_n)} \leq \| - \Delta p\|_{L^2(S_n)} \leq C_2 \|p\|_{L^2(S_n)} \tag{3.29}$$

holds for all $p \in Q(n; 2 + j)$.

We observe that for $p \in P(n; k)$ by $\|p\|_{L^2(S_n)}$ in general $(k \geq 2)$ no longer a norm is defined. But we observe that on this finite dimensional space any two norms are equivalent. We then easily deduce

Theorem 3.1. *Let $\emptyset \neq G_0 \subset\subset \mathbb{R}^n$, let $k \in \mathbb{N}, k \geq 1$ and let $1 < q < \infty$. Then*

$$-\Delta : \tilde{Q}(n; k+1) \rightarrow P(n; k-1)$$

is bijective and there are constants $C_i(k; q; G_0) > 0$ such that (compare (3.3))

$$C_1 |p|_{k+1, q; G_0} \leq | - \Delta p|_{k-1, q; G_0} \leq C_2 |p|_{k+1, q; G_0} \text{ for } p \in \tilde{Q}(n; k+1). \tag{3.30}$$

For our construction of solutions we need a further result. If $\emptyset \neq G_0 \subset \mathbb{R}^n$ and $m \in \mathbb{N}$, $1 < q < \infty$ then we put

$$\hat{H}^{m,q}_{loc}(\mathbb{R}^n; G_0) := \{ \ g : \mathbb{R}^n \rightarrow \mathbb{R} \text{ such that } q|_{G_1} \in H^{m,q}(G_1)$$
$$\text{for all } G_1 \subset\subset \mathbb{R}^n \text{ and}$$
$$\int_{G_0} D^\beta g = 0 \text{ for } |\beta| \leq m - 1 \ \}$$

Lemma 3.4. *Let $\emptyset \neq G_0 \subset\subset \mathbb{R}^n$. Let $k \in \mathbb{N}_0$ and let $1 < q < \infty$. Let $f \in \hat{H}^{k,q}_{loc}(\mathbb{R}^n; G_0)$ and let $v \in \hat{H}^{2+k,q}_{loc}(\mathbb{R}^n)$ such that $-\Delta v = f$ a.e. in $\mathbb{R}^n$. Then there exists a unique $h \in \tilde{H}(n; k+1)$ such that*

$$\int_{G_0} D^\sigma(v - h) \, dx = 0 \quad , \text{ for } |\sigma| \leq k + 1 \tag{3.31}$$

(that is $u := (v - h) \in \hat{H}^{2+k,q}_{loc}(\mathbb{R}^n; G_0)$).

3.2.2. Solution of the equation $-\Delta u = f$ in $\mathbb{R}^n$

Our main result is

Theorem 3.2. *Let $k \in \mathbb{N}_0$ and let $1 < q < \infty$, let $\emptyset \neq G_0 \subset\subset \mathbb{R}^n$.*

i) $-\Delta : \hat{H}^{2+k,q}(\mathbb{R}^n; G_0) \rightarrow \hat{H}^{k,q}(\mathbb{R}^n; G_0)$ is bijective and there exist constants $C_i = C_i(q, k, G_0) > 0$ $(i = 1, 2)$ such that

$$C_1 |u|_{2+k, q} \leq | - \Delta u|_{k, q} \leq C_2 |u|_{2+k, q} \text{ for } u \in \hat{H}^{2+k,q}(\mathbb{R}^n; G_0). \tag{3.32}$$

94

ii) In the sense of a direct decomposition holds (compare (3.9), (3.23) - (3.25))

$$\hat{H}^{2+k,q}(\mathbb{R}^n) = \hat{H}^{2+k,q}(\mathbb{R}^n; G_0) \oplus \tilde{Q}(n; k+1) \oplus \tilde{H}(n; k+1) \qquad (3.33)$$

and there are constants $D_i = D_i(k, q; G_0) > 0$ $(i = 1, 2)$ such that

$$D_1 \left(\|\|u_1\|\|^q_{2+k,q;G_0} + |u_2|^q_{k+1,q;G_0} + |u_3|^q_{k+1,q;G_0} \right)^{\frac{1}{q}} \leq \qquad (3.34)$$

$$\leq \|\|u_1 + u_2 + u_3\|\|_{2+k,q;G_0} \leq$$

$$\leq D_2 \left(\|\|u_1\|\|^q_{2+k,q;G_0} + |u_2|^q_{k+1,q;G_0} + |u_3|^q_{k+1,q;G_0} \right)$$

holds for $(u_1, u_2, u_3) \in \hat{H}^{2+k,q}(\mathbb{R}^n; G_0) \times \tilde{Q}(n; k+1) \times \tilde{H}(n; k+1)$ (observe that for $p \in P(n; k+1)$ holds $\|\|p\|\|_{2+k,q;G_0} = |p|_{1+k,q;G_0}$ (see (3.3),(3.4)) and that $|.|_{1+k,q;G_0}$ is a norm on $P(n; k+1)$).

iii) Let $N^{2+k,q}(\mathbb{R}^n) := \{u \in \hat{H}^{2+k,q}(\mathbb{R}^n) : -\Delta u = 0\}$. Then $N^{2+k,q}(\mathbb{R}^n) = \tilde{H}(n; k+1)$ and $\dim N^{2+k,q}(\mathbb{R}^n) = \binom{n+k-1}{n-1} \frac{(n+2k+1)}{k+1}$. Further

$$\begin{array}{rcl} -\Delta & : & \hat{H}^{2+k,q}(\mathbb{R}^n; G_0) \oplus \tilde{Q}(n; k+1) \to \hat{H}^{k,q}(\mathbb{R}^n) \qquad (3.35) \\ -\Delta & : & \tilde{Q}(n; k+1) \to P(n; k-1) \\ -\Delta & : & \hat{H}^{2+k,q}(\mathbb{R}^n; G_0) \to \hat{H}^{k,q}(\mathbb{R}^n; G_0) \end{array}$$

are bijections. Further there are constants $\tilde{C}_i = \tilde{C}_i(C_i, D_i) > 0$ $(i = 1, 2)$ (where C_i, D_i denote the constants according to i.), ii.)) such that for $u_1 \in \hat{H}^{2+k,q}(\mathbb{R}^n; G_0)$ and $u_2 \in \tilde{Q}(n; k+1)$ holds

$$\tilde{C}_1 \left(\|\|u_1\|\|^q_{2+k,q;G_0} + |u_2|^q_{1+k,q;G_0} \right) \leq \|\| - \Delta(u_1 + u_2) \|\|_{2+k,q;G_0} \qquad (3.36)$$

$$\leq \tilde{C}_2 \left(\|\|u_1\|\|^q_{2+k,q;G_0} + |u_2|^q_{1+k,q;G_0} \right)^{\frac{1}{q}} .$$

We want to sketch the proof:

i) First let $f \in C_0^\infty(\mathbb{R}^n) \subset L^q(\mathbb{R}^n)$. Let $v \in \hat{H}^{2,q}(\mathbb{R}^n; G_0)$ be the unique solution of $-\Delta v = f$ according to Theorem 2.2 satisfying (2.14). For $\varepsilon > 0$ let $j_\varepsilon \in C_0^\infty(B_\varepsilon)$ denote the mollifier kernel and put $v_\varepsilon := j_\varepsilon * v$. Observe $-\Delta v_\varepsilon = (-\Delta v)_\varepsilon = f_\varepsilon$ and $v_\varepsilon \in C^\infty(\mathbb{R}^n) \cap \hat{H}^{2+k,q}(\mathbb{R}^n)$. There exists a unique $P^{(\varepsilon)} \in P(n; k-1)$ such that $g^{(\varepsilon)} := (f_\varepsilon - P^{(\varepsilon)}) \in \hat{H}^{k,q}(\mathbb{R}^n; G_0)$. By Theorem 3.1 there exists a unique $p^{(\varepsilon)} \in \tilde{Q}(n; k+1)$ with $-\Delta p^{(\varepsilon)} = P^{(\varepsilon)}$. Then $-\Delta(v_\varepsilon - p^{(\varepsilon)}) = g^{(\varepsilon)}$. Choose now $h^{(\varepsilon)} \in \tilde{H}(n; k+1)$ such that $u^{(\varepsilon)} := (v_\varepsilon - p^{(\varepsilon)} - h^{(\varepsilon)}) \in \hat{H}^{2+k,q}(\mathbb{R}^n; G_0)$, $-\Delta u^{(\varepsilon)} = g^{(\varepsilon)}$. If $|\beta| = k$ then $D^\beta u^{(\varepsilon)} \in \hat{H}^{2,q}(\mathbb{R}^n; G_0)$ and satisfies $-\Delta(D^\beta u^{(\varepsilon)}) = D^\beta g^{(\varepsilon)} = D^\beta f_\varepsilon \in L^q(\mathbb{R}^n)$. Again by Theorem 2.2 we conclude that $D^\beta u_\varepsilon$ is the unique solution of the last equation and (2.14) applies to u replaced by $D^\beta u^{(\varepsilon)}$ for each $|\beta| = k$. Then we get from (2.14) for $\varepsilon', \varepsilon'' > 0$

$$|u^{(\varepsilon')} - u^{(\varepsilon'')}|_{2+k,q} \leq C \sum_{|\beta|=k} |D^\beta \left(u^{(\varepsilon')} - u^{(\varepsilon'')} \right)|_{2,q} \leq C \cdot C(q, G_0)$$

$$\sum_{|\beta|=k} \| -\Delta D^\beta \left(u^{(\varepsilon')} - u^{(\varepsilon'')} \right) \|_{2,q} \leq$$

$$\leq C' \cdot C(q; G_0) |g^{(\varepsilon')} - g^{(\varepsilon'')}|_{k,q} = C' \cdot C(q; G_0) |f_{\varepsilon'} - f_{\varepsilon''}|_{k,q} \to 0.$$

By completeness of $\hat{H}^{2+k,q}(\mathbb{R}^n; G_0)$ there exists a $u \in \hat{H}^{2+k,q}(\mathbb{R}^n; G_0)$ such that $|u - u^{(\varepsilon)}|_{2+k,q} \to 0$. Further there exists a $P \in P(n; k-1)$ such that $g :=$ $(f - P) \in \hat{H}^{k,q}(\mathbb{R}^n; G_0)$ and $|g_\varepsilon - g|_{k,q} \to 0$. Therefore we solved the equation $-\Delta u = g = f - P$ for $f \in C_0^\infty(\mathbb{R}^n)$, $P \in P(n; k-1)$ and the estimate

$$|u|_{2+k,q} \leq C|g|_{k,q}. \tag{3.37}$$

ii) If $f \in \hat{H}^{k,q}(\mathbb{R}^n; G_0)$ is given then by Lemma 2.1 there exists a sequence $(f_i) \subset C_0^\infty(\mathbb{R}^n)$ such that $|f - f_i|_{k,q} \to \infty$. With unique $P_i \in P(n; k-1)$ we put $g_i := (f_i - P_i) \in \hat{H}^{k,q}(\mathbb{R}^n; G_0)$, $\||f - g_i\||_{k,q;G_0} \leq C_1 |f - g_i|_{k,q} = C_1 |f - f_i|_{k,q}$ (compare (3.8)). Then by step i.) we solve the equation $-\Delta u_i = g_i$ and using (3.37) we again may pass to the limit $i \to \infty$ and we finally get the solution $u \in \hat{H}^{2+k,q}(\mathbb{R}^n; G_0)$ of the equation $-\Delta u = f$ (observe $|P_i|_{k-1,q;G_0} \to 0$).

iii) The remaining assertions are clear via (3.9) and Theorem 3.1. $\qquad\square$

All the results achieved until now are depending on the choice of the set $\emptyset \neq G_0 \subset\subset \mathbb{R}^n$ acting like as a parameter. In the case of strong solutions and for $n \geq 3$ and $1 < q < \frac{n}{2}$ (that is $2q < n$) we achieved more natural results via the Sobolev embedding theorem. The following result corresponds to Theorem 2.3, part b) in case $k = 0$.

Definition and Theorem 3.3. *Let $k \in \mathbb{N}_0$ and let $1 < q < \infty$ satisfy $(2 + k)q < n$ (that means $n > 2 + k$). For $j = 0, 1, \ldots, 2 + k$ let $q_j := \frac{nq}{n - jq}$.*

i) We put for $m = k$ or $m = 2 + k$

$$\hat{H}_0^{m,q}(\mathbb{R}^n) = \{ \ f \in \hat{H}^{m,q}(\mathbb{R}^n) : D^\beta f \in L^{q_j}(\mathbb{R}^n) \ for \tag{3.38}$$
$$|\beta| = m - j \ and \ j = 0, \ldots, m \},$$

and for $f \in \hat{H}_0^{m,q}(\mathbb{R}^n)$

$$|f|_{m-j,q_j} := \left(\sum_{|\beta|=m-j} \|D^\beta f\|_{q_j}^{q_j} \right)^{\frac{1}{q_j}} \ for \ j = 0, \ldots, m \tag{3.39}$$

$$|f|_{m,q} := \sum_{j=0}^m |f|_{m-j,q_j}. \tag{3.40}$$

Then there exists a constant $C_{m,q} = C(m, q, n) > 0$ such that

$$C_{m,q} |f|_{m,q} \leq |f|_{m,q} \leq |f|_{m,q}. \tag{3.41}$$

$\left(\hat{H}_0^{m,q}(\mathbb{R}^n), |.|_{m,q} \right)$ is a reflexive Banach space and

$$\hat{H}_0^{m,q}(\mathbb{R}^n) = \overline{C_0^\infty(\mathbb{R}^n)}^{|.|_{m,q}}. \tag{3.42}$$

Further as direct decompositions hold

$$\hat{H}^{m,q}(\mathbb{R}^n) = \hat{H}_0^{m,q}(\mathbb{R}^n) \oplus P(n; m-1) \tag{3.43}$$
$$= \hat{H}_0^{m,q}(\mathbb{R}^n) \oplus \tilde{Q}(n; m-1) \oplus \tilde{H}(n; m-1)$$

ii)

$$-\Delta : \quad \hat{H}_0^{2+k,q}(\mathbb{R}^n) \oplus \tilde{Q}(n; k+1) \;\to\; \hat{H}_0^{k,q}(\mathbb{R}^n) \oplus P(n; k-1)$$
$$-\Delta : \hat{H}_0^{2+k,q}(\mathbb{R}^n) \to \hat{H}_0^{k,q}(\mathbb{R}^n) \tag{3.44}$$
$$-\Delta : \tilde{Q}(n; k+1) \to P(n; k-1)$$

are bijective maps. Further there exist constants $C_i = C_i(n, k, q) > 0$ *(i = 1,2)*
such that for $u \in \hat{H}_0^{2+k,q}(\mathbb{R}^n)$ *holds*

$$C_1 |u|_{2+k,q} \quad \leq \quad |-\Delta u|_{k,q} \quad \leq \quad C_2 |u|_{2+k,q} \tag{3.45}$$

To sketch the proof: i) Iterated application of the Sobolev embedding theorem shows
$\overline{C_0^\infty(\mathbb{R}^n)}^{|\cdot|_{m,q}} \subset \hat{H}^{m,q}(\mathbb{R}^n)$. The converse inclusion is easily proved by cut-off and molli-
fying techniques. The inequalities (3.41) are then a consequence of Sobolev's theorem.
The decomposition (3.43) and the results of (3.44),(3.45) are then an easy consequence
of Lemma 2.1.

In the contrary if $q \geq n$ we have no "natural decomposition" like (3.43). If we put
for $\emptyset \neq G_0 \subset\subset \mathbb{R}^n$ and $m = 2 + k$ *or* $m = k$ in case $k \geq 1$

$$C_0^\infty(\mathbb{R}^n; m; G_0) := \left\{ \phi \in C_0^\infty(\mathbb{R}^n) : \int_{G_0} D^\beta \phi \, dx = 0 \text{ for } |\beta| \leq m-1 \right\} \tag{3.46}$$

then by cut-off techniques we easily show in case $q > n$ and with some additional
functional-analytic considerations in case $q = n$, too, that

$$\hat{H}^{m,q}(\mathbb{R}^n; G_0) = \overline{C_0^\infty(\mathbb{R}^n; m; G_0)}^{|\cdot|_{m,q}} = \overline{C_0^\infty(\mathbb{R}^n; m; G_0)}^{\|\cdot\|_{m,q;G_0}}. \tag{3.47}$$

Therefore we gain for $q \geq n$ no better informations than Theorem 3.2 tells us. In
case $1 < q < n$ but $(2 + k)q \geq n$ we have to consider that integer $r \in \mathbb{N}_0$ such that
$r < \frac{n}{q} \leq r + 1$. Depending as well on r as on a "parameter set" $\emptyset \neq G_0 \subset \mathbb{R}^n$ we derive
analogous results to Theorem 2.3 part a), but we have to distinguish a variety of cases.

3.3. (2+k)-solutions in exterior domains

3.3.1. The null spaces of $-\Delta$ in $\hat{H}_D^{2+k,q}(G)$

For $k \in \mathbb{N}_0$ let us choose now any basis in $\tilde{H}(n; k+1) = N^{2+k,q}(\mathbb{R}^n)$ and denote it
by $\tilde{h}_i, i = 1, \ldots, m(n, k)$, where $m(n, k) = \dim N^{2+k,q}(\mathbb{R}^n)$. Now with this meaning of
$\tilde{h}_i$ we solve the equations (2.25) and put $h_i := p_i + (1 - \tilde{\eta})\tilde{h}_i$. If we observe the regularity
result Theorem 1.4 we clearly expect that the harmonic functions h_i $(i = 1, \ldots, m(n, k))$
have the properties listed subsequently. It remains to show that they form a basis of
$N^{2+k,q}(G)$.

Theorem 3.4. *Let $k \in \mathbb{N}$ and let $1 < q < \infty$. Suppose $\partial G \in C^{2+k}$. We define*

$$N^{2+k,q}(G) := \left\{ h \in \hat{H}_D^{2+k,q}(G) : \quad \Delta h = 0 \right\}.$$

Then

$$\dim N^{2+k,q}(G) = m(n,k) = \binom{n+k-1}{n-1} \frac{(n+2k+1)}{k+1}. \tag{3.48}$$

There exists a basi

$$h_i \in \bigcap_{1 < t < \infty} \hat{H}_D^{2+k,t}(G), \quad i = 0, 1, \ldots, m(n,k) - 1$$

with the additional property $h_0 \in \hat{H}_\bullet^{1,t}(G)$ for $\frac{n}{n-1} < t < \infty$ but $h_i \notin \hat{H}_\bullet^{1,t}(G)$ for all $1 < t < \infty$ and $1 \le i \le m(n,k) - 1$.

3.3.2. Solutions of the equation $-\Delta u = f$ in $\hat{H}_D^{2+k,q}(G)$

We proceed exactly like as in section 2.4. First we consider again A_R defined by (2.28) and we put $G_0 := A_R$. Given $f \in \hat{H}_D^{k,q}(G)$ we consider $\tilde{f} := E_{k,q} f \in \hat{H}^{k,q}(\mathbb{R}^n)$ (see 3.12). According to Theorem 3.2 iii) there exists a unique $\tilde{v} \in \hat{H}^{2+k,q}(\mathbb{R}^n; G_0) \oplus \tilde{Q}(n; k+1)$ with $-\Delta \tilde{v} = \tilde{f}$ such that (3.36) holds. We put $v := \tilde{v}|_G$ and observe $-\Delta \tilde{v}|_G = f$. Then we proceed again as in section 2.4. First we solve (2.33). Again we observe Theorem 1.4 with $j = 2 + k$. From (1.18) we get estimates for the derivatives of p up to order $2 + k$. Combining the estimates our construction leads to solutions of $-\Delta u = f$ such that

$$\|u\|_{2+k,q;G_0} \le C(k, q; G_0) \|f\|_{k,q;G_0}$$

holds. Uniqueness of this construction again depends on the way of construction. To get now a realization of $\hat{H}_D^{2+k,q}(G)/N^{2+k,q}(G)$ instead of (2.36) we may introduce on $\hat{H}_D^{2+k,2}(G)$ the inner product $(u, v \in \hat{H}_D^{2+k,2}(G))$

$$< u, v >_{2+k} := \sum_{|\alpha| = 2+k} \int D^\alpha u \, D^\alpha v. \tag{3.49}$$

By (3.49) a continuous sesquilinear form on $\hat{H}_D^{2+k,q}(G) \times \hat{H}_D^{2+k,q'}(G)$ is defined. Then we can formulate similar statements to Theorem 2.6. Sharper results can be achieved like in Theorems 2.1 and 2.6 in case of strong solutions if $1 < q < n$. But even formulating the results becomes more and more complicated and we omit this here.

4. $(2+k)$-SOLUTIONS OF STOKES' SYSTEM

4.1. Stokes' System in $\hat{\mathbf{H}}^{2+k,q}(\mathbb{R}^n; G_0)$

Let $k \in \mathbb{N}_0$, $n \ge 2$ and $1 < q < \infty$. Let $\emptyset \ne G_0 \subset\subset \mathbb{R}^n$. Then for $m \in \mathbb{N}_0$ we define

$$\hat{\mathbf{H}}^{m,q}(\mathbb{R}^n; G_0) := \hat{H}^{m,q}(\mathbb{R}^n; G_0)^n = \tag{4.1}$$

$$= \{ \mathbf{g} = (g_1, \ldots, g_n) : g_i \in \hat{H}^{m,q}(\mathbb{R}^n; G_0), i = 1, \ldots, n \}$$

and for $\mathbf{g} \in \hat{\mathbf{H}}^{m,q}(\mathbb{R}^n; G_0)$ we put

$$|\mathbf{g}|_{m,q} := \left(\sum_{i=1}^{n} |g_i|_{m,q}^q \right)^{\frac{1}{q}}. \tag{4.2}$$

(Let us put for $m = 0$ $\hat{\mathbf{H}}^{0,q}(\mathbb{R}^n; G_0) := L^q(\mathbb{R}^n)^n$). Analogously we define $|\mathbf{g}|_{m-1,q;G_0}$ and $\|\|\mathbf{g}\|\|_{m,q;G_0}$ (compare (3.3),(3.4)) and we get an estimate similar to (3.8). Further let

$$\mathbf{P}(n; m-1) \; := \; P(n; m-1)^n = \tag{4.3}$$
$$= \; \{\mathbf{p} = (p_1, \ldots, p_n) : p_i \in P(n; m-1), i = 1, \ldots, n\}$$

Then the decomposition

$$\hat{\mathbf{H}}^{m,q}(\mathbb{R}^n) = \hat{\mathbf{H}}^{m,q}(\mathbb{R}^n; G_0) \oplus \mathbf{P}(n; m-1) \tag{4.4}$$

holds and an inequality similar to (3.10). In case of Stokes' system the result parallel to the corresponding Theorem 3.2 for the Laplacian is

Theorem 4.1. *Let $k \in \mathbb{N}_0$, let $1 < q < \infty$ and let $\emptyset \neq G_0 \subset\subset \mathbb{R}^n$ ($n \geq 2$). We put for $k \geq 1$*

$$\mathbf{D}^{k,q}(\mathbb{R}^n; G_0) := \{\mathbf{v} \in \hat{\mathbf{H}}^{k,q}(\mathbb{R}^n; G_0) : div\,\mathbf{v} = 0\} \tag{4.5}$$

and for $k \geq 0$

$$\mathbf{E}^{k,q}(\mathbb{R}^n; G_0) := \{\nabla p : p \in \hat{H}^{k+1,q}(\mathbb{R}^n; G_0)\} \subset \hat{\mathbf{H}}^{k,q}(\mathbb{R}^n; G_0) \tag{4.6}$$

i) Then for $k \geq 1$ the Helmholtz decomposition

$$\hat{\mathbf{H}}^{k,q}(\mathbb{R}^n; G_0) = \mathbf{D}^{k,q}(\mathbb{R}^n; G_0) \oplus \mathbf{E}^{k,q}(\mathbb{R}^n; G_0) \tag{4.7}$$

holds in the sense of a direct ($q = 2$: orthogonal) sum. There is a constant $C_1 = C_1(k, q, G_0) > 0$ such that for $\mathbf{v} \in \mathbf{D}^{k,q}(\mathbb{R}^n; G_0)$ and $\nabla p \in \mathbf{E}^{k,q}(\mathbb{R}^n; G_0)$ holds

$$C_1(|\mathbf{v}|_{k,q} + |\nabla p|_{k,q}) \leq |\mathbf{v} + \nabla p|_{k,q} \leq |\mathbf{v}|_{k,q} + |\nabla p|_{k,q} \tag{4.8}$$

ii)

$$-\Delta : \; \mathbf{D}^{2+k,q}(\mathbb{R}^n; G_0) \; \rightarrow \; \mathbf{D}^{k,q}(\mathbb{R}^n; G_0) \; and \tag{4.9}$$
$$-\Delta : \; \mathbf{E}^{2+k,q}(\mathbb{R}^n; G_0) \; \rightarrow \; \mathbf{E}^{k,q}(\mathbb{R}^n; G_0)$$

are bijective maps for $k \geq 0$. Furthermore there are constants $D_i = D_i(k, q; G_0) > 0$ and $E_i = E_i(k, q; G_0) > 0$ ($i = 1, 2$) such that

$$D_1|\mathbf{v}|_{2+k,q} \leq \; |-\Delta\mathbf{v}|_{k,q} \leq \; D_2\,|\mathbf{v}|_{2+k,q} \tag{4.10}$$
$$\text{for } \mathbf{v} \in \mathbf{D}^{2+k,q}(\mathbb{R}^n; G_0)$$

$$E_1|\nabla p|_{2+k,q} \leq \; |\nabla(-\Delta p)|_{k,q} \leq \; E_2\,|\nabla p|_{2+k,q}$$
$$\text{for } \nabla p \in \mathbf{E}^{2+k,q}(\mathbb{R}^n; G_0)$$

iii) For $\mathbf{f} \in \hat{\mathbf{H}}^{k,q}(\mathbb{R}^n; G_0)$ $(k \geq 0)$ *given there exists a unique pair*
$(\mathbf{u}, \pi) \in \mathbf{D}^{2+k,q}(\mathbb{R}^n; G_0) \times \mathbf{E}^{k,q}(\mathbb{R}^n; G_0)$ *such that*

$$\left.\begin{array}{r} -\Delta\mathbf{u} + \nabla\pi = \mathbf{f} \\ div\ \mathbf{u} = 0 \end{array}\right\} \quad Stokes'\ system \qquad (4.11)$$

Further there are constants $S_i = S_i(k, q; G_0) > 0$ $(i = 1, 2)$ *such that*

$$S_1(|\mathbf{u}|_{2+k,q} + |\nabla\pi|_{k,q}) \leq |-\Delta\mathbf{u} + \nabla\pi|_{k,q} \leq S_2(|\mathbf{u}|_{2+k,q} + |\nabla\pi|_{k,q}) \qquad (4.12)$$

holds for all $(\mathbf{u}, \pi) \in \mathbf{D}^{2+k,q}(\mathbb{R}^n; G_0) \times \mathbf{E}^{k,q}(\mathbb{R}^n; G_0)$.

On the basis of Theorem 3.2 and elementary facts on harmonic functions the proof of Theorem 4.1 is most simple.

Proof:

a) Given $\mathbf{f} \in \hat{\mathbf{H}}^{k,q}(\mathbb{R}^n; G_0)$ then div $\mathbf{f} \in \hat{\mathbf{H}}^{k-1,q}(\mathbb{R}^n; G_0)$. We observe $k - 1 \geq 0$. By Theorem 3.2 there exists a unique $p \in \hat{\mathbf{H}}^{k+1,q}(\mathbb{R}^n; G_0)$ such that $-\Delta p = $ div $\mathbf{f}$ and by (3.32) there are constants $\tilde{C}_i = \tilde{C}_i(q, k, G_0) > 0$ $(i = 1, 2)$ such that

$$\tilde{C}_1|p|_{1+k,q} \leq |-\Delta p|_{k-1,q} = |\text{div } \mathbf{f}|_{k-1,q} \leq \tilde{C}_2|p|_{1+k,q}. \qquad (4.13)$$

Since $\nabla p \in \hat{\mathbf{H}}^{k,q}(\mathbb{R}^n; G_0)$ we see

$$\mathbf{v} := \mathbf{f} - \nabla p \in \hat{\mathbf{H}}^{k,q}(\mathbb{R}^n; G_0) \text{ and div } \mathbf{v} = 0. \qquad (4.14)$$

b) Then (4.7) and (4.8) are immediate consequences of (4.13) and (4.14) if we show $\mathbf{D}^{k,q}(\mathbb{R}^n; G_0) \cap \mathbf{E}^{k,q}(\mathbb{R}^n; G_0) = \{0\}$. Suppose that $\mathbf{g}$ belongs to the intersection mentioned last. Then $\mathbf{g} = \nabla p \in \mathbf{E}^{k,q}(\mathbb{R}^n; G_0)$. Since $\mathbf{g} \in \mathbf{D}^{k,q}(\mathbb{R}^n; G_0)$ we see $0 = $ div $\mathbf{g} = \Delta p$ and therefore $p \in \hat{\mathbf{H}}^{k+1,q}(\mathbb{R}^n; G_0)$ is harmonic. Since for $|\sigma| = k+1$ holds $D^\sigma p \in L^q(\mathbb{R}^n)$ and $0 = D^\sigma \Delta p = \Delta D^\sigma p$ we conclude by the L^q-version $(1 < q < \infty)$ of Liouville's theorem that $D^\sigma p = 0$ for all $|\sigma| = k + 1$. Therefore p is a polynomial of degree at most k. On the other hand $p \in \hat{H}^{k+1}(\mathbb{R}^n; G_0)$ tells us $\int_{G_0} D^\beta p\, dx = 0$ for $|\beta| \leq k$ and therefore $p \equiv 0$.

c) If $\mathbf{w} \in \mathbf{D}^{k,q}(\mathbb{R}^n; G_0)$ is given then by application of Theorem 3.2 (applied to each component w_i) we get $\mathbf{v} \in \hat{\mathbf{H}}^{2+k,q}(\mathbb{R}^n; G_0)$ such that $-\Delta\mathbf{v} = \mathbf{w}$. Since $0 = $ div $\mathbf{w} = -\Delta(\text{div } \mathbf{v})$ and div $\mathbf{v} \in \hat{H}^{k+1,q}(\mathbb{R}^n; G_0)$ we conclude as in part b) of this proof that div $\mathbf{v} = 0$. That means $\mathbf{v} \in \mathbf{D}^{2+k,q}(\mathbb{R}^n; G_0)$.

d) If $\nabla p \in \mathbf{E}^{k,q}(\mathbb{R}^n; G_0)$ with $p \in \hat{H}^{k+1,q}(\mathbb{R}^n; G_0)$ then by Theorem 3.2 there exists $\pi \in \hat{H}^{k+3,q}(\mathbb{R}^n; G_0)$ such that $-\Delta\pi = p$. Therefore $\nabla\pi \in \mathbf{E}^{2+k,q}(\mathbb{R}^n; G_0)$ and $-\Delta(\nabla\pi) = \nabla p$. The estimates (4.9) are clear by Theorem 3.2 and our construction.

e) For $\mathbf{f} \in \hat{\mathbf{H}}^{k,q}(\mathbb{R}^n; G_0)$ there exists $\mathbf{w} \in \hat{\mathbf{H}}^{2+k,q}(\mathbb{R}^n; G_0)$ such that $-\Delta\mathbf{w} = f$. By (4.7) applied to $\mathbf{w}$ there is a unique $(u, \nabla p) \in \mathbf{D}^{2+k,q}(\mathbb{R}^n; G_0) \times \mathbf{E}^{2+k,q}(\mathbb{R}^n; G_0)$ such that $\mathbf{w} = \mathbf{u} + \nabla p$. If we put $\nabla\pi := -\nabla(\Delta p) \in \mathbf{E}^{k,q}(\mathbb{R}^n; G_0)$ then all properties of part iii) of the theorem are clear.

Let us remark for part i) of the preceeding theorem that in case $k = 0$ (that is $\hat{\mathbf{H}}^{0,q}(\mathbb{R}^n; G_0) \equiv L^q(\mathbb{R}^n)^n$ the Helmholtz decomposition of $L^q(\mathbb{R}^n)^n$ still holds true. But in this case we have to understand the condition "div $\mathbf{v} = 0$" for the first component of the decomposition in a weak sense. Modified in a suitable way the Helmholtz decomposition for the whole space $\mathbb{R}^n$ holds in case $k = 0$ and $1 < q < \infty$, too (see e.g. Simader and Sohr, 1992, Remark 5.1, Theorems 3.2, Lemma 3.3 and Theorem 1.4).

As in the case of Poisson's equation by means of the Sobolev embedding theorem we get in case $(2 + k)q < n$ results similar to Theorem 3.3. Further the remarks following Theorem 3.3 apply literally to the underlying case of Stokes' system.

4.2. Polynomial solutions of Stokes' system

We put for $j \in \mathbb{N}_0$

$$\mathbf{A}(n; j) := A(n; j)^n, \quad \mathbf{H}(n; j) := H(n; j)^n \quad \text{and} \quad \mathbf{Q}(n; j) := Q(n; j)^n \tag{4.15}$$

Further let for $j \geq 1$

$$\mathbf{D}(n; j) := \{\mathbf{v} \in \mathbf{A}(n; j) : \text{div } \mathbf{v} = 0\} \tag{4.16}$$

To calculate the dimension of $\mathbf{D}(n; j)$ we proceed as follows. For $\mathbf{v} \in \mathbf{D}(n; j)$ let $g := \sum_{i=1}^{n-1} \partial_i v_i \in A(n; j - 1)$. Then v_n satisfies

$$v_n \in A(n; j) \text{ and } \partial_n v_n = -g. \tag{4.17}$$

If conversely arbitrary $v_i \in A(n; j)$ $(i = 1, \ldots, n - 1)$ are given and if we find v_n satisfying (4.17) then $\mathbf{v} = (v_1, \ldots, v_{n-1}, v_n) \in \mathbf{D}(n; j)$. To construct v_n let us write first $(x = (x', x_n), x' = (x_1, \ldots, x_{n-1}) \in \mathbb{R}^{n-1})$

$$g(x) = g(x', x_n) = \sum_{r=0}^{j-1} g_{j-1-r}(x')x_n^r \tag{4.18}$$

with uniquely determined $g_{j-1-r} \in A(n-1; j-1-r)$ for $r = 0, \ldots, j-1$. Analogously for $v_n \in A(n; j)$ holds $v_n(x) = \sum_{s=0}^{j} w_{j-s}(x')x_n^s$ with $w_{j-s} \in A(n-1; j-s)$ for $s = 0, \ldots, j$. Then

$$(\partial_n v_n)(x', x_n) = \sum_{s=0}^{j} w_{j-s}(x')sx_n^{s-1} = \sum_{r=0}^{j-1} w_{j-1-r}(x')(r+1)x_n^r \tag{4.19}$$

Comparison of coefficients in (4.18), (4.19) shows that (4.17) holds if and only if

$$w_{j-s}(x') = -\frac{1}{s}g_{j-s}(x') \text{ for } s = 1, \ldots, j \tag{4.20}$$

101

Then we conclude from (4.17) and (4.20) that with

$$v_n(x', x_n) := w_j(x') - \sum_{s=1}^{j} \frac{1}{s} g_{j-s}(x') x_n^s \qquad (4.21)$$

for given functions $v_i \in A(n; j)$ $(i = 1, \ldots, n - 1)$ holds $\mathbf{v} = (v_1, \ldots, v_n) \in \mathbf{D}(n; j)$ if and only if (4.21) is satisfied with *arbitrary* $w_j \in A(n - 1; j)$. Therefore

$$\begin{aligned}
\dim \mathbf{D}(n; j) &= (n-1) \dim A(n; j) + \dim A(n-1; j) = \qquad (4.22) \\
&= (n-1)\binom{n-1+j}{n-1} + \binom{n-2+j}{n-2} = \\
&= (n+j)\binom{n-2+j}{j}.
\end{aligned}$$

Further we see immediately

$$\mathbf{D}(n; 0) = \mathbf{A}(n; 0), \quad \dim \mathbf{D}(n; 0) = \dim \mathbf{A}(n; 0) = n. \qquad (4.23)$$

By Lemma 3.3 we easily get

Lemma 4.1. (polynomial Helmholtz decomposition) *Let* $j \in \mathbb{N}_0$

$$\mathbf{E}(n; j) := \{\nabla p : \quad p \in Q(n; j+1)\} \quad \subset \mathbf{A}(n; j) \qquad (4.24)$$

Then in the sense of a direct decomposition holds for $j \geq 1$

$$\begin{aligned}
\mathbf{A}(n; j) &= \mathbf{D}(n; j) \quad \oplus \quad \mathbf{E}(n; j) \\
\mathbf{w} &= \mathbf{v} \quad + \quad \nabla p
\end{aligned} \qquad (4.25)$$

Let for $j \geq 1$

$$\begin{aligned}
\mathbf{N}(n; j) := \quad \{ \quad &(\mathbf{v}, \nabla p) \in \mathbf{D}(n, j) \times E(n; j) : \qquad (4.26) \\
&-\Delta \mathbf{v} + \nabla(-\Delta p) = 0 \quad \}.
\end{aligned}$$

Then by $(\mathbf{v}, \nabla p) \to (\mathbf{v} + \nabla p)$ *a bijection from* $\mathbf{N}(n; j)$ *on* $\mathbf{H}(n; j)$ *is defined. Therefore for* $j \geq 1$

$$\dim \mathbf{N}(n; j) = \dim \mathbf{H}(n; j) = n \binom{n-3+j}{n-2} \frac{n-2+2j}{j} \qquad (4.27)$$

Clearly for $j = 0$ *we have* $\mathbf{N}(n; 0) = \{(\mathbf{v}, \nabla p) : \mathbf{v} \in \mathbf{A}(n; 0), p \in A(n; 1)\}$.

Proof. For $\mathbf{w} \in \mathbf{A}(n; j)$ holds $\operatorname{div} \mathbf{w} \in A(n; j - 1)$ and by Lemma 3.3 there exists a unique $p \in Q(n; j + 1)$ such that $-\Delta p = -\operatorname{div} \mathbf{w}$. Then $\mathbf{v} := (\mathbf{w} - \nabla p) \in \mathbf{A}(n; j)$ and $\mathbf{w} = \mathbf{v} + \nabla p$. If $\mathbf{v}_i \in \mathbf{D}(n; j)$, $\nabla p_i \in \mathbf{E}(n; j)$ such that $\mathbf{w} = \mathbf{v}_i + \nabla p_i$ $(i = 1, 2)$ then $(p_1 - p_2) \in Q(n; j + 1)$, $0 = \operatorname{div}(\mathbf{v}_2 - \mathbf{v}_1) = \Delta(p_1 - p_2)$ and therefore $(p_1 - p_2) = 0$. Let $\mathbf{h} \in \mathbf{A}(n; j)$, $\mathbf{h} = \mathbf{v} + \nabla p$. Then $\mathbf{h} \in \mathbf{H}(n; j)$ if and only if $(\mathbf{v}, \nabla p) \in \mathbf{N}(n; j)$. $\square$

A further easy consequence is

Corollary 4.1. *Let now $j \geq 2$. We put (compare (3.21))*

$$\mathbf{Q}(n;j) \;:= \{ \;\; (\mathbf{v}, \nabla p) \in \mathbf{D}(n;j) \times \mathbf{E}(n;j) : \tag{4.28}$$
$$(\mathbf{v} + \nabla p) \in Q(n;j)^n \;\}$$

Then for every $\mathbf{f} \in \mathbf{A}(n; j-2)$ there exists a unique $(\mathbf{v}, \nabla p) \in Q(n;j)$ such that $-\Delta \mathbf{v} + \nabla(-\Delta p) = \mathbf{f}$.

Proof. By Lemma 3.3 there exists a unique $\mathbf{w} \in \mathbf{Q}(n;j)^n$ such that $-\Delta \mathbf{w} = f$. By Lemma 4.2 $\mathbf{w} = \mathbf{v} + \nabla p$. Uniqueness is clear by Lemma 4.1. $\qquad\square$

Let now $\mathbb{N} \ni k \geq 1$. We put

$$\begin{aligned}
\mathbf{P}(n; k-1) &:= P(n; k-1)^n, \\
\tilde{\mathbf{D}}(n; k+1) &:= \overset{k+1}{\underset{j=0}{\oplus}} \mathbf{D}(n;j), \qquad \tilde{\mathbf{E}}(n; k+1) := \overset{k+1}{\underset{j=0}{\oplus}} \mathbf{E}(n;j), \\
\tilde{\mathbf{N}}(n; k+1) &:= \overset{k+1}{\underset{j=0}{\oplus}} \mathbf{N}(n;j), \qquad \tilde{\mathbf{Q}}(n; k+1) := \overset{k+1}{\underset{j=2}{\oplus}} \mathbf{Q}(n;j)
\end{aligned} \tag{4.29}$$

Then by Lemma 4.1 and Corollary 4.1 we get a result analoguous to Theorem 3.1.

Theorem 4.2. *Let $k \geq 1$. The Stokes operator defined as the map*

$$\begin{aligned}
ST: \quad \tilde{\mathbf{D}}(n; k+1) \times \tilde{\mathbf{E}}(n; k+1) \;\; &\rightarrow \;\; \mathbf{P}(n; k-1) \\
(\mathbf{v}, \nabla p) \qquad\qquad &\mapsto \;\; -\Delta \mathbf{v} + \nabla(-\Delta p)
\end{aligned}$$

is surjective. Its null space is $\tilde{\mathbf{N}}(n; k+1)$ and restricted to $\tilde{\mathbf{Q}}(n; k+1)$ it forms a bijection on $\mathbf{P}(n; k-1)$.

It is now completely clear how a theorem analogous to Theorem 3.2 can be established and clearly in case of e.g. $(2+k)q < n$ results similar to Theorem 3.3 may be achieved.

4.3. Stokes' system in exterior domains

As we have seen in the preceeding sections the construction of strong and $(2+k)$-solutions of the Dirichlet problem for the Laplacian was based on the existence of weak solutions (Theorem 1.2) and certain regularity results (Theorem 1.4). Weak solutions of the Stokes' system in $L^q(G)$ where $G \subset \mathbb{R}^n$ is an exterior domain had been considered in Galdi and Simader (1990). In this paper the solenoidal part $\mathbf{u}$ was considered in the smaller space $\hat{H}_0^{1,q}(G)^n$ (see (1.2)) instead of $\hat{H}_\bullet^{1,q}(G)^n$. This demanded a compatibility condition for the given data (see Galdi and Simader, 1990). For the Laplacian compare the remark following Lemma 1.2. But if we replace in Galdi and Simader (1990) the spaces $\hat{H}_0^{1,q}(G)^n$ by $\hat{H}_\bullet^{1,q}(G)^n$ we get a result similar to Theorem 1.2 for Stokes' system, too. Further, regularity results like Theorem 1.4 can be derived. Then as we proceeded in Theorem 3.4 from Theorem 4.2 we easily can determine the null spaces of $(2+k)$-solutions ($k \geq 0$) of Stokes' system and construct all solutions of Stokes' system in an exterior domain. Clearly we cannot present all details in this overview.

REFERENCES

AGMON, SH. (1965), *Lectures on elliptic boundary value problems.* (Princeton: Van Nostrand).

GALDI, G.P. and C.G. SIMADER (1990), Existence, uniqueness and L^q-estimates for the Stokes' problem in an exterior domain, *Arch. Rational Mech. Anal.* **112**, p. 291-318.

MÜLLER, C. (1966), Spherical harmonics, *Lecture Notes in Mathematics* **17** (Berlin, Heidelberg, New York: Springer).

SCHECHTER, M. (1963a), On L^p estimates and regularity, *Amer. J. Math.* **85**, p. 1-13.

SCHECHTER, M. (1963b), Coerciviness in L^p, *Trans. Amer. Math. Soc.* **107**, p. 10-29.

SIMADER, C.G. (1972), On Dirichlet's boundary value problem, *Lecture Notes in Mathematics* **268**, (Berlin, Heidelberg, New York: Springer).

SIMADER, C.G. (1990), The weak Dirichlet and Neumann problem for the Laplacian in L^q for bounded and exterior domains, *in:* KRBEC, M., KUFNER, A., OPIC, B. and J. RÁKOSNIK (eds.), Nonlinear analysis, function spaces and applications **4**, p. 180-223, *Teubner-Texte zur Mathematik* **110** (Leipzig: Teubner).

SIMADER, C.G. (1992), Mean value formulas, Weyl's lemma and Liouville theorems for Δ^2 and Stokes' system, *Results in Mathematics* **22**, p. 761-780.

SIMADER, C.G. and H. SOHR (1992), A new approach to the Helmholtz decomposition and the Neumann problem in L^q-spaces for bounded and exterior domains, *in:* GALDI, G.P. (ed.), Mathematical problems relating to the Navier-Stokes equations, *Series on advances in mathematics for applied sciences* **11**, p. 1-35, (Singapore: World Scientific).

SIMADER, C.G. and H. SOHR (1995), The weak and strong Dirichlet problem for the Laplacian in L^q in bounded and exterior domains, *Pitman Research Notes in Mathematics Series*, (to appear).

THE WEAK NEUMANN PROBLEM AND THE HELMHOLTZ DECOMPOSITION OF TWO-DIMENSIONAL VECTOR FIELDS IN WEIGHTED L^r-SPACES

Maria Specovius-Neugebauer

University of Paderborn, Fachbereich 17
Warburger Straße 100, D–33098 Paderborn, Germany

ABSTRACT

Let $\Omega \subseteq \mathbb{R}^2$ be an exterior domain with smooth boundary $\partial\Omega$. We construct the Helmholtz decomposition of weighted L^r-Spaces $L_\delta^r(\Omega)$ defined by

$$L_\delta^r(\Omega) = \{u, \|u\|_{r,\delta,\Omega} = \left(\int_\Omega (1+|x|^2)^{\delta r/2} |u(x)|^r dx \right)^{1/r} < \infty\}$$

for any $\delta \neq -n/r \underline{+} k, k \in \mathbb{N}$. For $-n/r < \delta < n-n/r$ we get the "classical" Helmholtz decomposition: Let $X_\delta^r(\Omega)$ denote the closure of all smooth solenoidal test functions in $L_\delta^r(\Omega)$ and $G_\delta^r(\Omega) = \{\nabla p \in L_\delta^r(\Omega), p \in L_{\delta-1}^r(\Omega)\}$. Then $L_\delta^r(\Omega) = X_\delta^r(\Omega) \oplus G_\delta^r(\Omega)$. This can be extended to the cases $\delta < -n/r$ and $\delta > n-n/r$ with appropriate modifications.

1. INTRODUCTION

A useful tool in the theory of hydrodynamical and electromagnetical equations is the decomposition of a given vector field u into a gradient field and a solenoidal vector field. In this paper the decomposition of weighted L^r-spaces in two-dimensional exterior domains is constructed. Let $\Omega \subseteq \mathbb{R}^2$ be an unbounded domain with smooth compact boundary $\partial\Omega$. For simplicity we assume that $\mathbb{R}^2 \backslash \Omega$ is a domain, too. We define a basic weight function $w(x) = (1+|x|^2)^{1/2}$. For $\delta \in \mathbb{R}$, let $L_\delta^r(\Omega)$ be the space of all measurable functions f with $fw \in L^r(\Omega)$. Furthermore, let $X_\delta^r(\Omega)$ denote the closure of all smooth solenoidal vector fields with compact support in $(L_\delta^r(\Omega))^2$. For any $\delta \in \mathbb{R}\backslash D$, where D is a discrete set of exceptional points, we construct a projection

$$P : (L_\delta^r(\Omega))^2 \rightarrow X_\delta^r(\Omega).$$

If $-2/r < \delta < 2/r'$, $\delta \neq -2/r + 1$, we obtain the analogue to the classical Helmholtz

Navier-Stokes Equations and Related Nonlinear Problems
Edited by A. Sequeira, Plenum Press, New York, 1995

decomposition of $(L^r(\Omega))^2$

$$(L^r_\delta(\Omega))^2 = X^r_\delta(\Omega) \oplus \{\nabla p; p \in L^r_{\delta-1}(\Omega), \nabla p \in (L^r_\delta(\Omega))^2\}. \tag{1.1}$$

Decomposition results of this type for $L^r(\Omega))^n$ were proved by many authors (see, e.g. Fujiwara and Morimoto, 1977; Miyakawa, 1982; Simader and Sohr, 1992). The basic tool is the investigation of the following Laplace problems in suitable Sobolev spaces

$$\Delta u = f \quad \text{in } \mathbb{R}^2 \quad \text{and} \tag{1.2}$$

$$\Delta u = 0 \quad \text{in } \Omega, \quad \frac{\partial}{\partial \nu} u|_{\partial\Omega} = h, h \in H^{-1/r,r}(\partial\Omega), \tag{1.3}$$

where ν is the exterior normal vector (with respect to Ω) in $x \in \partial\Omega$.

The two-dimensional case differs from the case $n \geq 3$, because the logarithmic singularity of the Laplace fundamental solution causes special difficulties. The present paper fills a gap in the results of Specovius-Neugebauer (1990), where the Helmholtz decomposition of weighted L^r-spaces is constructed for $n-$dimensional exterior domains with $n \geq 3$.

The problems (1.2), (1.3) will be solved in weighted Sobolev spaces by means of potential theory. The regularity of the boundary layer potentials is investigated on the boundary $\partial\Omega$. Here we follow an idea of v. Wahl (1990), who examined in detail the integral equation of the weak Neumann problem (1.3) for $n = 3$. Furthermore, we obtain explicit decay rates for the solution of the problem (1.3) which may be interesting for themselves. The construction of the projection can be done similar as in Specovius-Neugebauer (1990).

2. NOTATIONS AND AUXILIARY RESULTS

Let $\Omega \subset \mathbb{R}^2$ be an exterior domain, i.e. an open unbounded connected subset with closure $\overline{\Omega}$ and compact boundary $\partial\Omega$ of class C^2. We further assume that $\mathbb{R}^2 \backslash \overline{\Omega} \neq \emptyset$ is a bounded domain. We recall the notations of some function spaces. Let $C(\Omega)$, $C^m(\Omega)$, $m = 1, 2, \ldots \infty$, denote the space of all continuous and of all m-times differentiable functions in Ω, respectively. $C_0^m(\Omega)$ is the subspace of all $f \in C^m(\Omega)$ with compact support in Ω. For $1 < r < \infty$, $m \in \mathbb{N}$, $H^{m,r}(\Omega)$ denotes the Sobolev space of order m, i.e. $H^{m,r}(\Omega) = \{f \in L^r(\Omega); D^\alpha f \in L^r(\Omega), 0 \leq |\alpha| \leq m\}$. The norm on $H^{m,r}(\Omega)$ is indicated by $\|\cdot\|_{m,r,\Omega}$.[1]

As already mentioned, $w(x) = (1 + |x|^2)^{1/2}$ is the basic weight function in Ω. For $\delta \in \mathbb{R}$, $1 < r < \infty$ we have $L^r_\delta(\Omega) = \{f \text{ measurable}; fw^\delta \in L^r(\Omega)\}$. $L^r_\delta(\Omega)$ is a reflexive Banach space normed by

$$\|f\|_{r,\delta,\Omega} = \|fw^\delta\|_{r,\Omega}.$$

The dual space can be identified with $L^{r'}_{-\delta}(\Omega)$, $r' = r/(r-1)$ by the bilinear form $\langle f, g \rangle = \int_\Omega fg \, dx$, where $f \in L^r_\delta(\Omega)$, $g \in L^{r'}_{-\delta}(\Omega)$. For $m \in \mathbb{N}$ we define corresponding Sobolev spaces by

$$H^{m,r}_\delta = \{f \in L^r_\delta(\Omega); D^\alpha \in L^r_{\delta+|\alpha|}(\Omega), 0 \leq |\alpha| \leq m\}$$

with the norm

$$\|f\|_{m,r,\delta,\Omega} = \left(\sum_{|\alpha|=0}^m \|D^\alpha f\|^r_{r,\delta+|\alpha|,\Omega}\right)^{1/r}.$$

---

[1] Here we use the customary multi-index terminology: $|\alpha| = \alpha_1 + \alpha_2$, $D^\alpha = \frac{\partial^{\alpha_1}}{\partial x_1} \frac{\partial^{\alpha_2}}{\partial x_2}$, $\alpha \in \mathbb{N}_0^2$.

By $\overset{\circ}{H}^{m,r}_\delta(\Omega)$ we mean the closure of $C_0^\infty(\Omega)$ in $H_\delta^{m,r}(\Omega)$. We introduce Sobolev spaces of negative order by $H_\delta^{-m,r} = (\overset{\circ}{H}^{m,r'}_{-\delta}(\Omega))^*$, here $(\ )^*$ expresses the strong dual space. From the definition it follows $H_\delta^{m,r}(\Omega) \subset H_{loc}^{m,r}(\overline{\Omega})$ (i.e. for $u \in H_\delta^{m,r}(\Omega)$ we have $u \in H^{m,r}(B \cap \Omega)$ for any open ball B with $B \cap \Omega \neq \emptyset$). Hence we have a continuous trace operator $\gamma : H_\delta^{m,r}(\Omega) \to H^{m-1/r,r}(\partial\Omega)$, with $\gamma u = u|_{\partial\Omega}$ for smooth functions u (see Adams, 1975, p.216, e.g.). The Sobolev spaces $H^{s,r}(\partial\Omega)$ of noninteger order are defined as usual by means of local coordinates and interpolation methods (Adams, 1975, p.204-216).

For $x \in \partial\Omega$ let $\nu(x)$ denote the exterior (unit) normal vector (with respect to Ω). We set $Y_\delta^r(\Omega) = \{u \in (L_\delta^r(\Omega))^2; \operatorname{div} u \in L_{\delta+1}^r(\Omega)\}$. Then from the results of Fujiwara and Morimoto (1977), p.686 and Simader and Sohr, 1992, p.31, we have a continuous trace operator $\gamma_\nu : Y_\delta^r(\Omega) \to H^{-1/r,r}(\partial\Omega)$ with $\gamma_\nu u = u \cdot \nu|_{\partial\Omega}$ for any smooth vector field u. Moreover, since the smooth functions with compact support in $\overline{\Omega}$ are dense in $Y_\delta^r(\Omega)$ and in $H_{-\delta-1}^{1,r'}(\Omega)$, we have Green's formula

$$\int_\Omega \operatorname{div} u\, p \,\mathrm{d}x = \langle \gamma_\nu u, p\rangle_{\partial\Omega} - \int_\Omega u \cdot \nabla p \,\mathrm{d}x \tag{2.1}$$

for $u \in Y_\delta^r(\Omega)$, $p \in H_{-\delta-1}^{1,r'}(\Omega)$, $\langle \cdot, \cdot\rangle_{\partial\Omega}$ indicates the duality $\langle H^{-1/r,r}(\partial\Omega), H^{1/r,r'}(\partial\Omega)\rangle$.

3. THE MAIN THEOREM

For $k \in \mathbb{N}_o$ we set $I_k = (k, k+1)$. Furthermore, let $E_\delta^r(\Omega)$ denote the space of all gradient fields ∇p with $p \in H_{\delta-1}^{1,r}(\Omega)$. With the notations of the previous sections we have the following theorem.

Theorem 3.1. Let $\Omega \subseteq \mathbb{R}^2$ be an exterior domain with $\partial\Omega \in C^2$. For $1 < r < \infty$, $\delta \in -2/r \overset{+}{} I_k$, $k \in \mathbb{N}_0$ there exists a continuous projection $P : (L_\delta^r(\Omega))^2 \to X_\delta^r(\Omega)$. The following decompositions hold:
(i) For $-2/r < \delta < 2/r'$, $\delta \neq -2/r + 1$,

$$(L_\delta^r(\Omega))^2 = X_\delta^r(\Omega) \oplus E_\delta^r(\Omega).$$

(ii) For $\delta \in -I_k - 2/r$, $k = 0, 1, \ldots$

$$(L_\delta^r(\Omega))^2 = X_\delta^r(\Omega) + E_\delta^r(\Omega).$$

The intersection $X_\delta^r(\Omega) \cap E_\delta^r(\Omega) = \nabla\mathcal{H}_k(\frac{\partial}{\partial\nu})$ is a finite dimensional subspace of $(L_\delta^r(\Omega))^2$ generated by the gradients of the nontrivial solutions to the exterior Neumann problem (1.3). $\nabla\mathcal{H}_k(\frac{\partial}{\partial\nu})$ is independent of δ on the intervall $-I_k - 2/r$.
(iii) For $\delta \in I_k - 2/r$, $k = 2, 3, \ldots$

$$(L_\delta^r(\Omega))^2 = X_\delta^r(\Omega) \oplus E_\delta^r(\Omega) \oplus G_k(\Omega)$$

where $G_k(\Omega)$ is a finite dimensional subspace of $(L_\delta^r(\Omega))^2$ with $\dim G_k(\Omega) = \dim \nabla\mathcal{H}_k(\frac{\partial}{\partial\nu})$. Moreover,

$$\int_\Omega u \cdot \nabla p = 0$$

for any $u \in X_\delta^r(\Omega)$, $p \in E_{-\delta}^{r'}(\Omega)$.

Remark 3.1. It is possible to remove the condition $\delta \neq -2/r + 1$ if we substitute the space $E_\delta^r(\Omega)$ by $\overline{E}_\delta^r(\Omega) = \{p \in H_{loc}^{1,r}(\overline{\Omega}), \nabla p \in (L_\delta^r(\Omega))^2\}$ in the Helmholtz decomposition. This can be proved similar as in Simader and Sohr (1992) by using Stein's theorem (see e.g. McOwen, 1979) instead of the Calderon Zygmund inequality.

4. THE WEAK NEUMANN PROBLEM VIA INTEGRAL EQUATIONS

First we sum up the results for the problem (1.2).

Lemma 4.1. *Let* $1 < r < \infty$. *For* $k \in \mathbb{N}_0$ *let* $\mathcal{H}_k$ *denote the set of all harmonic polynomials with degree* $\leq k$. *Then it holds:*
(i) *For* $\delta \in -I_k - 2/r$, *the mapping*

$$\text{(L)} \qquad \Delta : H_\delta^{1,r}(\mathbb{R}^2) \to H_{\delta+2}^{-1,r}(\mathbb{R}^2)$$

is surjective with finite dimensional null space $\ker(\Delta) = \mathcal{H}_k$.
(ii) *For* $\delta \in I_k - 2/r$, *the mapping* (L) *is injective with closed range* $R(\Delta) = \{f : \langle f, h \rangle = 0$ *for all* $h \in \mathcal{H}_k\}$. *For* $f \in L_{\delta+2}^r$ *we get* $\Delta^{-1} f = K * f$, *where* $K(x) = 1/2\pi \ln |x|$ *is the fundamental solution of the Laplace operator.*

Proof: For the proof we refer the reader to the proof of Lemma 2 in Specovius-Neugebauer (1990). Using Lemma 6 of McOwen (1979), p.793, part (i) can be proved in the same way as in Specovius-Neugebauer (1990), since the arguments do not depend on the dimension. Part (ii) follows from (i) using the closed range theorem (Yoshida, 1965, p. 205), because the adjoint mapping is defined by

$$\text{(L)}^* \qquad \Delta : H_{-\delta-2}^{1,r'}(\mathbb{R}^2) \to H_{-\delta}^{-1,r'}(\mathbb{R}^2).$$

The representation formula follows from Lemma 6 in McOwen (1979).   □

Next we consider the weak Neumann problem. Let $h \in H^{-1/r,r}(\partial\Omega)$ be given. Suppose $u \in H_\delta^{1,r}(\Omega)$ for some $\delta \in \mathbb{R}$ with $\Delta u = 0$, then $\nabla u \in Y_{\delta+1}^r(\Omega)$ and $\gamma_\nu \nabla u$ is well defined by the trace theorem of Section 2. Hence it makes sense to look for a solution u of the problem (1.3) with $u \in H_\delta^{1,r}(\Omega)$ for suitable δ. If $\delta > -2/r$, we may substitute $p \equiv 1$ into Green's formula (1.4) and obtain the necessary condition

$$\langle h, 1 \rangle_{\partial\Omega} = 0. \tag{4.1}$$

Assume for a moment that $h \in C^1(\partial\Omega)$ and $\int_{\partial\Omega} h \, ds = 0$. The classical potential theory yields a solution of the Neumann problem (1.3) as a single layer potential

$$u = Vf = \int_{\partial\Omega} K(x - y) f(y) \, ds_y, \tag{4.2}$$

(see Folland, 1976, p. 171, ff.). The density $f \in C(\partial\Omega)$ is the solution of the integral equation

$$-\frac{1}{2} f + D^* f = h, \tag{4.3}$$

where D^* is the adjoint operator to the double layer potential Df,

$$D^* = \int_{\partial\Omega} \frac{\partial}{\partial\nu_x} K(x - y) f(y) \, ds_y.$$

To extend this setup to the case of arbitrary data $h \in H^{-1/r,r}(\partial\Omega)$, we follow an idea of von Wahl (1990). In the following lemmas we estimate the integral operators V and D on the boundary $\partial\Omega$. We recall an often used notation: For two Banach spaces X and Y let $L(X, Y)$ denote the space of all continuous linear mappings from X to Y.

Lemma 4.2. *Assume that* $\partial\Omega \subset \mathbb{R}^2$ *is a closed curve of class* C^2. *Let* V *be defined as in* (4.2). *Then* $V \in L(W^{-\alpha}(\partial\Omega), W^{1-\alpha}(\partial\Omega))$ *for any* $\alpha \in [0, 1]$.

Proof: We fix $f \in C^1(\partial\Omega)$ and start with $\alpha = 0$, i.e. we show the estimate

$$\|Vf\|_{1,r,\partial\Omega} \leq C\|f\|_{r,\partial\Omega} \tag{4.4}$$

Since $\ln \in L^1_{loc}(\mathbb{R})$ we have (see also Folland, 1976, p.158)

$$\|Vf\|_{r,\partial\Omega} \leq \|f\|_{r,\partial\Omega}. \tag{4.5}$$

To estimate the derivatives of Vf, we choose a parametric representation of $\partial\Omega$ and a smooth cut-off procedure. Since $\partial\Omega$ is a closed C^2-curve, there exist a finite number of open intervals $I_j = (0, a_j)$, and C^2–mappings $\psi_j : I_j \to \partial\Omega$, such that $\partial\Omega = \cup_{j=1}^d \psi_j(I_j)$ and $|\psi'(s)| = 1$ for all $s \in I_j, j = 1, \ldots, d$ [2]. Moreover, there exist constants $\varepsilon_0, \bar{C}_1, \bar{C}_2$, such that

$$\bar{C}_1|s - \sigma| \leq |\psi(s) - \psi(\sigma)| \leq \bar{C}_2|s - \sigma|, \tag{4.6}$$

for $|s - \sigma| \leq \varepsilon_0$. Using the partition of unity $\{w_j\}_{j=1,\ldots,d}$ subordinated to the covering $\{\psi_j(I_j)\}_j$, we see that it is sufficient to show that

$$\|V\tilde{f}_j\|_{1,r,I_j} \leq C\|\tilde{f}_j\|_{r,I_j}.$$

Here $\tilde{f}_j(s) = (fw_j) \circ \psi_j(s)$ for $s \in I_j$, and $\tilde{f}(s) = 0$ elsewhere, hence $\tilde{f}_j \in C^1_0(\mathbb{R})$, and

$$V\tilde{f}_j(s) = \int_{I_j} \ln |(\psi_j(s) - \psi_j(\sigma)|\tilde{f}(\sigma)\mathrm{d}\sigma = \int_{-\infty}^{\infty} \ln |(\psi_j(s) - \psi_j(\sigma)|\tilde{f}_j(\sigma)\mathrm{d}\sigma.$$

In the following calculations we omit the index j. To use the cut-off procedure now, we fix $\varphi \in C^\infty_0(\mathbb{R})$ with $0 \leq \varphi \leq 1$, $\varphi(r) = 0$ for $|r| \geq 2$, $\varphi(r) = 1$ for $|r| \leq 1$. We set

$$K_\varepsilon(s, \sigma) = \ln |\psi(s) - \psi(\sigma)|(1 - \varphi((s - \sigma)/\varepsilon))$$

and $V_\varepsilon\tilde{f}(s) = \int_{\mathbb{R}} K_\varepsilon(s, \sigma)\tilde{f}(\sigma)\mathrm{d}\sigma$, for any $s \in I$. We have $\lim_{\varepsilon \to 0} V_\varepsilon\tilde{f} = V\tilde{f}(s)$, for any $s \in I$, and $\|V_\varepsilon\tilde{f}\|_{r,I} \leq C\|f\|_{r,\partial\Omega}$, where C is independent of f and ε. It remains to show that

$$\|(V_\varepsilon\tilde{f})'\|_{r,I} \leq C\|f\|_{r,\partial\Omega}, \tag{4.7}$$

where C is again independent of f and ε. Then $(\tilde{V}f)' \in L^r(I)$ and $\|(\tilde{V}f)'\|_{r,I} \leq \|f\|_{r,\partial\Omega}$ by the uniform boundedness theorem (Yoshida, 1965, p.68). Since $K_\varepsilon(s, \sigma)$ is smooth, we may differentiate under the integral sign and obtain

$$(V_\varepsilon\tilde{f})'(s) = \int_{\mathbb{R}} \frac{\partial}{\partial s} K_\varepsilon(s, \sigma)\tilde{f}(\sigma)\mathrm{d}\sigma$$

$$= \int_{\mathbb{R}} \frac{\psi(s) - \psi(\sigma)}{|\psi(s) - \psi(\sigma)|^2}\psi'(s)(1 - \varphi((s - \sigma)/\varepsilon))\tilde{f}(\sigma)\mathrm{d}\sigma$$

$$+ \int_{\mathbb{R}} \ln |\psi(s) - \psi(\sigma)|\tilde{f}(\sigma)\frac{1}{\varepsilon}\varphi'((s - \sigma)/\varepsilon))\mathrm{d}\sigma =: I_{1,\varepsilon}f + I_{2,\varepsilon}.$$

To treat the first integral we use Taylor's formula and obtain

$$\psi(s) - \psi(\sigma) = \psi'(s)(s - \sigma) - \frac{1}{2}\psi''(s_o)(s - \sigma)^2 \tag{4.8}$$

---

[2] We have $|\psi'(s)| = 1$, if the parameter s is the length of the curve between a fixed point $x_o \in \psi_j(I_j)$ and $\psi_j(s)$.

with some suitable s_0 depending continuously an s and σ. Using (4.8) and observing that $|\psi'(s)| \equiv 1$ we get after some elementary calculations

$$\frac{\psi(s) - \psi(\sigma)}{|\psi(s) - \psi(\sigma)|^2}\psi'(s) = \frac{1}{s - \sigma} + \tilde{k}(s, \sigma),$$

where $\tilde{k}(s, \sigma)$ is a continuous kernel with $|\tilde{k}(s, \sigma)| \leq C|s - \sigma|^{-2}$ as s, σ tend to infinity. Hence the results about the Hilbert transform (Morrey, 1966, p.55) lead to

$$\|I_{1,\varepsilon}\tilde{f}\|_{r,I} \leq C\|\tilde{f}\|_{r,\mathbb{R}} \leq C\|f\|_{r,\partial\Omega}.$$

To estimate $I_{2,\varepsilon}\tilde{f}$, we use supp $\varphi(x/\varepsilon) \subset [\varepsilon, 2\varepsilon]$ and inequality (4.6), for $2\varepsilon < \min\{1, \varepsilon_0\}$, thus

$$|I_{2,\varepsilon}\tilde{f}| \leq \frac{C}{\varepsilon} \int\limits_{\varepsilon \leq |s-\sigma| \leq 2\varepsilon} |\ln|\psi(s) - \psi(\sigma)||\tilde{f}(\sigma)|d\sigma \leq \int\limits_{\varepsilon \leq |s-\sigma| \leq 2\varepsilon} |\ln \bar{C}_1|s - \sigma|| |\tilde{f}(\sigma)|d\sigma.$$

With Young's inequality we obtain

$$\|I_{2,\varepsilon}\tilde{f}\|_{r,I} \leq \frac{2C}{\varepsilon} \int\limits_{\varepsilon}^{2\varepsilon} |\ln \bar{C}_1\tau|d\tau \, \|\tilde{f}\|_{r,\mathbb{R}} \leq C\|f\|_{r,\partial\Omega},$$

and inequality (4.4) is proved.

Since the operator V is formally self adjoint, we find that

$$\|Vf\|_{r,\partial\Omega} \leq C\|f\|_{-1,r,\partial\Omega}$$

by duality. To prove the lemma for arbitrary $\alpha \in (0, 1)$, we need some facts from the interpolation theory.

Assume that A_o, A_1, B_o, B_1 are complex Banach spaces with continuous embeddings $A_1 \hookrightarrow A_o$, $B_1 \hookrightarrow B_o$, $[A_o, A_1]_{\theta,r}$ the interpolation space defined by the K-method (Triebel, 1978, p.23). Then $[A_o, A_1]_{\theta,r} = [A_1, A_o]_{1-\theta,r}$, $[A_o, A_1]_{\theta,r}^* = [A_o^*, A_1^*]_{\theta,r'}$ (see Triebel, 1978, p.69). If $K \in L(B_o, A_0)$ such that the restriction $K|_{B_1} \in L(B_1, A_1)$, then $K|_{[B_o, B_1]_{\theta,r}} \in L([B_o, B_1]_{\theta,r}, [A_o, A_1]_{\theta,r})$.

We apply these facts to $A_o = L^r(\partial\Omega)$, $A_1 = H^{1,r}(\partial\Omega)$, $B_o = H^{-1,r}(\partial\Omega)$, $B_1 = L^r(\partial\Omega)$, $\theta = 1 - \alpha$. We have $[L^r(\partial\Omega), H^{1,r}(\partial\Omega)]_{1-\alpha,r} = W^{1-\alpha,r}(\partial\Omega)$, (Triebel, 1978, p.168, p.185 ff.) and $W^{-\alpha,r}(\partial\Omega) = [H^{-1,r}(\partial\Omega), L^r(\partial\Omega)]_{1-\alpha,r}$. $\qquad\square$

For the double layer potential we get a weaker result

Lemma 4.3. *Let $\partial\Omega$ be as in Lemma 4.2. For $f \in L^r(\partial\Omega)$ we set*

$$Df = \int_{\partial\Omega} \frac{1}{2\pi} \frac{(x - y)\nu(y)}{|x - y|^2} f(y)ds_y.$$

Then $D \in L(L^r(\partial\Omega), H^{1-1/r,r}(\partial\Omega))$, hence $D^* \in L(H^{-1/r',r'}(\partial\Omega), L^{r'}(\partial\Omega))$.

Proof: We abbreviate the kernel of the double layer potential by $d(x, y)$. Then it holds:

$$|d(x, y)| \leq C \qquad \text{for } x, y \in \partial\Omega, \quad \text{(Folland, 1976, p.163)} \tag{4.9}$$

and by an elementary calculation

$$|d(x, y) - d(x', y)| \leq C\frac{|x - x'|}{R}, \quad R = \min(|x - y|, |x' - y|). \tag{4.10}$$

From (4.9) it follows immediately that

$$\|Df\|_{r,\partial\Omega} \le C\|f\|_{r,\partial\Omega}.$$

If we use the direct representation of the norm $\|\cdot\|_{1-1/r,r,\partial\Omega}$ on $\partial\Omega$, it remains to show that

$$\int_{\partial\Omega}\int_{\partial\Omega} \frac{|Df(x)-Df(x')|^r}{|x-x'|^r}\mathrm{d}s_x\mathrm{d}s_{x'} \le C\|f\|_{r,\partial\Omega}. \tag{4.11}$$

For this purpose we choose $\varepsilon < \min\{1/r, 1/r'\}$. By Hölder's inequality we obtain

$$|Df(x)-Df(x')|^r \le \left(\int_{\partial\Omega}|d(x,y)-d(x',y)|^{1/r-\varepsilon+1/r'+\varepsilon}|f(y)|\mathrm{d}s_y\right)^r$$

$$\le \int_{\partial\Omega}|d(x,y)-d(x',y)|^{1-\varepsilon r}|f(y)|^r\mathrm{d}s_y \left(\int_{\partial\Omega}|d(x,y)-d(x',y)|^{1+\varepsilon r'}\mathrm{d}s_y\right)^{r-1}$$

$$=: A_1(A_2)^{r-1} \tag{4.12}$$

Let $\Sigma_x = \partial\Omega \cap \{y, |y-x| \le 2|x-x'|\}$. We estimate A_2, using the triangle inequality, (4.9), (4.10) and polar coordinates

$$A_2 = \int_{\Sigma_x}\ldots + \int_{\partial\Omega\setminus\Sigma_x}\ldots \le C_1|x-x'| + C_2|x-x'|\int_{|x-x'|}^{\infty}\frac{1}{r^{1+\varepsilon r'}}\mathrm{d}r \le C|x-x'|. \tag{4.13}$$

Substituting (4.12), (4.13) into the left hand side of (4.11) we get

$$\int_{\partial\Omega}\int_{\partial\Omega}\frac{Df(x)-Df(x')|^r}{|x-x'|^r}\mathrm{d}s_x\mathrm{d}s_{x'}$$

$$\le C\int_{\partial\Omega}\int_{\partial\Omega}\int_{\partial\Omega}\frac{1}{|x-x'|}|d(x,z)-d(x',y)|^{1-\varepsilon r}\mathrm{d}s_x\mathrm{d}s_{x'}|f(y)|^r\mathrm{d}s_y.$$

Using once more (4.10) and $\frac{1}{R} \le \frac{1}{|x-y|} + \frac{1}{|x'-y|}$ we arrive at

$$\int_{\partial\Omega}\int_{\partial\Omega}\frac{Df(x)-Df(x')|^r}{|x-x'|^r}\mathrm{d}s_x\mathrm{d}s_{x'} \le C\int_{\partial\Omega}|f(y)|^r\int_{\partial\Omega}\frac{1}{|x-y|^{1-\varepsilon r}}\int_{\partial\Omega}|x-x'|^{-\varepsilon r}\mathrm{d}s_{x'}\mathrm{d}s_x\mathrm{d}s_y$$

$$+ \int_{\partial\Omega}|f(y)|^r\int_{\partial\Omega}\frac{1}{|x'-y|^{1-\varepsilon r}}\int_{\partial\Omega}|x-x'|^{-\varepsilon r}\mathrm{d}s_x\mathrm{d}s_{x'}\mathrm{d}s_y \le \bar{C}\|f\|_{r,\partial\Omega}^r$$

which proves Lemma 4.3. $\qquad\square$

The next lemma characterizes the solution of the weak Neumann problem in the weighted spaces $H_\delta^{m,r}(\Omega)$.

Lemma 4.4. *Let* $\Omega \subseteq \mathbb{R}^2$ *be an exterior domain with* $\partial\Omega \in C^2$. *Suppose* $h \in H^{-1/r,r}(\partial\Omega)$ *with*

$$\langle h, 1\rangle_{\partial\Omega} = 0. \tag{4.14}$$

Then there exists a unique solution u_h *of the problem (1.3), with* $u_h \in H_\delta^{1,r}(\Omega)$ *for any* $\delta < 1 - 2/r$. *Moreover,* u_h *obeys the following decay properties as* $|x| \to \infty$

$$|u_h(x)| = O(|x|^{-1}), \quad |\nabla u(x)| = O(|x|^{-2}). \tag{4.15}$$

For $\delta \in -I_k - 2/r$, the solutions of the homogeneous problem, i.e. $h = 0$, form a finite-dimensional subspace $\mathcal{H}_{k-1}(\frac{\partial}{\partial \nu})$ of $H_\delta^{1,r}(\Omega)$, with $\dim \mathcal{H}_{k-1}(\frac{\partial}{\partial \nu}) = \dim \mathcal{H}_k$.

Moreover, if

$$\langle h, v|_{\partial\Omega}\rangle_{\partial\Omega} = 0 \ \text{for any } v \in \mathcal{H}_k(\frac{\partial}{\partial \nu}), \tag{4.16}$$

then

$$|u_h(x)| = O(|x|^{-1-k}), \qquad |\nabla u_h(x)| = O(|x|^{-2-k}) \tag{4.17}$$

as $|x| \to \infty$. In this case, $u_h \in H_\delta^{1,r}(\Omega)$ for $\delta < k + 1 - 2/r$.

Proof: First we prove the uniqueness. Let $u \in H_\delta^{1,r}(\Omega)$ for some $\delta > -2/r$ with $\Delta u = 0$, $\gamma_\nu \nabla u = 0$. Since $\partial\Omega \in C^2$, we have $u \in H_{loc}^{2,r}(\overline{\Omega})$ and $u \in H_\delta^{2,r}(\Omega)$ by inequality (2.7) in Nirenberg and Walker (1973). Let $\phi \in C_0^\infty(\mathbb{R}^2)$ be a neighbourhood of Ω. We decompose $u = \phi u + (1 - \phi)u$. Since supp ϕu is compact, we may use the representation formula for bounded domains and obtain

$$\phi u(x) = \int_{\partial\Omega} \frac{(x - y) \cdot \nu(y)}{|x - y|^2} u(y) \mathrm{d}s_y + \int_\Omega K(x - y)\Delta(\phi u)\mathrm{d}y.$$

For $(1 - \phi)u$ we have by Lemma 4.1

$$(1 - \phi)u(x) = \int_{\mathbb{R}^2} K(x - y)\Delta((1 - \phi)u(y)\mathrm{d}y = \int_\Omega K(x - y)\Delta((1 - \phi)u(y))\mathrm{d}y.$$

Adding both formulas leads to the representation

$$u(x) = \int_{\partial\Omega} \frac{(x - y) \cdot \nu(y)}{|x - y|^2} u(y)\mathrm{d}s_y,$$

hence $|u(x)| = O(|x|^{-1})$ as $|x| \to \infty$. This implies $u \equiv 0$ by the classical uniqueness result (Folland, 1976, p.194).

Next we prove the existence of a solution with the decay properties (4.15). Then $u \in H_\delta^{1,r}(\Omega)$ for any $\delta < -2/r + 1$ is obvious. Assume first that $h \in C^1(\partial\Omega)$ with $\int_{\partial\Omega} h = 0$. There exists a solution $u = Vf$, where f is the unique solution of the integral equation (4.3). Moreover, $\int_{\partial\Omega} f \mathrm{d}s = 0$, too (see Folland, 1976, p.178). It is clear that $u \in H_{loc}^{1,r}(\overline{\Omega})$. To show the decay properties (4.15), we use the expansion of $K(x - y)$ into Gegenbauer polynomials (Du Plessis, 1970, p.27). For $|x| > \max\{|y|, y \in \partial\Omega\}$ we have

$$K(x - y) = K(x) + \sum_{l=1}^\infty p_l(\hat{x}, y)\frac{1}{|x|^l} \tag{4.18}$$

where $\hat{x} = x/|x|$ and for any $l \in \mathbb{N}$, $x \in \mathbb{R}^2$ the expression $p_l(\hat{x}, y)$ is a harmonic polynomial of degree l in y. Substituting the equation (4.18) into the equation $u(x) = \int_{\partial\Omega} K(x - y)f(y)\mathrm{d}s_y$, the first term vanishes and we obtain the decay properties (4.15).

Now we use an approximation argument. We fix $h \in H^{-1/r,r}(\partial\Omega)$ with $\langle h, 1\rangle_{\partial\Omega} = 0$ and choose a sequence $h_n \in C^1(\partial\Omega)$ with $\int_{\partial\Omega} h_n \mathrm{d}s = 0$ and $\lim_{n\to\infty} h_n = h$ in $H^{-1/r,r}(\partial\Omega)$. Let $(f_n)_{n\in\mathbb{N}}$ be the corresponding solutions of the integral equation (4.3). Using the compactness of the embedding $L^r(\partial\Omega) \hookrightarrow H^{-1/r,r}(\partial\Omega)$ and Lemma 4.3 we see that the mapping $-\frac{1}{2}I + D^*$ is an isomorphism on $H^{-1/r,r}(\partial\Omega)$, hence $\lim_{n\to\infty} f_n = f$, where $f = (-\frac{1}{2}I + D^*)^{-1}h$. To derive "reasonable" convergence in Ω, we first look at the points far away from $\partial\Omega$. Let $R_o \geq \max\{|y|, y \in \partial\Omega\}$ and $x \in \Omega$

with $|x| > 2R_0$. We may substitute the expansion (4.18) into Vf_n and, by exploiting the properties of p_l (see Du Plessis, 1970, p.27 for the explicit form), we obtain

$$|Vf_n(x)| \leq \frac{R_o}{|x|} \int_{\partial\Omega} \sum_{l=1}^{\infty} \frac{|y|^l}{2^l R_o^l} |f_n(y)| \mathrm{d}s_y \leq \frac{C}{|x|} \|f_n\|_{-1/r,r,\partial\Omega}. \tag{4.19}$$

Analogously,

$$|\nabla Vf_n(x)| \leq \frac{C}{|x|^2} \|f_n\|_{-1/r,r,\partial\Omega}. \tag{4.20}$$

Hence

$$\sup_{|x| \leq 2R_o} \left(|x||Vf_n(x) - Vf(x)| + |x|^2 |\nabla Vf_n(x) - \nabla Vf(x)| \right) \to 0$$

as $n \to \infty$.

Near the boundary, we apply the classical estimates of the Dirichlet problem to Vf_n (Lions and Magenes, 1961, III, Thm. 8.2, p.80, Simader,1990, Thm. 4.2, p.12). Let $B_o = \{x, |x| < 2R_o\}$. Then

$$\|Vf_n\|_{1,r,\Omega\cap B_o} \leq C \left(\|Vf_n|_{\partial\Omega}\|_{1-1/r,r,\partial\Omega} + \|Vf_n|_{\partial B_o}\|_{1-1/r,r\partial\Omega} \right) \leq C\|f_n\|_{1-1/r,r,\partial\Omega}. \tag{4.21}$$

For the second inequality we used Lemma 4.2 and (4.19). From (4.21) it follows

$$\lim_{n\to\infty} \|Vf_n - Vf\|_{1,r,\Omega\cap B_o} = 0.$$

By the trace theorem of Section 2 we deduce that Vf is the solution of the Neumann problem (1.3), and that the decay properties (4.15) are a consequence of (4.19) and (4.20).

Now we fix $\delta \in -I_k - 2/r$. Let $v \in H_\delta^{1,r}(\Omega)$ with $\Delta u = 0$, $\gamma_\nu \nabla = 0$. With the same arguments as in the first part of the proof it follows $v \in H_\delta^{2,r}(\Omega)$. We extend v to $\tilde{v} \in H_\delta^{2,r}(\mathbb{R}^2)$. Then supp $\Delta\tilde{v} \subset \mathbb{R}\backslash\Omega$. We set $v_1 = \tilde{v} - K * \Delta\tilde{v}$. From the decay properties of K we deduce that $K * \Delta\tilde{v} =: v_2 \in H_\alpha^{2,r}(\mathbb{R}^2)$ for any $\alpha < -2/r$, hence $v_1 \in H_\delta^{2,r}(\mathbb{R}^2)$. Lemma 4.1 implies that v_1 is a polynomial of degree $\leq k$. Obviously, we have $\gamma_\nu \nabla v_1 = -\gamma_\nu \nabla v_2|_{\partial\Omega}$. Furthermore, it holds

$$\int_{\mathbb{R}^2\backslash\Omega} \Delta\tilde{v} \, \mathrm{d}x = -\int_{\partial\Omega} \gamma_\nu \nabla v \mathrm{d}s = 0, \tag{4.22}$$

so with (4.18) we obtain $|v_2(x)| = |K * \Delta\tilde{v}(x)| = O(|x|^{-1})$ as $|x| \to \infty$. Therefore v_2 is the unique solution of the Neumann problem (1.3) with data $h = -\gamma_\nu \nabla v_1$ and the decay properties (4.15). On the other hand, there exists for any $v_1 \in \mathcal{H}_k$ a unique v_2 with these properties. The necessary condition (4.14) can be derived from (4.22) by replacing $\tilde{v}$ through v_1, hence $\dim\mathcal{H}_{k-1}(\frac{\partial}{\partial\nu}) = \dim\mathcal{H}_k$.

The decay properties (4.17) under the condition (4.16) can be proved exactly as in Specovius-Neugebauer (1990) by using the representation formula for u_h, Green's formula (2.4) and the expansion (4.18).

5. PROOF OF THE MAIN THEOREM

First we observe that $(\mathrm{div}\,, \gamma_\nu) : Y_\delta^r(\Omega) \to L_{\delta+1}^r(\Omega) \times H^{-1/r,r}(\partial\Omega)$ is a continuous linear operator. For $u \in Y_\delta^r(\Omega)$ with $\mathrm{div}\,u = 0$ we have $\|u\|_{Y_\delta^r(\Omega)} = \|u\|_{r,\delta,\Omega}$, hence $\ker(\mathrm{div}\,, \gamma_\nu) = \{u \in (L_\delta^r(\Omega))^2, \mathrm{div}\,u = 0, \gamma_\nu u = 0\} =: N_\delta^r(\Omega)$ is a closed subspace of $(L_\delta^r(\Omega))^2$ which contains $X_\delta^r(\Omega)$. To show the other inclusion, we prove the following

Proposition 5.1. *For $\delta \neq -2/r + 1$, $X_\delta^r(\Omega)^\perp = N_\delta^r(\Omega)^\perp = E_{-\delta}^{r'}(\Omega)$, where $^\perp$ denotes the annihilator in $L_{-\delta}^{r'}(\Omega)^2$.*

As an immediate consequence we get the following

Corollary 5.1. *$X_\delta^r(\Omega) = N_\delta^r(\Omega)$ for $\delta \neq -2/r + 1$.*

Proof: $E_{-\delta}^{r'}(\Omega) \subset N_\delta^r(\Omega)^\perp \subset X_\delta^r(\Omega)^\perp$ follows from Green's formula (2.4). So we fix $v \in (L_{-\delta}^{r'}(\Omega))^2$ with $\int_\Omega v \cdot u \, dx = 0$, for any $u \in X_\delta^r(\Omega)$. From Theorem 1.1 and Corollary 1.2 in Simader and Sohr (1992) we deduce $v = \nabla p$ with $p \in H_{loc}^{1,r'}(\overline{\Omega})$. Let $R = \max\{|y|, y \in \partial\Omega\}$ and $\psi \in C_0^\infty(\mathbb{R}^2)$ be a cut-off function with $\psi(x) = 0$ for $|x| \leq 2R$ and $\psi(x) = 1$ for $|x| \geq 3R$. Then $\psi p|_{\partial B(0,2R)} = 0$ and $\nabla(\psi p) \in L_{-\delta}^{r'}(\Omega)$. Hence by Hardy's inequality (see e.g. Kufner, 1980, p.33) $\psi p \in H_{-\delta-1}^{1,r'}(\mathbb{R}^2 \backslash B(0, 2R))$ which implies $\nabla p \in E_{-\delta}^{r'}(\Omega)$. $\square$

Proof of the main theorem: (i). Let $\delta \in (-2/r, 2/r')$, $\delta \neq -2/r + 1$. For $u \in (L_\delta^r(\Omega))^2$ let $\tilde{u}$ denote the extension of u to $\mathbb{R}^2$ with $\tilde{u} = 0$ for $x \notin \Omega$. Then div $\tilde{u} \in H_{\delta+1}^{-1,r}(\mathbb{R}^2)$ and we find a solution

$$p_1 \in H_{\delta+1}^{1,r}(\mathbb{R}^2) \text{ with } \Delta p_1 = \text{div } \tilde{u} \tag{5.1}$$

by Lemma 4.1. Since div $(u - \nabla p_1) = 0$ in Ω, we have $\gamma_\nu(u - \nabla p_1) \in H^{-1/r,r}(\partial\Omega)$. Green's formula (2.4) applied to the bounded domain $\mathbb{R}^2 \backslash \Omega$ leads to $\langle \gamma_\nu(u - \nabla p_1), 1 \rangle_{\partial\Omega} = 0$. Thus, by Lemma 4.4, there exists a solution $p_2 \in H_{\delta-1}^{1,r}(\Omega)$ of the exterior Neumann problem

$$\Delta p_2 = 0 \text{ in } \Omega, \quad \gamma_\nu \nabla p_2 = \gamma_\nu(u - \nabla p_1). \tag{5.2}$$

We set $Pu = u - \nabla(p_1 + p_2)$, hence $(L_\delta^r(\Omega))^2 = X_\delta^r(\Omega) + E_\delta^r(\Omega)$. $X_\delta^r(\Omega) \cap E_\delta^r(\Omega) = \emptyset$ follows from the uniqueness results of the Neumann problem in Lemma 4.4, which proofs part (i).

(ii). We fix $k \in \mathbb{N}$ and $\delta \in -I_k - 2/r$. From Lemma 4.1 and Lemma 4.4 we obtain solutions of (5.1) and (5.4), hence $(L_\delta^r(\Omega))^2 = X_\delta^r(\Omega) + E_\delta^r(\Omega)$. Furthermore, by Proposition 5.1 and Part (ii) of Lemma 4.4,

$$E_\delta^r(\Omega) \cap X_\delta^r(\Omega) = \{\nabla p; p \in H_{\delta-1}^{1,r}(\Omega), \Delta p = 0, \gamma_\nu \nabla p = 0\} = \nabla \mathcal{H}_k(\frac{\partial}{\partial \nu}).$$

Let P_k denote the projection of $(L_\delta^r(\Omega))^2$ onto $\nabla \mathcal{H}_k(\frac{\partial}{\partial \nu})$. For $u \in (L_\delta^r(\Omega))^2$ we set $Pu = u - \nabla(p_1 + p_2) + P_k \nabla(p_1 + p_2)$, where p_1, p_2 are solutions of (5.1) and (5.2), respectively. Pu is well defined: Let $p_1, \bar{p}_1$ be different solutions of the problem (5.1), $p_2, \bar{p}_2$ the corresponding solutions of the exterior Neumann problem (5.2), and $\bar{P}u = u - \nabla(\bar{p}_1 + \bar{p}_2) + P_k \nabla(\bar{p}_1 + \bar{p}_2)$. Then $p_1 - \bar{p}_1 \in \mathcal{H}_{k+1}$ due to Lemma 4.1, hence $p_1 - \bar{p}_1 + p_2 - \bar{p}_2 \in \nabla \mathcal{H}_k(\frac{\partial}{\partial \nu})$, which implies $Pu - \bar{P}u = 0$, and Part (ii) is proved.

(iii). Here we use duality arguments, which we summarize in the following proposition

Proposition 5.2. *Suppose X is a Banach space, X_1, X_2 are closed subspaces with $X_1 + X_2 = X$, $X_1 \cap X_2 = F$ is a finite dimensional subspace with basis $\{x_1, \ldots, x_d\}$. Let $P_i : X \to X_i$ denote the projection onto X_i. Then the (strong) dual space X^* can be decomposed in the following form:*

$$X^* = X_1^\perp \oplus X_2^\perp \oplus F^*,$$

where $X_i^\perp = \{x^* \in X^*; \langle x^*, x \rangle = 0, x \in X_i\}$ and $F^* = span[x_1^*, \ldots, x_d^*]$ with $\langle x_i^*, x_j \rangle = \delta_{ij}, j = 1, \ldots, d$.

Proof: For $x \in X$ we set $P_F = \sum_{i=1}^{d} \langle x_i^*, x \rangle x_i$ the projection of X onto F. Then $I - P_1 = P_2 - P_F$ hence

$$x = (I - P_1)x + (I - P_2)x + P_F x. \tag{5.3}$$

Let $y \in X^*$. Then $y_i := y \circ (I - P_i) \in X_i^\perp$, $i = 1, 2$, and $y_F = y \circ P_F \in F^*$. Due to (5.3), $\langle y, x \rangle = \langle y_1, x \rangle + \langle y_2, x \rangle + \langle y_F, x \rangle$ for any $x \in X$.

For $y \in X_1^\perp \cap X_2^\perp$, $x \in X$, we have

$$\langle y, x \rangle = \langle y, P_1 x + P_2 x - P_d x \rangle = \langle y, P_1 x \rangle + \langle y, P_2 x - P_d x \rangle = 0.$$

Similarly, for $y \in X_i^\perp \cap F^*$, we get

$$\langle y, x \rangle = \sum_{i=1}^{d} \langle y, x_i \rangle \langle x_i^* \rangle = \langle y, P_F x \rangle = 0$$

since $P_F x \in X_1 \cap X_2$. This shows that the sum is direct. □

Now let $\delta \in \mathbb{R}$ be fixed with $-2/r + k < \delta < -2/r + k + 1$, $k \geq 2$. We set $X = (L_{-\delta}^{r'}(\Omega))^2$, $X_1 = X_{-\delta}^{r'}(\Omega)$, $X_2 = E_{-\delta}^{r'}(\Omega)$, $F = \nabla \mathcal{H}_k(\frac{\partial}{\partial \nu})$. Then assertion (iii) of the main theorem follows from Proposition 1 and 2 and Part (ii) of the main theorem.

REFERENCES

ADAMS, R.A. (1975), *Sobolev spaces*, (Academic Press, New York).

BENCI, V. and D. FORTUNATO (1979), Weighted Sobolev spaces and the nonlinear Dirichlet problem, *Ann. Mat. Pur. Appl.* **121**, 319-336.

DU PLESSIS, N. (1970), *An Introduction to Potential Theory*, (Oliver & Boyd, Edingbourgh).

FOLLAND, G.B. (1976), *Introduction to Partial Differential Equations*, (Princeton University Press, Princeton).

FUJIWARA, D. and H. MORIMOTO (1977), An L_r theorem of the Helmholtz-decomposition of Vector fields, *J. Fac. Sci. Univ. Tokyo* **24**, 685-699.

GALDI, G.P. and C.G. SIMADER (1990), Existence, uniqueness and L^q-estimates for the Stokes problem in an exterior domain, *Arch. Rat. Mech. Anal.* **112**, 291-318.

KUFNER, A. (1980), *Weighted Sobolev spaces*, (Teubner, Leipzig).

LADYZHENSKAYA, O.A. (1966), *The Mathematical Theory of Viscous Incompressible Flow*, (Gordon & Breach, New York).

LIONS, J.L. and E. MAGENES (1961), Probleme ai limiti non omogeni III, *Ann. Scuola Norm. Sup. Pisa* **25**, 41 -103.

LIONS, J.L. and E. MAGENES (1962), Probleme ai limiti non omogeni V, *Ann. Scuola Norm. Sup. Pisa, Sci.Mat.* **16**, 1-44.

LIONS, J.L. and E. MAGENES (1963), Problèmes aux limites non homogènes VI, *J. d'Analyse Math.* **11**, 165-188.

MCOWEN, R.C. (1979), The behavior of the Laplacian in weighted Sobolev spaces, *Comm. Pure Appl. Math* **32**, 783-795.

MIYAKAWA, T. (1982), On nonstationary solutions of the Navier-Stokes equations in an exterior domain , *Hiroshima Mat. J* **12**, 115-140.

MORREY, C.B. Jr. (1966), *Multiple Integrals in the Calculus of Variations*, (Springer Verlag, Berlin).

NIRENBERG, L. and H. WALKER (1973), The null spaces of elliptic partial differential operators in $I\!R^n$, *J. Math. Anal. App.* **42**, 271-301.

NERI, U. (1977), *Singular integrals*, (Springer, New York).

SIMADER, C.G. (1990), *The weak Dirichlet and Neumann problem for the Laplacian in L^q for bounded and exterior domains. Applications.* Nonlinear Analysis, function spaces and applications, Vol. 44, (Teubner Texte zur Mathematik 119, Leipzig).

SIMADER, C.G. H. SOHR (1992), A new approach to the Helmholtz decomposition and the Neumann problem in L^q spaces for bounded and exterior domains, in: *Mathematical problems relating to the Navier-Stokes equations* (ed. G.P. Galdi), (World Scientific, Singapore).

SPECOVIUS-NEUGEBAUER, N. (1990), The Helmholtz decomposition of weighted L^r-spaces, *Comm. Part. Diff. Eq.* **15**, 273-288.

TRIEBEL, (1978) H. *Interpolation Theory, Function spaces and Differential Operators*, (North Holland Publishing Company, Berlin).

VON WAHL, W. (1990), *Abschätzungen für das Neumann-Problem und die Helmholtz-Zerlegung von L^p*, Nachr. Akad. Wiss. Göttingen, Math.-Phys. Kl. II.

YOSHIDA, K. (1965), *Functional Analysis*, (Springer, New York).

BOUNDARY INTEGRAL EQUATIONS FOR THE STOKES PROBLEM IN EXTERIOR DOMAINS OF $I\!R^n$

Werner Varnhorn

Institute of Numerical Mathematics
Technical University Dresden
01062 Dresden, Germany

1. INTRODUCTION

The aim of this paper is the construction and representation of solutions u, p to the homogeneous Stokes equations

$$- \Delta u + \nabla p = 0 \quad \text{in} \quad G_e, \qquad \nabla \cdot u = 0 \quad \text{in} \quad G_e, \qquad u = \Phi \quad \text{on} \quad \Gamma, \qquad (1.1)$$

with methods of hydrodynamical potential theory. Here $G_e \subset I\!R^n$ $(n \geq 2)$ is an exterior domain with boundary $\Gamma = \partial G_e \in C^2$, and $\Phi \in C^0(\Gamma)$ is some prescribed boundary value.

The hydrodynamical potential theory for the solution of the stationary Stokes equations in three-dimensional domains was developed even in the twenties of this century. Lichtenstein (1928) and Odquist (1930) independently of each other constructed hydrodynamical potentials, studied their properties, and used them for the solution of the Stokes equations. This theory has several advantages compared to the merely functional analytical approach. It is constructive and it leads to explicit representations of the solution. The differentiability properties of this solution can be investigated in Hölder or Lebesgue spaces, not only in the interior of the domain but also close to the boundary. From the numerical point of view the boundary element method certainly has also contributed to revive this theory and encouraged studies of new potential formulations also for numerical purposes.

The theory of hydrodynamical potentials only differs from the classical potential theory of the Laplace equation by the definite analytical form of the potentials. However, due to the same singular behaviour of the potential kernels, the properties of these potentials are mainly identical to the properties of the corresponding classical potentials. Therefore we will only list these properties below for the general n–dimensional case $(n \geq 2)$. A complete convergence analysis of the various weakly and strongly singular surface integrals as well as a precise deduction of the continuity properties and jump

Navier-Stokes Equations and Related Nonlinear Problems
Edited by A. Sequeira, Plenum Press, New York, 1995

relations of the hydrodynamical potentials is not given here (a detailed representation for the case $n = 3$ has been given by Deuring, von Wahl and Weidemaier, 1988, and Deuring and von Wahl, 1989).

2. HYDRODYNAMICAL POTENTIAL THEORY

As in the classical potential theory we require Green formulas as a starting point, the hydrodynamical version of which has the following form. For sufficiently smooth, solenoidal vector functions u, v and scalar functions p, q in a bounded domain $A \subset \mathbb{R}^n$ $(n \geq 2)$ with the boundary $\partial A \in C^1$ we have Green's first and second formula

$$\int_A \left(S_p^u\right) \cdot \binom{v}{q} \, dy = \int_{\partial A} (T_p^u N) \cdot v \, do + 2 \int_A Du : Dv \, dy, \tag{2.1}$$

$$\int_A \left\{ \left(S_p^u\right) \cdot \binom{v}{q} - \binom{u}{p} \cdot \left(S'^v_q\right) \right\} dy = \int_{\partial A} \left\{ (T_p^u N) \cdot v - u \cdot (T'^v_q N) \right\} do. \tag{2.2}$$

Here

$$S : \binom{u}{p} \longrightarrow S_p^u := \begin{pmatrix} -\Delta u + \nabla p \\ \nabla \cdot u \end{pmatrix} \tag{2.3}$$

stands for the formal Stokes operator,

$$S' : \binom{u}{p} \longrightarrow S'^u_p := \begin{pmatrix} -\Delta u - \nabla p \\ -\nabla \cdot u \end{pmatrix} \tag{2.4}$$

for the corresponding adjoint operator, and

$$T : \binom{u}{p} \longrightarrow T_p^u := -2Du + pI_n,$$

$$\tag{2.5}$$

$$T' : \binom{u}{p} \longrightarrow T'^u_p := -2Du - pI_n$$

denote the stress tensors adjoint to each other. In (2.5) the deformation tensor is defined by

$$Du := \tfrac{1}{2}(\nabla u + (\nabla u)^T) \tag{2.6}$$

with $(\nabla u)^T$ as the matrix transposed to $\nabla u := (\partial_i u_k)_{k,i=1,\dots,n}$. Here and in the following, I_n is always the $n \times n$ identity matrix and $N = N(y)$ the exterior (with respect to the bounded domain A) unit surface normal vector in $y \in \partial A$. For vectors $a, b \in \mathbb{R}^n$ and $n \times n$ matrices $C = (C_{ij})$, $D = (D_{ij})$ we set

$$a \cdot b := \sum_{i=1}^n a_i b_i \quad \text{and} \quad C : D := \sum_{i,j=1}^n C_{ij} D_{ij}.$$

Now the following uniqueness statement for classical solutions u, p of the equations (1.1) (i.e. $u \in C^2(G_e) \cap C^1(\overline{G_e})$, $p \in C^1(G_e) \cap C^0(\overline{G_e})$) can be proved using Green's first formula (2.1):

Lemma 2.1. Let $G_e \subset \mathbb{R}^n$ $(n \geq 2)$ be an exterior domain with boundary $\Gamma \in C^2$ and $\Phi \in C^0(\Gamma)$. Then there exists at most one classical solution u, p to the equations (1.1), which satisfies the decay conditions

$$\nabla^k u(x) = \mathcal{O}(|x|^{2-n-k}), \quad k = 0, 1, \tag{2.7}$$

$$p(x) = \mathcal{O}(|x|^{1-n}), \tag{2.8}$$

as $|x| \to \infty$.

Let $a_\infty \in I\!\!R^2$ be given. Then the condition (2.7) in the case $n = 2$, $k = 0$ may be weakened by

$$u(x) - a_\infty \ln |x| = \mathcal{O}(1). \tag{2.9}$$

Proof: We define $u = u^1 - u^2$ and $p = p^1 - p^2$ as the difference between two solutions u^1, p^1 and u^2, p^2 of (1.1). If $K_R(0)$ denotes an open ball centered at zero with a sufficiently large radius R so that $\Gamma \subset K_R(0)$, then Green's first formula (2.1) in $G_R = G_e \cap K_R(0)$ provides for the relation

$$2 \int_{G_R} |Du|^2 dx + \int_{\partial K_R} (T_p^u N) \cdot u \, do = 0. \tag{2.10}$$

Here the corresponding boundary integral on Γ disappears due to $u = 0$ on Γ. In the case $n \geq 3$ it now results from the decay conditions (2.7) and (2.8) that the boundary integral in (2.10) vanishes as $R \to \infty$. Hence $u = 0$ in $\overline{G}_e$, and the differential equations (1.1) imply $p = c$, whereas the constant c has to disappear due to the decay (2.8) of the pressure. In the case $n = 2$ the conditions (2.7) and (2.8) at first are not sufficient to guarantee that the boundary integral in (2.10) vanishes as $R \to \infty$. However, due to $u(x) = \mathcal{O}(1)$ for $|x| \to \infty$ it follows from the representation formula of Chang and Finn (1961) [p. 393] that $\nabla u(x)$ and $p(x)$ even decay like $\mathcal{O}(|x|^{-2})$ as $|x| \to \infty$. This is sufficient to imply the asserted uniqueness also in the case $n = 2$. □

Apart from the Green formulas (2.1) and (2.2), the explicit form of the Stokes fundamental tensor $E = (E_{jk})_{j,k=1,\ldots,n+1}$ is of a great importance. The following representation is well-known (see Hsiao and Kress, 1985, in the case $n = 2$ and Fabes, Kenig and Verchota, 1988 for $n \geq 3$).

$n \geq 2$ $(j, k = 1, \ldots, n)$:

$$E_{jk}(x) = \frac{1}{2\omega_n} \left\{ \frac{x_j x_k}{|x|^n} + \delta_{jk} \left\{ \begin{array}{ll} \ln \frac{1}{|x|} & (n = 2) \\ \frac{|x|^{2-n}}{n-2} & (n \geq 3) \end{array} \right\} \right\},$$

$$E_{n+1,k}(x) = E_{k,n+1}(x) = \frac{x_k}{\omega_n |x|^n}, \tag{2.11}$$

$$E_{n+1,n+1}(x) = \delta(x).$$

Here and in future, ω_n always stands for the surface area of the $(n - 1)$ dimensional unit sphere in $I\!\!R^n$ $(n \geq 2)$.

Now let u, p denote a solution to $S_p^u = \binom{0}{0}$ in A. Insert for v, q the column vector $E_k(x - \cdot) := (E_{jk}(x - \cdot))_{j=1,\ldots,n+1}$ as a function of y in Green's second formula (2.2), one after another for each $k = 1, \ldots, n + 1$. Then the result is the following row representation of u and p, depending on the position of the point x:

$$\int_{\partial A} \langle\!\langle T_p^u N(y), E^{(r)}(x - y) \rangle\!\rangle \, do_y - \int_{\partial A} \langle\!\langle u(y), T_y' E(x - y) N(y) \rangle\!\rangle \, do_y$$

$$= \left\{ \begin{array}{ll} -(u_1(x), \ldots, u_n(x), p(x)) & , \ x \in A, \\ 0 & , \ x \notin \overline{A}. \end{array} \right. \tag{2.12}$$

Here the multiplication $\langle\langle .,.\rangle\rangle$ is defined for $a \in \mathbb{R}^n$ and matrices $A = (A_{jk}) \in \mathbb{R}^n \times \mathbb{R}^m$ by the m-componental row

$$\langle\langle a, A\rangle\rangle := \left(\sum_{j=1}^{n} a_j A_{j1}, \ldots, \sum_{j=1}^{n} a_j A_{jm}\right).$$

The tensor $E^{(r)}$ is the $n \times (n+1)$ matrix which results from cancelling the last row ($\sim r$) of E. The term $T_y' EN$ in the last boundary integral is to be understood as follows. Applying T' to the $n+1$ columns of E yields $(n+1)$ $n \times n$ matrices, which result in some $n \times (n+1)$ matrix if multiplied by the unit normal vector N. The index y in T_y' means that the differentiation has to be carried out with respect to y.

A representation of a solution u, p to $S_p^u = \binom{0}{0}$ in A in the form of columns follows by transposing (2.12):

$$\int_{\partial A} E^{(c)}(x-y)\, T_p^u N(y)\, do_y - \int_{\partial A} D(x,y)\, u(y)\, do_y = \begin{cases} -\binom{u}{p}(x) & , \ x \in A, \\[2mm] 0 & , \ x \notin \overline{A}. \end{cases} \tag{2.13}$$

Here the $(n+1) \times n$ matrix $E^{(c)}$ is obtained from E by cancelling the last column ($\sim c$) (note that $E = E^T$), and the $(n+1) \times n$ matrix $D(x,y)$ is defined by (use $T_y' E(x-y) = -T_x E(x-y)$)

$$D(x,y) = (-T_x E(x-y)\, N(y)) = \left((-T_x E_k(x-y))_{ij} N_j(y)\right)_{ki}. \tag{2.14}$$

Setting $z := x - y$ and $N := N(y)$, the double layer tensor D has the following form. $n \geq 2$ $(k, i = 1, \ldots, n)$:

$$D_{ki}(x,y) = -\frac{n}{\omega_n} \frac{z_k z_i z \cdot N}{|z|^{n+2}},$$

$$\tag{2.15}$$

$$D_{n+1,i}(x,y) = -\frac{2}{\omega_n} \left(\frac{n z_i z \cdot N}{|z|^{n+2}} - \frac{N_i}{|z|^n}\right).$$

The representations given in (2.12) and (2.13) are so-called direct representations of a solution u, p to $S_p^u = \binom{0}{0}$ by its boundary values u and $T_p^u N$. This kind of representation is not unique, of course, since it depends on the corresponding Green formula. Besides the direct representations there are indirect representations by hydrodynamical surface potentials with general vector-valued source densities. For such densities $\Psi \in C^0(\Gamma)$ we therefore define the hydrodynamical single layer potential

$$(E_n \Psi)(x) := \int_{\Gamma} E^{(c)}(x-y)\, \Psi(y)\, do_y, \quad x \notin \Gamma, \tag{2.16}$$

and the hydrodynamical double layer potential

$$(D_n \Psi)(x) := \int_{\Gamma} D(x,y)\, \Psi(y)\, do_y, \quad x \notin \Gamma. \tag{2.17}$$

Both potentials are smooth functions outside the surface Γ and there they satisfy the homogeneous Stokes equations, i.e. we have

$$S E_n \Psi = 0, \quad S D_n \Psi = 0.$$

In the following, we will supply that part of these potentials referring to the velocity field u with a dot. This leads to the n–componental column vectors

$$(E_n^{\bullet}\Psi)(x) \;=\; \int_{\Gamma} E^{(r,c)}(x-y)\,\Psi(y)\,do_y, \quad x \notin \Gamma, \tag{2.18}$$

$$(D_n^{\bullet}\Psi)(x) \;=\; \int_{\Gamma} D^{(r)}(x,y)\,\Psi(y)\,do_y, \quad x \notin \Gamma. \tag{2.19}$$

Here the $n \times n$ matrices $E^{(r,c)}$ and $D^{(r)}$ are defined by cancelling the last row $(\sim r)$ in $E^{(c)}$ and D, respectively.

Besides the single layer potential $E_n\Psi$, in the following we also need its normal stresses $T(E_n\Psi)N$, which are defined in a neighbourhood $U \subset I\!\!R^n$ of the surface Γ for all $x \in U\backslash\Gamma$ and $\Psi \in C^0(\Gamma)$ by the n–componental column vector

$$(H_n^{\bullet}\Psi)(x) \;=\; \int_{\Gamma} T_x\Big(E^{(c)}(x-y)\,\Psi(y)\Big)\,N(\tilde{x})\,do_y$$

$$=: \; \int_{\Gamma} H(x,y)\,\Psi(y)\,do_y, \quad x \notin \Gamma. \tag{2.20}$$

Here $\tilde{x} \in \Gamma$ is the uniquely determined projection of $x \in U$ onto Γ (for boundaries of class C^2 the construction of parallel surfaces is possible, see Smirnow, 1979 [§200]), and for the $n \times n$ matrix H we find

$$H(x,y) = \Big(D^{(r)}(y,x)\Big)^T = D^{(r)}(y,x) \quad \text{on} \quad \Gamma \times \Gamma. \tag{2.21}$$

Next we need some statements regarding the behaviour of the potentials in the point $x \in I\!\!R^n\backslash\Gamma$, if x passes through the surface Γ. Therefore let

$$G_i := I\!\!R^n\backslash\overline{G_e}$$

always denote the bounded complement of G_e. Then the following analogue to the Gauss integral formula of the classical potential theory holds (see Ladyzhenskaya, 1969 [p. 56] in the case $n = 3$):

Lemma 2.2. For the double layer potential $D_n^{\bullet}b$ with some constant density $b \in I\!\!R^n$ $(n \geq 2)$ we have

$$(D_n^{\bullet}b)(x) = \begin{cases} b, & x \in G_i, \\ \tfrac{1}{2}b, & x \in \Gamma, \\ 0, & x \in G_e. \end{cases} \tag{2.22}$$

Let us now assume $z \in \Gamma$. With the definitions

$$w^i(z) \;:=\; \lim_{\substack{x \to z \\ x \in G_i}} w(x) \qquad \text{(limiting value from the interior } G_i\text{)},$$

$$w^e(z) \;:=\; \lim_{\substack{x \to z \\ x \in G_e}} w(x) \qquad \text{(limiting value from the exterior } G_e\text{)}, \tag{2.23}$$

we have the following continuity and jump relations (see Ladyzhenskaya, 1969 [p. 56ff] in the case $n = 3$).

Lemma 2.3. Let $\Psi \in C^0(\Gamma)$ and let $E_n^\bullet \Psi, D_n^\bullet \Psi, H_n^\bullet \Psi$ be the potentials defined by (2.18), (2.19), and (2.20). Then:

$$(E_n^\bullet \Psi)^i \ = \ (E_n^\bullet \Psi) = (E_n^\bullet \Psi)^e, \tag{2.24}$$

$$(D_n^\bullet \Psi)^i - D_n^\bullet \Psi \ = \ +\tfrac{1}{2}\Psi \ = \ D_n^\bullet \Psi - (D_n^\bullet \Psi)^e, \tag{2.25}$$

$$(H_n^\bullet \Psi)^i - H_n^\bullet \Psi \ = \ -\tfrac{1}{2}\Psi \ = \ H_n^\bullet \Psi - (H_n^\bullet \Psi)^e. \tag{2.26}$$

This means that the hydrodynamical potentials behave exactly as the classical potentials for the scalar Laplace equation $-\Delta u = 0$. The single layer potential corresponding to the velocity field is continuous in the entire space $\mathbb{R}^n$, whereas the normal stresses of all the single layer potential as well as the double layer potential jump when they pass through the surface. The decay behaviour at infinity is also identical in both cases. Whereas the single layer potential regarding the velocity increases logarithmically ($n = 2$) or decays like $\mathcal{O}(|x|^{2-n})$ ($n \geq 3$), respectively, the double layer potential regarding the velocity has the same decay behaviour $\mathcal{O}(|x|^{1-n})$ as the classical double layer potential of the Laplace equation for all $n \geq 2$ (compare Günther, 1957).

3. THE BOUNDARY INTEGRAL EQUATIONS

In the following we will prove the existence of a solution u, p to the Stokes equations (1.1) using the method of boundary integral equations. In contrast to the classical proceeding as described in the case $n = 3$ by Odquist (1930) or in the third chapter of Ladyzhenskaya's book (1969), for this purpose we use for all $n \geq 2$ a combined potential ansatz, i.e. a sum of double layer potential and single layer potential with unknown source densities (Odquist's formulation requires $\frac{n(n+1)}{2}$ single layer potentials, the densities of which span the null space of the integral operator corresponding to the interior Neumann problem of the Stokes equations, see Ladyzhenskaya, 1969 [p. 63] in the case $n = 3$). Formulations of this kind are originally due to Brakhage and Werner (1965), Leis (1964) for the Helmholtz equation, as well as to Panich (1965) for Maxwell's equation. In the connection with boundary element methods for the Stokes equations they were apparently used for the first time by Hsiao and Kress (1985), in the case $n = 2$ and by Hebeker (1986) in the case $n = 3$.

As in Hsiao and Kress (1985) (also see Borchers and Varnhorn, 1993a, 1993b) we choose in the case $n = 2$ in $x \in G_e$ the potential ansatz ($\eta, \alpha \in \mathbb{R}$)

$$\binom{u}{p}(x) \ = \ -4\pi(E_2 a_\infty)(x) + (D_2\Psi)(x) - \eta(E_2 M_2 \Psi)(x) - \alpha \int_\Gamma \binom{\Psi}{0} do. \tag{3.1}$$

Here $\Psi \in C^0(\Gamma)$ is an unknown source density, and the constant $a_\infty \in \mathbb{R}^2$ has to be selected identically to the prescribed constant a_∞ from the condition (2.9). The projector $M_n : C^0(\Gamma) \to C^0(\Gamma)$ is defined by

$$\Psi \to M_n\Psi \ := \ \Psi - \Psi_M \tag{3.2}$$

with the surface mean value

$$\Psi_M := \tfrac{1}{|\Gamma|} \int_\Gamma \Psi \, do, \qquad |\Gamma| = \int_\Gamma 1 \, do.$$

In the case $n \geq 3$ a simpler formulation is sufficient. Here we choose in $x \in G_e$ ($\eta \in \mathbb{R}$) the potential ansatz

$$\binom{u}{p}(x) = (D_n \Psi)(x) - \eta(E_n \Psi)(x). \tag{3.3}$$

Using the continuity and jump relations of the potentials according to Lemma 2.3, now the following systems of boundary integral equations result from the formulations (3.1) and (3.3), respectively:

$$\Phi + 4\pi E_2 a_\infty = K_2 \Psi := \left(-\tfrac{1}{2} I_2 + D_2^\bullet - \eta E_2^\bullet M_2 - \alpha |\Gamma|(I_2 - M_2)\right)\Psi, \tag{3.4}$$

$$\Phi = K_n \Psi := \left(-\tfrac{1}{2} I_n + D_n^\bullet - \eta E_n^\bullet\right)\Psi, \quad n \geq 3. \tag{3.5}$$

Here $\Phi \in C^0(\Gamma)$ is the boundary value prescribed in (1.1). The solvability of these systems in $C^0(\Gamma)$ (see also Borchers and Varnhorn, 1993a in the case $n = 2$ and Hebeker, 1986, for $n = 3$) is described by the following theorem.

Theorem 3.1. Let $\Phi \in C^0(\Gamma)$. Then:

1. Let $a_\infty \in \mathbb{R}^2$ be given. Then for $\eta > 0$, $\alpha \neq 0$ there exists exactly one solution $\Psi = (\Psi_1, \Psi_2) \in C^0(\Gamma)$ of the boundary integral equations' system (3.4).

2. Let $3 \leq n \in \mathbb{N}$. Then for $\eta > 0$ there exists exactly one solution $\Psi = (\Psi_1, \ldots, \Psi_n) \in C^0(\Gamma)$ of the boundary integral equations' system (3.5).

Proof: The operators $K_n : C^0(\Gamma) \to C^0(\Gamma)$ defined in (3.4) and (3.5) are Fredholm operators. Therefore their adjoint operators K_n^* with respect to the dual system

$$\langle \Phi, \Psi \rangle := \int_\Gamma \Phi \cdot \Psi \, do = \int_\Gamma (\Phi_1 \Psi_1 + \cdots + \Phi_n \Psi_n) \, do$$

have to be investigated. Consequently, due to

$$E_n^{\bullet *} = E_n^\bullet, \qquad\qquad D_n^{\bullet *} = H_n^\bullet,$$

$$M_n^* = M_n, \qquad (E_n^\bullet M_n)^* = M_n E_n^\bullet,$$

we have to consider the homogeneous boundary integral equations' systems $(n = 2)$

$$0 = K_2^* \Psi := \left(-\tfrac{1}{2} I_2 + H_2^\bullet - \eta M_2 E_2^\bullet - \alpha |\Gamma|(I_2 - M_2)\right)\Psi \tag{3.6}$$

and $(n \geq 3)$

$$0 = K_n^* \Psi := \left(-\tfrac{1}{2} I_n + H_n^\bullet - \eta E_n^\bullet\right)\Psi \tag{3.7}$$

on Γ.

1. Let $0 \neq \Psi \in C^0(\Gamma)$ be a solution of (3.6). Then from the relations (2.24) and (2.26) we find

$$(H_2^\bullet \Psi)^i = -\tfrac{1}{2}\Psi + H_2^\bullet \Psi = \eta M_2 E_2^\bullet \Psi + \alpha |\Gamma| \Psi_M \tag{3.8}$$

123

on Γ. Since for every constant vector $b \in \mathbb{R}^2$ we have

$$M_2 b = 0, \quad (I_2 - M_2)b = b \quad \text{and} \quad (-\tfrac{1}{2}I_2 + D_2^\bullet)b = 0$$

(see (2.22)), hence

$$0 = \langle b, K_2^* \Psi \rangle = \langle K_2 b, \Psi \rangle = -\langle \alpha |\Gamma| \, b, \Psi \rangle = -\alpha |\Gamma| \, b \cdot \int_\Gamma \Psi \, do,$$

this implies

$$\Psi_M = \tfrac{1}{|\Gamma|} \int_\Gamma \Psi \, do = 0 \tag{3.9}$$

due to $\alpha \neq 0$. Let us consider the single layer potential $\binom{u}{p} = E_2 \Psi$ with source density Ψ in the bounded domain $G_i = \mathbb{R}^2 \backslash \overline{G_e}$. Then from Green's first formula (2.1) it follows by (2.20), (3.8) and (3.9) that

$$2 \int_{G_i} |Du|^2 dx = - \int_\Gamma (H_2^\bullet \Psi)^i \cdot u^i \, do$$

$$= \int_\Gamma (-\eta M_2 u) \cdot u \, do = -\eta \int_\Gamma |M_2 u|^2 do. \tag{3.10}$$

Here the last equation is correct since $\int_\Gamma M_2 u \, do = 0$. Due to $\eta > 0$ as assumed above it follows $M_2 u = M_2 E_2^\bullet \Psi = 0$ on Γ, hence $u = u_M$ in $\overline{G_i}$ according to (3.10), and thus $(H_2^\bullet \Psi)^i = (pN)^i = 0$ by (3.8). This implies

$$(H_2^\bullet \Psi)^e = \Psi \tag{3.11}$$

using the jump relations $(H_2^\bullet \Psi)^e - (H_2^\bullet \Psi)^i = \Psi$ (see (2.26)).

Now, as $M_2 u = 0$ on Γ, the functions $v = u - u_M$ and p represent a classical solution to the equations

$$-\Delta v + \nabla p = 0 \quad \text{in} \quad G_e, \qquad \nabla \cdot v = 0 \quad \text{in} \quad G_e, \qquad v = 0 \quad \text{on} \quad \Gamma.$$

Note that $E_2 \Psi$ with Ψ as a solution to the homogeneous equations (3.6) is sufficiently regular. Therefore v and p fulfill the decay conditions (2.7) (with v instead of u) and (2.8). This follows directly from the representation of the single layer potentials (2.11) for $n = 2$, with one exception: we still have to prove $v(x) = \mathcal{O}(1)$ for $|x| \to \infty$ (condition (2.7) in the case $n = 2$, $k = 0$). This condition, however, is obtained from (2.11) since the average value Ψ_M of the density Ψ of the single layer potential $u = E_2^\bullet \Psi$ vanishes according to (3.9). Hence from

$$E_2^\bullet \Psi(x) = \frac{1}{2\omega_2} \int_\Gamma \left(-\ln |x - y| \Psi(y) + \frac{(x-y) \cdot \Psi(y)}{|x-y|^2}(x-y) \right) do_y$$

$$= \frac{1}{2\omega_2} \int_\Gamma \ln \left| \frac{x}{x-y} \right| \Psi(y) \, do_y + \mathcal{O}(1)$$

$$= o(1) + \mathcal{O}(1) = \mathcal{O}(1), \qquad |x| \to \infty,$$

it follows $v(x) = u(x) - u_M = \mathcal{O}(1)$ for $|x| \to \infty$. According to the uniqueness statement from Lemma 2.1 now it results $v = 0$ (i.e. $u = u_M = \text{const.}$) and $p = 0$ in G_e. Due to

$H_2^\bullet \Psi = 0$ in G_e it follows $(H_2^\bullet \Psi)^e = 0$ on Γ, and (3.11) implies $\Psi = 0$, in contradiction to the assumption. This proves the first part.

2. Analogously, let $0 \neq \Psi \in C^0(\Gamma)$ be a solution to the homogeneous adjoint integral equations (3.7). From (2.24) and (2.26) (see (3.8)) it follows

$$(H_n^\bullet \Psi)^i = -\tfrac{1}{2}\Psi + H_n^\bullet \Psi = \eta E_n^\bullet \Psi \tag{3.12}$$

on Γ. With Green's first formula (2.1) we know that the single layer potential $\binom{u}{p} = E_n \Psi$ with density Ψ satisfies

$$2 \int\limits_{G_i} |Du|^2 dx = -\eta \int\limits_{\Gamma} |u|^2 do,$$

similar to (3.10). Due to $\eta > 0$ according to the assumption we find $u = 0$ on Γ, hence $u = 0$ in $\overline{G_i}$, and (3.12) implies

$$(H_n^\bullet \Psi)^i = 0. \tag{3.13}$$

On the other hand, from the uniqueness statement (Lemma 2.1) it finally follows $u = 0$ in the entire space $\mathbb{R}^n$ and $p = 0$ in G_e, as the operator $H_n^\bullet - \eta E_n^\bullet : C^0(\Gamma) \to C^0(\Gamma)$ from (3.7) regularizes, and the validity of the decay conditions (2.7) and (2.8) results for $n \geq 3$ directly from the representations (2.11) of the single layer potential kernels. Therefore

$$(H_n^\bullet \Psi)^e = 0,$$

and from the jump relations (2.26) with (3.13) it follows

$$\Psi = (H_n^\bullet \Psi)^e - (H_n^\bullet \Psi)^i = 0.$$

This also proves the second part. □

Let us now summarize the result of this section in a theorem.

Theorem 3.2. Let $G_e \subset \mathbb{R}^n$ ($n \geq 2$) be an exterior domain with boundary $\Gamma \in C^2$. Let $\Phi \in C^0(\Gamma)$ and in the case $n = 2$ also $a_\infty \in \mathbb{R}^2$ be given. Let $\Psi \in C^0(\Gamma)$ be the uniquely determined solution of the boundary integral equations (3.4) for $n = 2$ or (3.5) for $n \geq 3$, according to Theorem 3.1 ($\eta > 0$, $\alpha \neq 0$). Then (3.1) or (3.3), respectively, represents a solution $u \in C^\infty(G_e) \cap C^0(\overline{G_e})$, $p \in C^\infty(G_e)$ to the Stokes equations (1.1) which satisfies the decay conditions from Lemma 2.1.

In some special cases the resulting integral equations can be simplified considerably by a suitable choice of parameters in the potential ansätze. If e.g.

$$G_e := \mathbb{R}^n \setminus \overline{K_R(0)} \ (n \geq 3)$$

is the complement of the closed ball with centre at zero and radius $R > 0$, the ansatz

$$\binom{u}{p} := D_n \Psi - \eta E_n \Psi, \qquad \eta = \frac{n}{R} \tag{3.14}$$

for the solution of

$$S_p^u = \binom{0}{0} \quad \text{in} \quad G_e, \qquad u = \Phi \quad \text{on} \quad \Gamma$$

leads to a completely decoupled system of boundary integral equations. Namely, for $x, y \in \Gamma = \partial K_R$ we have

$$(x - y) \cdot N(y) = \tfrac{1}{2} 2(x - y) \cdot \tfrac{1}{R} y = \tfrac{1}{2R}(2x \cdot y - x^2 - y^2) = -\tfrac{1}{2R}|x - y|^2,$$

hence (2.15) and (2.11) imply

$$D_{ki}(x,y) - \frac{n}{R} E_{ki}(x-y) = \frac{n}{R} \cdot \frac{1}{2\omega_n} \left(-\delta_{ki} \frac{|x-y|^{2-n}}{n-2} \right).$$

This leads to n uniquely solvable *scalar* boundary integral equations of the form

$$\Phi = -\tfrac{1}{2}\Psi - \frac{n}{2(n-2)R\omega_n} \int_\Gamma |x-y|^{2-n} \Psi(y)\, do_y.$$

REFERENCES

BORCHERS, W. and W. VARNHORN (1993a), On the Boundedness of the Stokes Semigroup in Two–Dimensional Exterior Domains, *Math. Z.* **213**, p. 275–300.

BORCHERS, W. and W. VARNHORN (1993b), Die Stokes–Resolvente in Außengebieten des $I\!R^2$, *Z.A.M.M.* **73**, p. T849–T852.

BRAKHAGE, H. and P. WERNER (1965), Über das Dirichlet'sche Außenraumproblem für die Helmholtz'sche Schwingungsgleichung, *Arch. Math.* **16**, p. 325–329.

CHANG, I–DEE and R. FINN (1961), On the Solutions of a Class of Equations Occurring in Continuum Mechanics with Application to the Stokes Paradox, *Arch. Rat. Mech.* **7**, p. 389–401.

DEURING, P., VON WAHL, W. and P. WEIDEMAIER (1988), Das lineare Stokes–System in $I\!R^3$ (1. Vorlesungen über das Innenraumproblem), *Bayreuther Mathematische Schriften* **27**, p. 1–252.

DEURING, P. and W. VON WAHL (1989), Das lineare Stokes–System in $I\!R^3$ (Das Außenraumproblem), *Bayreuther Mathematische Schriften* **28**, p. 1–109.

FABES, E.B., KENIG, C.E. and G.C. VERCHOTA (1988), The Dirichlet Problem for the Stokes System on Lipschitz Domains, *Duke Math. J.* **57**, p. 769–793.

GÜNTHER, N.M. (1957), *Die Potentialtheorie und ihre Anwendungen auf Grundaufgaben der mathematischen Physik* (Verlagsgesellschaft, Leipzig).

HEBEKER, F.K. (1986), Efficient Boundary Element Methods for Three–Dimensional Exterior Viscous Flow, *Numer. Meth. Part. Diff. Equa.* **2**, p. 273–297.

HSIAO, G.C. and R. KRESS (1985), On an Integral Equation for the Two–Dimensional Exterior Stokes Problem, *Appl. Num. Math.* **1**, p. 77–93.

LADYZHENSKAYA, O.A. (1969), *The Mathematical Theory of Viscous Incompressible Flow* (Gordon and Breach, New York et al.).

LEIS, R. (1964), Zur Eindeutigkeit der Randwertaufgaben der Helmholtz'schen Schwingungsgleichung, *Math. Z.* **85**, p. 141–153.

LICHTENSTEIN, L. (1928), Über einige Existenzprobleme der Hydrodynamik, *Math. Z.* **28**, p. 387–415.

ODQUIST, F.K.G. (1930), Über die Randwertaufgaben in der Hydrodynamik zäher Flüssigkeiten, *Math. Z.* **32**, p. 329–375.

PANICH, O.I. (1965), On the Question of the Solvability of the Exterior Boundary Value Problem for the Wave Equation and Maxwell's Equation, *Russ. Math. Surv.* **20**, p. 221–226.

SMIRNOW, W.I. (1979), *Lehrgang der höheren Mathematik 4* (Deutscher Verlag der Wissenschaften, Berlin).

Flows in Bounded Domains

REGULAR SOLUTIONS TO THE STEADY NAVIER-STOKES EQUATIONS

Jens Frehse and Michael Růžička

University Bonn, Institute of Applied Mathematics
Beringstraße 6, 53115 Bonn, Germany

ABSTRACT

The existence of regular solutions to the periodic steady Navier-Stokes equations in dimension N, $10 \leq N \leq 15$, is established.

1. INTRODUCTION

In a previous paper Frehse and Růžička (1995) proved the existence of regular solutions to the steady Navier-Stokes equations

$$
\begin{aligned}
-\Delta \mathbf{u} + \mathbf{u} \cdot \nabla \mathbf{u} + \nabla p &= \mathbf{f} \\
\operatorname{div} \mathbf{u} &= 0
\end{aligned}
\quad \text{in } \Omega
\tag{1.1}
$$

with periodic boundary conditions, where $\Omega = (0, L)^N$, $L > 0$, $5 \leq N < 10$. Note also the paper of Struwe (1995) who proves the existence of a smooth solution to (1.1) for $\Omega = \mathbb{R}^5$ by a different method. The proof in Frehse and Růžička (1995) is based on the apriori estimate

$$
\int_\Omega \left(\frac{\mathbf{u}^2}{2} + p \right)_+^q \, dx \leq c(\mathbf{f}), \qquad q > N/2,
\tag{1.2}
$$

which was previously established by Frehse, Růžička (1994b) only for $N < 10$. Here, we prove that it is possible to shift the bound for N up to $N = 15$ and hence the same arguments as in Frehse and Růžička (1995) can be applied also here. Note that (1.2) expresses a weak version of the maximum principle for the head pressure $\frac{\mathbf{u}^2}{2} + p$, which was observed by Gilbarg and Weinberger (1974) in a different context.

2. PRELIMINARY RESULTS

Let us first introduce some notation. By $(L^q(\Omega), \|\cdot\|_{0,q})$, $(W^{k,q}(\Omega), \|\cdot\|_{k,q})$ we denote the usual Lebesgue, resp. Sobolev, spaces of periodic functions. The seminorm in the space of Hölder continuous functions $C^{0,\alpha}(\Omega)$ is denoted by $[\cdot]_{\alpha,\Omega}$. We denote the mean value of some function g by $\bar{g}$. By $\dot{X}$ we denote the subspace of functions from X with zero mean value. By c we denote a generic positive constant. The dependence of c on certain quantities is sometimes specified in brackets, e.g. $c = c(\mathbf{f})$ means that the constant depends only on the L^∞-norm of $\mathbf{f}$. Further we use $B_R(x_0) = \{x \in \mathbb{R}^N, |x - x_0| < R\}$, $T_R(x_0) = B_{2R}(x_0) \backslash B_R(x_0)$. Let $\zeta \in C_0^\infty(B_1(0))$ be a given monotonous function with $\zeta = \zeta(|x|)$, $0 \le \zeta \le 1$ and $\zeta \equiv 1$ on $B_{1/2}(0)$. For $x_0 \in \mathbb{R}^N$, $R > 0$ we define

$$\zeta_{R,x_0}(x) \equiv \zeta\left(\frac{x - x_0}{R}\right). \tag{2.1}$$

We shall omit the subscripts R and x_0 when there is no danger of confusion. Note that

$$\frac{\partial \zeta_{R,x_0}(x)}{\partial x_i} = \zeta'_{R,x_0}(x)\frac{(x - x_0)}{|x - x_0|}, \tag{2.2}$$

with

$$\left|\zeta'_{R,x_0}(x)\right| \le \frac{c}{R}$$
$$\text{supp}\,\zeta'_{R,x_0} \subseteq T_{R/2}(x_0). \tag{2.3}$$

The summation convention over repeated indices is used.

The arguments used here are similiar to that in Frehse and Růžička (1995), and thus we only briefly repeat the main points. Let $\Omega = (0, L)^N$, $10 \le N \le 15$ be a cube of length $L > 0$ and let us denote $\Gamma_j = \partial\Omega \cap \{x_j = 0\}$, $\Gamma_{j+N} = \partial\Omega \cap \{x_j = L\}$. We consider the steady Navier-Stokes equations

$$-\Delta\mathbf{u} + \mathbf{u} \cdot \nabla\mathbf{u} + \nabla p = \mathbf{f}$$
$$\text{div}\,\mathbf{u} = 0 \tag{2.4}$$

with periodic boundary conditions

$$\mathbf{u}|_{\Gamma_j} = \mathbf{u}|_{\Gamma_{j+N}}; \qquad p|_{\Gamma_j} = p|_{\Gamma_{j+N}} \qquad \forall j = 1, \dots, N$$
$$\frac{\partial\mathbf{u}}{\partial x_k}\Big|_{\Gamma_j} = \frac{\partial\mathbf{u}}{\partial x_k}\Big|_{\Gamma_{j+N}} \qquad \forall j, k = 1, \dots, N, \tag{2.5}$$

and subject to the requirement of zero average flow

$$\int_\Omega \mathbf{u}\,dx = 0. \tag{2.6}$$

We also require that the pressure p has mean value zero. For simplicity we assume that the mean value $\bar{\mathbf{f}}$ of the external force $\mathbf{f}$ is zero and that

$$\mathbf{f} \in L^\infty(\Omega).$$

By standard methods one easily proves the following

Proposition 2.1. *Let* $\mathbf{f} \in \dot{L}^\infty(\Omega)$. *Then there exists a weak solution* $\mathbf{u}, p$ *to* (2.4)–(2.6), *i.e.* $\mathbf{u} \in \dot{W}^{1,2}(\Omega)$ *with* $\operatorname{div} \mathbf{u} = 0$, *and* $p \in \dot{W}^{1,\frac{N}{N-1}}(\Omega)$ *satisfying for all* $\varphi \in C^\infty(\overline{\Omega})$

$$\int_\Omega \frac{\partial u_i}{\partial x_j} \frac{\partial \varphi_i}{\partial x_j} + u_j \frac{\partial u_i}{\partial x_j} \varphi_i + \frac{\partial p}{\partial x_i} \varphi_i \, dx = \int_\Omega f_i \, \varphi_i \, dx \tag{2.7}$$

such that

$$\|\mathbf{u}\|_{1,2} \le c(\mathbf{f})$$

$$\|p\|_{1,\frac{N}{N-1}} \le c(\mathbf{f}).$$

Because the final result is based on the Leray-Schauder theory we are only interested in apriori estimates and hence we shall assume in the sequel that $\mathbf{u}, p$ are smooth solutions to (2.4)–(2.6). Let us denote the head pressure ω by

$$\omega = \frac{\mathbf{u}^2}{2} + p. \tag{2.8}$$

Further, we have $g = g_+ + g_-$, where g_+ is the positive part of g and $g_- \le 0$ is the negative one. We will now study the properties of the Green type function $G = G_\delta$ solving

$$-\Delta G - \mathbf{u} \cdot \nabla G = \frac{\omega_+^s}{(1 + \delta\omega_+)^s} + k_0$$
$$\int_\Omega G \, dx = 0, \tag{2.9}$$

with periodic boundary conditions. The value $s \in \mathbb{R}^+$ will be specified later on. The weak formulation of (2.9) reads

$$\int_\Omega \nabla G \nabla \varphi \, dx - \int_\Omega \mathbf{u} \cdot \nabla G \varphi \, dx = \int_\Omega \frac{\omega_+^s}{(1 + \delta\omega_+)^s} \varphi \, dx + k_0 \int_\Omega \varphi \, dx, \quad \forall \varphi \in C^\infty(\Omega), \tag{2.10}$$

where

$$k_0 = -\frac{1}{|\Omega|} \int_\Omega \frac{\omega_+^s}{(1 + \delta\omega_+)^s} \, dx. \tag{2.11}$$

In the same way as in Frehse and Růžička (1995), one can prove

Proposition 2.2. *There exists a weak solution* $G = G_\delta \in \dot{W}^{1,2}(\Omega)$ *satisfying* (2.10), *together with the estimates*

$$\|G_-\|_{0,\infty} \le c(1 + |k_0|),$$
$$\|\nabla G\|_{0,N/(N-1)} \le c(1 + |k_0|), \tag{2.12}$$

where the constants are independent on δ.

3. A PRIORI ESTIMATES FOR ω

Now we want to use the Green type function G in order to obtain higher integrability for ω_+ in $L^r(\Omega)$, $r > N/2$. It is easy to derive from (2.7) the weak formulation for the head pressure $\omega = \frac{\mathbf{u}^2}{2} + p$

$$\int_\Omega \nabla \left(\frac{\mathbf{u}^2}{2} + p \right) \nabla \varphi \ + \ \mathbf{u} \cdot \nabla \left(\frac{\mathbf{u}^2}{2} + p \right) \varphi \, dx \tag{3.1}$$

$$= \int_\Omega (\nabla \mathbf{u} \circ \nabla \mathbf{u}) \, \varphi \, dx + \int_\Omega \mathbf{f} \cdot \nabla \varphi + \mathbf{f} \cdot \mathbf{u} \, \varphi \, dx \, ,$$

where $\nabla \mathbf{u} \circ \nabla \mathbf{u} = \dfrac{\partial u_i}{\partial x_j} \dfrac{\partial u_j}{\partial x_i}$. Inserting $\varphi = G$ into (3.1) and using $\nabla \mathbf{u} \circ \nabla \mathbf{u} - |\nabla \mathbf{u}|^2 \leq 0$ we get

$$\int_\Omega \nabla \omega \, \nabla G \ - \ \mathbf{u} \cdot \nabla G \, \omega \, dx$$

$$\leq \int_\Omega (\nabla \mathbf{u} \circ \nabla \mathbf{u} - |\nabla \mathbf{u}|^2) G_- \, dx + \int_\Omega \mathbf{f} \cdot \nabla G + \mathbf{f} \cdot \mathbf{u} \, G \, dx$$

Taking into account the weak formulation, (2.10), for G and the estimates (2.12), we obtain

$$\int_\Omega \frac{\omega_+^{s+1}}{(1 + \delta \omega_+)^s} \, dx \ \leq c + c|k_0| + \|\mathbf{f}\|_{0,\infty} \|\nabla G\|_{0,1} + |k_0| \int_\Omega |\omega| \, dx + \|\mathbf{f}\|_{0,\infty} \int_\Omega |\mathbf{u}| \, |G| \, dx$$

$$\leq c(1 + |k_0|) + c(\mathbf{f}) \int_\Omega |\mathbf{u}| \, |G| \, dx \, .$$

For k_0 defined in (2.11) one gets the estimate

$$|k_0| \leq c \left(\int_\Omega \frac{\omega_+^{s+1}}{(1 + \delta \omega_+)^s} \, dx \right)^{s/(s+1)} \left(\int_\Omega \frac{1}{(1 + \delta \omega_+)^s} \, dx \right)^{1/(s+1)} \, ,$$

which can be absorbed by the left–hand side by means of the Young inequality. Thus we have proved

$$\int_\Omega \frac{\omega_+^{s+1}}{(1 + \delta \omega_+)^{s+1}} \, dx \leq c(\mathbf{f}) \left(1 + \int_\Omega |\mathbf{u}| \, |G| \, dx \right) \, . \tag{3.2}$$

If we would already have proved an estimate of ω_+ in $L^r(\Omega)$, $r > N/2$, we would finally obtain an estimate of $\mathbf{u} \in L^q(\Omega)$, $q < 4$; see e.g. Frehse and Růžička (1994a). Thus we will try to derive an estimate of $\|G\|_{0,q'}$, $q' > 4/3$, in terms of $\|\mathbf{u}\|_{0,q}$ only. Let us therefore start with $\varphi = G|G|^\lambda$, $\lambda > -1$, in (2.10) in order to get first an estimate of G in terms of $\dfrac{\omega_+}{1 + \delta \omega_+}$. We obtain

$$\left(\int_\Omega |G|^{(\lambda+2)\frac{N}{N-2}} \, dx \right)^{\frac{N-2}{N}} \leq c \int_\Omega \frac{\omega_+^s}{(1 + \delta \omega_+)^s} |G|^{\lambda+1} \, dx$$

$$\leq c \left(\int_\Omega \left[\frac{\omega_+}{(1 + \delta \omega_+)} \right]^{s\frac{N(\lambda+2)}{N+2(\lambda+1)}} \right)^{\frac{N+2(\lambda+1)}{N(\lambda+2)}} \left(\int_\Omega |G|^{(\lambda+2)\frac{N}{N-2}} \, dx \right)^{\frac{(\lambda+1)(N-2)}{(\lambda+2)N}}$$

134

$$\leq \frac{K}{\varepsilon}\left(\int_{\Omega}\left[\frac{\omega_+}{(1+\delta\omega_+)}\right]^{s\frac{N(\lambda+2)}{N+2(\lambda+1)}}\right)^{\frac{N+2(\lambda+1)}{N}} + \varepsilon\left(\int_{\Omega}|G|^{(\lambda+2)\frac{N}{N-2}}\,dx\right)^{\frac{N-2}{N}}$$

and consequently

$$\|G\|_{0,\frac{N(\lambda+2)}{N-2}} \leq c\left\|\frac{\omega_+}{1+\delta\omega_+}\right\|_{s\cdot\frac{N(\lambda+2)}{N+2(\lambda+1)}}^s \quad . \tag{3.3}$$

However here we get a new restriction for λ from below, namely

$$\frac{N(\lambda+2)}{N-2} > \frac{4}{3}$$

or equivalently

$$\lambda > -\frac{2(N+4)}{3N} . \tag{3.4}$$

Using (3.3) together with the Hölder inequality in (3.2) we get

$$\left\|\frac{\omega_+}{1+\delta\omega_+}\right\|_{0,s+1}^{s+1} \leq c(\mathbf{f})\left(1 + \|\mathbf{u}\|_{\frac{N(\lambda+2)}{N(\lambda+1)+2}}\left\|\frac{\omega_+}{1+\delta\omega_+}\right\|_{s\cdot\frac{N(\lambda+2)}{N+2(\lambda+1)}}^s\right) . \tag{3.5}$$

Requiring now

$$s+1 = s\frac{N(\lambda+2)}{N+2(\lambda+1)}$$

we obtain

$$s = \frac{N+2(\lambda+1)}{(N-2)(\lambda+1)}$$

$$s+1 = \frac{N(\lambda+2)}{(N-2)(\lambda+1)}$$

and hence from (3.5) and Young's inequality

$$\left\|\frac{\omega_+}{1+\delta\omega_+}\right\|_{\frac{N(\lambda+2)}{(N-2)(\lambda+1)}} \leq c(\mathbf{f})(1 + \|\mathbf{u}\|_{\frac{N(\lambda+2)}{N(\lambda+1)+2}}) .$$

Denoting now for simplicity

$$\frac{N(\lambda+2)}{N(\lambda+1)+2} = q,$$

we get

$$\frac{N(\lambda+2)}{(N-2)(\lambda+1)} = \frac{Nq}{N-2q},$$

for which we want that

$$\frac{Nq}{N-2q} > \frac{N}{2} \tag{3.6}$$

or equivalently

$$q > \frac{N}{4} . \tag{3.7}$$

Rewriting also (3.4) in terms of q we get

$$q < 4 \,. \tag{3.8}$$

Both (3.7) and (3.8) immediately imply $N < 16$. Letting $\delta \to 0^+$ we therefore proved

Proposition 3.3. *Let* $10 \leq N \leq 15$ *and let* $\mathbf{u}, p$ *be a smooth solution to* (2.4)–(2.6). *Then we have for* $q \in (N/4, 4)$

$$\|\omega_+\|_{0, \frac{Nq}{N-2q}} \leq c(\mathbf{f})(1 + \|\mathbf{u}\|_{0,q}) \,. \tag{3.9}$$

4. REGULARITY FOR u AND p

Based on the estimate (3.9) we can now derive weighted L^q–estimates for ω. Namely we have

Proposition 4.1. *Let* $10 \leq N \leq 15$ *and let* $\mathbf{u}, p$ *be a smooth solution to* (2.4)–(2.6). *Then we have for all* $x_0 \in \Omega$ *and all* $q \in (N/4, 4)$

$$\int_{\Omega} \left| \frac{\mathbf{u}^2}{2} + p \right| \frac{1}{|x - x_0|^{N-2}} \, dx \leq c(\mathbf{f})(1 + \|\omega_+\|_{0, \frac{Nq}{N-2q}}) \,. \tag{4.1}$$

Proof: Using in (2.7) the test function

$$\varphi_i = \frac{\partial}{\partial x_i} \frac{\zeta^2}{|x - x_0|^{N-4}} \,,$$

where $\zeta = \zeta_{R,x_0}$ we get after some rearrangements, using also

$$\left(\frac{\mathbf{u}^2}{2} + p \right) = 2 \left(\frac{\mathbf{u}^2}{2} + p \right)_+ - \left| \frac{\mathbf{u}^2}{2} + p \right| \,,$$

the relation

$$(N-4) \cdot (N-2) \cdot \int_{\Omega} \frac{(\mathbf{u} \cdot (x - x_0))^2}{|x - x_0|^N} \zeta^2 \, dx + 2(N-4) \int_{\Omega} \left| \frac{\mathbf{u}^2}{2} + p \right| \frac{\zeta^2}{|x - x_0|^{N-2}} \, dx =$$

$$= 4(N-4) \int_{\Omega} \left(\frac{\mathbf{u}^2}{2} + p \right)_+ \frac{\zeta^2}{|x - x_0|^{N-2}} \, dx + \int_{\Omega} u_i \, u_j \frac{\partial^2 \zeta^2}{\partial x_i \partial x_j} \frac{1}{|x - x_0|^{N-4}} \, dx$$

$$+ 2(N-4) \int_{\Omega} u_i \, u_j \frac{(x - x_0)_i}{|x - x_0|^{N-2}} \frac{\partial \zeta^2}{\partial x_j} \, dx - \int_{\Omega} p \Delta \zeta^2 \frac{1}{|x - x_0|^{N-4}} \, dx$$

$$+ 2(N-4) \int_{\Omega} p \frac{(x - x_0)_i}{|x - x_0|^{N-2}} \frac{\partial \zeta^2}{\partial x_i} \, dx - \int_{\Omega} f_i \frac{\partial \zeta^2}{\partial x_i} \frac{1}{|x - x_0|^{N-4}} \, dx$$

$$+ (N-4) \int_{\Omega} f_i \frac{(x - x_0)_i}{|x - x_0|^{N-2}} \zeta^2 \, dx \,.$$

Using (3.9) we get

$$\int_\Omega \frac{(\mathbf{u} \cdot (x - x_0))^2}{|x - x_0|^N} \zeta^2 \, dx + \int_\Omega \left| \frac{\mathbf{u}^2}{2} + p \right| \frac{\zeta^2}{|x - x_0|^{N-2}} \, dx \leq c(\mathbf{f}, R) \left(1 + \|\omega_+\|_{0, \frac{Nq}{N-2q}} \right),$$

which is a local version of (4.1). Due to the periodicity of the problem we immediately get (4.1) by a covering argument. □

Using instead of

$$\varphi_i = \frac{\partial}{\partial x_i} \frac{\zeta^2}{|x - x_0|^{N-4}}$$

the test function

$$\varphi_i = \frac{\partial}{\partial x_i} \frac{\zeta^2}{|x - x_0|^{N-4-\varepsilon}} \quad \varepsilon > 0 \,,$$

in the weak formulation (2.7) we immediately get

Corollary. Let $10 \leq N \leq 15$ and let $\mathbf{u}, p$ be a smooth solution to (2.4)–(2.6). Then we have for all $x_0 \in \Omega$ and all $r \in [1, N-2)$

$$\int_\Omega |p| \frac{1}{|x - x_0|^r} \, dx \;\leq\; c(\mathbf{f}) \left(1 + \|\omega_+\|_{0, \frac{Nq}{N-2q}} \right), \tag{4.2}$$

$$\int_\Omega \mathbf{u}^2 \frac{1}{|x - x_0|^r} \, dx \;\leq\; c(\mathbf{f}) \left(1 + \|\omega_+\|_{0, \frac{Nq}{N-2q}} \right). \tag{4.3}$$

Now we have all prepared for the following

Theorem 4.1. Let $10 \leq N \leq 15$ and let $\mathbf{u}, p$ be a smooth solution to (2.4)–(2.6). Then we have for all $q \in (N/4, 4)$

$$\int_\Omega |\mathbf{u}|^q \, dx \leq c(\mathbf{f}) \,. \tag{4.4}$$

Proof: First, notice that φ_0 given by

$$\varphi_0(x_0) = \int_\Omega \frac{1}{|x - x_0|^{N-2}} |p|^{\frac{q-2}{2}} \operatorname{sgn} p \, dx \,, \tag{4.5}$$

satisfies

$$-\Delta \varphi_0 = |p|^{\frac{q-2}{2}} \operatorname{sgn} p \,.$$

Due to (4.2), (3.9) and the Hölder inequality, we get for $r \in ((N-2)/2, N-2)$

$$|\varphi_0(x_0)| \leq \left(\int_\Omega \frac{|p|}{|x - x_0|^r} \, dx \right)^{\frac{q-2}{2}} \left(\int_\Omega \frac{1}{|x - x_0|^{[N-2-r\frac{q-2}{2}]\frac{2}{4-q}}} \right)^{\frac{4-q}{2}}$$

$$\leq c(\mathbf{f}) \left(1 + \|\mathbf{u}\|_{0,q}^{\frac{q-2}{2}} \right)$$

and thus

$$\|\varphi_0\|_{0,\infty} \leq c(\mathbf{f}) \left(1 + \|\mathbf{u}\|_{0,q}^{\frac{q-2}{2}} \right).$$

Further we have $\|p\|_{0, N/(N-2)} \leq c(\mathbf{f})$ and hence

$$\|\nabla \varphi_0\|_{0,1} \leq c(\mathbf{f}) \,.$$

Using in (2.7) $\varphi = \nabla\varphi_0$ we get also using the definition of φ_0 (4.5)

$$\int_\Omega |p|^{q/2}\, dx = \int_\Omega \nabla\mathbf{u}\circ\nabla\mathbf{u}\,\varphi_0\, dx + \int_\Omega \mathbf{f}\cdot\nabla\varphi_0\, dx$$

$$\leq c(\mathbf{f})\|\varphi_0\|_{0,\infty} + \|\mathbf{f}\|_{0,\infty}\|\nabla\varphi_0\|_{0,1}$$

$$\leq c(\mathbf{f})\left(1 + \|\mathbf{u}\|_{0,q}^{\frac{q-2}{2}}\right).$$

Since $\frac{\mathbf{u}^2}{2} \leq \omega_+ + |p|$ we get

$$\int_\Omega |\mathbf{u}|^q\, dx \leq c\int_\Omega |p|^{q/2}\, dx + c\int_\Omega \omega_+^{q/2}\, dx$$

$$\leq c(\mathbf{f})\left(1 + \|\mathbf{u}\|_{0,q}^{\frac{q-2}{2}}\right) + c\|\omega_+\|_{0,\frac{Nq}{N-2q}}^{q/2}. \tag{4.6}$$

From (4.6), (3.9) and Young's inequality we finally get

$$\|\mathbf{u}\|_{0,q} \leq c(\mathbf{f})\left(1 + \|\mathbf{u}\|_{0,q}^{\frac{q-2}{2q}}\right) + c(\mathbf{f})\left(1 + \|\mathbf{u}\|_{0,q}^{\frac{1}{2}}\right)$$

$$\leq c(\mathbf{f}) + \tfrac{1}{2}\|\mathbf{u}\|_{0,q},$$

which proves the theorem. $\qquad\qquad\qquad\qquad\qquad\qquad\qquad\qquad\square$

Combining now Theorem 4.1., Proposition 3.3. and (3.6) we obtain

Theorem 4.2. *Let* $10 \leq N \leq 15$ *and let* $\mathbf{u}, p$ *be a smooth solution to* (2.4)–(2.6). *Then we have for some* $r > N/2$

$$\|\omega_+\|_r \leq c(\mathbf{f}). \tag{4.7}$$

With the estimate (4.7) in hand we can repeat the arguments from Frehse, Růžička (1995). Therefore, we finally obtain

Theorem 4.3. *Let* $10 \leq N \leq 15$ *and let* $\mathbf{u}, p$ *be a smooth solution to* (2.4)–(2.6). *Then there exists a* $\alpha > 0$, *such that*

$$[\mathbf{u}]_{\alpha,\Omega} \leq c(\mathbf{f}),$$

$$[p]_{2\alpha,\Omega} \leq c(\mathbf{f}).$$

Now the Leray-Schauder theory gives

Theorem 4.4. *Let* $10 \leq N \leq 15$ *and let* $\mathbf{f} \in \dot{L}^\infty(\Omega)$. *Then there exists a regular solution* $\mathbf{u}, p$ *to* (2.4)–(2.6), *i.e. for all* $q \in [1,\infty)$

$$\mathbf{u} \in W^{2,q}(\Omega),$$
$$p \in W^{1,q}(\Omega).$$

REFERENCES

FREHSE, J. and M. RŮŽIČKA (1994a), On the regularity of the stationary Navier–Stokes equations, *Ann. Scu. Norm. Pisa* **21**, 63-95.

FREHSE, J. and M. RŮŽIČKA (1994b), Weighted Estimates for the Stationary Navier–Stokes Equations, *Acta Appl. Mathematicae* **37**.

FREHSE, J. and M. RŮŽIČKA (1995), Existence of Regular Solutions to the Stationary Navier–Stokes Equations, MATH. ANN., (to appear).

GILBARG, D. and H.F. WEINBERGER (1974), Asymptotic properties of Leray's solution of the stationary two-dimensional Navier-Stokes equations, *Russ. Math. Surv.* **29**, 109-123.

STRUWE, M. (1995), Regular Solutions of the Stationary Navier–Stokes equations on R^5, MATH. ANN., (to appear).

A NOTE ON DERIVATIVE ESTIMATES FOR A HOPF SOLUTION TO THE NAVIER-STOKES SYSTEM IN A THREE-DIMENSIONAL CUBE

J. Málek[1], M. Padula[2], M. Růžička[3]

[1]Mathematical Institute of Charles University
Sokolovská 83, 18600 Prague 8, Czech Republic

[2] Università di Ferrara, Dipartimento di Matematica
Via Machiavelli, 35, 44100 Ferrara, Italy

[3] University of Bonn, Institute of Applied Mathematics
Beringstraße 4–6, 53115 Bonn, Germany

ABSTRACT

In this paper we derive apriori estimates for the first (and second) derivatives of a weak solution to the three–dimensional Navier–Stokes equations.

1. INTRODUCTION

Let L and T be positive constants. We denote by $\Omega \equiv [0, L]^3$ a cube in $\mathbb{R}^3$ and by $I \equiv (0, T)$ a time interval. For $\mathbf{u} : I \times \Omega \mapsto \mathbb{R}^3$ and $\pi : I \times \Omega \mapsto \mathbb{R}$, let us consider the Navier–Stokes system in the form

$$\operatorname{div} \mathbf{u} = 0 \qquad (1)_1$$

$$\frac{\partial \mathbf{u}}{\partial t} + [\nabla \mathbf{u}]\mathbf{u} = -\nabla \pi + \Delta \mathbf{u} \qquad (1)_2$$

completed partly by space periodic requirements

$$\begin{aligned} \mathbf{u}(t, \mathbf{x} + L\mathbf{e}^i) &= \mathbf{u}(t, \mathbf{x}) \\ \pi(t, \mathbf{x} + L\mathbf{e}^i) &= \pi(t, \mathbf{x}) \end{aligned} \qquad \forall\, i = 1, 2, 3 \qquad (1)_3$$

($\mathbf{e}^i$ being the unit vector at the i-th direction) and by an initial condition

Navier-Stokes Equations and Related Nonlinear Problems
Edited by A. Sequeira, Plenum Press, New York, 1995

$$\mathbf{u}(0, \mathbf{x}) = \mathbf{u}_0(\mathbf{x}) \tag{1}_4$$

where $\mathbf{u}_0 : \Omega \mapsto I\!\!R^3$ is supposed to be a smooth divergence–free function.

Since Leray's paper (1934), the global existence of a weak (Hopf) solution to (1)

$$\mathbf{u} \in C_w(I; V^0) \cap L^2(I; V^1) \tag{2}$$

is well known[1]. However, the question whether this solution is smooth for all $t \in I$ remains open.

The opposite assertions are valid for a so–called strong solution $\mathbf{v}$ to (1) for which

$$\mathbf{v} \in C_w(I; V^1) \cap L^2(I; V^2) \,. \tag{3}$$

Such $\mathbf{v}$ has to be smooth in all times of its existence, but this existence is known only for a certain (short) interval $(0, t^*)$, where $t^* = t^*(\|\nabla \mathbf{u}_0\|_2)$. As far as $\mathbf{v}(0, \cdot) = \mathbf{u}(0, \cdot) = \mathbf{u}_0(\cdot)$, then the strong solution $\mathbf{v}$ coincides with the weak solution $\mathbf{u}$ on $[0, t^*)$. We refer to Constantin and Foias (1988), Chap. 8–10, for proofs of above mentioned statements.

Various attempts have been done to obtain global in time estimates for the second derivatives of $\mathbf{u}$. Firstly, Foias, Guillopé and Temam (1981) have proved that

$$\int_0^T \|D^{(2)}\mathbf{u}(t)\|_2^{2/3} dt < \infty \,. \tag{4}_1$$

This result has been obtained also in Málek, Nečas and Růžička (1993) as a by–product of a method providing the global existence of solution for a certain class of non–Newtonian fluids (so–called fluids with shear dependent viscosities). Recently, Constantin (1990) has constructed a weak solution satisfying

$$D^{(2)}\mathbf{u} \in L^q(I; L^q(\Omega)^{3\times3\times3}) \qquad \forall q < \frac{4}{3} \,. \tag{4}_2$$

Finally, Giga and Sohr (1991) have revealed that for any weak solution

$$\int_0^\infty \|D^{(2)}\mathbf{u}(t)\|_q^{\frac{2q}{4q-3}} dt < C \,, \tag{4}_3$$

whenever $q \in (1, 3/2)$.

Regarding the first derivatives of $\mathbf{u}$, it is known that

$$\nabla \mathbf{u} \in L^\infty(I; L^1(\Omega)^{3\times3}) \,, \tag{5}_1$$

which is a consequence of the result due to Coifman, Lions, Meyer and Semmes (1993) saying that $D^{(2)}\pi \in L^1(I; \mathcal{H}^1(\Omega)^{3\times3})$, where $\mathcal{H}^1$ is the Hardy space. From (2), we know also

$$\nabla \mathbf{u} \in L^\infty(I; L^2(\Omega)^{3\times3}) \,, \tag{5}_2$$

and it follows from $(4)_1$ that

$$\int_0^T \|\nabla \mathbf{u}(t)\|_6^{2/3} dt < \infty \,. \tag{5}_3$$

---

[1] Here, V^k denotes the closure of smooth space–periodic divergence–free functions, having mean value zero, with respect to the norm of $W^{k,2}(\Omega)^3$, $k = 0, 1, 2, \ldots$.

If X is a Hilbert space, then $C_w(I; X)$ consists of functions $u : I \mapsto X$ such that for almost all $t_0 \in I$ and for all $h \in X$ $(u(t), h)_X \to (u(t_0), h)_X$ as $t \to t_0$.

By $\| \cdot \|_q$ we mean the norm of the Lebesgue spaces $L^q(\Omega)$ or $L^q(\Omega)^3$ or $L^q(\Omega)^{3\times3}$. The symbol $(\cdot, \cdot)$ denotes the scalar product in $L^2(\Omega)$ or its multidimensional extensions.

The objective of this paper is to show that for arbitrary $q \in [2, \infty)$ there exists a constant $C(q)$ independent on T such that

$$\int_0^T \|\nabla \mathbf{u}(t)\|_q^{\frac{q}{2q-3}} \, dt < C(q) . \tag{6}$$

Till now, we do not know if the estimates (6) can be handled independently of q. Assuming that C in (6) does not depend on q, we obtain (letting $q \longrightarrow \infty$)

$$\int_0^T \|\nabla \mathbf{u}(t)\|_\infty^{\frac{1}{2}} dt < C(q) . \tag{7}$$

Let us finally note that our approach is based on the energy method, where we use the nonlinear p–Laplace operator as a test function.

2. THEOREM AND ITS PROOF

Let $p \geq 1$. We denote

$$\Delta_p \mathbf{u} = \mathrm{div}\,(|\nabla \mathbf{u}|^{p-2}\nabla \mathbf{u}) = \sum_k \frac{\partial}{\partial x_k}(|\nabla \mathbf{u}|^{p-2}\frac{\partial \mathbf{u}}{\partial x_k}),$$

$$I_p(\mathbf{u}) = \int_\Omega |\nabla \mathbf{u}|^{p-2}|D^{(2)}\mathbf{u}|^2 \, dx .$$

Then for $q \in [1, 2]$ there exists $C > 0$ such that

$$\|\nabla \mathbf{u}\|_{\frac{3p}{3-q}} \leq C \left(\frac{p}{q}\right)^{\frac{q}{p}} \{I_p(\mathbf{u})\}^{\frac{q}{2p}} \|\nabla \mathbf{u}\|_p^{\frac{2-q}{2}} . \tag{8$_1$}$$

Putting $q = 2$ in (8)$_1$, we get

$$\|\nabla \mathbf{u}\|_{3p}^p \leq C \, I_p(\mathbf{u}) \tag{8$_2$}$$

Also, it holds for $p \leq 2$

$$\|D^{(2)}\mathbf{u}\|_p^2 \leq C(1 + \|\nabla \mathbf{u}\|_p)^{2-p} I_p(\mathbf{u}) . \tag{9}$$

The proofs of (8)–(9) can be found in Málek, Nečas, Rokyta and Růžička (1995), for example.

Theorem 1.1. *Let $\mathbf{u}$ be a Hopf solution to the Navier–Stokes system (1). Then (6) holds for all $q \in [2, \infty)$.*

Proof: Let us take $\mathbf{u}_0$ (which is smooth, for example from V^1) arbitrary, but fixed. Then there exists $t^* = t^*(\|\nabla \mathbf{u}_0\|)$ such that $\mathbf{u}$ is smooth on $[0, t^*)$. Considering (1)$_2$ on $[0, t^*)$, we can multiply each term by $-\Delta_p \mathbf{u}$ and integrate over Ω. For $p > 1$, the resulting inequality reads

$$\frac{1}{p}\frac{d}{dt}\|\nabla \mathbf{u}\|_p^p + C(p-1)I_p(\mathbf{u}) \leq C\|\nabla \mathbf{u}\|_{p+1}^{p+1}, \tag{10}$$

where we have applied many times Green's theorem and used the pressure equation

$$-\Delta \pi = \sum_{i,j} \frac{\partial u_i}{\partial x_j}\frac{\partial u_j}{\partial x_i} . \tag{11}$$

143

More precisely,

$$-(\frac{\partial \mathbf{u}}{\partial t}, \Delta_p \mathbf{u}) = (\frac{\partial \nabla \mathbf{u}}{\partial t}, |\nabla \mathbf{u}|^{p-2} \nabla \mathbf{u}) = \frac{1}{p} \frac{d}{dt} \|\nabla \mathbf{u}\|_p^p, \tag{12}_1$$

$$(\Delta \mathbf{u}, \Delta_p \mathbf{u}) = \sum_k (\frac{\partial}{\partial x_k} \nabla \mathbf{u}, \frac{\partial}{\partial x_k} |\nabla \mathbf{u}|^{p-2} \nabla \mathbf{u}) \geq C(p-1) I_p(\mathbf{u}), \tag{12}_2$$

$$-([\nabla \mathbf{u}]\mathbf{u}, \Delta_p \mathbf{u}) = \sum_{i,j,k} \int_\Omega \frac{\partial u_i}{\partial x_j} \frac{\partial u_j}{\partial x_k} |\nabla \mathbf{u}|^{p-2} \frac{\partial u_i}{\partial x_k} \, dx. \tag{12}_3$$

The last equality holds since

$$\sum_{i,j,k} \int_\Omega \frac{\partial^2 u_i}{\partial x_k \partial x_j} u_j |\nabla \mathbf{u}|^{p-2} \frac{\partial u_i}{\partial x_k} \, dx = \frac{1}{p} \sum_j \int_\Omega \frac{\partial}{\partial x_j} |\nabla \mathbf{u}|^p u_j = -\frac{1}{p} \int_\Omega |\nabla \mathbf{u}|^p \operatorname{div} \mathbf{u} \, dx = 0.$$

Further,

$$(\nabla \pi, \Delta_p \mathbf{u}) = \sum_{i,k} (\frac{\partial^2 \pi}{\partial x_i \partial x_k}, |\nabla \mathbf{u}|^{p-2} \frac{\partial u_i}{\partial x_k}) \equiv Y.$$

By Hölder's inequality

$$|Y| \leq \|D^{(2)} \pi\|_{\frac{p+1}{2}} \|\nabla \mathbf{u}\|_{p+1}^{p-1}.$$

From the regularity of $\mathbf{u}$ on $(0, t^*)$ and (11), we know that for $p > 1$

$$\|D^{(2)} \pi\|_{\frac{p+1}{2}} \leq C \|\nabla \mathbf{u}\|_{p+1}^2.$$

Hence

$$|(\nabla \pi, \Delta_p \mathbf{u})| \leq C \|\nabla \mathbf{u}\|_{p+1}^{p+1}. \tag{12}_4$$

The inequality (10) then easily follows from $(12)_{1-4}$.

Using two interpolation inequalities

$$\|\nabla \mathbf{u}\|_{p+1}^{p+1} \leq \|\nabla \mathbf{u}\|_2^{\frac{2(2p-1)}{3p-2}} \|\nabla \mathbf{u}\|_{3p}^{\frac{3p(p-1)}{3p-2}}, \tag{13}_1$$

$$\|\nabla \mathbf{u}\|_{p+1}^{p+1} \leq \|\nabla \mathbf{u}\|_p^{\frac{2p-1}{2}} \|\nabla \mathbf{u}\|_{3p}^{\frac{3}{2}}, \tag{13}_2$$

we can estimate the right–hand side of (10) by means of (13) as follows: if $\alpha \in [0,1]$ then

$$
\begin{aligned}
\|\nabla \mathbf{u}\|_{p+1}^{p+1} &= \|\nabla \mathbf{u}\|_{p+1}^{\alpha(p+1)+(1-\alpha)(p+1)} \\
&\leq \|\nabla \mathbf{u}\|_2^{\frac{2\alpha(2p-1)}{3p-2}} \|\nabla \mathbf{u}\|_p^{(1-\alpha)\frac{2p-1}{2}} \|\nabla \mathbf{u}\|_{3p}^{\frac{3(1-\alpha)}{2}+\frac{3\alpha p(p-1)}{3p-2}} \\
&\leq C \|\nabla \mathbf{u}\|_2^{\frac{2\alpha(2p-1)}{3p-2}} \|\nabla \mathbf{u}\|_p^{(1-\alpha)\frac{2p-1}{2}} \{I_p(\mathbf{u})\}^{\frac{3(1-\alpha)}{2p}+\frac{3\alpha(p-1)}{3p-2}},
\end{aligned}
\tag{14}
$$

where the last inequality is due to $(8)_2$. The strategy is clear: we combine (14) with (10) and we want to apply Young's inequality in order to move $I_p(\mathbf{u})$ onto the left–hand side of (10). Simultaneously, we demand that the exponent at $\|\nabla \mathbf{u}\|_2$ is equal to 2. The exponent at $\|\nabla \mathbf{u}\|_p^p$, say λ, is already determined from these requirements. Thus, we have

$$
\begin{aligned}
\frac{2\alpha(2p-1)}{3p-2} \delta &= 2, \\
(\frac{3(1-\alpha)}{2p} + \frac{3\alpha(p-1)}{3p-2}) \delta' &= 1, \\
\frac{1}{\delta} + \frac{1}{\delta'} &= 1
\end{aligned}
\tag{15}
$$

and
$$\lambda p = \frac{2p-1}{2}(1-\alpha)\,\delta\,. \tag{16}$$

Elementary calculations in (15) imply
$$\alpha = \frac{(2p-3)(3p-2)}{(5p-6)(2p-1)}\,, \qquad 1-\alpha = \frac{4p(p-1)}{(5p-6)(2p-1)}$$

and
$$\delta = \frac{5p-6}{2p-3}\,, \qquad \delta' = \frac{1}{3}\frac{5p-6}{p-1}\,.$$

Notice that p has to be greater than $\frac{3}{2}$ in order to $\alpha \in [0,1]$ and $\delta,\ \delta' > 1$. Finally, we obtain from (16)
$$\lambda = 2\frac{p-1}{2p-3}\,,$$

and (10) is transformed into
$$\frac{1}{p}\frac{d}{dt}(1+\|\nabla\mathbf{u}\|_p^p) + c\,I_p(\mathbf{u}) \leq C(1+\|\nabla\mathbf{u}\|_p^p)^\lambda\|\nabla\mathbf{u}\|_2^2\,. \tag{17}$$

Note that $\lambda \in [2,\infty)$ for $p \in (\frac{3}{2},2]$. Multiplying (17) by $(1+\|\nabla\mathbf{u}\|_p^p)^{-\lambda}$, integrating between 0 and t, $t \leq t^*$, we get
$$\int_0^t \frac{I_p(\mathbf{u})}{(1+\|\nabla\mathbf{u}\|_p^p)^\lambda}\,d\tau \leq C\int_0^t \|\nabla\mathbf{u}\|_2^2 d\tau + \frac{2}{p} \leq \int_0^T \|\nabla\mathbf{u}\|_2^2 d\tau + \frac{2}{p} \equiv C_0\,. \tag{18}$$

Since C_0 is independent of t, we can extend the validity of (18) for all $t \in \overline{I}$. Particularly,
$$\int_0^T \frac{I_p(\mathbf{u})}{(1+\|\nabla\mathbf{u}\|_p^p)^\lambda}\,d\tau \leq C_0\,. \tag{19}$$

Taking $(8)_2$ into consideration and using the interpolation inequality $(p > 2)$
$$\|\nabla\mathbf{u}\|_p \leq \|\nabla\mathbf{u}\|_2^{\frac{4}{3p-2}}\|\nabla\mathbf{u}\|_{3p}^{\frac{3(p-2)}{3p-2}}\,, \tag{20}$$

we can conclude from (19) that there exists C (depending on p) such that
$$\int_0^T \|\nabla\mathbf{u}(t)\|_{3p}^{\frac{p}{2p-1}}\,dt < C \tag{21}$$

Indeed, we observe first, using $(8)_2$ and (20), that
$$\int_0^T \|\nabla\mathbf{u}\|_{3p}^{p\beta}\,dt \leq C\int_0^T \left\{\frac{I_p(\mathbf{u})}{(1+\|\nabla\mathbf{u}\|_p^p)^\lambda}\right\}^\beta (1+\|\nabla\mathbf{u}\|_p^p)^{\lambda\beta}\,dt$$
$$\leq C\int_0^T \left\{\frac{I_p(\mathbf{u})}{(1+\|\nabla\mathbf{u}\|_p^p)^\lambda}\right\}^\beta\,dt$$
$$+C\int_0^T \left\{\frac{I_p(\mathbf{u})}{(1+\|\nabla\mathbf{u}\|_p^p)^\lambda}\right\}^\beta \|\nabla\mathbf{u}\|_2^{\lambda\beta\frac{4p}{3p-2}}\|\nabla\mathbf{u}\|_{3p}^{\lambda\beta\frac{3p(p-2)}{3p-2}}\,dt\,.$$

It suffices to estimate the last integral, increased by means of $(8)_2$, by Hölder's inequality requiring
$$1 = \frac{1}{\delta} + \frac{1}{\delta'} + \frac{1}{\delta''}\,, \quad \beta\delta = 1\,, \quad \lambda\beta\frac{4p}{3p-2}\delta' = 2\,, \quad \lambda\beta\frac{3p(p-2)}{3p-2}\delta'' = p\beta\,.$$

An elementary calculus gives

$$\beta = \frac{1}{2p - 1},$$

which is (21). Setting $q = 3p$, we get (6) for $q > 6$. For $q \in [2, 6]$ (6) is a consequence (by the interpolation) of $(5)_2$ and $(5)_3$.

The proof of the Theorem is complete. $\qquad\square$

Remark 1.1. It might seem that the use of $(8)_1$ instead of $(8)_2$ could lead to some improvements. But this is not true. In such a case, though α depends on q, the other parameters as δ, δ', λ and β are independent of q and coincide with those calculated above.

Remark 1.2. Taking (9) and $(5)_1$ into consideration, it is possible to derive from (19) that whenever $p \in (\frac{3}{2}, 2]$

$$\int_0^T \|D^{(2)}\mathbf{u}(t)\|_p^{\frac{2p(2p-3)}{7p^2 - 14p + 6}} \, dt < \infty. \tag{22}$$

But (22) is worse than the result obtained by a straightforward interpolation of $(4)_1$ and $(4)_2$.

Conclusion. We have derived new estimates (6) for a Hopf solution to the Navier–Stokes system for space period problem in three dimensions. Despite of our effort, we did not extend $(4)_2$ for $q = \frac{4}{3}$. None of the estimates (4)–(6) implies the regularity of a weak solution.

REFERENCES

COIFMAN, R., LIONS, P.L., MEYER, Y. and S. SEMMES (1993), Compensated Compactness and Hardy Spaces, *J. Math. Pures Appl.* **72**, 247-286.

CONSTANTIN, P. (1990), Remarks on the Navier-Stokes equations, in: *New Perspectives in Turbulence*, L.Sirovich (ed.), (Springer, New York).

CONSTANTIN, P. and C. FOIAS (1988), Navier-Stokes Equations, (The University of Chicago Press, Chicago).

FOIAS, C., GUILLOPÉ, C. and R. TEMAM (1981), New a priori estimates for Navier-Stokes equations in dimension 3, *Comm. in PDE* **6**, 329-359.

GIGA, Y. and H. SOHR (1991), Abstract L^p Estimates for the Cauchy Problem with Applications to the Navier–Stokes Equations in Exterior Domains, *J. Funct. Anal.* **102**, 72-94

LERAY, J. (1934), Sur le mouvement d'un liquide visqueux emplissant l'espace, *Acta Mathematica* **63**, 193-248.

MÁLEK, J., NEČAS, J. and M. RŮŽIČKA (1993), On the Non-Newtonian Incompressible Fluids, M^3AS **3**, 35-63.

MÁLEK, J., NEČAS, J., ROKYTA, M. and M. RŮŽIČKA (1995), *Weak and Measure–valued Solutions to Evolutionary Partial Differential Equations*, Applied Mathematics and Mathematical Computation, (Chapman and Hall), to appear.

CLASSICAL SOLUTION TO THE NAVIER-STOKES SYSTEM

Paolo Maremonti

Università della Basilicata, Dipartimento di Matematica
Via Nazario Sauro, 85, 85100 - Potenza, Italy

1. INTRODUCTION

In this paper we consider in $\Omega \subset \mathbb{R}^n$ ($n = 2, 3$) the boundary-value problem for the stationary Navier-Stokes system. The domain Ω is a bounded domain, whose boundary $\partial\Omega$ is assumed to be of class C^2. So far (cf. Galdi, 1994; Ladyzhenskaya, 1969) the boundary-value problem has been considered assuming that the boundary data $\mathbf{a}(x)$ has suitable properties of differentiability on $\partial\Omega$ (usually it is assumed that $\mathbf{a}(x) \in H^{\frac{1}{2}}(\partial\Omega)$). Here we try to give an existence theorem without making these assumptions on $\mathbf{a}(x)$. In fact, we prove that the boundary-value problem admits smooth solutions $(\mathbf{v}(x), \pi(x)) \in C^2(\Omega) \cap C(\overline{\Omega}) \times C^1(\Omega)$ under the sole hypothesis that $\mathbf{a}(x) \in C(\partial\Omega)$.

2. STATEMENT OF THE PROBLEM

In order to give our statement, we need to make some preliminary assumptions. We consider the following Navier-Stokes system

$$\mathbf{v}(x) \cdot \nabla\mathbf{v}(x) = \nu\Delta\mathbf{v}(x) + \nabla\pi(x), \quad \nabla \cdot \mathbf{v}(x) = 0, \qquad \text{in } \Omega. \tag{1}$$

To system (1) we append the boundary condition

$$\mathbf{v}(x)_{|\partial\Omega} = \mathbf{a}(x) \qquad \text{with} \qquad \int_{\partial\Omega} \mathbf{a}(x) \cdot \vec{n} d\sigma = 0. \tag{2}$$

A *classical solution* of system (1) is a pair $(\mathbf{v}(x), \pi(x)) \in C^2(\Omega) \cap C(\overline{\Omega}) \times C^1(\Omega)$.

Condition $(2)_2$ must be strenghtened in the case of multiconnected domains. If $\Omega^c = \cup_{h=1}^m \Omega_h$, with Ω_h connected components of the domain Ω, then it is usually assumed that $(2)_2$ is satisfied by means of $\int_{\partial\Omega_h} \mathbf{a}(x) \cdot \vec{n} d\sigma = 0, \forall h = 1, ..., m$. As is well known, this assumption is an undesired fact from the physical point of view. Recently some authors (see Borchers and Pileckas, 1994; Galdi, 1994) have modified the above assumption making the following one

$$\sum_{h=1}^m C_h | \int_{\partial\Omega_h} \mathbf{a}(x) \cdot \vec{n} d\sigma | < \nu, \qquad \text{with} \qquad \mathbf{a}(x) \in H^{\frac{1}{2}}(\partial\Omega). \tag{3}$$

For further details on this question we refer the reader to Galdi and Simader (1990). Here we only stress that our theorem works under an assumption which is analogous to relation (3). More precisely we have the following theorem

Theorem 2.1. *Let* $\mathbf{a}(x) \in C(\partial\Omega)$ *and assume that one of the following alternative conditions is valid for* $\mathbf{a}(x)$:

$$i) \quad Cd\nu^{-1} \max_{\partial\Omega} |\mathbf{a}(x)| < 1,$$

with d *denoting the diameter of* Ω *and for a suitable constant* C.

$$ii) \quad C\nu^{-1}m \max_{h=1,\dots,m} \left| \int_{\partial\Omega_h} \mathbf{a}(x) \cdot \vec{n} d\sigma \right| < 1,$$

for a suitable constant C.

Then there exists a classical solution of system (1).

Remark 2.1. In the case of the linear system (Stokes problem), Theorem 2.1 becomes the maximum modulus theorem (see Theorem 3.1, below). The maximum modulus theorem for Stokes problem is also an auxiliary result for the proof of our theorem. However, since this result is proved only for a two-dimensional domain Ω and in a particular geometry of three-dimensional domains, (see Maremonti and Russo, 1994; Maremonti, 1993, respectively), Theorem 2.1 has partial validity related with the choice of Ω.

Finally, but not least, there is the question of the uniqueness of a classical solution. In this regard we consider only the two-dimensional case. The uniqueness of classical solutions $(\mathbf{v}(x), \pi(x))$ is certainly achieved when the following assumptions are satisfied

$\mathcal{U}_1$) Condition i) is verified as $\max_{\partial\Omega} |\mathbf{a}(x)| < \nu/2Cd$; or ii) is verified as $\max_{h=1,\dots,m} |\int_{\partial\Omega_h} \mathbf{a}(x) \cdot \vec{n} d\sigma < \nu/2Cm$;

$\mathcal{U}_2$) $\nabla \mathbf{u}(x) = \nabla(\mathbf{v}(x) - \mathbf{a}(x)) \in L^2(\Omega)$, where $\mathbf{a}(x)$ is the solution of the Stokes problem with $\mathbf{A}(x)_{|\partial\Omega} = \mathbf{a}(x)$.

We conclude saying that under assumption $\mathcal{U}_1$ it is possible to speak of a sort of "conditioned maximum modulus theorem."

For the sake of brevity we confine our proof to the case i) of the Theorem 2.1.

3. TWO PRELIMINARY RESULTS

Let us consider the Stokes problem:

$$\Delta \mathbf{A}(x) + \nabla \Pi(x) = \mathbf{F}(x), \qquad \nabla \cdot \mathbf{A}(x) = 0, \quad \text{in } \Omega, \qquad \mathbf{A}(x)_{|\partial\Omega} = \mathbf{a}(x). \qquad (4)$$

For system (4) we have the following theorem

Theorem 3.1. *Let* $\mathbf{F}(x) = 0$ *and assume that* $\mathbf{a}(x) \in C(\partial\Omega)$ *in system (4). Then there exists a unique classical solution to system (4). Moreover*

$$\max_{\Omega} |\mathbf{A}(x)| \leq C \max_{\partial\Omega} |\mathbf{a}(x)|, \quad \max_{K} \sum_{|\alpha|=1}^{2} |D^\alpha \mathbf{A}(x)| \leq C(K) \max_{\partial\Omega} |\mathbf{a}(x)|, \qquad (5)$$

where $K \subset \Omega$ is any arbitrary compact set. Here Ω can be an arbitrary bounded domain of R^2 or a three-dimensional domain of the type $\Omega = S(0, R_0) - \cup_{l=1}^m S(x^l, R_l)$ $(m \geq 0)$ with $l \geq 1$, $S(x^l, R_l) = \left\{ x \in I\!R^3 : |x - x^l| \leq R_l \right\}$, $\partial S(0, R_0) \cap \partial S(x^l, R_l) = \emptyset$, $\forall l = 1, ..., m$ and where the balls $S(x^l, R_l)$ are mutually disjoint.

Proof: See Maremonti, (1993); Maremonti and Russo (1994).

Theorem 3.2. Let $\mathbf{F}(x) \in H^{-1,p}(\Omega)$, $(p > 1)$ and assume that $\mathbf{a}(x) = 0$ in system (4). Then there exists a unique weak solution to system (4) such that

$$|\mathbf{A}|_p + |\nabla \mathbf{A}|_p + |\Pi|_p \leq |\mathbf{F}|_{H^{-1,p}}. \tag{6}$$

Proof: This theorem is a particular case of Theorem 2.1 of Galdi and Simader (1990). See also Theorem 6.1 of Galdi (1994).

4. PROOF OF THEOREM 2.1

Now we are in a position to prove Theorem 2.1 under the assumption i) for $\mathbf{a}(x)$. We start by assuming that $\mathbf{a}(x) \in C^1(\partial\Omega)$; then we relax the assumption on the data requiring only the continuity. Corresponding to $\mathbf{a}(x)$ we try to obtain a solution $\mathbf{v}(x)$ in the form

$$\mathbf{v}(x) = \mathbf{u}(x) + \mathbf{A}(x).$$

Function $\mathbf{A}(x)$ is the solution of (4) with $\mathbf{F}(x) = 0$ in Ω, function $\mathbf{u}(x)$ is the solution to the following boundary-value problem

$$\begin{cases} \nu \Delta \mathbf{u}(x) + \nabla \pi_1(x) = (\mathbf{u}(x) + \mathbf{A}(x)) \cdot \nabla(\mathbf{u}(x) + \mathbf{A}(x)) & \text{in } \Omega \\ \nabla \cdot \mathbf{u}(x) = 0 \\ \mathbf{u}(x)_{|\partial\Omega} = 0. \end{cases} \tag{7}$$

For system (7) we investigate a weak solution of the type

$$\mathbf{u}(x) \in J^1(\Omega) \cap C(\overline{\Omega}),$$
$$\nu(\nabla \mathbf{u}, \nabla \varphi) = ((\mathbf{u} + \mathbf{A}) \cdot \nabla \varphi, \mathbf{u} + \mathbf{A}), \, \forall \varphi(x) \in J^1(\Omega). \tag{8}$$

It is well known that to obtain such a solution in $J^1(\Omega)$ there exist several techniques. We refer to Galdi (1994), Ladyzhenskaya (1969) for details. Here, having in mind to prove the existence by the Galerkin method, we only recall the crucial point in the existence theorem, which is to obtain an estimate of the type

$$|\nabla \mathbf{u}|_2 \leq M. \tag{9}$$

Multiplying (7)$_1$ by $\mathbf{u}(x)$ and integrating over Ω we get

$$\nu |\nabla \mathbf{u}|^2 = (\mathbf{u} \cdot \nabla \mathbf{u}, \mathbf{A}) + (\mathbf{A} \cdot \nabla \mathbf{u}, \mathbf{A}).$$

Applying the Schwarz inequality and taking into account (5)$_1$ we arrive at

$$\nu \left(1 - \frac{C(\Omega)}{\nu} \max_{\partial\Omega} |\mathbf{a}(x)|\right) |\nabla \mathbf{u}|_2 \leq C(\Omega) \max_{\partial\Omega} |\mathbf{a}(x)|^2.$$

149

This last inequality and hypothesis i) imply (9). Now we prove that

$$\mathbf{u}(x) \in C(\overline{\Omega}) \qquad \text{and} \qquad \max_{\overline{\Omega}} |\mathbf{u}(x)| \leq M_1. \tag{10}$$

Since the right hand side of (7) belongs to $L^2(\Omega)$ (we have the assumption $\mathbf{a}(x) \in C^1(\partial\Omega)$), by virtue of theorems of regularity (cf. Galdi, 1994; Ladyzhenskaya, 1969) we have $\mathbf{u}(x) \in J^1(\Omega) \cap W^{2,2}(\Omega) \subset C(\overline{\Omega})$. On the other hand, setting

$$\mathbf{F}(x) = (\mathbf{u}(x) + \mathbf{a}(x)) \cdot \nabla(\mathbf{u}(x) + \mathbf{A}(x))$$

it is easy to verify that $\mathbf{F}(x) \in H^{-1,p}(\Omega)$ for suitable exponent p. From Theorem 3.2 we have

$$|\pi_1|_p + |\nabla\mathbf{u}|_p + |\mathbf{u}|_p \leq C|\mathbf{F}|_{H^{-1,p}(\Omega)}. \tag{11}$$

Now, we try to find a bound for $|\mathbf{F}|_{H^{-1,p}(\Omega)}$. To this end it is not difficult to deduce that

$$|\mathbf{F}|_{H^{-1,p}(\Omega)} \leq \{|\mathbf{u}|_{2p}^2 + 2\max_{\overline{\Omega}}|\mathbf{A}(x)||\mathbf{u}|_p + C(\Omega)\max_{\overline{\Omega}}|\mathbf{A}(x)|^2\}. \tag{12}$$

We consider $p > n$ ($n = 2, 3$). By virtue of Sobolev inequality

$$|\mathbf{u}|_{2p} \leq C|\nabla\mathbf{u}|_2^{1-a}|\mathbf{u}|_\infty^a, \qquad \begin{cases} a = 0 & \text{if } n = 2, \\ a = \frac{p-3}{p} & \text{if } n = 3. \end{cases}$$

Now we require $p < 6$. Then from (9), (11), (12) we have, in particular

$$\begin{aligned} \max_{\overline{\Omega}} |\mathbf{u}(x)| \ &\leq \ C\{|\nabla\mathbf{u}|_2^{\frac{2-2a}{1-2a}} + \max_{\overline{\Omega}}|\mathbf{A}(x)||\nabla\mathbf{u}|_2 \\ &+ \max_{\overline{\Omega}}|\mathbf{A}(x)||\nabla\mathbf{u}|_2 + \max_{\overline{\Omega}}|\mathbf{A}(x)|^2\} = M_1. \end{aligned} \tag{13}$$

Constants M and M_1 only depend on $\max_{\partial\Omega}|\mathbf{a}(x)|$ and Ω. We conclude the proof considering the case $\mathbf{a}(x) \in C(\partial\Omega)$. There exists a sequence $\{\mathbf{a}_n(x)\}_{n\in N} \subset C^1(\partial\Omega)$ such that $\mathbf{a}_n(x) \to \mathbf{a}(x)$ in $C(\partial\Omega)$ and any element of the sequence satisfies i) and $(2)_2$. By virtue of the above result and inequalities (9) and (10) we find a sequence $\{\mathbf{u}_n(x)\}_{n\in N}$ of weak solutions such that

$$|\nabla\mathbf{u}_n|_2 \leq M = M(\max_{\partial\Omega}|\mathbf{a}(x)|)$$

$$\max_{\overline{\Omega}} |\mathbf{u}_n(x)| \leq M_1 = M_1(\max_{\partial\Omega}|\mathbf{a}(x)|).$$

The first estimate implies that
   $\mathbf{u}_n(x)$ converges weakly to a limit $\mathbf{u}(x)$ in $H^1(\Omega)$;
   $\mathbf{u}_n(x)$ converges strongly in $J^p(\Omega)$ to the function $\mathbf{u}(x)$.
We terminate proving that $\mathbf{u}_n(x)$ converges to $\mathbf{u}(x)$ in $C(\overline{\Omega})$. We define $\mathbf{w}_{n,m}(x) = \mathbf{u}_n(x) - \mathbf{u}_m(x)$ and $\pi_{n,m}^1 = \pi_n^1(x) - \pi_m^1(x)$. Then $(\mathbf{w}_{n,m}(x), \pi_{n,m}^1(x))$ is a solution to the Stokes problem (4) with null boundary data and $\mathbf{F}_{n,m}(x) = \mathbf{F}(\mathbf{u}_n, \mathbf{u}_m, \mathbf{A}_n, \mathbf{A}_m)$. Since $\mathbf{F}_{n,m}(x) \in H^{-1,p}(\Omega)$, we have again

$$|\mathbf{w}_{n,m}|_p + |\nabla\mathbf{w}_{n,m}|_p \leq C|\mathbf{F}_{n,m}|_{H^{-1,p}}.$$

Considerations quite analogous to the ones already employed in obtaining (13), imply that, for some $p > n$

$$\begin{aligned} \max_{\overline{\Omega}} |\mathbf{w}_{n,m}(x)| \ &\leq \ C\{|\mathbf{w}_{n,m}|_p(\max_{\overline{\Omega}}|\mathbf{u}_n| + \max_{\overline{\Omega}}|\mathbf{u}_m| + \max_{\overline{\Omega}}|\mathbf{A}_n| + \max_{\overline{\Omega}}|\mathbf{A}_m|) \\ &+ \max_{\overline{\Omega}}|\mathbf{A}_n(x) - \mathbf{A}_m(x)|(\max_{\overline{\Omega}}|\mathbf{A}_n(x)| + \max_{\overline{\Omega}}|\mathbf{A}_m(x)|. \end{aligned} \tag{14}$$

In view of compactness of $\mathbf{w}_{n,m}$ in $L^p(\Omega)$, for $p \in (n, 6)$, and $(5)_1$ applied to difference $\mathbf{A}_n(x) - \mathbf{A}_m(x)$, we deduce from (13) that $\mathbf{w}_{n,m}(x)$ is converging in $C(\overline{\Omega})$. The existence of a weak solution to system (7) is proved. As far as the regularity of $\mathbf{u}(x)$ is concerned, it is sufficient to follow the usual arguments of local regularity, see Galdi (1994), Ladyzhenskaya (1969). Therefore $\mathbf{v}(x) = \mathbf{u}(x) + \mathbf{A}(x)$, $\pi(x) = \pi_1(x) + \Pi(x)$ is a classical solution to system (1). The existence theorem is completely proved.

5. UNIQUENESS OF CLASSICAL SOLUTIONS

In the following we consider $\Omega \subset \mathbb{R}^2$. For any pair $(\mathbf{v}(x), \pi(x))$ and $(\mathbf{v}_1(x), \pi_1(x))$ of classical solutions we set $\mathbf{u}(x) = \mathbf{v}(x) - \mathbf{v}_1(x)$, $\tilde{\pi}(x) = \pi(x) - \pi_1(x)$. Then we have

$$\begin{cases} \Delta\mathbf{u}(x) + \nabla\tilde{\pi}(x) = \mathbf{u}(x) \cdot \nabla\mathbf{u}(x) + \mathbf{u}(x) \cdot \nabla\mathbf{v}(x) + \mathbf{v}(x) \cdot \nabla\mathbf{u}(x), & \text{in } \Omega \\ \nabla \cdot \mathbf{u}(x) = 0, & \text{in } \Omega \\ \mathbf{u}(x)_{|\partial\Omega} = 0. \end{cases} \quad (15)$$

Now we prove that $\mathbf{u}(x) \in H^{1,p}(\Omega), \forall p > 1$. Set the right hand side of $(15)_1$ equal to $\mathbf{F}(\mathbf{u}, \mathbf{v})$. Let us consider the following Stokes problem

$$\begin{cases} \Delta\mathbf{w}(x) + \nabla\hat{\pi}(x) = \mathbf{F}(\mathbf{u}(x), \mathbf{v}(x)), & \text{in } \Omega \\ \nabla \cdot \mathbf{w}(x) = 0, & \text{in } \Omega \\ \mathbf{w}(x)_{|\partial\Omega} = 0. \end{cases} \quad (16)$$

Since $\mathbf{F}(\mathbf{u}, \mathbf{v}) \in H^{-1,p}(\Omega)$, from Theorem 3.2 we have the existence of a unique $(\mathbf{w}(x), \hat{\pi}(x))$ solution to system (16) belonging to $H^{1,p}(\Omega)$ and being smooth inside of Ω. Setting $\mathbf{U}(x) = \mathbf{u}(x) - \mathbf{w}(x)$ we have

$$\begin{cases} \Delta\mathbf{U}(x) + \nabla\tilde{\Pi}(x) = 0, & \text{in } \Omega \\ \nabla \cdot \mathbf{U}(x) = 0, & \text{in } \Omega \\ \mathbf{U}(x)_{|\partial\Omega} = 0, \end{cases} \quad (17)$$

with $\mathbf{u}(x) \in C(\overline{\Omega}) \cap C^2(\Omega)$. It is not difficult to prove that $\mathbf{U}(x) = 0$. In fact from (17) we see that there exists a biharmonic stream function $\psi(x)$ such that

$$U_1(x) = \frac{\partial\psi(x)}{\partial x_2}, \quad U_2(x) = -\frac{\partial\psi(x)}{\partial x_1}, \quad \nabla\psi(x)_{|\partial\Omega} = 0.$$

Such $\psi(x)$ has also $D^2\psi(x) \in L^2(\Omega)$. Therefore it follows that $\nabla\mathbf{U}(x) \in L^2(\Omega)$. This last property is sufficient to prove that $\mathbf{U}(x) = 0$ for system (17). As a consequence we have $\mathbf{u}(x) = \mathbf{w}(x)$, which implies, in particular, that $\nabla\mathbf{u}(x) \in L^2(\Omega)$. Now, taking into account assumption $\mathcal{U}_1$, it is possible to prove that $\mathbf{u}(x) = 0$ in system (15) making the usual consideration about uniqueness of weak solutions to Navier-Stokes system.

REFERENCES

BORCHERS, W. and K. PILECKAS (1994), Note on flux problem for stationary incompressible Navier-Stokes equations in domains with multiply-connected boundary, *Acta Applicandae Mathematicae* **37**, 21-30.

GALDI, G.P. (1994), *An Introduction to the Mathematical Theory of the Navier-Stokes equations* Vol. I, Springer Tracts in Natural Philosophy (Springer, Heidelberg).

GALDI, G.P. and C. SIMADER (1990), Existence, uniqueness and L^q-estimates for the Stokes problem in an exterior domain,*Arch. Rational Mech. Anal.* **112**, 291-318.

LADYZHENSKAYA, O.A. (1969), *The Mathematical Theory of Viscous Incompressible Flow*, (Gordon and Breach, New York).

MAREMONTI, P. (1993), On Stokes flow: the maximum modulus theorem, *Proc. Waves and Stability*, (World Scientific).

MAREMONTI, P. and R. RUSSO (1994), On the maximum modulus theorem, *Ann. Sc. Norm. Sup. Pisa*, S. IV, **22**, 629-643.

Compressible Fluids

THE MAXIMUM PRINCIPLE FOR THE EQUATION OF CONTINUITY OF COMPRESSIBLE MEDIA

V.N. Maslennikova and M.E. Bogovskii

Russian People's Friendship University
Mikluho Maklay, 6, 117198 Moscow, Russia

1. INTRODUCTION

The maximum principle is established for classical and generalized solution $\rho(\mathbf{x}, t)$ for the problem

$$\frac{\partial \rho}{\partial t} + \operatorname{div}(\rho \mathbf{v}) = 0, \qquad \mathbf{x} \in \Omega \subset \mathbf{R}^n, \ t > 0 \tag{1}$$

$$\rho(\mathbf{x}, t)|_{t=0} = \rho_0(\mathbf{x}), \quad (\mathbf{v}, \mathbf{n})|_{\partial\Omega} = 0 \tag{2}$$

for a given flow $\mathbf{v}(\mathbf{x}, t)$, where $\mathbf{n}$ is the unit normal to $\partial\Omega$, c.f. Maslennikova and Bogovskii (1994).

For simplicity let Ω be a bounded domain. The conditions on the flow $\mathbf{v}(\mathbf{x}, t)$ which are sufficient for the realization of these conditions are illustrated by the examples of flows for which the maximum principle loses its sense, i.e. the density $\rho(\mathbf{x}, t)$ either vanishes (the expanding bubble is formed in the continuous medium) or the generalized solution $\rho(\mathbf{x}, t)$ does not exist.

2. CLASSICAL SOLUTION FOR THE PROBLEM (1)–(2)

Let us denote

$$f(t) = \inf_{\mathbf{x} \in \Omega} \operatorname{div} \mathbf{v}(\mathbf{x}, t) \tag{3}$$

$$g(t) = \sup_{\mathbf{x} \in \Omega} \operatorname{div} \mathbf{v}(\mathbf{x}, t) \tag{4}$$

Let $F(t) = \displaystyle\int_0^t f(\tau)\, \mathrm{d}\tau$, $G(t) = \displaystyle\int_0^t g(\tau)\, \mathrm{d}\tau$.

Theorem 1. *If $\rho_0(x) \geq 0$, $\mathbf{v}(\mathbf{x}, t)$ is continuously differentiable with respect to $\mathbf{x}$, $f(t)$, $g(t) \in L^1(0, T)$, then for all $\mathbf{x} \in \Omega$, $t \in [0, T]$, the inequalities*

$$\inf_{\mathbf{x} \in \Omega} \rho_0(\mathbf{x}) e^{-G(t)} \leq \rho(\mathbf{x}, t) \leq \sup_{\mathbf{x} \in \Omega} \rho_0(\mathbf{x}) e^{-F(t)} \tag{5}$$

are satistied.

Proof: To obtain the above bound, we introduce the cut function

$$\eta(\xi) = \begin{cases} \xi - 1, & \xi > 1 \\ 0, & \xi \leq 1. \end{cases}$$

Let $s > 0$ be any number. We multiply the equation (1) by $(1/s)\eta'(\rho/s)$ and integrate over Ω:

$$\frac{\partial}{\partial t} \int_\Omega \eta\left(\frac{\rho}{s}\right) d\mathbf{x} + \int_\Omega \left(\mathbf{v}, \nabla \eta\left(\frac{\rho}{s}\right)\right) d\mathbf{x} + \int_\Omega \frac{\rho}{s} \eta'\left(\frac{\rho}{s}\right) \operatorname{div} \mathbf{v} \, d\mathbf{x} = 0.$$

Integrating by parts in the second integral, we obtain

$$\frac{\partial}{\partial t} \int_\Omega \eta\left(\frac{\rho}{s}\right) d\mathbf{x} - \int_\Omega \eta\left(\frac{\rho}{s}\right) \operatorname{div} \mathbf{v} \, d\mathbf{x} + \int_\Omega \frac{\rho}{s} \eta'\left(\frac{\rho}{s}\right) \operatorname{div} \mathbf{v} \, d\mathbf{x} = 0.$$

Now we integrate over s from some $\lambda > 0$ to ∞:

$$\frac{\partial}{\partial t} \int_\lambda^\infty ds \int_\Omega \eta\left(\frac{\rho}{s}\right) d\mathbf{x} - \int_\lambda^\infty ds \int_\Omega \eta\left(\frac{\rho}{s}\right) \operatorname{div} \mathbf{v} \, d\mathbf{x} + $$
$$+ \int_\Omega \operatorname{div} \mathbf{v} \, d\mathbf{x} \int_\lambda^\infty \frac{\rho}{s} \eta'\left(\frac{\rho}{s}\right) = 0. \tag{6}$$

We suppose the classical solution is bounded, therefore $\rho/s \to 0$, when $s \to 0$. Integrating by parts over s in the last integral (6), we have

$$\frac{\partial}{\partial t} \int_\lambda^\infty ds \int_\Omega \eta\left(\frac{\rho}{s}\right) d\mathbf{x} + \lambda \int_\Omega \eta\left(\frac{\rho}{\lambda}\right) \operatorname{div} \mathbf{v} \, d\mathbf{x} = 0. \tag{7}$$

If we set

$$u(\lambda, t) = \int_\lambda^\infty ds \int_\Omega \eta\left(\frac{\rho}{s}\right) d\mathbf{x}, \tag{8}$$

then (7) assumes the form

$$\frac{\partial u}{\partial t}(\lambda, t) = -\lambda \int_\Omega \eta\left(\frac{\rho}{\lambda}\right) \operatorname{div} \mathbf{v} \, d\mathbf{x}$$

We assume that $\operatorname{div} \mathbf{v} \geq f(t)$, therefore

$$\frac{\partial u}{\partial t} \leq -\lambda f(t) \int_\Omega \eta\left(\frac{\rho}{\lambda}\right) d\mathbf{x} = \lambda f(t) \frac{\partial u}{\partial \lambda}. \tag{9}$$

We define the function $h(\lambda, t)$ with the help of the equality

$$\frac{\partial u}{\partial t}(\lambda, t) - \lambda f(t) \frac{\partial u}{\partial \lambda}(\lambda, t) = h(\lambda, t), \tag{10}$$

where $h(\lambda, t) \le 0$. By changing the variables $(\lambda, t) \to (\xi, \sigma)$:

$$\xi = \lambda\, e^{F(t)}, \quad \sigma = t,$$

we obtain

$$\frac{\partial u}{\partial t} = \frac{\partial u}{\partial \xi} \lambda f(t) \exp\left\{ \int_0^t f(\tau)\, d\tau \right\} + \frac{\partial u}{\partial \sigma}$$

$$\frac{\partial u}{\partial \lambda} = \frac{\partial u}{\partial \xi} \exp\left\{ \int_0^t f(\tau)\, d\tau \right\} \quad \text{and}$$

$$\frac{\partial u}{\partial t} - \lambda f(t) \frac{\partial u}{\partial \lambda} = \frac{\partial \tilde{u}}{\partial \sigma}(\xi, \sigma) = \tilde{h}(\xi, \sigma) \le 0$$

whence

$$\tilde{u}(\xi, \sigma) \le \tilde{u}(\xi, 0).$$

But

$$\tilde{u}(\xi, \sigma) = u\left(\lambda(\xi, \sigma), t(\xi, \sigma)\right) = u\left(\xi\, e^{-F(\sigma)}, \sigma\right).$$

Therefore, by changing back to the variables (λ, t), we get

$$u(\lambda, t) \le u\left(\lambda e^{F(t)}, 0\right), \qquad \forall \lambda > 0,$$

which imply together with (8) that

$$u(\lambda, t) \le \int_{\lambda e^{F(t)}}^{\infty} ds \int_{\Omega} \eta\left(\frac{\rho_0(x)}{s}\right) dx \qquad \forall \lambda > 0.$$

If we choose

$$\lambda \ge e^{-F(t)} \sup_{x \in \Omega} \rho_0(x),$$

then $(\rho_0(x)/s) \le 1$ and, consequently, in view of the definition of the function η, $\eta(\rho_0(x)/s) = 0$. Therefore $u(\lambda, t) \le 0$.

On the other hand, in view of (8) and of the nonnegativity of the function η, one has $u(\lambda, t) \ge 0$.

Thus, we conclude that

$$u(\lambda, t) = 0, \quad \forall \lambda \ge e^{-F(t)} \sup_{x \in \Omega} \rho_0(\mathbf{x}).$$

This means, again in view of (8) and nonnegativity of $\eta(\xi)$ that

$$\eta\left(\frac{\rho(\mathbf{x}, t)}{s}\right) = 0, \qquad \forall s \ge \lambda,$$

from which and from the definition of the function η we obtain

$$\frac{\rho(\mathbf{x}, t)}{s} \le 1, \quad \forall s \ge \lambda, \quad \forall \lambda \ge e^{-F(t)} \sup_{x \in \Omega} \rho_0(\mathbf{x}).$$

In particular, we can take

$$s = e^{-F(t)} \sup_{x \in \Omega} \rho_0(\mathbf{x}), \text{ that is}$$

$$\rho(\mathbf{x}, t) \le e^{-F(t)} \sup_{\mathbf{x} \in \Omega} \rho_0(\mathbf{x}).$$

In order to get a lower bound for $\rho(\mathbf{x}, t)$, we choose the cut function in the form of

$$\eta(\xi) = \begin{cases} 1 - \xi, & 0 < \xi < 1 \\ 0, & \xi \ge 1 \end{cases}$$

and integrate over s from zero to $\lambda > 0$. Theorem 1 is proved.

Consequence. The classical solution $\rho(\mathbf{x}, t)$ of the problem (1)–(2) is unique.

3. THE MAXIMUM PRINCIPLE

The maximum principle for the conjugate of equation (1) is simpler. We consider the solutions for the problem

$$\frac{\partial \rho}{\partial t} + (\mathbf{v}, \nabla \rho) = 0, \quad \mathbf{x} \in \Omega \subset \mathbf{R}^n, \ t > 0 \tag{11}$$

$$\rho(\mathbf{x}, t)|_{t=0} = \rho_0(\mathbf{x}), \quad (\mathbf{v}, \mathbf{n})|_{\partial \Omega} = 0, \tag{12}$$

where the vector field $\mathbf{v}(\mathbf{x}, t)$ may be nonsolenoidal and derivatives in (11) are understood in Sobolev sense.

Theorem 2. *If $g(t) = \sup_{\mathbf{x} \in \Omega} \operatorname{div} \mathbf{v}(x, t) \in L^1(0, T)$, then for any $\mathbf{x} \in \Omega$, $t \in [0, T]$ the inequalities*

$$\inf_{\mathbf{x} \in \Omega} \rho_0(\mathbf{x}) \le \rho(\mathbf{x}, t) \le \sup_{\mathbf{x} \in \Omega} \rho_0(\mathbf{x})$$

are satisfied.

Proof: Let $\eta(\xi)$ be a continuously differentiable function:

$$\begin{cases} \eta(\xi) > 0, & \xi \notin [\xi_1, \xi_2], \\ \eta(\xi) = 0, & \xi \in [\xi_1, \xi_2], \end{cases}$$

where the numbers ξ_1 and ξ_2 are defined by

$$\xi_1 = \inf_{\mathbf{x} \in \Omega} \rho_0(\mathbf{x}), \quad \xi_2 = \sup_{\mathbf{x} \in \Omega} \rho_0(\mathbf{x}).$$

Multiplying the equation (11) by $\eta'(\rho)$ and integrating over Ω by parts, we get

$$\frac{\partial}{\partial t} \int_\Omega \eta(\rho) \, d\mathbf{x} \le g(t) \int_\Omega \eta(\rho) \, d\mathbf{x}. \tag{13}$$

Multiplying (13) by the integrating factor $e^{-G(t)}$, where $G(t) = \int_0^t g(\tau) \, d\tau$, and integrating in t, we obtain

$$e^{-G(t)} \int_\Omega \eta(\rho) \, d\mathbf{x} \le \int_\Omega \eta(\rho_0(\mathbf{x})) \, d\mathbf{x}.$$

But the values of $\rho_0(\mathbf{x})$ are contained between ξ_1 and ξ_2, where $\eta(\xi) = 0$, therefore $\eta(\rho_0(\mathbf{x})) = 0$. It follows that

$$\int_\Omega \eta(\rho) \, d\mathbf{x} \le 0.$$

Now, η is a non-negative function and therefore $\eta(\rho) = 0$, from which we conclude that

$$\xi_1 \leq \rho(\mathbf{x}, t) \leq \xi_2.$$

Theorem 2 is proved. $\square$

The maximum principle for generalized solutions loses its sense when the flow $\mathbf{v}(\mathbf{x}, t)$ is smooth, if div $\mathbf{v}(\mathbf{x}, t)$ is not bounded at least in one point.

Example. Let $\Omega = \{\mathbf{x} \in \mathbf{R}^3 \mid |\mathbf{x}| < 1\}$. We consider the flow $\mathbf{v}(\mathbf{x}) = \alpha \mathbf{x}(|\mathbf{x}|^{-\beta} - 1) \in C(\bar{\Omega}; \mathbf{R}^3) \cap C^\infty(\bar{\Omega} \setminus \{0\}; \mathbf{R}^3)$, where $\alpha \neq 0$, $0 < \beta < 1$. Evidently $\mathbf{v}|_{\partial\Omega} = 0$, div $\mathbf{v} = \alpha(3 - \beta)|\mathbf{x}|^{-\beta} - 3\alpha \in L^1(\Omega)$, $\sup_{\mathbf{x} \in \Omega}$ div $\mathbf{v} = \infty$. Then the solution of problem (1)–(2) is

$$\rho(\mathbf{x}, t) = e^{\alpha\beta t}|\mathbf{x}|^{\beta-3}\left[1 - (1 - |\mathbf{x}|^\beta)\, e^{\alpha\beta t}\right]^{(3-\beta)/\beta} \times$$
$$\times \rho_0\left(\frac{\mathbf{x}}{|\mathbf{x}|}\left[1 - (1 - |\mathbf{x}|^\beta)\, e^{\alpha\beta t}\right]^{1/\beta}\right), \tag{14}$$

if $(1 - e^{-\alpha\beta t})^{1/\beta} < |\mathbf{x}| < 1$; $\rho(\mathbf{x}, t) = 0$, if $|\mathbf{x}| \leq (1 - e^{-\alpha\beta t})^{1/\beta}$ in the case $\alpha > 0$. The uniqueness of the solutions (14) will be proved in Theorem 4.

Defining $\rho(\mathbf{x}, t)$ to be zero outside Ω, we note that

$$\lim_{t \to \infty} \rho(\mathbf{x}, t) = \mu\, \delta_s(x) \quad \text{in } S'(\mathbf{R}^3),$$

where $\delta_s(x)$ is δ-function concentrated on unit sphere $S \subset \mathbf{R}^3$ with constant density $\mu > 0$.

Thus, our example shows that an extending emptiness is formed in the ball and it occupies the whole ball when $t \to \infty$.

In the case $\alpha > 0$ the generalized solution $\rho(\mathbf{x}, t)$ of the problem (1)–(2) does not exists, even in the class $L^p(Q_T)$, $Q_T = \Omega \times [0, T]$, $1 < p < \infty$, where Ω is the ball.

Definition. *We call $\rho(\mathbf{x}, t) \in L^p(Q_T)$ ($1 < p < \infty$) a generalized solution to the problem (1)–(2) if*

$$\int_{Q_T} \left[\rho\,\Psi_t + (\rho\mathbf{v}, \nabla\Psi)\right] d\mathbf{x}\, dt = \int_\Omega \rho_0(\mathbf{x})\,\Psi(\mathbf{x}, 0)\, d\mathbf{x} \tag{15}$$

for any $\Psi(\mathbf{x}, t) \in C^1(\bar{Q}_T)$, $\Psi(\mathbf{x}, T) = 0$. We assume that the vector field $\mathbf{v}(\mathbf{x}, t) \in L^1(Q_T; \mathbf{R}^n)$ and that a generalized divergence exists for $\mathbf{v}(\mathbf{x}, t)$ with nonflowing condition $(\mathbf{v}, \mathbf{n})|_{\partial\Omega} = 0$, i. e. the function $\omega(\mathbf{x}, t) \in L^1(Q_T)$ exists for which

$$\int_{Q_T} (\mathbf{v}, \nabla_x \varphi(\mathbf{x}, t))\, d\mathbf{x}\, dt = -\int_{Q_T} \omega\, \varphi\, d\mathbf{x}\, dt, \quad \forall \varphi \in C^1(\bar{Q}_T). \tag{16}$$

We do not assume on $\mathbf{v}(\mathbf{x}, t)$ the existence of separate derivatives of the first order with respect to $\mathbf{x}$.

Choosing in (16) the test functions $\varphi = \varphi(t)$, we get

$$\int_\Omega \omega(\mathbf{x}, t)\, d\mathbf{x} = 0 \tag{17}$$

159

for almost all $t \in (0, T)$. We denote

$$f(t) = \operatorname*{ess\,inf}_{\mathbf{x} \in \Omega} \omega(\mathbf{x}, t) > -\infty,$$

$$g(t) = \operatorname*{ess\,sup}_{\mathbf{x} \in \Omega} \omega(\mathbf{x}, t) < +\infty,$$

$$F(t) = \int_0^t f(\tau)\, d\tau, \quad G(t) = \int_0^t g(\tau)\, d\tau.$$

Theorem 3. *Let $\Omega \subset \mathbf{R}^n, n \geq 2$, be a bounded domain satisfying the cone condition and let $1 < q < \infty$ and $T > 0$. Let $\mathbf{v}(\mathbf{x}, t) \in L^1(Q_T; \mathbf{R}^n)$, $\omega(\mathbf{x}, t) \in L^q(Q_T)$, $f(t) \in L^1(0, T)$, $\rho_0(\mathbf{x}) \in L^\infty(\Omega)$, $\rho_0(\mathbf{x}) > 0$. Then there exists a generalized solution $\rho(\mathbf{x}, t) \in L^\infty(Q_T)$ to the problem (15). This solution satisfies the inequalities*

$$\operatorname*{ess\,inf}_{\mathbf{x} \in \Omega} \rho_0(\mathbf{x})\, e^{-G(t)} \leq \rho(\mathbf{x}, t) \leq \operatorname*{ess\,sup}_{\mathbf{x} \in \Omega} \rho_0(\mathbf{x})\, e^{-F(t)} \tag{18}$$

for almost all $(\mathbf{x}, t) \in Q_T$.

Proof: First we approximate $\omega(\mathbf{x}, t)$ in $L^q(Q_T)$ by smooth functions $\omega^n(\mathbf{x}, t) \in C^\infty(Q_T)$, which satisfy (17) and for which

$$\min_{\mathbf{x} \in \Omega} \omega^n(\mathbf{x}, t) \geq \operatorname*{ess\,inf}_{\mathbf{x} \in \Omega} \omega(\mathbf{x}, t)$$
$$\max_{\mathbf{x} \in \Omega} \omega^n(x, t) \leq \operatorname*{ess\,sup}_{\mathbf{x} \in \Omega} \omega(\mathbf{x}, t) \tag{19}$$

After that we solve the sequence of problems

$$\operatorname{div} \mathbf{u}^n = \omega^n(\mathbf{x}, t)$$
$$\mathbf{u}^n|_{\partial \Omega} = 0$$

Using the result of Bogovskii (1979), we obtain

$$\|\mathbf{u}^n - \mathbf{u}\|_{W^{1,0}_{q,x,t}(Q_T; \mathbf{R}^n)} \leq c \|\omega^n - \omega\|_{L^q(Q_T; \mathbf{R}^n)} \to 0,$$

then $n \to \infty$; $\mathbf{u}(\mathbf{x}, t) \in W^{1,0}_{q,x,t}(Q_T; \mathbf{R}^n)$ is solution of the problem

$$\operatorname{div} \mathbf{u} = \omega(\mathbf{x}, t)$$
$$\mathbf{u}|_{\partial \Omega} = 0.$$

For flow $\mathbf{v}(\mathbf{x}, t) \in L^1(Q_T; \mathbf{R}^n)$ we have $\mathbf{w} = \mathbf{v} - \mathbf{u} \in \mathring{J}_1(Q_T; \mathbf{R}^n)$, where $\mathring{J}_1$ is the closure of

$$\mathring{J}_\infty(\Omega) = \{\mathbf{w}(\mathbf{x}, t) \mid \mathbf{w}(\mathbf{x}, \cdot) \in \mathring{C}^\infty(\Omega), \ \operatorname{div} \mathbf{w} = 0\}$$

in the norm $L^1(\Omega)$.

Hence there exists a sequence

$$\{\mathbf{w}^n(\mathbf{x}, t)\} \in \mathring{J}_1(Q_T; \mathbf{R}^n) \text{ such that } \lim_{n \to \infty} \|\mathbf{v} - \mathbf{u} - \mathbf{w}^n\|_{L^1(Q_T; \mathbf{R}^n)} = 0.$$

We take by definition the approximation for $\mathbf{v}(\mathbf{x},t)$: $\mathbf{v}^n = \mathbf{u}^n + \mathbf{w}^n$; then $\mathbf{v}^n(\mathbf{x},t) \in \mathring{C}^\infty(Q_T; \mathbf{R}^n)$, and

$$\lim_{n\to\infty} \|\mathbf{v} - \mathbf{v}^n\|_{L^1(Q_T; \mathbf{R}^n)} = 0\,.$$

By such method we constructed the sequence $\mathbf{v}^n(\mathbf{x},t) \in \mathring{C}^\infty(Q_T)$, which approximates $\mathbf{v}(\mathbf{x},t)$ in $L^1(Q_T; \mathbf{R}^n)$ and for which div $\mathbf{v}^n = \omega^n$ in Q_T.

Let $\rho_0^n(\mathbf{x}) \in C^\infty(\bar{\Omega})$ be smooth approximation of $\rho_0(\mathbf{x})$ in $L^p(\Omega)$, $1 < p < \infty$, which is constructed with the help of averaging.

Solving the problem

$$\frac{\partial \rho^n}{\partial t} + \operatorname{div}(\mathbf{v}^n \rho^n) = 0$$

$$\rho^n(\mathbf{x},t)|_{t=0} = \rho_0^n(\mathbf{x})\,,$$

with smooth $\mathbf{v}^n(\mathbf{x},t)$ and $\rho_0^n(\mathbf{x})$, we obtain a smooth solution $\rho^n(\mathbf{x},t)$, which satisfies Theorem 1, i.e.:

$$\inf_{\mathbf{x}\in\Omega} \rho_0^n(\mathbf{x}) \exp\left\{-\int_0^t g^n(\tau)\,d\tau\right\} \le \rho^n(\mathbf{x},t) \le \sup_{\mathbf{x}\in\Omega} \rho_0^n(\mathbf{x}) \exp\left\{-\int_0^t f^n(\tau)\,d\tau\right\},$$

where $f^n(t) = \min_{\mathbf{x}\in\Omega}\omega^n(\mathbf{x},t)$, $g^n(t) = \max_{\mathbf{x}\in\Omega}\omega^n(\mathbf{x},t)$.

Since $f(t) \in L^1(0,T)$ the sequence $\rho^n(\mathbf{x},t)$ is bounded in $L^\infty(Q_T)$. Choosing a subsequence, which weakly converges to $\rho(\mathbf{x},t)$ in $L^p(Q_T)$ for some $1 < p < \infty$, we obtain the generalized solution of the problem (15), for which

$$
\begin{aligned}
\liminf_{n\to\infty} \operatorname*{ess\,sup}_{\mathbf{x}\in\Omega} \rho^n(\mathbf{x},t) &\ge \operatorname*{ess\,sup}_{\mathbf{x}\in\Omega} \rho(\mathbf{x},t) \\
\limsup_{n\to\infty} \operatorname*{ess\,inf}_{\mathbf{x}\in\Omega} \rho^n(\mathbf{x},t) &\le \operatorname*{ess\,inf}_{\mathbf{x}\in\Omega} \rho(\mathbf{x},t)
\end{aligned}
\tag{20}
$$

for almost all $t \in (0,T)$. From (20) and (19) we get (18).

Remark 1. If $g(t) \in L^1(0,T)$, then the estimate (18) for $\rho(\mathbf{x},t)$ guarantees the impossibility of formation of vacuums in the continuous medium.

Therefore we naturally assume that sufficient conditions for realization of maximum principle are:

$$f(t) \in L^1(0,T)\,, \quad g(t) \in L^1(0,T)\,.$$

The exactness of these conditions is verified by the above-mentioned examples.

Theorem 4. (Uniqueness of generalized solution) *Let $\Omega \subset \mathbf{R}^n$ be a bounded domain, $n \ge 2$, $1 < p < \infty$. For the uniqueness of generalized solution $\rho(\mathbf{x},t) \in L^p(Q_T)$ in the sense of (15), it is sufficient for $\rho(\mathbf{x},t)$ to have $\partial\rho/\partial x_k \in L^1(Q_T)$, $k = 1,\dots,n$ and for the vector field $\mathbf{v}(\mathbf{x},t) \in L^\infty(Q_T; \mathbf{R}^n)$ to have generalized divergence $\omega(\mathbf{x},t) \in L^{p'}(Q_T)$ in the sense of (16) with the nonflowing condition $(\mathbf{v},\mathbf{n})|_{\partial\Omega} = 0$, where p' is the conjugate index of p.*

Proof: First we prove the existence of generalized derivative $\rho_t(\mathbf{x},t) \in L^1(Q_T)$. Then the generalized solution in the sense of (15) is the solution of the homogeneous problem

$$\frac{\partial\rho}{\partial t} + (\mathbf{v},\nabla\rho) + \rho\omega = 0\,, \quad (\mathbf{x},t) \in Q_T$$

$$\rho|_{t=0} = 0\,, \qquad \mathbf{x} \in \Omega\,.
\tag{21}$$

This problem has only a trivial solution. In fact, choosing the cut function $\eta(\xi) \in \overset{\circ}{C}{}^\infty(\mathbf{R}^1)$ in such a way that

$$\eta(\xi) > 0, \qquad \xi \in \left(\tfrac{1}{2}, 1\right)$$

$$\eta(\xi) = 0, \qquad \xi \notin \left(\tfrac{1}{2}, 1\right)$$

we multiply the equation (21) by $(2\rho/s^2)\,\eta'(\rho^2/s^2)$ and integrate over Ω and in s from $\lambda > 0$ to some $M > \lambda$.

After integration by parts and in t from 0 to T we get

$$\int_\lambda^M ds \int_\Omega \eta\left(\frac{\rho^2}{s^2}\right) d\mathbf{x} = \int_\lambda^M ds \int_\Omega \eta\left(\frac{\rho_0^2}{s^2}\right) d\mathbf{x} +$$

$$+ M \int_{Q_T^M} \eta\left(\frac{\rho^2}{M^2}\right) \operatorname{div} \mathbf{v}\, d\mathbf{x}\, d\tau - \lambda \int_{Q_T} \eta\left(\frac{\rho^2}{s^2}\right) \operatorname{div} \mathbf{v}\, d\mathbf{x}\, d\tau, \tag{22}$$

where

$$Q_T^M = \left\{(\mathbf{x}, t) \in Q_T \mid \frac{M}{\sqrt{2}} < |\rho(\mathbf{x}, t)| < M\right\},$$

and

$$\operatorname{mes} Q_T^M \leq \left(\frac{\sqrt{2}}{M}\right)^p \|\rho\|_{l^p(Q_T)}^p \quad \forall\, M > 0 \tag{23}$$

since $\eta(\rho^2/M^2) \neq 0$ only for $\tfrac{1}{2} < \rho^2/M^2 < 1$.

Using (23) and passing to the limit: $\lambda \to 0$, $M \to \infty$ in (22), we obtain

$$\int_0^\infty ds \int_\Omega \eta\left(\frac{\rho^2}{s^2}\right) d\mathbf{x} = 0$$

for almost all $t \in (0, T)$. Hence $\eta(\rho^2/s^2) = 0$, $\forall\, s \in (0, \infty)$. From the last equality and the maximum principle we have $\rho(\mathbf{x}, t) \equiv 0$ almost everywhere in Q_T.

Remark 2. We proved all Theorems for bounded domain Ω, but we used its boundedness only for integration by parts, where we used the condition $(\mathbf{v}, \mathbf{n})|_{\partial\Omega} = 0$. Therefore if we impose the condition $\mathbf{v}(\mathbf{x}, t) \to 0$ when $|\mathbf{x}| \to \infty$, then all results will remain valid in an unbounded domain Ω.

Remark 3. Generalized solution of the problem (1)–(2) was introduced here with regard to the maximum principle. For the generalized solution itself see Di Perna and Lions (1989).

Remark 4. To establish the maximum principle for the problem (1)–(2) in case of smooth flow $\mathbf{v}(\mathbf{x}, t)$ one may start from a well-known general form of a solution of the first order partial differential equation. This paper displays a new approach to the proof of the maximum principle via integration by parts which was applied here to the problem (1)–(2). Being applied to general elliptic or parabolic equations, this approach results in an immediate and elementary proof of the maximum principle.

REFERENCES

BOGOVSKII, M.E. (1979), *Soviet Math. Dokl.* **248**, 1037–1040; *Outlines Sobolev Seminar,* Novosibirsk (1980), n. **1**, 5–40.

DI PERNA, R.J. and P.L. LIONS (1989), Ordinary differential equations, Transport Theory and Sobolev Spaces, *Inventions Math.* **98**, 511–547.

MASLENNIKOVA, V.N. and M.E. BOGOVSKII (1994), The maximum principle for the equation of compressible medium, *Doklady of Russian Academy of Sciences* **339**, n. **4**.

UNIQUENESS OF WEAK SOLUTIONS OF UNSTEADY MOTIONS OF VISCOUS COMPRESSIBLE FLUIDS (II)

Šárka Matušu-Nečasová

Czech Technical University, Department of Technical Mathematics
Karlovo nám. 13, Prague 2, Czech Republic

1. INTRODUCTION

The study of the uniqueness of the flow of compressible viscous fluid presents mathematical and physical interest. We recall the equations governing general motions of compressible viscous fluids

$$\begin{cases} \rho[v_{,t} + v \cdot \nabla v] - \mu \Delta v - (\lambda + \mu)\nabla\nabla \cdot v = -\nabla\rho + \rho h \\[2mm] \rho_{,t} + \operatorname{div}(\rho v) = 0 \\[2mm] v(x,0) = v_0(x), \quad \rho(x,0) = \rho_0(x) \qquad\qquad\qquad \text{in } \Omega \\[2mm] v(x,t) = 0 \qquad\qquad\qquad\qquad\qquad\qquad \text{at } \partial\Omega \times (0,T) \end{cases} \qquad (1)$$

where $\Omega \subseteq R^3$, Ω is a sufficiently smooth domain, p is the pressure, ρ is the density, v is the velocity, μ, λ are viscosity coefficients, $h(x,t)$ is the external specific body force, $v_0(x)$ is the initial velocity and $\rho_0(x)$ is the initial density.

The first results about uniqueness of viscous compressible fluids trace back to Graffi and Serrin. They presented results for strictly positive density in bounded domain (see Graffi, 1961; Serrin, 1959). After them, Graffi and Russo extended these results for non-negative densities in unbounded domains (see Graffi, 1960; Russo, 1980). The question of summability of the densities in unbounded domains was left open. The uniqueness of the weak solution of compressible viscous fluids under more general assumptions of nonnegativeness on the density when density can vanish is proved in Matušu-Nečasová (1995).

The aim of this paper is to extend these results to the case where density can blow up. We shall assume that the viscosities μ and λ are bigger than the density which can blow up. Under this assumption we can prove uniqueness for weak solutions.

Section 2 deals with preliminary lemmas. Section 3 is devoted to the uniqueness proof in the case where density can blow up.

Navier-Stokes Equations and Related Nonlinear Problems
Edited by A. Sequeira, Plenum Press, New York, 1995

2. MATHEMATICAL PRELIMINARIES

We set $Q_T = \Omega \times (0, T)$, $u = (u_1, u_2, u_3)$ and introduce some function spaces. Let $D(\Omega) = C_0^\infty$, and denote by H the completion of $D(\Omega)$ in L^2-norm. $(H_p^m(\Omega))^3$ is the space of vector functions having finite L^p-norm of $|D^m \phi|$. Moreover, $(.\,,.)$ denotes the usual scalar product in $L^2(\Omega)$ and

$$|u|_{g,p,C} = \left(\int_C g u^p \, dx \right)^{1/p},$$

$$|u|_p \equiv |u|_{1,p,\Omega}, \qquad |u|_{g,p} \equiv |u|_{g,p,\Omega}, \qquad |u|_\infty = \operatorname*{ess\,sup}_\Omega u,$$

(2)

where g is a measurable non-negative scalar function in Ω. Moreover $H_p^{0,1}(\Omega)$ is the completion of $D(\Omega)$ in the norm $|D\phi|_p$. If Ω is bounded, then $H_p^{0,1}(\Omega)$ coincides with the Sobolev space $W_0^{1,p}(\Omega)$. We define $W^{-1,p'}(\Omega)$ as the dual of $W^{1,p'}(\Omega)$. We shall consider the following case

$$\Omega \text{ bounded in } \mathbb{R}^3, \quad 0 \le g(x, t).$$

Now we recall some important lemmas.

Lemma 2.1. *Let $u \in C_0^\infty(\Omega)$. Then it holds*

$$|u|_{\sqrt{g},4} \le c |u|_{g,2}^{1/4} |\nabla u|_2^{3/4}.$$

(3)

Proof: see Padula (1990). □

Definition 2.1. *The space $W(Q_T)$ is a space with the norm*

$$|||u|||^2 = |u|_{g,2}^2 + |\nabla u|_2^2 + \int_0^T \left(|u_t|_{g,2}^2 + |D^2 u|_2^2 \right) dt < \infty.$$

(4)

with functions u having zero trace.

Lemma 2.2. *Let $\nabla \cdot v$ and $v\, D(\ln f)$ belong to $L^1(0, T; L^\infty(\Omega))$ and let*

$$m f(x) \le \rho_0(x) \le M f(x), \quad m, M > 0.$$

(5)

Then there exists a constant c_1 depending only on the integral

$$\int_0^T \left(|v\, D(\ln f)|_\infty + 2|\nabla \cdot v|_\infty \right) ds,$$

(6)

such that

$$m c^{-1} \le \rho(x, t)(f(x))^{-1} \le M c_1.$$

(7)

Proof: Matušu-Nečasová (1995). □

We shall also use Young's inequality

$$ab \le \delta |a|^p / p + \delta^{-\frac{p'}{p}} |b|^{p'} / p.$$

(8)

Remark 2.1. This lemma gives us, under assumptions on the initial density, on the velocity and on the given function f, the behaviour of density in $(x, t) \in Q_T$. We apply this lemma to the case in which Ω is bounded and $f = 1/|x|^\alpha, \alpha > 0$. Another application was described in Matušu-Nečasová (1995), for the case of vanishing density.

3. UNIQUENESS OF ISOTHERMAL CASE WITH DENSITY WHICH CAN BLOW UP

We consider Ω as a bounded domain, with sufficiently smooth boundary. Our control function f will be $f = 1/|x|^\alpha, \alpha > 0$. It means that ρ doesn't belong to $L^\infty(Q_T)$.

Definition 3.1. *A pair (ρ, v) is said to be a generalized solution to (1) if and only if, for all $T > 0$*

(i) $\rho \in L^\infty(0, T; L^6(\Omega))$, $\rho \geq 0$;

(ii) $\rho_{,t} \in L^2(0, T; W^{-1,2}(\Omega))$;

(iii) $v \in W(Q_T)$;

(iv) (ρ, v) satisfy the following identities

$$\int_0^T \int_\Omega \phi[\rho(v_{,t} + v \cdot \nabla v) - \rho h] \, dx \, dt$$

$$+ \int_{Q_T} \mu \nabla \phi \nabla v + (\lambda + \mu) \operatorname{div} \phi \operatorname{div} v = - \int_{Q_T} K \rho \nabla \cdot \phi \quad (9)$$

$$\int_{Q_T} (\rho_{,t} \psi - \rho v \cdot \nabla \psi) \, dx \, dt = 0$$

(v) $\lim_{t \to 0} |\rho - \rho_0|_2 = 0$ and $\lim_{t \to 0} |\rho v - \rho_0 v_0|_2 = 0$, where ϕ and ψ are functions which satisfy

(vi) $\phi \in L^2(0, T; L^2_\rho(\Omega)) \cap L^2(0, T; H_2^{0,1}(\Omega))$, $\psi \in L^2(0, T; L^6(\Omega)) \cap L^2(0, T; W^{1,2}(\Omega))$ and initial ρ_0, v_0 satisfy

(vii) $\rho_0 \in L^6(\Omega)$, $v_0 \in L^2_{\rho_0}(\Omega)$.

Definition 3.2. *A scalar function $\psi(x)$ in Ω is be said to be in the class $I_{K,\alpha}$ if*

(i) $m < \psi(x)|x|^K < M$;

(ii) $|\nabla \psi(x)| \cdot |x|^\alpha < M_1$.

A vector function $\varphi(x, t)$ in Q_T is be said to belong to the class V if

(i) $\nabla \varphi \in L^2(0, T; L^\infty(\Omega))$;

(ii) $\nabla \varphi_{,t} \in L^2(Q_T)$.

Theorem 3.1. *Let $(\tilde{\rho}, \tilde{v})$ and (ρ, v) be two generalized solutions to (9) and let assumptions of Lemma 2 be satisfied with $\tilde{\rho}_{,t} \in L^2(Q_T)$ and corresponding to the same data $h \in L^\infty(0, T; L^4(\Omega))$ and $v_0 \in L^2 \rho_0(\Omega)$. Moreover, let $\tilde{v} \in V$ and let $\rho_0 \in I_K$, $\tilde{\rho} \in I_{K,\alpha}$, $\alpha < \frac{K}{2}$ (I_K means that only the first condition of Definition 3 is satisfied). Finally, assume that $\tilde{\rho}$ and ρ have the same speed (it means that $\rho, \tilde{\rho} \in I_K$) and $(\lambda + \mu)^2 > \tilde{\rho}^2$. Then $\tilde{\rho} = \rho$, $\tilde{v} = v$.*

Proof: We set $\sigma = \tilde{\rho} - \rho$ and $u = \tilde{v} - v$. Since both solutions satisfy system (9) we deduce for perturbations (σ, u) that

$$\int_\Omega (\phi\rho u_{,t} + \phi\rho v \cdot \nabla u) \; dx \; dt + \int_\Omega \mu\nabla\phi\nabla u +$$

$$\int_\Omega (\lambda + \mu)\nabla \cdot \phi\nabla \cdot u = -\int_\Omega \phi\sigma\tilde{v}_{,t} + \phi\sigma\tilde{v} \cdot \nabla\tilde{v} - \tag{10}$$

$$\int_\Omega \phi\rho u \cdot \nabla\tilde{v} + \int_\Omega \sigma h\phi + K\int_\Omega \sigma\nabla \cdot \phi$$

and

$$\int_\Omega \Big(\frac{\partial\sigma}{\partial t} + \nabla\sigma v\Big)\psi = -\int_\Omega (\nabla \cdot \tilde{\rho}u + \sigma\nabla \cdot v + \tilde{\rho}\nabla \cdot u)\psi. \tag{11}$$

We assume that $u(x,0) = \sigma(x,0) = 0$ and we set $\phi = u$ and $\psi = \sigma$. We obtain

$$\int_\Omega \rho\frac{d}{dt}\Big(\frac{u^2}{2}\Big) + \int_\Omega \mu\nabla u\nabla u + (\lambda + \mu)\,\mathrm{div}\,u\,\mathrm{div}\,u$$

$$= -\int_\Omega \sigma(\tilde{v}_{,t} + \tilde{v} \cdot \nabla\tilde{v} - h)u - \int_\Omega \rho u \cdot \nabla\tilde{v}u + K\int_\Omega \sigma\nabla \cdot u. \tag{12}$$

Using (3) and (8) we have

$$\int_\Omega \sigma\tilde{v}_{,t}u \; dx \quad \leq \quad |\sigma|_2|\tilde{v}_{,t}|_{\sqrt{\rho},4}|u|_{\sqrt{\rho},4}|\rho^{-\frac{1}{4}}|_\infty$$

$$\leq \quad \eta^{(\frac{3}{8}+\frac{1}{8}-\frac{1}{2})}|\sigma|_2|\tilde{v}_{,t}|_{\rho,2}^{\frac{1}{4}}|\nabla\tilde{v}_{,t}|_2^{\frac{3}{4}}|u|_{\rho,2}^{\frac{1}{4}}|\nabla u|_2^{\frac{3}{4}}|\rho^{-\frac{1}{4}}|_\infty$$

$$\leq \quad \Big(\frac{\eta}{2}|\sigma|_2^2|\nabla\tilde{v}_{,t}|_2^{\frac{3}{2}} + \frac{\eta^{-4}}{8}|u|_{\rho,2}^2|\tilde{v}_{,t}|_{\rho,2}^2 + \frac{3}{8}|\nabla u|_2^2\Big)|\rho^{-\frac{1}{4}}|_\infty,$$

$$\int_\Omega \sigma\tilde{v} \cdot \nabla\tilde{v}u \; dx \quad \leq \quad |\sigma|_2|\tilde{v} \cdot \nabla\tilde{v}|_{\sqrt{\rho},4}|u|_{\sqrt{\rho},4}|\rho^{-\frac{1}{4}}|_\infty$$

$$\leq \quad |\sigma|_2|\tilde{v}|_{\rho,2}^{\frac{1}{4}}|\nabla\tilde{v}|_2^{\frac{3}{4}}|\nabla\tilde{v}|_\infty|u|_{\rho,2}^{\frac{1}{4}}|\nabla u|_2^{\frac{3}{4}}|\rho^{-\frac{1}{4}}|_\infty \cdot \eta^{(\frac{3}{8}+\frac{1}{8}-\frac{1}{2})}$$

$$\leq \quad \Big(\frac{\eta}{2}|\sigma|_2^2|\nabla\tilde{v}|_\infty^2 + \frac{3}{8}|\nabla u|_2^2 +$$

$$\frac{\eta^{-4}}{8}|u|_{\rho,2}^2|\tilde{v}|_{\rho,2}^2|\nabla\tilde{v}|_2^6\Big) \cdot |\rho^{-\frac{1}{4}}|_\infty,$$

$$\int_\Omega \sigma hu \; dx \quad \leq \quad |\sigma|_2|u|_{\sqrt{\rho},4}|h|_4|\rho^{-\frac{1}{8}}|_\infty$$

$$\leq \quad |\sigma|_2|u|_{\rho,2}^{\frac{1}{4}}|\nabla u|_2^{\frac{3}{4}}|h|_4|\rho^{-\frac{1}{8}}|_\infty \cdot \eta^{(\frac{3}{8}+\frac{1}{8}-\frac{1}{2})}$$

$$\leq \quad |\rho^{-\frac{1}{8}}|_\infty\Big(\frac{\eta}{2}|\sigma|_2^2 + \frac{3}{8}|\nabla u|_2^2 + \frac{\eta^{-4}}{8}|u|_2^2\Big)|h|_4,$$

$$\int_\Omega \rho u\nabla\tilde{v}u \; dx \quad \leq \quad |u|_{\rho,2}^2|\nabla\tilde{v}|_\infty,$$

$$K\int_\Omega \sigma\nabla \cdot u \; dx \quad \leq \quad K|\sigma|_2|\nabla \cdot u|_2 \leq \Big(\frac{\eta^{-1}}{2}|\sigma|_2^2 + \frac{\eta}{2}|\nabla \cdot u|_2^2\Big)K.$$

From (11) we obtain

$$\int_\Omega \frac{1}{2}\frac{d}{dt}\sigma^2 = -\int_\Omega (\nabla\tilde\rho u + \sigma\nabla\cdot v + \tilde\rho\nabla\cdot u)\sigma. \tag{13}$$

Applying (3), (8) we get

$$\int_\Omega \nabla\tilde\rho u\sigma \le |u|_{\rho,2}|\sigma|_2|\nabla\tilde\rho\rho^{-\frac{1}{2}}|_\infty \le |\nabla\tilde\rho\rho^{-\frac{1}{2}}|_\infty\left(\frac{\eta}{2}|u|^2_{\rho,2} + \frac{\eta^{-1}}{2}|\sigma|^2_2\right),$$

$$\int_\Omega \sigma^2\nabla\cdot v \le |\sigma|^2_2|\nabla\cdot v|_\infty,$$

$$\int_\Omega \tilde\rho\sigma\nabla\cdot u \le |\sigma|_2|\nabla\cdot u|_{\tilde\rho^2,2} \le \frac{\eta^{-1}}{2}|\sigma|^2_2 + \frac{\eta}{2}|\nabla\cdot u|^2_{\tilde\rho^2,2}.$$

So

$$|u|^2_{\rho,2} + |\sigma|^2_2 + \int_0^T |\nabla\cdot u|^2_2(\lambda+\mu) + \mu|\nabla\cdot u|^2_2 \le$$

$$\le \int_0^T 2\left\{\left(\frac{\eta}{2}|\nabla\tilde v_{,t}|^{\frac{3}{2}}_2 + \frac{\eta}{2}|\nabla\tilde v|^2_\infty\right)|\rho^{-\frac{1}{4}}|_\infty + |\rho^{-\frac{1}{8}}|_\infty\frac{\eta}{2}|h|_4 + \right.$$

$$\left. + \frac{K\eta^{-1}}{2} + \frac{\eta^{-1}}{2}|\nabla\tilde\rho\rho^{-\frac{1}{2}}|_\infty + |\nabla v|_\infty + \frac{\eta^{-1}}{2}\right\}|\sigma|^2_2 +$$

$$+ \left\{|\rho^{-\frac{1}{4}}|_\infty\left(\frac{\eta^{-4}}{8}|\tilde v_{,t}|^2_{\rho,2} + \frac{\eta^{-4}}{8}|\tilde v|^2_{\rho,2}|\nabla\tilde v|^6_2\right) + \right.$$

$$\left. + |\rho^{-\frac{1}{8}}|_\infty\left(|h|_4\frac{\eta^{-4}}{8} + |\nabla\tilde v|_\infty + \frac{\eta}{2}|\nabla\tilde\rho\rho^{-\frac{1}{2}}|_\infty\right)\right\}|u|^2_{\rho,2} + \int_0^T \frac{\eta}{2}|\nabla\cdot u|^2_{\tilde\rho^2,2}.$$

Observing that $|\nabla\tilde\rho\rho^{-\frac{1}{2}}|_\infty \le \frac{1}{|x|^\alpha}\frac{1}{|x|^{-K/2}} \le c$ under assumption $\alpha < \frac{K}{2}$, and applying the assumption of large viscosity which satisfyies $(\lambda+\mu)^2 > \tilde\rho^2$, we obtain

$$|u|^2_{\rho,2} + |\sigma|^2_2 \le |u|^2_{\rho,2} + |\sigma|^2_2 + \int_0^T\int_\Omega |\nabla.u|^2_2\{(\lambda+\mu)^2 - \tilde\rho^2\} + \int_0^T \mu|\nabla.u|^2_2 \le$$

$$\le \int_0^T 2(a(t)|\sigma|^2_2 + b(t)|u|^2_{\rho,2})\,dt,$$

where

$$a(t) = \{\left(\frac{\eta}{2}|\nabla\tilde v_{,t}|^{\frac{3}{2}}_2 + \frac{\eta}{2}|\nabla\tilde v|^2_\infty\right)|\rho^{-\frac{1}{4}}|_\infty + |\rho^{-\frac{1}{8}}|_\infty\frac{\eta}{2}|h|_4 +$$

$$+ \frac{K\eta^{-1}}{2} + \frac{\eta^{-1}}{2}|\nabla\tilde\rho\rho^{-\frac{1}{2}}|_\infty + |\nabla v|_\infty + \frac{\eta^{-1}}{2}\}$$

and

$$b(t) = \left\{|\rho^{-\frac{1}{4}}|_\infty\left(\frac{\eta^{-4}}{8}|\tilde v_{,t}|^2_{\rho,2} + \frac{\eta^{-4}}{8}|\tilde v|^2_{\rho,2}|\nabla\tilde v|^6_2\right) + \right.$$

$$\left. + |\rho^{-\frac{1}{8}}|_\infty\left(|h|_4\frac{\eta^{-4}}{8} + |\nabla\tilde v|_\infty + \frac{\eta}{2}|\nabla\tilde\rho\rho^{-\frac{1}{2}}|_\infty\right)\right\}.$$

From regularity assumptions on $\tilde\rho$, $\tilde v$, ρ, v, we see that $a(t)$ and $b(t)$ belong to $L^1(0,T)$. Using Gronwall's lemma we get $\rho = \tilde\rho$, $v = \tilde v$. $\qquad\square$

Remark 3.1. We can generalize this result for $\rho \in L^\infty(0, T; L^p)$.

ACKNOWLEDGEMENTS

The author wishes to express her gratitude to Profs. M. Padula and G.P. Galdi for their help and support during her stay at the University of Ferrara. This research was supported by Italian CNR and by Grant nr. 201/93/2177 of the Grant Agency of Czech Republic.

REFERENCES

ADAMS, R.A. (1975), *Sobolev spaces*, (New York, Academic Press).

GRAFFI, D. (1961), Ancora sul teorema di unicità per le equazioni dei moto dei fluidi, *Atti Accad. Sci. Ist. Bologna (II)* **8**, 7-14.

GRAFFI, D. (1960), Sul teorema di unicità per le equazioni del moto dei fluidi compressibili in un dominio illimitato, *Atti. Accad. Sci. Ist. Bologna* **7**, 1-8.

MATUŠU-NEČASOVÁ, S. (1995), Uniqueness of weak solutions of unsteady motions of viscous compressible fluids, (in preparation).

NEČAS, J. (1967), *Les méthodes directes en théorie des équations elliptiques*, (Prague).

PADULA, M. (1990), On the existence and uniqueness of non-homogeneous motions in exterior domains, *Mathematische Zeitschrift* **203**, 581-604.

RUSSO, R. (1978), On the uniqueness of viscous compressible fluid motions in bounded domains, *Meccanica* 13, 78-82.

RUSSO, R. (1980), Continuous dependence for viscous compressible fluids in unbounded domains, *Riv Math. Univ. Parma* **4**, 333-345.

SERRIN, J. (1959), On the uniqueness of compressible fluid motions, *Arch. Rat. Mech. Anal.* **3**, 272-288.

ON THE EXISTENCE OF STEADY MOTIONS OF A VISCOUS ISOTHERMAL FLUID IN A PIPE

Mariarosaria Padula[1] and Konstantin Pileckas[2]

[1] Università di Potenza, Dipartimento di Matematica
via N. Sauro 85, 85100 Potenza, Italy

[2] Universität GH Paderborn, Fachbereich Mathematik-Informatik
Warburger Str. 100, 33098 Paderborn, Germany

ABSTRACT

We study the problem of existence of steady flows of compressible fluids filling a cylinder, subject to a compact support body force and prescribed flux ϕ. We are not able to prove existence of solutions which decay exponentially to zero for the original problem. However, we solve the problem by adding a control force $\mathbf{f}_1$ with compact support.

1. INTRODUCTION

The paper deals with the existence of solutions to the steady Poisson-Stokes (*alias* compressible Navier-Stokes) equations governing the three-dimensional flows of a viscous compressible fluid in a cylinder Ω, with two outlets, in the direction x_3. We recall that the Poisson-Stokes equations governing the steady flows of a barotropic gas, in Ω are given by

$$\rho\mathbf{v}\cdot\nabla\mathbf{v} = -\nabla p + \mu\Delta\mathbf{v} + (\lambda+\mu)\nabla\nabla\cdot\mathbf{v} + \rho\mathbf{f},$$

$$\nabla\cdot(\rho\mathbf{v}) = 0. \tag{1.1}$$

where the unknowns ρ and $\mathbf{v}$ denote the density and the kinetic field. Moreover, the constants λ and μ are the bulk and shear viscosities, while for an ideal isothermal gas the pressure is given by $p = k\rho$ where the constant k is related to the (uniform) temperature of the gas. The external specific force is denoted by $\mathbf{f}$. Since the domain is unbounded in the direction of the x_3-axis, it seems reasonable to consider, together with the usual adherence condition on the boundary $\partial\Omega$, a condition at infinity, too.

Navier-Stokes Equations and Related Nonlinear Problems
Edited by A. Sequeira, Plenum Press, New York, 1995

Since the condition $(1.1)_2$ implies the constancy of the flux of the momentum $\rho\mathbf{v}$ over the cross section S, in the wake of what is done for incompressible fluids, we are lead to add to (1.1) the following side conditions

$$\mathbf{v}|_{\partial\Omega} = 0$$

$$\int_S \rho v_3 dx' = \beta. \tag{1.2}$$

For incompressible fluids, problem (1.1), (1.2) was investigated by several authors from different points of view, see, e.g., Amick (1977), Ladyzhenskaja and Solonnikov (1983), Horgan and Wheeler (1978), Galdi (1994), and the references quoted therein. Here, we want to indagate the more unexplored field of compressible fluids, where one finds, at once, completely new kinds of problems. To begin with, we notice that in order to prescribe the flux, here we must fix conditions on two variables. It is mathematically unclear in which way one has to fix them, and it constitutes just the object of Novotný, Padula and Penel (1994). Precisely, it was proved in this paper that, in the class of regular solutions having bounded density and a fixed growth condition for the velocity, corresponding to potential external forces summable in the L^2-norm, the flux over the cross section must be necessarily zero. More recently, in Novotný, Padula, and Penel (1995), it was studied again the problem of a correct formulation of the problem, in the class of regular solutions having bounded density and a fixed growth condition for the velocity, corresponding to external forces summable in the L^2-norm. Precisely, they showed the existence of the limit value for the mean (over the cross section S) density ρ,

$$\lim_{x_3 \to 0} \int_S \rho dx' = L|S|, \tag{1.3}$$

$|S|$ denoting the measure of S. Then, provided L has a given value, they proved a uniqueness theorem and the exponential decay for $\mathbf{v}$, and ρ to 0, and L, respectively.

On the other hand, in Padula and Pileckas (1995) we considered the problem (1.1)–(1.2) with external forces of the form $\mathbf{f} = \mathbf{f}_1 + \mathbf{f}_2$, where $\mathbf{f}_1 = h\mathbf{e}_3$, $h = $ const., $\mathbf{e}_3 = (0,\,0,\,1)$ and where the perturbation $\mathbf{f}_2$ vanishes exponentially as $|x_3| \to \infty$. The basic force $\mathbf{f} = \mathbf{f}_1$ has the form $(\mathbf{f}_1)_i(x) = 0$, $i = 1, 2$, $(\mathbf{f}_1)_3 = h$. Therefore, it is natural to suppose that the solution $(\mathbf{v},\ \rho)$ of the perturbed problem should exponentially approach the exact solution $(\mathbf{v}_E,\ \rho_E)$ corresponding to the force $\mathbf{f}_1$. However, because of technical difficulties we are not able to prove the existence result for the perturbed problem in the case of arbitrary exponentially vanishing perturbation $\mathbf{f}_2$. In order to overcome this difficulty, we modify the problem. Precisely, in Padula and Pileckas (1995) we proved an existence theorem by prescribing the quantity $\int_S v_3 dx'$ and by adding a control force. Namely, instead of (1.1)–(1.2), in Padula and Pileckas (1995) we considered the following *control problem*

$$
\begin{cases}
-\mu\Delta\mathbf{v} - (\lambda + \mu)\nabla\mathrm{div}\,\mathbf{v} = -k\nabla\rho + \rho(\mathbf{f}_1 + \mathbf{f}_2) - \rho\mathbf{v}\cdot\nabla\mathbf{v} + \rho\nabla(\alpha\psi), \\[4pt]
\mathrm{div}(\mathbf{v}\rho) = 0, \qquad\qquad x \in \Omega, \\[4pt]
\mathbf{v} = 0, \qquad\qquad\qquad x \in \partial\Omega, \\[4pt]
\displaystyle\int_S \rho\mathbf{v}\cdot\mathbf{k}\,dx' = \beta.
\end{cases}
\tag{1.4}
$$

where ψ is a given function, depending only of x_3, such that $\nabla\psi \in C_0^\infty(\mathbb{R}^1)$ and $supp\,\nabla\psi \subseteq [d,\ d+1]$.

172

Here, we give a short version of the control problem analyzed in Padula and Pileckas (1995). Precisely, in Section 2 we introduce some exact solutions to (1.1)–(1.2), and some functional space. Thus, in Section 3 we formulate the well-posedness question for some elliptic problems and for the transport equation in pipes. Next, in Section 4 we prove existence, uniqueness and asymptotic behavior of steady flows for the control problem (1.4).

2. EXACT SOLUTIONS AND AUXILIARY RESULTS

In this section, we wish to give exact solutions to problem (1.1), (1.2) in the following two cases:

(a) $\mathbf{f} = (f_1(x'), f_2(x'), f_3(x'))$ and $\beta \neq 0$;

(b) $\mathbf{f} = (0, 0, f_3(x_3))$ and $\beta = 0$.

Case (a): We look for solutions in the form

$$\mathbf{v}_E = (0, 0, q_E), \quad \rho_0 = \alpha_0 \rho_E(x'). \tag{2.1}$$

Then, the first two components of $(1.1)_1$ furnish

$$k \nabla' \rho_0 = \rho_0 \mathbf{f}',$$

with $\mathbf{f}' = (f_1, f_2), \nabla' = \nabla_{x'}$,

$$k \nabla' \ln \rho_0 = \mathbf{f}'$$

and $\mathbf{f}'$ should satisfy the compatibility condition

$$(\operatorname{curl} \mathbf{f})_3 = 0$$

Therefore, if ∂S is "good", it results $\mathbf{f}' = \nabla' \varphi_0$ and

$$\rho_0(x', x_3) = \kappa_0(x_3) \exp(\varphi_0(x')/k), \qquad \kappa_0 > 0. \tag{2.2}$$

Hence, a precise expression for the density is completely given. In particular, if $\mathbf{f}(x') = (0, 0, h)$, the exact solution $(\mathbf{v}_E, \rho_E)$ is defined by the relations

$$\mathbf{v}_E(x') = (0, \ 0, \ q_E(x')), \ q_E(x') = (h\beta)^{1/2} k_1^{-1/2} Q_1(x'),$$

$$\rho_E(x') = \rho_E = (\beta h^{-1})^{1/2} k_1^{-1/2} \tag{2.3}$$

Remark 2.1. The flow defined by (2.3) coincides with the classical Hagen-Poiseuille flow.

Remark 2.2. We recall that the incompressible Navier-Stokes equations for laminar Poiseuille flow have the form

$$-\Delta q_E = \nabla p_E, \quad x \in S,$$

$$q_E = 0, \quad x \in \partial S. \tag{2.4}$$

For constant pressure drop $\nabla p_0 = h$ we deduce that

$$q_E(x') = h Q_1(x'), \qquad p_0(x) = h x_3, \qquad \beta = \int_S q_E(x')\, dx' = h k_1.$$

Therefore, in the incompressible case, it is physically equivalent to fix either the kinetic flux β or the pressure drop h. For compressible fluids, on the contrary, the meaning of h is essentially different because for compressible fluids h corresponds to the intensity of the external force f_3 which is fixed. As a consequence, the flux must be in agreement with the external force $\mathbf{f}$. In particular, we have the condition

$$\beta h > 0. \tag{2.5}$$

Case (b): For $\mathbf{f} = (0,\ 0,\ f_3(x_3))$ the force is potential and the non-homogeneous rest state is possible. Precisely, an exact solution is

$$\mathbf{v}_E = (0,0,0), \quad \rho_E = \rho_E(x_3).$$

with

$$\rho_E(x_3) = \kappa_0 \exp\!\left(\int_0^{x_3} f_3(s)/k\right), \quad \kappa_0 > 0,$$

positive and eventually vanishing at infinity. For example, if

$$f_3(x_3) = \ln |x_3|^{-k\alpha},$$

we obtain the solutions

$$\mathbf{v}_E = (0,\ 0,\ 0), \quad \rho_E(x_3) = \kappa_0 |x_3|^{-\alpha}.$$

In Novotný, Padula, and Penel (1995) it has been proved that there exists

$$\lim_{z \to \infty} \int_S \rho \, dx' = L. \tag{2.6}$$

Moreover, in a class of solutions which are sufficiently regular at infinity, and satisfying (2.6) also uniqueness and asymptotic decay to $(0,L)$ was obtained. In what follows, we shall introduce some functional spaces which will be used in the sequel. Precisely, denoting by $W^m(\Omega)$, $W^{m-1/2}(\partial\Omega)$ the usual Sobolev spaces in L^2, we set

$W_\eta^m(\Omega)$ is a space of functions with a finite norm

$$\|u; W_\eta^m(\Omega)\| := \left(\sum_{|\alpha| \le m} \int_\Omega |D^\alpha u|^2 \exp\left(2\eta\sqrt{1+x^2}\right) dx \right)^{1/2};$$

$L_\eta^2(\Omega) := W_\eta^0(\Omega)$;

$\dot{H}_2^1(\Omega)$ is the completion of $C_0^\infty(\Omega)$ with respect to the norm $\|\cdot\,; W^1(\Omega)\|$;

$H_\eta^m(\Omega)$ is the space of functions with a norm

$$\|u; H_\eta^m(\Omega)\| := \|u \exp(\eta x_n); W_2^m(\Omega)\|;$$

$H_\eta^{m-1/2}(\partial\Omega)$ has the norm

$$\|u; W_\eta^{m-1/2}(\partial\Omega)\| := \|u \exp(\eta x_n); W^{m-1/2}(\partial\Omega)\|;$$

$\mathcal{C}_\eta^h(\Omega)$ is a space of functions with

$$\|u; \mathcal{C}_\eta^h(\Omega)\| := \sum_{|\alpha| < h} \sup_{x \in \Omega} \left(|D^\alpha u| \exp\left(\eta \sqrt{1 + x^2} \right) \right) +$$

$$+ \sum_{|\alpha| = [h]} \sup_{x,y \in \Omega} \frac{|D^\alpha u \exp\left(\eta \sqrt{1 + x^2} \right) - D^\alpha u \exp\left(\eta \sqrt{1 + y^2} \right)|}{|x - y|^{h - [h]}};$$

$$\mathcal{W}_\eta^m(\Omega) = \{ u \in W_\eta^m(\Omega) : (\partial u / \partial n)|_{\partial\Omega} = \Delta u|_{\partial\Omega} = 0 \}.$$

Of course, one has

$$W_\eta^m(\Omega) \cong H_\eta^m(\Omega) \cap H_{-\eta}^m(\Omega),$$

$$W_\eta^{m+2-(1/2)}(\partial\Omega) := H_\eta^{m+2-(1/2)}(\partial\Omega) \cap H_{-\eta}^{m+2-(1/2)}(\partial\Omega).$$

Now, we recall some imbedding theorem in weighted spaces.

Lemma 2.1. *Let* $u \in W_\eta^m(\Omega), m \geq 1$. *Then*

(i) If $qm \leq n$, *then* $u \in L_\eta^s(\Omega)$ *with* $2 \leq s \leq 2n/(n - 2m)$ *and*

$$\|u; L_\eta^s(\Omega)\| \leq c\|u; W_\eta^m(\Omega)\|.$$

(ii) If $qm > n$, *then* $u \in \mathcal{C}_\eta^h(\Omega)$ *with* $h \leq (2m - n)/2$ *and*

$$\|u; \mathcal{C}_\eta^h(\Omega)\| \leq c\|u; W_\eta^m(\Omega)\|.$$

Lemma 2.2. *Let* $u \in W_\eta^m(\Omega)$; $\varphi = u|_{\partial\Omega}$, *then* $\varphi \in W_\eta^{m-1/2}(\partial\Omega)$ *and*

$$\|\varphi; W_\eta^{m-1/2}(\partial\Omega)\| \leq c\|u; W_\eta^m(\Omega)\|. \tag{2.7}$$

3. LINEAR PROBLEMS

In this section we recall some main results known for linear systems.

3.1. Linear elliptic problem in the cylinder

Consider in the cylinder Ω an elliptic boundary value problem (ADN-elliptic, see Agmon, Douglis and Nirenberg, 1964)

$$\begin{aligned}
\mathcal{L}(x', D_{x'}, D_{x_n})u(x', x_n) &= f(x', x_n) & x \in \Omega; \\
\mathcal{B}(x', D_{x'}, D_{x_n})u(x', x_n) &= g(x', x_n) & x \in \partial\Omega;
\end{aligned} \tag{3.1}$$

Denote by $\mathcal{A}(D_{x_n})$ the operator of the problem (3.1). Consider the operator $\mathcal{A}(\lambda)$. Let

$$\mathcal{A}(\lambda_0)\varphi^{(0)} = 0, \quad \varphi^{(0)} \neq 0.$$

Then λ_0 is called an eigenvalue of $\mathcal{A}(\lambda)$, and $\varphi^{(0)}$ is an eigenfuction of $\mathcal{A}(\lambda)$. The associated vectors $\varphi^{(j)}$ corresponding to the eigenvalue λ_0 and the eigenfunction $\varphi^{(0)}$ are defined as solutions of the equations

$$\sum_{n=0}^{k} \frac{1}{n!} \frac{\partial^n}{\partial\lambda^n} \mathcal{A}(\lambda)\varphi^{(k-n)}|_{\lambda=\lambda_0} = 0, \qquad k = 1, \ldots, k_0 - 1.$$

175

The system $\{\varphi^{(0)}, \dots, \varphi^{(k_0-1)}\}$ is called a Jordan chain corresponding to λ_0, (see Gohberg and Krein, 1969). It is known (see Kondrat'ev, 1967; Maz'ya and Plamenevskii, 1977) that the function

$$u_\nu^{(\sigma,\tau)}(x', x_n) = e^{i\lambda_\nu x_n} \sum_{k=0}^{\sigma} \frac{1}{k!} (ix_n)^k \varphi_\nu^{(\sigma-k,\tau)}(x') \tag{3.2}$$

satisfies the problem (3.1) with $f = 0$, $g = 0$ if and only if λ_ν is an eigenvalue of $\mathcal{A}(\lambda)$ and $\{\varphi_\nu^{(j,\tau)}, \tau = 1, \dots, I_\nu; j = 0, \dots, k_{\tau\nu}-1\}$ are corresponding systems of Jordan chains.

Assume that the problem (3.1) has the following systems of "elliptic weights": $\{s_h\}, \{t_j\}, \{r_\gamma\}$, i.e. $\mathcal{L}$ is an elliptic $k \times k$-matrix operator with elements $\mathcal{L}_{hj}$, having ord $\mathcal{L}_{hj} = s_h + t_j$, $s_{\max} = 0$, $t_{\min} > 0$ and

$$s_1 + t_1 + \dots + s_k + t_k = 2M$$

$\mathcal{B}$ is $m \times k$-matrix with elements $\mathcal{B}_{\gamma j}$, ord $\mathcal{B}_{\gamma j} = r_\gamma + t_j$, $\gamma = 1, \dots, M$, $j = 1, \dots, k$, see Agmon, Douglis and Nirenberg (1964). It holds

Theorem 3.1. (i) *The operator $\mathcal{A}$ of problem (3.1)*

$$\mathcal{A}(x', D_{x'}, D_{x_n}) : \prod_{j=1}^{k} H_{q,\eta}^{m+t_j}(\Omega) \equiv \mathcal{D}_\eta^m \to$$
$$\prod_{h=1}^{k} H_\eta^{m-s_h}(\Omega) \times \prod_{\gamma=1}^{M} H_\eta^{m-r_\gamma/2}(\partial\Omega) \equiv \mathcal{R}_\eta^m, \tag{3.3}$$

realizes an isomorfism if and only if the line $\Re^1 + i\eta = \{\lambda \in \mathcal{C} : \mathrm{Im}\lambda = \eta\}$ is free of the eigenvalues of the operator-function $\mathcal{A}(x', Dx', \lambda)$. The solution u to (3.1) admits the estimate

$$\|u; \mathcal{D}_\eta^m\| \le \frac{c}{d} \|(f,g); \mathcal{R}_\eta^m\|$$

where d is a distance from the line $\{\lambda : \mathrm{Im}\lambda = \eta\}$ to the spectrum of $\mathcal{A}(\lambda)$. If the eigenvalues are situated on this line the map (3.3) has no Fredholm property.

(ii) *Let $\{f,g\} \in \mathcal{R}_{\eta_1}^m \cap \mathcal{R}_{\eta_2}^m$, $\eta_1 > \eta_2$. Assume that the lines $\Re^1 + i\eta_j = \{\lambda \in \mathcal{C} : \mathrm{Im}\lambda = \eta_j\}$, $j = 1, 2$, are free of eigenvalues of $\mathcal{A}(\lambda)$. Moreover, suppose that N eigenvalues $\lambda_1, \dots, \lambda_N$ are in the strip $\{\lambda \in \mathcal{C} : \eta_2 < \mathrm{Im}\lambda < \eta_1\}$. Then the solutions $u^{(j)} \in \mathcal{D}_{q,\eta_j}^m$, $j = 1, 2$, of the problem (3.1) are related by the equality*

$$u^{(1)}(x', x_n) = \sum_{\nu=1}^{N} \sum_{\tau=1}^{I_\nu} \sum_{\sigma=0}^{k_{\tau\nu}-1} c_\nu^{(\sigma,\tau)} u_\nu^{(\sigma,\tau)}(x', x_3) + u^{(2)}(x', x_3), \tag{3.4}$$

where $c_\nu^{(\sigma,\tau)}$ are constants and $K := k_{\tau\nu}-1$ and $u_\nu^{(\sigma,\tau)}$ are defined by (3.2). Furthermore, $c_\nu^{(\sigma,\tau)}$ can be functions of data only.

Remark 3.1. We get $\mathcal{A}(\lambda)$ from $\mathcal{A}(\dots, x_n)$ by Fourier transform

$$\mathcal{F}_{x_n} \to \lambda.$$

Remark 3.2. In the simpler case of scalar elliptic boundary value problems of second order, the space $\mathcal{D}_\eta^m$, and $\mathcal{R}_\eta^m$ reduce to $H_\eta^{m+1}(\Omega)$ and $H_\eta^m(\Omega) \times H_\eta^{m+2-(1/2)}(\partial\Omega)$, respectively.

Remark 3.3. Let $\eta_1 = \eta$, $\eta_2 = -\eta$, $\eta > 0$. If

$$(f, g) \in \mathcal{H}_\eta^m(\Omega) \cup \mathcal{R}_{-\eta}^m(\Omega)$$

176

and the strip $\{\lambda \in \mathcal{C} : |Im\lambda| < \eta\}$ does not contain the eigenvalues of $\mathcal{A}(\lambda)$ or all constants $c_\nu^{\sigma,\tau} = 0$, we have

$$u^{(1)} = u^{(2)} \in \mathcal{D}_\eta^m(\Omega) \cup \mathcal{D}_{-\eta}^m(\Omega).$$

Notice that in some cases the constants $c_\nu^{\sigma,\tau}$ can be computed as functionals of the data. Then, under certain "compatibility" conditions on the data, it results $c_\nu^{\sigma,\tau} = 0$.

Now, we shall apply Theorem 3.1 to some simple second order elliptic boundary value problems.

(a) *Dirichlet problem for the Poisson equation.*

$$-\Delta u = f, \quad \text{in } \Omega$$
$$u|_{\partial\Omega} = 0. \tag{3.5}$$

In this case there are no eigenvalues of $\mathcal{A}(\lambda)$ in the strip $\{Im(\lambda) \in (-\eta_1^*, \eta_1^*)\}$ where η_1^* is the Poincaré constant given by

$$\eta_1^* := \sup_{u \in H_2^1(S)} \frac{\|u; L^2(S)\|}{\|\nabla u; L^2(S)\|} \tag{3.6}$$

Applying the part (i) of Theorem 3.1, we have

Lemma 3.1. *Let* $f \in W_\eta^{m+2}(\Omega)$ *with* $|\eta| < \eta_1^*$ *and* $m \geq -1$. *Then, there exists only one solution* $u \in W_\eta^{m+2}(\Omega)$ *to the problem (3.5) which satisfies the estimate*

$$\|\nabla u; W_\eta^{m+2}(\Omega)\| \leq c(\eta)\|f; W_\eta^m(\Omega)\|. \tag{3.7}$$

(b) *Neumann problem for the Poisson equation.*

Considering in Ω the following problem

$$-\Delta u = f, \quad \text{in } \Omega$$
$$\left.\frac{\partial u}{\partial \nu}\right|_{\partial\Omega} = 0 \tag{3.8}$$

It is well known that (3.8) is the ADN-elliptic problem. We put

$$\eta_2^* := \sup \frac{\|u; L^2(S)\|}{\|\nabla u; L^2(S)\|} \tag{3.9}$$

where the supremum is taken over all functions $u \in H_2^1(S)$ with

$$\left.\frac{\partial u}{\partial n}\right|_{\partial S} = 0; \quad \int_S u \, dx' = 0.$$

So, in the strip $\{\lambda : Im\lambda \in (-\eta_2^*, \eta_2^*)\}$ there is only one isolated eigenvalue $\lambda_0 = 0$ of $\mathcal{A}(\lambda)$ to which corresponds one eigenfunction $u^{(0)} = 1$ and one associated function $u^{(1)} = ix_3$. Thus, we can state the following (see, e.g. Nazarov and Plamenevskii, 1993; Pileckas, 1981, 1994)

Lemma 3.2. *Let $|\eta| \le \eta_2^*$, $f \in W_\eta^m(\Omega)$, $m \ge -1$. If the compatibility condition*

$$\int_\Omega f dx = 0, \tag{3.10}$$

is satisfied, there exists a solution u to problem (3.8) with $\nabla u \in W_\eta^{m+1}(\Omega)$ such that

$$\|\nabla u; W_\eta^{m+1}(\Omega)\| \le c(\eta)\|f; W_\eta^m(\Omega)\|. \tag{3.11}$$

If, in addition, it is satisfied,

$$\int_\Omega f x_n dx = 0 \tag{3.12}$$

we have $u \in W_\eta^{m+2}(\Omega)$ and

$$\|u; W_\eta^{m+2}(\Omega)\| \le c(\eta)\|f; W_\eta^m(\Omega)\|. \tag{3.13}$$

(c) *Dirichlet problem for the Stokes equations.*

Consider in Ω the Stokes problem

$$\begin{cases} -\Delta \mathbf{u} + \nabla p = \mathbf{f}, & \text{in } \Omega \\ \operatorname{div} \mathbf{u} = 0, & \text{in } \Omega \\ \mathbf{u}|_{\partial\Omega} = \mathbf{g}, & \text{on } \partial\Omega. \end{cases} \tag{3.14}$$

We now derive existence and estimates for solutions to the Stokes problem. In order to do this, we must apply Theorem 3.1 and follow the same reasoning outlined for the Neumann problem. For sake of simplicity, we limit ourselves here to state only the main results, the full proofs are given elsewhere. In this case, we deal with an elliptic system which is more complicated. Precisely, the part (ii) of the Theorem 3.1 furnishes

Lemma 3.3. *Let $\mathbf{f} \in W_\eta^m(\Omega)$, $\mathbf{g} \in W_\eta^{(m+2-(1/2))}(\partial\Omega)$, with $0 < \eta < \eta_3^*$, $m \ge -1$. Assume that the following compatibility condition holds*

$$\int_{\partial\Omega} \mathbf{g} \cdot \mathbf{n}\, dS = 0. \tag{3.15}$$

Then the Stokes problem (3.14) has a unique solution $(\mathbf{u}, p)$ with $\mathbf{u} \in W_\eta^{m+2}(\Omega)$, $\nabla p \in W_\eta^m(\Omega)$ and the following estimate is valid

$$\|\mathbf{u};\ W_\eta^{m+2}(\Omega)\| + \|\nabla p;\ W_\eta^m(\Omega)\| \le c \left(\|\mathbf{f};\ W_\eta^m(\Omega)\| + \right.$$
$$\left. + \|\mathbf{g}; W_\eta^{m+3/2}(\partial\Omega)\| \right). \tag{3.16}$$

Moreover, there exists the "pressure drop" $p_* = \lim_{x_3 \to +\infty} p(x) - \lim_{x_3 \to -\infty} p(x)$ given by the formula

$$p_* = M(\mathbf{f}, \mathbf{g}) := (\chi_1)^{-1} \left[\int_{\partial\Omega} \mathbf{g} \cdot \mathbf{n} x_3\, dS - \int_{\partial\Omega} g_3 \nabla' Q_1 \cdot \mathbf{n}' dS + \int_\Omega f_3\, Q_1\, dx \right],$$
$$\chi_1 := \int_S Q_1(x')\, dx'. \tag{3.17}$$

If, in addition, the second compatibility condition

$$M(\mathbf{f}, \mathbf{g}) = 0 \tag{3.18}$$

holds, then $(\mathbf{u}, p) \in W_\eta^{m+2}(\Omega) \times W_\eta^{m+1}(\Omega)$ *and*

$$\|\mathbf{u}; W_\eta^{m+2}(\Omega)\| + \|p; W_\eta^{m+1}(\Omega)\| \le c \left(\|\mathbf{f}; W_\eta^m(\Omega)\| + \right.$$
$$\left. + \|\mathbf{g}; W_\eta^{m+3/2}(\partial\Omega)\| \right). \tag{3.19}$$

Lemma 3.4. *The compatibility conditions* (3.10), (3.12) *(for the Neumann problem* (3.8)*) and* (3.15), (3.18) *(for the Stokes problem* (3.14)*) follow by computing the constants in the asymptotic representation* (3.4) *of the solutions.*

Lemma 3.5. *The Laplacian of the pressure p can be estimated only in terms of the external force* $\mathbf{f}$ *and if* $m \ge 1$*, we get the inequality*

$$\|\Delta p; W_\eta^{m-1}(\Omega)\| \le c\|\mathbf{f}; W_\eta^{m-1}(\Omega)\|.$$

To prove this, it is enough to apply the operator div *to the equations* (3.14). *Then we obtain*

$$\Delta p = \operatorname{div} \mathbf{f}.$$

3.2. Linear symmetric problem in the cylinder

Consider in Ω the following equation which is symmetric positive in the sense of Friedrichs (1958)

$$\sigma + \chi_0 \operatorname{div}(\sigma \mathbf{v}) = p \tag{3.20}$$

with

$$\chi_0 := \frac{\lambda + 2\mu}{k\rho_0}, \quad \mathbf{v} = \mathbf{v}_E + \mathbf{u}.$$

Assume that

$$\mathbf{v} = \mathbf{u} + \mathbf{v}_E, \quad \mathbf{v} \cdot \mathbf{n}|_{\partial\Omega} = 0,$$

$$\mathbf{u} \in W_\eta^{m+2}(\Omega), \quad \mathbf{v}_E = \mathbf{v}_E(x') \in W^{m+2}(S), \quad \eta > 0,$$

$$\|\mathbf{u}; W^{m+2}(\Omega)\| < \epsilon, \quad \epsilon \text{ sufficiently small.}$$

If

$$\|p; W^{m+1}(\Omega)\| < \infty, \quad \|\mathbf{v}; W^{m+2}(\Omega)\| < \epsilon,$$

for Ω bounded, or $\Omega = \mathcal{R}^n$, or Ω exterior, the problem (3.20) has a unique solution σ in $W^{m+1}(\Omega)$, see Beirão da Veiga (1987), Padula (1987, 1992, 1993a, 1993b) Novotný (1995) and the bibliography there quoted. In particular, the following estimate holds

$$\|\sigma; W^{m+1}(\Omega)\| \le c\|p; W_q^{m+1}(\Omega)\|.$$

In our case, it is still $\mathbf{v}|_{\partial\Omega} = 0$, but now Ω has a non compact boundary. It holds

$$\|\mathbf{v}_E; W^{m+1}(\Omega_R)\| = R^{1/2}\|\mathbf{v}_E; W^{m+1}(S)\|,$$

with $\Omega_R = S \times (-R, R)$. Therefore, also the norm of $\mathbf{v}$ in $W^{m+2}(\Omega)$ is not bounded. In Padula and Pileckas (1995) we proved existence in weighted spaces under assumptions weaker than summability. In particular, we stated the following result.

179

Lemma 3.6. *Let $p \in W^{m+1}_\eta(\Omega) \cap L^2(\Omega)$ with $m \geq 1, \eta \in (0, \gamma_*)$, where $\gamma_* := \min\{\eta_1^*, \eta_2^*, \eta_3^*\}$. Set*

$$\|v_E; W^{m+2}(S)\| + \|u; W^{m+2}_\eta(\Omega)\| := \gamma_\eta.$$

If γ_η is sufficiently small, then there exists a unique solution $\sigma \in W^{m+1}_\eta(\Omega)$ to the problem (3.20) satisfying the estimates

$$\ll \sigma; W^{m+1}_\eta(\Omega) \gg \;\leq\; c\|p; W^{m+1}_\eta(\Omega)\|,$$

$$\ll \sigma; W^{-1}_\eta(\Omega) \gg \;\leq\; c\|p; W^{-1}_\eta(\Omega)\|, \tag{3.21}$$

$$< \nabla div(\mathbf{v}\sigma) >_{-1} \;\leq\; c < \nabla p >_{-1}$$

where

$$\ll \sigma; X \gg := \|\sigma; X\| + \|div(\mathbf{v}\sigma); X\|.$$

4. NONLINEAR PROBLEMS

Consider in the cylinder $\Omega = S \times (-\infty, +\infty)$ the modified Poisson-Stokes equations

$$\begin{cases} -\mu\Delta\mathbf{v} - (\lambda + \mu)\nabla div\,\mathbf{v} = -k\nabla\rho + \rho\mathbf{f} - \rho\mathbf{v}\cdot\nabla\mathbf{v} + \rho\nabla(\alpha\psi), \\ div(\mathbf{v}\rho) = 0, \\ \mathbf{v}|_{\partial\Omega} = 0, \\ \int_S \rho\,\mathbf{v}\cdot\mathbf{n}\,dx' = \beta \end{cases} \tag{4.1}$$

where $x' = (x_1, x_2)$. Moreover, $\mathbf{f} = \mathbf{f}^{(1)} + \mathbf{f}^{(2)}$, with $\mathbf{f}^{(1)} = h\mathbf{e}_3$ and with $\mathbf{f}^{(2)}$ exponentially vanishing, $\mathbf{f}^{(2)} \in W^m_\eta(\Omega)$. Moreover, ψ is a smooth function, of x_3 only, such that $\psi \in C_o^\infty(\mathbb{R}^1)$, $supp\,\psi \subseteq [d, d+1]$,

$$\lim_{x_3 \to +\infty} \psi(x_3) = 1, \quad \lim_{x_3 \to -\infty} \psi(x_3) = 0,$$

$$\int_\Omega \psi(x_3)q_0(x')dx'dx_n = 1.$$

The exact solution corresponding to $\mathbf{f}^{(2)} = 0, \alpha = 0$ has the form

$$v_E = (0, 0, q_0(x')), \quad \rho_E = const.$$

The norm of $\mathbf{f}^{(2)}$ will be assumed sufficiently small compared with h, i.e.

$$\|\mathbf{f}^{(2)}; W^m(\Omega)\| \leq \delta_0 \ll h. \tag{4.2}$$

Furthermore, we know that

$$\beta \sim h \quad \text{as} \quad h \to 0.$$

Then

$$\|\mathbf{v}_E; W^{m+2}(S)\| \sim h, \quad \rho_E \sim 1 \quad \text{as} \quad h \to 0. \tag{4.2*}$$

We look for a velocity vector $\mathbf{v}$, density function ρ and constant α such that the solution $(\mathbf{v}, \rho, \alpha)$ to (4.1) has the prescribed flux β. Here we use the decomposition method introduced by Novotný and Padula (1994).

Let us represent $(\mathbf{v}, \sigma)$ in the form

$$\mathbf{v} = \mathbf{u} + \mathbf{v}_E, \quad \rho = \sigma + \rho_E.$$

For the perturbation $\mathbf{u}, \sigma$ we derive

$$\begin{cases} -\mu\Delta\mathbf{u} - (\lambda + \mu)\nabla\mathrm{div}\,\mathbf{u} = -k\nabla\sigma + \mathbf{F}(\mathbf{u}, \sigma, \alpha) + \rho_E\nabla(\alpha\psi), \\ \mathrm{div}[(\sigma + \rho_E)(\mathbf{u} + \mathbf{v}_E)] = 0, \\ \mathbf{u}|_{\partial\Omega} = 0, \\ \int_S [(\sigma + \rho_E)\mathbf{u}\cdot\mathbf{n} + \sigma\mathbf{v}_E\cdot\mathbf{n}\,]dx' \end{cases} \tag{4.3}$$

where

$$\mathbf{F}(\mathbf{u}, \sigma, \alpha) := (\sigma + \rho_E)\mathbf{f}_2 + \sigma\mathbf{f}_1 - \mathbf{N}(\mathbf{u}, \sigma) + \sigma\nabla(\alpha\psi),$$
$$\mathbf{N}(\mathbf{u}, \sigma) := (\sigma + \rho_E)[(\mathbf{u} + \mathbf{v}_E)\cdot\nabla(\mathbf{u} + \mathbf{v}_E)] =$$
$$= (\sigma + \rho_E)[\mathbf{u}\cdot\nabla\mathbf{v}_E + \mathbf{u}\cdot\nabla\mathbf{u} + \mathbf{v}_E\cdot\nabla\mathbf{u}].$$

We look for $\mathbf{u}$ in the form

$$\mathbf{u} = \mathbf{w} + \epsilon\nabla\varphi \tag{4.4}$$

with

$$\mathrm{div}\,\mathbf{w} = 0, \quad \left.\frac{\partial\varphi}{\partial n}\right|_{\partial\Omega} = 0, \quad 0 < \epsilon < 1.$$

It is not difficult to check that setting

$$p = k\sigma - (\lambda + 2\mu)\epsilon\Delta\varphi \tag{4.5}$$

we obtain three problems which together are equivalent to the system (4.1),

$$-\mu\Delta\mathbf{w} + \nabla p = \mathbf{F}^{(1)}(\mathbf{w}, \nabla\varphi, \sigma, \alpha) + \rho_E\nabla(\alpha\psi),$$
$$\mathrm{div}\,\mathbf{w} = 0, \tag{4.6}$$
$$\mathbf{w}|_{\partial\Omega} = -\epsilon\nabla\varphi|_{\partial\Omega},$$

$$\sigma + \frac{(\lambda+2\mu)}{k\rho_E}\mathrm{div}[\sigma(\mathbf{w} + \epsilon\nabla\varphi + \mathbf{v}_E)) = \overline{p}, \tag{4.7}$$

$$\epsilon\Delta\varphi = G(\mathbf{w}, \nabla\varphi, \sigma),$$
$$\frac{\partial\varphi}{\partial n}|_{\partial\Omega} = 0, \tag{4.8}$$

and where

$$\mathbf{F}^{(1)}(\mathbf{w}, \nabla\varphi, \sigma) := \mathbf{N}(\mathbf{w} + \epsilon\nabla\varphi, \sigma) + (\rho_E + \sigma)\mathbf{f}^{(2)} + \sigma\mathbf{f}^{(1)},$$

$$G(\mathbf{w}, \nabla\varphi, \sigma) := -(1/\rho_E)\mathrm{div}[\sigma(\mathbf{w} + \epsilon\nabla\varphi + \mathbf{v}_E)], \tag{4.9}$$
$$\overline{p} = p/k.$$

We solve the nonlinear problem (4.1) by the method of successive approximations. Assume that the functions $\mathbf{w}_{l-1} \in W_\eta^{m+2}(\Omega)$, $\nabla\varphi_{l-1} \in W_\eta^{m+2}(\Omega)$, $\sigma_{l-1} \in W_\eta^{m+1}(\Omega)$, $m > 1$, and the constant $\alpha_{l-1} \in \mathbb{R}^1$ are already defined. Moreover, let

$$(\partial\varphi_{l-1}/\partial n)|_{\partial\Omega} = 0, \quad (\Delta\varphi_{l-1})|_{\partial\Omega} = 0. \tag{4.10}$$

We put

$$\alpha_l = -(\rho_E \chi_1)^{-1} \left[\int_{\partial\Omega} \varepsilon \nabla' \varphi_{l-1} \cdot \mathbf{n}' x_3 \, dS - \mu \int_{\partial\Omega} \varepsilon \frac{\partial \varphi_{l-1}}{\partial x_3} \nabla' Q_1 \cdot \mathbf{n}' dS + \right.$$
$$\left. + \int_{\partial\Omega} F_3(\mathbf{w}_{l-1}, \nabla \varphi_{l-1}, \sigma_{l-1}, \alpha_{l-1}) Q_1 \, dx \right]. \tag{4.11}$$

Because of the condition (4.10) we have

$$\int_{\partial\Omega} \varepsilon \nabla \varphi_{l-1} \cdot \mathbf{n} x_3 dS = 0.$$

Moreover,

$$\int_{\partial\Omega} \varepsilon \frac{\partial \varphi_{l-1}}{\partial x_3} \nabla' Q_1 \cdot \mathbf{n}' dS =$$

$$= \varepsilon \left(\lim_{x_3 \to +\infty} \int_{\partial S} \varphi_{l-1}(x) \nabla' Q_1 \cdot \mathbf{n}' dS - \lim_{x_3 \to -\infty} \int_{\partial S} \varphi_{l-1}(x) \nabla' Q_1 \cdot \mathbf{n}' \, dS \right) =$$

$$= \varepsilon(\varphi_{l-1}^+ - \varphi_{l-1}^-) \int_{\partial S} \nabla' Q_1 \cdot \mathbf{n}' \, dS = \varepsilon \varphi_{l-1}^* \int_{\partial S} \nabla' Q_1 \cdot \mathbf{n}' \, dS,$$

where $\varphi_{l-1}^* = \varphi_{l-1}^+ - \varphi_{l-1}^-$. From the last formula it is easy to verify that

$$|\alpha_l| \le c \left(\varepsilon |\varphi_{l-1}^*| + \| \mathbf{F}(\mathbf{w}_{l-1}, \nabla \varphi_{l-1}, \sigma_{l-1}, \alpha_{l-1}); W_\eta^m(\Omega) \| \right). \tag{4.12}$$

Now, we define $\mathbf{w}_l$, p_l as a solution to the Stokes problem

$$-\mu \Delta \mathbf{w}_l + \nabla p_l = \mathbf{F}(\mathbf{w}_{l-1}, \nabla \varphi_{l-1}, \sigma_{l-1}, \alpha_{l-1}) + \rho_E \nabla(\alpha_l \psi), \qquad x \in \Omega,$$
$$\mathrm{div}\, \mathbf{w}_l = 0, \qquad\qquad\qquad x \in \Omega, \tag{4.13}$$
$$\mathbf{w}_l = -\varepsilon \nabla \varphi_{l-1}, \qquad\qquad\qquad x \in \partial\Omega.$$

Since $(\partial \varphi_{l-1} / \partial n)|_{\partial\Omega} = 0$, the first compatibility condition (3.15) is valid. Moreover, because of (4.11), we satisfy the second compatibility condition (3.18). Therefore, by Lemma 3.3, problem (4.13) has a unique solution $(\mathbf{w}_l,\ p_l)$ satisfying the estimates

$$\|\mathbf{w}_l;\ W_\eta^{m+2}(\Omega)\| + \|p_l;\ W_\eta^{m+1}(\Omega)\|$$
$$\le c \left(\| \mathbf{F}(\mathbf{w}_{l-1}, \nabla \varphi_{l-1}, \sigma_{l-1}, \alpha_{l-1}); W_\eta^m(\Omega) \| + |\alpha_l| + \varepsilon \| \nabla \varphi_{l-1}; W_\eta^{m+2}(\Omega) \| \right)$$
$$\le c \left(\| \mathbf{F}(\mathbf{w}_{l-1}, \nabla \varphi_{l-1}, \sigma_{l-1}, \alpha_{l-1}); W_\eta^m(\Omega) \| + \varepsilon |\varphi_{l-1}^*| + \right. \tag{4.14}$$
$$\left. + \varepsilon \| \nabla \varphi_{l-1}; W_\eta^{m+2}(\Omega) \| \right),$$

$$\| \Delta p_l;\ W_\eta^{m-1}(\Omega) \| \le c \left(\| \mathrm{div}\, \mathbf{F}(\mathbf{w}_{l-1}, \nabla \varphi_{l-1}, \sigma_{l-1}, \alpha_{l-1}); W_\eta^{m-1}(\Omega) \| + |\alpha_l| \right)$$
$$\le c \left(\| \mathbf{F}(\mathbf{w}_{l-1}, \nabla \varphi_{l-1}, \sigma_{l-1}, \alpha_{l-1}); W_\eta^m(\Omega) \| + \varepsilon |\varphi_{l-1}^*| \right). \tag{4.15}$$

Let us assume that the functions $\mathbf{w}_{l-1}, \nabla \varphi_{l-1}, \sigma_{l-1}$ and the constant α_{l-1} admit the estimates

$$\|\mathbf{w}_{l-1};\ W_\eta^{m+2}(\Omega)\| \le r_0, \quad \|\nabla \varphi_{l-1};\ W_\eta^{m+2}(\Omega)\| \le r_0,$$
$$\|\sigma_{l-1};\ W_\eta^{m+1}(\Omega)\| \le r_0, \quad |\alpha_{l-1}| \le r_0, \quad |\varphi_{l-1}^*| \le \gamma r_0, \tag{4.16}$$

with sufficiently small r_0 and γ. By using the weighted imbedding theorem (see Lemma 3.1) it is easy to verify that

$$\|\mathbf{F}(\mathbf{w}_{l-1}, \nabla\varphi_{l-1}, \sigma_{l-1}, \alpha_{l-1}); W_\eta^{m,2}(\Omega)\| \le c(\delta_0 + r_0^2 + hr_0). \tag{4.17}$$

Then, by (4.16) the inequalities (4.12), (4.14) and (4.15) imply that

$$|\alpha_l| \le c_1(\delta_0 + r_0^2 + hr_0 + \varepsilon\gamma r_0),$$

$$\|\mathbf{w}_l; W_\eta^{m+2,2}(\Omega)\| + \|p_l; W_\eta^{m+1,2}(\Omega)\| \le c_1(\delta_0 + r_0^2 + hr_0 + \varepsilon r_0), \tag{4.18}$$

$$\|\Delta p_l; W_\eta^{m-1,2}(\Omega)\| \le c_1(\delta_0 + r_0^2 + hr_0 + \varepsilon\gamma r_0).$$

Now we solve the transport equation:

$$\sigma_l + \frac{(\lambda + 2\mu)}{k\rho_E}\operatorname{div}[\sigma_l(\mathbf{w}_l + \varepsilon\nabla\varphi_{l-1} + \mathbf{v}_E)] = \overline{p_l}, \qquad x \in \Omega. \tag{4.19}$$

By virtue of (4.16) and (4.18), we have

$$\|\mathbf{v}_E; W^{m+2}(S)\| + \|\mathbf{w}_l; W_\eta^{m+2}(\Omega)\| + \varepsilon\|\nabla\varphi_{l-1}; W_\eta^{m+2}(\Omega)\|$$

$$\le (h + c_*(\delta_0 + r_0^2 + hr_0 + \varepsilon r_0) + \varepsilon r_0) := \gamma_1.$$

Assume that

$$\gamma_1 < \gamma_*, \quad \gamma_1 < \gamma, \tag{4.20}$$

where γ_* is defined in Lemma 3.7. Then by this Lemma there exists a unique solution $\sigma_l \in W_\eta^{m+1,2}(\Omega)$ to the problem (4.19) satisfying the estimates (3.21) which together with (4.14), (4.15) imply that

$$\|\sigma_l; W_\eta^{m+1,2}(\Omega)\| + \|\operatorname{div}[(\mathbf{w}_l + \varepsilon\nabla\varphi_{l-1} + \mathbf{v}_E)\sigma_l]; W_\eta^{m+1,2}(\Omega)\|$$

$$\le c_2(\delta_0 + r_0^2 + hr_0 + \varepsilon r_0),$$

$$\|\Delta\sigma_l; W_\eta^{m-1,2}(\Omega)\| + \|\Delta\left(\operatorname{div}[(\mathbf{w}_l + \varepsilon\nabla\varphi_{l-1} + \mathbf{v}_E)\sigma_l]\right); W_\eta^{m-1,2}(\Omega)\| \tag{4.21}$$

$$\le c_2(\delta_0 + r_0^2 + hr_0 + \varepsilon\gamma r_0).$$

Finally, we consider the Neumann problem

$$\varepsilon\Delta\varphi_l = -\rho_E^{-1}\operatorname{div}[(\mathbf{w}_l + \varepsilon\nabla\varphi_{l-1} + \mathbf{v}_E)\sigma_l], \qquad x \in \Omega,$$

$$\frac{\partial\varphi_l}{\partial n} = 0, \qquad x \in \partial\Omega. \tag{4.22}$$

It is easy to see that the right–hand side

$$G = -\rho_E^{-1}\operatorname{div}[(\mathbf{w}_l + \varepsilon\nabla\varphi_{l-1} + \mathbf{v}_E)\sigma_l]$$

of the problem (4.22) satisfies the compatibility condition (3.10), i.e.

$$\int_\Omega G\, dx = 0.$$

Therefore, by Lemma 3.2 there exists a solution φ_l of (4.20) with $\nabla\varphi_l \in W_\eta^{m+2}(\Omega)$ and

$$\varepsilon\|\nabla\varphi_l; W_\eta^{m+2}(\Omega)\| \le c\|G; W_\eta^{m+1}(\Omega)\|. \tag{4.23}$$

Since
$$(\Delta\varphi_{l-1})|_{\partial\Omega} = 0, \quad \mathbf{w}_l|_{\partial\Omega} = -\varepsilon(\nabla\varphi_{l-1})|_{\partial\Omega},$$

we have

$$G|_{\partial\Omega} = -\rho_E^{-1}\{(\mathbf{w}_l + \varepsilon\varphi_{l-1})\cdot\nabla\sigma_l + \sigma_l\varepsilon\Delta\varphi_{l-1} + \mathbf{v}_E\cdot\nabla\sigma_l\}|_{\partial\Omega} = 0. \tag{4.24}$$

Therefore, we can consider G as a solution of the Dirichlet problem for the Poisson equation

$$\Delta G = -\rho_E^{-1}\Delta\left(\text{div}[(\mathbf{w}_l + \varepsilon\nabla\varphi_{l-1} + \mathbf{v}_E)\sigma_l]\right),$$
$$G = 0, \quad x \in \partial\Omega.$$

According to Lemma 3.1, the solution G admits the estimate

$$\|G;\ W_\eta^{m+1}(\Omega)\| \le c\|\Delta\left(\text{div}[(\mathbf{w}_l + \varepsilon\nabla\varphi_{l-1} + \mathbf{v}_E)\sigma_l]\right);\ W_\eta^{m-1}(\Omega)\|$$

and by (4.14), (4.16), (4.21) we have

$$\varepsilon\|\nabla\varphi_l;\ W_\eta^{m+2}(\Omega)\| \le c\|G;\ W_\eta^{m+1}(\Omega)\|$$
$$\le c\|\Delta\left(\text{div}[(\mathbf{w}_l + \varepsilon\nabla\varphi_{l-1} + \mathbf{v}_E)\sigma_l]\right);\ W_\eta^{m-1}(\Omega)\| \tag{4.25}$$
$$\le c_3(\delta_0 + r_0^2 + hr_0 + \varepsilon\gamma r_0).$$

Let us estimate the difference $\varphi_l^* = \lim_{x_3\to+\infty}\varphi_l(x) - \lim_{x_3\to-\infty}\varphi_l(x)$. By (3.17)

$$\varepsilon\varphi_l^* = -\rho_E^{-1}\int_\Omega x_3\text{div}[(\mathbf{w}_l + \varepsilon\nabla\varphi_{l-1} + \mathbf{v}_E)\sigma_l)]\,dx$$
$$= \rho_E^{-1}\int_\Omega \sigma_l\left(w_{3,l} + \varepsilon(\frac{\partial\varphi_{l-1}}{\partial x_3} + v_{3,E})\right)\,dx.$$

Hence, in virtue of (4.18), (4.21), (4.25) we have

$$\varepsilon|\varphi_l^*| \le c\|\sigma_l;\ W_\eta^{m+1}(\Omega)\|\left(\|\mathbf{w}_l;\ W_\eta^{m+2}(\Omega)\| + \varepsilon\|\nabla\varphi_{l-1};\ W_\eta^{m+2}(\Omega)\| + \right.$$
$$\left. +\|\mathbf{v}_E;\ W^{m+2,2}(S)\|\right) \le c_4(\delta_0 + r_0^2 + hr_0 + \varepsilon r_0)(\delta_0 + r_0^2 + hr_0 + \varepsilon r_0 + h). \tag{4.26}$$

Let

$$\gamma \le \min\{\frac{1}{4c_3},\ 1\}, \quad \varepsilon \le \frac{1}{4c_*}, \quad r_0 \le \min\{\frac{1}{4c_*},\ \frac{\varepsilon}{4c_3}\},$$
$$h < \min\{\frac{1}{4c_*},\ \frac{\varepsilon}{4c_3}\}, \quad \delta_0 < \min\{\frac{r_0}{4c_*},\ \frac{\varepsilon r_0}{4c_3}\}, \tag{4.27}$$

where $c_* = \max\{c_1, c_2, c_3, c_4\}$. Then, using (4.27) the inequalities (4.24)–(4.18), (4.21), (4.25) furnish

$$\|\mathbf{w}_l;\ W_\eta^{m+2}(\Omega)\| \le r_0, \quad \|\nabla\varphi_l;\ W_\eta^{m+2}(\Omega)\| \le r_0,$$
$$\|\sigma_l;\ W_\eta^{m+1,2}(\Omega)\| \le r_0, \quad |\alpha_l| \le r_0.$$

Moreover, if in addition it holds

$$r_0 < \varepsilon\gamma, \quad h < \varepsilon\gamma, \tag{4.28}$$

from (4.27) it follows that

$$|\varphi_l^*| \le \gamma r_0.$$

Further, from the equation (4.22) we have
$$(\varepsilon \Delta \varphi_l)|_{\partial\Omega} = -\rho_E^{-1}(\text{div}[(\mathbf{w}_l + \varepsilon\nabla\varphi_{l-1} + \mathbf{v}_E)\sigma_l])|_{\partial\Omega} = 0. \tag{4.29}$$

Hence, the approximation $(\mathbf{w}_l, \nabla\varphi_l, \sigma_l, \alpha_l)$ satisfies the same conditions (4.10), (4.16), as it was assumed for the approximation $(\mathbf{w}_{l-1}, \nabla\varphi_{l-1}, \sigma_{l-1}, \alpha_{l-1})$.

Now, in order to prove the boundedness of the sequence, it is enough to consider the first term of it. We define the constant α_1 by the formula
$$\alpha_1 = -\chi_1^{-1} \int_\Omega f_3^{(1)} Q_1 \, dx$$

and $(\mathbf{w}_1, \varphi_1, \sigma_1)$ are the solutions of the equations
$$
\begin{cases}
-\mu\Delta\mathbf{w}_1 + \nabla p_1 = \rho_E b f f^{(1)} + \rho_E \nabla(\alpha_1 \psi), & x \in \Omega \\[2mm]
\text{div } \mathbf{w}_1 = 0, & x \in \Omega \\[2mm]
\mathbf{w}_1 = 0, & x \in \partial\Omega \\[2mm]
\sigma_1 + \dfrac{(\lambda+2\mu)}{k\rho_E}\text{div}[\sigma_1(\mathbf{w}_1 + \mathbf{v}_E)] = p_1/k, & x \in \Omega \\[2mm]
\varepsilon\Delta\varphi_1 = -(1/\rho_E)\text{div}[\sigma_1(\mathbf{w}_1 + \mathbf{v}_E)], & x \in \Omega \\[2mm]
\dfrac{\partial\varphi_1}{\partial n} = 0, & x \in \partial\Omega.
\end{cases}
$$

Then, the estimates (4.12), (4.14), (4.21), (4.23) and the conditions (4.27), (4.28) imply
$$|\alpha_1| \le r_0, \quad \|\mathbf{w}_1; \, W_\eta^{m+2}(\Omega)\| + \|p_1; \, W_\eta^{m+1}(\Omega)\| \le r_0,$$
$$\|\sigma_1; \, W_\eta^{m+1}(\Omega)\| \le r_0, \quad \|\nabla\varphi_1; \, W_\eta^{m+2}(\Omega)\| \le r_0, \quad |\varphi_1^*| \le \gamma r_0.$$

Thus, under the conditions (4.27), (4.28) the approximations $(\mathbf{w}_l, \nabla\varphi_l, \sigma_l, \alpha_l)$ satisfy the conditions (4.10), (4.16) for all l.

It remains to prove that $\{\mathbf{w}_l, \nabla\varphi_l, \sigma_l, \alpha_l\}$ is convergent. We will not make explicitly these calculations which are extensively written in Padula and Pileckas (1995). We only give a sketch of this proof. To this end we consider the differences between the terms
$$\alpha_l' = \alpha_l - \alpha_{l-1}, \quad \mathbf{w}_l' = \mathbf{w}_l - \mathbf{w}_{l-1}, \quad p_l' = p_l - p_{l-1},$$
$$\varphi_l' = \varphi_l - \varphi_{l-1}, \quad \sigma_l' = \sigma_l - \sigma_{l-1},$$

and write the equations (4.11), (4.13), (4.19), (4.22) for $(\mathbf{w}_l', \nabla\varphi_l', \sigma_l', \alpha_l')$. At right hand side there will appear $(\mathbf{w}_{l-1}', \nabla\varphi_{l-1}', \sigma_{l-1}', \alpha_{l-1}')$. Therefore, using again the estimates of Lemmas 3.1–3, and of Lemma 3.5, we prove that if
$$\varepsilon \le \min\{\frac{1}{8c_{**}}, 1\}, \quad r_0 \le \min\{\frac{1}{24c_{**}}, \frac{\varepsilon}{2}\},$$
$$h < \min\{\frac{1}{24c_{**}}, \frac{\varepsilon}{2}\}, \quad \delta_0 < \frac{1}{24c_{**}}, \tag{4.30}$$

then it holds
$$|\alpha_l - \alpha_{l-1}| + \|\mathbf{w}_l - \mathbf{w}_{l-1}; \, W_\eta^3(\Omega)\| + \|p_l - p_{l-1}; \, W_\eta^2(\Omega)\| +$$
$$+\|\sigma_l - \sigma_{l-1}; \, W_\eta^2(\Omega)\| + \|\nabla\varphi_l - \nabla\varphi_{l-1}; \, W_\eta^3(\Omega)\| + |\varphi_l^* - \varphi_{l-1}^*|$$
$$\le a\left(|\alpha_{l-1} - \alpha_{l-2}| + \|\mathbf{w}_{l-1} - \mathbf{w}_{l-2}; \, W_\eta^3(\Omega)\| + \|p_{l-1} - p_{l-2}; \, W_\eta^2(\Omega)\| + \right.$$
$$\left. +\|\sigma_{l-1} - \sigma_{l-2}; \, W_\eta^2(\Omega)\| + \|\nabla\varphi_{l-1} - \nabla\varphi_{l-2}; \, W_\eta^3(\Omega)\| + |\varphi_{l-1}^* - \varphi_{l-2}^*|\right)$$

with

$$0 < a < 1.$$

Therefore, the sequence $\{\alpha_l,\ \mathbf{w}_l,\ p_l,\ \sigma_l,\ \nabla\varphi_l,\ \varphi_l^*\}$ converges in the space $I\!R^1 \times W_\eta^3(\Omega) \times W_\eta^2(\Omega) \times W_\eta^2(\Omega) \times W_\eta^3(\Omega) \times I\!R^1$ to the limit point $\{\alpha,\ \mathbf{w},\ p,\ \sigma,\ \nabla\varphi,\ \varphi^*\}$. Because of the estimate (4.16) the limit point belongs to the more regular space $I\!R^1 \times W_\eta^{m+2}(\Omega) \times W_\eta^{m+1}(\Omega) \times W_\eta^{m+1}(\Omega) \times I\!R^1$.

Thus, we have proved the main result of the paper.

Theorem 4.1. *Let* $\mathbf{f}^{(1)} \in W_\eta^{m,2}(\Omega)$ *with*

$$m > 1,\ \eta \in (0,\ \min\{\eta_1^*,\ \eta_2^*,\ \eta_3^*,\ \gamma_*\}),$$

where $\eta_i^*, i = 1,2,3,$ *and* γ_* *are defined in Section 3. Assume that the condition* (4.2*) *is valid and that the numbers* δ_0, h, ε, r_0, γ *are sufficiently small (the values of these parameters are specified by the conditions* (4.30).*) Then the control problem* (4.1)*, with* $\beta \sim h$*, has a unique solution* $(\mathbf{v},\ \rho,\ \alpha)$*, satisfying the representation*

$$\mathbf{v} = \mathbf{w} + \varepsilon\nabla\varphi + \mathbf{v}_E,\ \ \rho = \rho_E + \sigma$$

and the estimates

$$\|\mathbf{w};\ W_\eta^{m+2}(\Omega)\| + \|\nabla\varphi;\ W_\eta^{m+2}(\Omega)\| + \|\sigma;\ W_\eta^{m+1}(\Omega)\| + |\alpha| \leq r_0,$$

$$|\varphi^*| \leq \gamma r_0.$$

AKNOWLEDGEMENTS

This research was made possible by the Italian C.N.R. which supported the visit of Professor K. Pileckas at the University of Ferrara. M. Padula thanks the G.N.F.M. of the italian C.N.R. and M.P.I. contracts for 60% and 40% respectively, at the University of Ferrara. The authors wish to thank Prof. G.P. Galdi for many helpful and constructive suggestions.

REFERENCES

AGMON, S., DOUGLIS, A., and L. NIRENBERG (1964), Estimates near the boundary for solutions of elliptic partial differential equations satisfying general boundary conditions II, *Commun. Pure Appl. Math.* **17**, 35–92.

AMICK, C.J. (1977), Steady solutions of the Navier- Stokes equations in unbounded channels and pipes, *Ann. Scuola Norm. Pisa* (4) **4**, 473-513.

BEIRÃO DA VEIGA, H. (1987), On L^p-theory for the n-dimensional, stationary compressible Navier-Stokes equations and the incompressible limit for compressible fluids. The equilibrium solutions, *Comm. Math. Phys.* **109**, 229–248.

FRIEDRICHS, K.O. (1958), Symmetric positive linear differential equations, *Comm. Pur. Appl. Math.* **11**, 333-418.

GALDI, G.P. (1994), *An Introduction to the Mathematical Theory of the Navier-Stokes Equations, Vol.I Linearized Stationary Problems*, Springer Tracts in Natural Philosophy **38**, (Springer, Heidelberg).

GOHBERG, I.C. and M.G. KREIN (1969), Introduction to the theory of linear non-selfadjoint operators. *Transl. of Math. M.* **18**, (Amer. Math. Soc., Providence, Rhode Island).

HORGAN, C.O. and L.T. WHEELER (1978), Spatial decay estimates for the Navier-Stokes equations with application to the problem of entry flow, *SIAM J. Appl. Math.* **35**, 97-116.

KONDRAT'EV, V.A. (1967), Boundary value problems for elliptic equations in domains with conical or corner points, *Trudy Moskov. Mat. Obshch.*, **16**, 209-292. English Transl.: *Trans. Moscow Math. Soc.* **16**.

LADYZHENSKAJA, O.A. and V.A. SOLONNIKOV (1983), Determination of the solutions of boundary value problems for stationary Stokes and Navier-Stokes equations having an unbounded Dirichlet integral, *J. Sov. Math.* **21**, 728 − 761.

MAZ'YA, V.G. and B.A. PLAMENEVSKII (1977), On the coefficients in the asymptotics of solutions of elliptic boundary value problems in domains with conical points, *Math. Nachr.* **76**, 29-60. English Transl.: *Amer. Math. Soc. Transl.* **123**(2), 57-88 (1984).

NAZAROV, S.A. and B.A. PLAMENEVSKII (1993), *Elliptic boundary value problems in domains with piecewise smooth boundary*, (Valter de Gruyter and Co., Berlin).

NOVOTNÝ, A. (1995), About the steady transport equation I $-L^p$ - approach in domains with sufficiently smooth boundaries, (to appear).

NOVOTNÝ, A. and M. PADULA (1994), L^p-approach to steady flows of viscous compressible fluids in exterior domains, *Arch. Rat. Mech. Anal.* **126**, 243-297.

NOVOTNÝ, A., PADULA, M. and P. PENEL (1994), A remark on the well posedness of the problem of a steady flow of a viscous barotropic gas in a pipe, (submitted).

NOVOTNÝ, A., PADULA, M. and P. PENEL (1995), On the spatial decay for steady flows of a viscous barotropic gas in a straight channel, (forthcoming).

PADULA, M. (1987), Existence and uniqueness for viscous steady compressible motions, *Arch. Rat. Mech. Anal.* **77**(2), 89–102.

PADULA, M. (1992), A representation formula for steady solutions of a compressible fluid moving at low speed, *Transp. Th. Stat. Phys.* **21**, 593–614.

PADULA, M. (1993a), On the exterior steady problem for the equations of a viscous isothermal gas, *Com. Mat. Univ. Carolinae* **34**(2), 275-293.

PADULA, M. (1993b), Mathematical properties of motions of viscous compressible fluids, in: G.P. Galdi, J. Malek and J. Nečas (eds.), *Progress in theoretical and computational fluid mechanics*, Proc. Winter School, Paseky, Pittman Research Notes in Math. Series, 128-172.

PADULA, M. and K. PILECKAS (1995), On the existence and asymptotic behavior of a steady flow of a viscous barotropic gas in a pipe, (forthcoming).

PILECKAS, K. (1981), Existence of solutions for the Navier–Stokes equations, having an infinite dissipation of energy, in a class of domains with noncompact boundaries, *Zapiski Nauchn. Sem. LOMI* **110**, 180–202. English Transl.: *J. Sov. Math.* **25**(1), 932–947 (1984).

PILECKAS, K. (1994), Weighted L^q–theory, uniform estimates and asymptotics for steady Stokes and Navier–Stokes equations in domains with noncompact boundaries, *Preprint–Universität–Gesamthochschule–Paderborn*.

VALLI, A. (1987), On the existence of stationary solutions to compressible Navier–Stokes equations, *Ann. Inst. H. Poincaré* **4**(1), 99–113 .

Free Boundary Problems

CLASSICAL SOLVABILITY OF THE PROBLEM   DESCRIBING THE EVOLUTION OF A DROP IN A LIQUID MEDIUM

I.V. Denisova

Institute of the Mechanical Engineering Problems
Russian Academy of Sciences
V.O., Bolshoy pr., 61,
199178 St.-Petersburg, Russia

ABSTRACT

A free boundary problem governing the unsteady motion of a drop in another viscous incompressible fluid is studied. One takes into account the surface tension on an unknown interface between the liquids. A local (in time) existence theorem for this problem is obtained in spaces of Hölder continuous functions with a power decay at infinity. The proof of the theorem is based on coercive estimates in Hölder weighted spaces for a linearized problem.

1. INTRODUCTION

The result of this paper is local (in time) solvability of the free boundary problem governing the unsteady motion of two viscous incompressible fluids separated by a closed unknown interface. A study of this problem in the Hölder spaces of functions was started in Denisova (1993) and in Denisova and Solonnikov (1991). It is based on the technique of I.Sh. Mogilevskii and V.A. Solonnikov developed for the investigation of the evolution of a drop in the vacuum (see Mogilevskii and Solonnikov, 1989, Mogilevskii and Solonnikov, 1992).

The peculiarity of the problem describing the motion of two liquids, besides the presence of the second fluid, consists in the unboundedness of the domain occupied by one of the fluids. This fact obliged us to analyze our problem in the classes of Hölder continuous functions with a power-like decay at infinity. Under the condition of fast enough decrease of the mass forces and of the initial field of velocities, we prove the existence of a finite time interval on which the problem has a unique solution with some power decay at infinity.

The solvability of the same problem in the Sobolev spaces (also on a finite time interval) was established in Denisova (1990, 1994), Denisova and Solonnikov (1989).

Let us give a mathematical formulation of the problem. It is required to find a closed surface $\Gamma_t = \partial \Omega_t^+$, $t > 0$, between a bounded domain $\Omega_t^+ \subset R^3$ occupied by a fluid with the viscosity $\nu^+ > 0$ and the density $\rho^+ > 0$ and the exterior domain $\Omega_t^- = R^3 \setminus \bar{\Omega}_t^+$ occupied by a fluid with the viscosity $\nu^- > 0$ and the density $\rho^- > 0$, as well as the velocity vector field $\boldsymbol{v} = (v_1, v_2, v_3)$ and the pressure function p as a solution of the initial-boundary value problem for the system of the Navier-Stokes equations

$$D_t \boldsymbol{v} - \nu^{\pm} \nabla^2 \boldsymbol{v} + (\boldsymbol{v} \cdot \nabla)\boldsymbol{v} + \frac{1}{\rho^{\pm}} \nabla p = \boldsymbol{f}, \quad \nabla \cdot \boldsymbol{v} = 0 \quad (x \in \Omega_t^- \cup \Omega_t^+, \ t > 0),$$

$$\boldsymbol{v}|_{t=0} = \boldsymbol{v}_0, \quad [\boldsymbol{v}]|_{\Gamma_t} = \lim_{\substack{x \to x_0 \in \Gamma_t \\ x \in \Omega_t^+}} \boldsymbol{v}(x) - \lim_{\substack{x \to x_0 \in \Gamma_t \\ x \in \Omega_t^-}} \boldsymbol{v}(x) = 0, \tag{1}$$

$$[\boldsymbol{T}\boldsymbol{n}]|_{\Gamma_t} = \sigma H \boldsymbol{n}$$

where $D_t = \partial/\partial t$, $\nabla = (\partial/\partial x_1, \partial/\partial x_2, \partial/\partial x_3)$, $\nu^{\pm}, \rho^{\pm}$ are step functions of viscosity and density, respectively, $\boldsymbol{f}$ is a given vector field of mass forces, $\boldsymbol{v}_0$ is the initial value of the velocity vector field, $\boldsymbol{T}$ is the stress tensor with the elements

$$T_{ik} = -\delta_i^k p + \mu^{\pm}(\partial v_i/\partial x_k + \partial v_k/\partial x_i), \qquad i, k, = 1, 2, 3;$$

$\mu^{\pm} = \nu^{\pm}\rho^{\pm}$, δ_i^k is the Kronecker symbol, $\sigma > 0$ is the coefficient of the surface tension, $\boldsymbol{n}$ is the outward normal to Ω_t^+, $H(x,t)$ is the doubled mean curvature of Γ_t ($H < 0$ at the points where Γ_t is convex towards Ω_t^-). We suppose that the Cartesian coordinate system $\{x\}$ is introduced in R^3. The centered dot denotes the Cartesian scalar product.

Moreover, to exclude the mass transport through the surface Γ_t, it is assumed that the liquid particles do not leave Γ_t. It means that Γ_t consists of the points $x(\xi, t)$ such that the corresponding vector $\boldsymbol{x}(\xi, t)$ solves the Cauchy problem

$$D_t \boldsymbol{x} = \boldsymbol{v}(x(t), t), \quad \boldsymbol{x}|_{t=0} = \boldsymbol{\xi}, \quad \xi \in \Gamma_0, \ t > 0 \tag{2}$$

with a given $\Gamma_0 = \Gamma_t|_{t=0}$. Hence, $\Omega_t^{\pm} = \{x(\xi, t) | \xi \in \Omega_0^{\pm}\}$.

Let us make use of a well known relation

$$H\boldsymbol{n} = \Delta(t)\boldsymbol{x} \tag{3}$$

where $\Delta(t)$ denotes the Beltrami-Laplace operator on Γ_t and pass from the Eulerian to Lagrangean coordinates by the formula

$$\boldsymbol{x} = \boldsymbol{\xi} + \int_0^t \boldsymbol{u}(\xi, \tau)d\tau \equiv X_{\boldsymbol{u}}(\xi, t) \tag{4}$$

(here $\boldsymbol{u}(\xi, t)$ is the velocity vector field in the Lagrangean coordinates). This transformation leads us to the problem for $\boldsymbol{u}$ and $q = p(X_{\boldsymbol{u}}, t)$ with a given interface $\Gamma \equiv \Gamma_0$. If the angle between $\boldsymbol{n}$ and the exterior normal $\boldsymbol{n}_0$ to Γ is acute this problem is equivalent to the system

$$D_t \boldsymbol{u} - \nu^{\pm} \nabla_{\boldsymbol{u}}^2 \boldsymbol{u} + \frac{1}{\rho^{\pm}} \nabla_{\boldsymbol{u}} q = \boldsymbol{f}(X_{\boldsymbol{u}}, t),$$

$$\nabla_{\boldsymbol{u}} \cdot \boldsymbol{u} = 0 \quad \text{in } Q_T^{\pm} = \Omega_0^{\pm} \times (0, T),$$

$$\boldsymbol{u}|_{t=0} = \boldsymbol{v}_0 \quad \text{on } \Omega_0^{\pm}, \quad \boldsymbol{u} \xrightarrow[|\xi| \to \infty]{} 0 \tag{5}$$

$$[\boldsymbol{u}]|_{\Gamma} = 0, \quad [\Pi_0 \Pi T_{\boldsymbol{u}}(\boldsymbol{u}, q)\boldsymbol{n}]|_{\Gamma} = 0,$$

$$[\boldsymbol{n}_0 \cdot \mathcal{T}_{\boldsymbol{u}}(\boldsymbol{u}, q)\boldsymbol{n}]|_\Gamma - \sigma \boldsymbol{n}_0 \cdot \Delta(t) X_{\boldsymbol{u}}|_\Gamma = 0.$$

Here the following notations are introduced: $\nabla_{\boldsymbol{u}} = \mathcal{A}\nabla$, $\mathcal{A}$ is the matrix of cofactors A_{ij} to the elements

$$a_{ij}(\xi, t) = \delta_i^j + \int_0^t \frac{\partial u_i}{\partial \xi_j} dt'$$

of the Jacobian matrix of the transformation (4), vector $\boldsymbol{n}(X_{\boldsymbol{u}})$ is related to $\boldsymbol{n}_0$ as $\boldsymbol{n} = \mathcal{A}\boldsymbol{n}_0/|\mathcal{A}\boldsymbol{n}_0|$; $\Pi\boldsymbol{\omega} = \boldsymbol{\omega} - \boldsymbol{n}(\boldsymbol{n} \cdot \boldsymbol{\omega})$, $\Pi_0\boldsymbol{\omega} = \boldsymbol{\omega} - \boldsymbol{n}_0(\boldsymbol{n}_0 \cdot \boldsymbol{\omega})$ are projections of the vector $\boldsymbol{\omega}$ onto the tangent plane to Γ_t and Γ, respectively. The tensor $\mathcal{T}_{\boldsymbol{u}}(\boldsymbol{w}, q)$ has the elements

$$(\mathcal{T}_{\boldsymbol{u}}(\boldsymbol{w}, q))_{ij} = -\delta_j^i q + \mu^\pm (A_{kj}\partial w_i/\partial \xi_k + A_{ki}\partial w_j/\partial \xi_k).$$

Repeated indices imply summation.

2. DEFINITION OF WEIGHTED SPACES

Now we define anisotropic weighted Hölder spaces. Let Ω be a domain in R^n, $n \in N$, $T > 0$, $\Omega_T = \Omega \times (0, T)$, and let $\alpha \in (0, 1)$, $\beta \geq 0$. We denote by $C_\beta^{\alpha, \alpha/2}(\Omega_T)$ the set of functions defined in Ω_T and having the finite norm

$$\|f\|_{\beta, \Omega_T}^{(\alpha)} = \|f\|_{\beta, \Omega_T} + \ll f \gg_{\beta, \Omega_T}^{(\alpha)}$$

where

$$\|f\|_{\beta, \Omega_T} = \max_{t \in (0, T)} \max_{x \in \Omega} (1 + |x|)^\beta |f(x, t)|,$$

$$\ll f \gg_{\beta, \Omega_T}^{(\alpha)} = \ll f \gg_{x, \beta, \Omega_T}^{(\alpha)} + \ll f \gg_{t, \beta, \Omega_T}^{\alpha/2},$$

$$\ll f \gg_{x, \beta, \Omega_T}^{(\alpha)} = \max_{t \in (0, T)} \max_{x, y \in \Omega} (1 + d_{x,y})^\beta |f(x, t) - f(y, t)||x - y|^{-\alpha},$$

$$\ll f \gg_{t, \beta, \Omega_T}^{(\mu)} = \max_{x \in \Omega} \max_{t, \tau \in (0, T)} (1 + |x|)^\beta |f(x, t) - f(x, \tau)||t - \tau|^{-\mu},$$

$\mu \in (0, 1)$, $d_{x,y} = \min(|x|, |y|)$.

Let

$$\|f\|_{x, \beta, \Omega_T}^{(\alpha)} = \|f\|_{\beta, \Omega_T} + \ll f \gg_{x, \beta, \Omega_T}^{(\alpha)},$$

$$\|f\|_{t, \beta, \Omega_T}^{(\mu)} = \|f\|_{\beta, \Omega_T} + \ll f \gg_{t, \beta, \Omega_T}^{(\mu)}.$$

We introduce the notation:

$$D_x^{\boldsymbol{r}} = \partial^{|\boldsymbol{r}|}/\partial x_1^{r_1}...\partial x_n^{r_n}, \quad \boldsymbol{r} = (r_1, ...r_n), \quad r_i > 0, \quad |\boldsymbol{r}| = r_1 + ... + r_n,$$

$$D_t^s = \partial^s/\partial t^s, \quad s \in N \cup \{0\},$$

and we fix $k \in N$. The space $C_\beta^{k+\alpha, (k+\alpha)/2}(\Omega_T)$ consists, by definition, of functions with the finite norm

$$\|f\|_{\beta, \Omega_T}^{(k+\alpha)} = \sum_{|\boldsymbol{r}|+2s \leq k} \|D_x^{\boldsymbol{r}} D_t^s f\|_{\beta, \Omega_T} + \ll f \gg_{\beta, \Omega_T}^{(k+\alpha)},$$

$$\ll f \gg_{\beta, \Omega_T}^{(k+\alpha)} = \sum_{|\boldsymbol{r}|+2s=k} \ll D_x^{\boldsymbol{r}} D_t^s f \gg_{\beta, \Omega_T}^{(\alpha)} + \sum_{|\boldsymbol{r}|+2s=k-1} \ll D_x^{\boldsymbol{r}} D_t^s f \gg_{t, \beta, \Omega_T}^{(1/2+\alpha/2)}.$$

We define $C_\beta^{k+\alpha}(\Omega)$, $k \in N \cup \{0\}$, as the space of functions $f(x)$, $x \in \Omega$, with the norm

$$\|f\|_{\beta, \Omega}^{(k+\alpha)} = \sum_{|\boldsymbol{r}| \leq k} \|D_x^{\boldsymbol{r}} f\|_{\beta, \Omega} + \ll f \gg_{\beta, \Omega}^{(k+\alpha)},$$

193

$$\ll f \gg_{\beta,\Omega}^{(k+\alpha)} = \sum_{|\boldsymbol{r}|=k} \ll D_x^{\boldsymbol{r}} f \gg_{\beta,\Omega}^{(\alpha)} = \sup_{x,y\in\Omega} (1 + d_{x,y})^\beta \sum_{|\boldsymbol{r}|=k} \|D_x^{\boldsymbol{r}} f(x) - D_y^{\boldsymbol{r}} f(y)\| |x - y|^{-\alpha}.$$

We also need the following semi-norm with $\alpha, \gamma \in (0,1)$:

$$|f|_{1,\gamma,\Omega_T}^{(1+\alpha,\gamma)} = \langle f \rangle_{1+\gamma,\Omega_T}^{(1+\alpha,\gamma)} + \ll f \gg_{t,1,\Omega_T}^{(\frac{1+\alpha-\gamma}{2})},$$

Here

$$\langle f \rangle_{1+\gamma,\Omega_T}^{(1+\alpha,\gamma)} = \max_{t,\tau\in(0,T)} \max_{x,y\in\Omega} (1 + d_{x,y})^{1+\gamma} \frac{|f(x,t) - f(y,t) - f(x,\tau) + f(y,\tau)|}{|x - y|^\gamma |t - \tau|^{(1+\alpha-\gamma)/2}}.$$

There exists the estimate

$$\langle f \rangle_{1+\gamma,\Omega_T}^{(1+\alpha,\gamma)} \le c_1 \ll f \gg_{1+\gamma,\Omega_T}^{(1+\alpha)} .$$

We consider that $f \in C_{1,\gamma}^{(1+\alpha,\gamma)}(\Omega_T)$ if

$$\|f\|_{1,\Omega_T} + |f|_{1,\gamma,\Omega_T}^{(1+\alpha,\gamma)} < \infty.$$

Finally, if function f has the finite norm

$$\|f\|_{1,\gamma,\Omega_T}^{(\gamma,\mu)} \equiv \ll f \gg_{x,1+\gamma,\Omega_T}^{(\gamma)} + \|f\|_{t,1,\Omega_T}^{(\mu)}, \qquad \gamma \in (0,1), \quad \mu \in [0,1),$$

then it belongs to the Hölder space $C_{1,\gamma}^{\gamma,\mu}(\Omega_T)$.

We suppose that a vector valued function is an element of a Hölder space if all its components belong to this space, and its norm is defined as the maximal norm of the components.

Let us set $D_T = Q_T^- \cup Q_T^+$ and

$$\|f\|_{\beta,D_T}^{(k+\alpha)} = \|f\|_{\beta,Q_T^-}^{(k+\alpha)} + \|f\|_{\beta,Q_T^+}^{(k+\alpha)},$$

$$\|f\|_{\beta,\Omega^\pm}^{(k+\alpha)} = \|f\|_{\beta,\Omega^-}^{(k+\alpha)} + \|f\|_{\beta,\Omega^+}^{k+\alpha)}.$$

As $\beta = 0$, all these spaces become ordinary Hölder spaces, for example, the space $C_0^{k+\alpha,(k+\alpha)/2}(\Omega_T)$ coinsides with $C^{k+\alpha,(k+\alpha)/2}(\Omega_T)$ and its norm is equal to $\|\cdot\|_{\Omega_T}^{(k+\alpha)}$. For functions defined on bounded manifolds (in particular, on Γ_T and on $G_T = \Gamma \times (0,T)$), all the weighted spaces introduced above are equivalent to ordinary Hölder spaces.

3. EXISTENCE THEOREM FOR PROBLEM (5)

We can now formulate the main result of this paper.

Theorem 3.1. *Suppose that* $\Gamma \in C^{3+\alpha}$, $\boldsymbol{f}, D_x\boldsymbol{f} \in C_2^{\alpha,(1+\alpha-\gamma)/2}(R^3 \times (0,T))$, $\boldsymbol{v}_0 \in C_{1+\gamma}^{2+\alpha}(\Omega^- \cup \Omega^+)$, $\sigma \in C^{1+\alpha}$, $\sigma \ge \sigma_0 > 0$ *with some* $\alpha, \gamma \in (0,1)$, $\gamma > \alpha$, $T < \infty$. *Moreover, let the compatibility conditions*

$$\nabla \cdot \boldsymbol{v}_0 = 0, \quad [\boldsymbol{v}_0]|_\Gamma = 0, \quad [\Pi_0 \mathcal{T}(\boldsymbol{v}_0)\boldsymbol{n}_0]|_\Gamma = 0,$$

$$[\Pi_0(\nu^\pm \nabla^2 \boldsymbol{v}_0 - \frac{1}{\rho^\pm}\nabla q_0)]|_\Gamma = 0$$

be fulfilled where $q_0(\xi) \equiv q(\xi,0)$ *is a solution of the diffraction problem*

$$\frac{1}{\rho^\pm}\nabla^2 q_0(\xi) = \nabla \cdot (\boldsymbol{f}(\xi,0) - D_t\mathcal{B}^*(\boldsymbol{v}_0)\boldsymbol{v}_0(\xi)) \qquad \text{in} \quad \Omega^- \cup \Omega^+,$$

$$[q_0]|_\Gamma = [2\mu^\pm \partial \boldsymbol{v}_0/\partial \boldsymbol{n}_0 \cdot \boldsymbol{n}_0]|_\Gamma,$$

$$[\frac{1}{\rho^\pm}\frac{\partial q_0}{\partial \boldsymbol{n}_0}]|_\Gamma = [\nu^\pm \boldsymbol{n}_0 \cdot \nabla^2 \boldsymbol{v}_0]|_\Gamma, \qquad (\frac{\partial}{\partial \boldsymbol{n}_0} = \boldsymbol{n}_0 \cdot \nabla).$$

Then there exists a positive constant $T_0 \leq T$ such that problem (5) has a unique solution $\boldsymbol{u} \in C_{1+\gamma}^{2+\alpha,1+\alpha/2}(D_{T_0})$, $q \in C_{1,\gamma}^{(1+\alpha,\gamma)}(D_{T_0})$ and $\nabla q \in C_{1+\gamma}^{\alpha,\alpha/2}(D_{T_0})$. The value of T_0 depends on the norms of $\boldsymbol{f}, \boldsymbol{v}_0$ and on the curvature of Γ.

We give only main steps of the proof of Theorem 3.1. The solvability of (5) is obtained by means of successive approximations. That is why one considers a linearized problem

$$D_t \boldsymbol{w} - \nu^\pm \nabla_{\boldsymbol{u}}^2 \boldsymbol{w} + \frac{1}{\rho_\pm}\nabla_{\boldsymbol{u}} s = \boldsymbol{f},$$

$$\nabla_{\boldsymbol{u}} \cdot \boldsymbol{w} = r \qquad in \ D_T,$$

$$\boldsymbol{w}|_{t=0} = \boldsymbol{w}_0, \qquad \boldsymbol{w} \xrightarrow[|\xi|\to\infty]{} 0, \ s \xrightarrow[|\xi|\to\infty]{} 0, \tag{6}$$

$$[\boldsymbol{w}]|_{G_T} = 0, \quad [\Pi_0 \Pi \mathcal{T}_{\boldsymbol{u}}(\boldsymbol{w},s)\boldsymbol{n}]|_{G_T} = \Pi_0 \boldsymbol{a},$$

$$[\boldsymbol{n}_0 \cdot \mathcal{T}_{\boldsymbol{u}}(\boldsymbol{w},s)\boldsymbol{n}]|_{G_T} - \sigma \boldsymbol{n}_0 \cdot \Delta(t)\int_0^t \boldsymbol{w}dt'|_{G_T} = b + \sigma\int_0^t Bdt'.$$

Here $G_T = \Gamma \times (0,T)$, and the functions $\boldsymbol{f}, r, \boldsymbol{w}_0, \boldsymbol{a}, b, B$ are known.

For this problem, we establish

Theorem 3.2. *Assume that for some $\alpha, \gamma \in (0,1)$, $\gamma > \alpha$, $T < \infty$, the surface $\Gamma \in C^{2+\alpha}$, the coefficient $\sigma \in C^{1+\alpha}(\Gamma)$, $\sigma \geq \sigma_0 > 0$, and the vector field $\boldsymbol{u} \in C_{1+\gamma}^{2+\alpha,1+\alpha/2}(D_T), [\boldsymbol{u}]|_{G_T} = 0$, satisfies the inequality*

$$(T + T^{\frac{\gamma}{2}})\|\boldsymbol{u}\|_{1+\gamma,D_T}^{(2+\alpha)} \leq \delta, \tag{7}$$

$\delta > 0$ being a sufficiently small number. In addition, suppose that four groups of conditions hold:

1) $\boldsymbol{f} \in C_{1+\gamma}^{\alpha,\alpha/2}(D_T)$, $\boldsymbol{f}(.,0) \in C_{1+\gamma,\alpha}^\alpha(\Omega^- \cup \Omega^+)$, $\boldsymbol{w}_0 \in C_{1+\gamma}^{2+\alpha}(\Omega^- \cup \Omega^+)$,

$$r \in C_{1+\gamma}^{1+\alpha,(1+\alpha)/2}(D_T), \qquad r = \nabla \cdot \boldsymbol{R}, \qquad [\boldsymbol{R} \cdot \boldsymbol{n}_0]|_\Gamma = 0,$$

$$\boldsymbol{R} \in C_{1+\gamma}^{(1+\alpha,\gamma)}(D_T), \ D_t \boldsymbol{R} \in C_{1+\gamma}^{\alpha,\alpha/2}(D_T), \ D_t \boldsymbol{R}(.,0) \in C_{1+\gamma,\alpha}^\alpha(\Omega^- \cup \Omega^+),$$

$$\boldsymbol{a} \in C^{1+\alpha,(1+\alpha)/2}(G_T), \ b \in C^{(1+\alpha,\gamma)}(G_T), \ B \in C^{\alpha,\alpha/2}(G_T);$$

2) $\nabla \cdot \boldsymbol{w}_0(\xi) = r(\xi,0) = 0$, $[\boldsymbol{w}_0]|_\Gamma = 0$,

$$[\Pi_0 \mathcal{T}(\boldsymbol{w}_0(\xi))\boldsymbol{n}_0]|_\Gamma = \Pi_0 \boldsymbol{a}(\xi,0)|_\Gamma,$$

$$[\Pi_0(\boldsymbol{f}(\xi,0) - \frac{1}{\rho^\pm}\nabla s_0 + \nu^\pm \nabla^2 \boldsymbol{w}_0(\xi))]|_\Gamma = 0;$$

3) $s_0 = s(\xi,0)$ is a solution of the following problem:

$$\frac{1}{\rho^\pm}\nabla^2 s_0(\xi) = \nabla \cdot (\boldsymbol{f}(\xi,0) - D_t \boldsymbol{R}(\xi,0) - D_t \mathcal{B}^*|_{t=0}\boldsymbol{w}_0(\xi)) \qquad in \quad \Omega^- \cup \Omega^+,$$

$$[s_0]|_\Gamma = [2\mu^\pm \partial \boldsymbol{w}_0/\partial \boldsymbol{n}_0 \cdot \boldsymbol{n}_0]|_\Gamma - b|_{t=0}, \tag{8}$$

$$\left[\frac{1}{\rho^\pm}\frac{\partial s_0}{\partial n_0}\right]\big|_\Gamma = [n_0 \cdot (f(\xi,0) + \nu^\pm \nabla^2 w_0)]|_\Gamma;$$

4) *There exist a vector* $h \in C_{1+\gamma}^{\alpha,\alpha/2}(D_T)$ *and a tensor* $H = \{H_{ik}\}_{k,i=1,2,3}$, $H \in C_{1,\gamma}^{(1+\alpha,\gamma)}(D_T)$, $H \in C_{1,\gamma}^{\gamma,0}(D_T)$ *such that two representation formulas hold:*

$$D_t r - \nabla u \cdot f = \nabla \cdot h \quad and \quad h_i = \partial H_{ik}/\partial \xi_k, \quad i = 1,2,3$$

(these equations are understood in a weak sense).

Under hypotheses 1)-4) problem (6) is uniquely solvable and its solution (w, s) has the properties:

$$w \in C_{1+\gamma}^{2+\alpha,1+\alpha/2}(D_T), \quad s \in C_{1,\gamma}^{(1+\alpha,\gamma)}(D_T), \quad \nabla s \in C_{1+\gamma}^{\alpha,\alpha/2}(D_T),$$

$$\|w\|_{1+\gamma,D_T}^{(2+\alpha)} + \|\nabla s\|_{1+\gamma,D_T}^{(\alpha)} + \|s\|_{t,1,D_T}^{((1+\alpha-\gamma)/2)} + \langle s\rangle_{1+\gamma,D_T}^{(1+\alpha,\gamma)} \leq$$

$$\leq c_2(T)\{\|f\|_{1+\gamma,D_T}^{(\alpha)} + \ll f(\cdot,0)\gg_{1+\gamma+\alpha,\Omega^\pm}^{(\alpha)} + \|r\|_{1+\gamma,D_T}^{(1+\alpha)} + \|w_0\|_{1+\gamma,\Omega^\pm}^{(2+\alpha)} +$$

$$+\|D_t R\|_{1+\gamma,D_T}^{(\alpha)} + \langle R\rangle_{1+\gamma,D_T}^{(1+\alpha,\gamma)} + \ll D_t R(.,0)\gg_{1+\gamma+\alpha,\Omega^\pm}^{(\alpha)} + |H|_{1,\gamma,D_T}^{(1+\alpha,\gamma)} + \|H\|_{1,\gamma,D_T}^{(\gamma,0)} +$$

$$+\|a\|_{G_T}^{(1+\alpha)} + |b|_{G_T}^{(1+\alpha,\gamma)} + \|b\|_{G_T} + \|\Pi_0\nabla b\|_{G_T}^{(\alpha)} + \|B\|_{G_T}^{(\alpha)} +$$

$$+(T^{\frac{1-\alpha}{2}}\|\nabla u\|_{1+\gamma,D_T} + \|\nabla u\|_{1+\gamma,D_T}^{(\alpha)})\|w_0\|_{1+\gamma,\Omega^\pm}^{(1)}\} \equiv c_2(T)Q[u,T], \quad (9)$$

$c_2(T)$ *being a non-decreasing function of* T.

The proof of Theorem 3.2 is also based on the method of successive approximations. We begin with a study of problem (6) in the case $u = 0$. The analogue of Theorem 3.2 in this case is proved on the basis of the solvability of the linear problem in ordinary Hölder spaces (see Denisova, 1993). Inequality (9) with $u = 0$ is obtained by means of considering local (in space) estimates for the solutions of corresponding Cauchy problem (see Solonnikov, 1976) and of linear problem with a drop (Denisova, 1993).

Next, we rewrite system (6) as

$$D_t w - \nu^\pm \nabla^2 w + \frac{1}{\rho^\pm}\nabla s = f + l_1(w,s) \equiv f_1,$$

$$\nabla \cdot w = r + l_2(w) \equiv r_1 \quad in \ D_T,$$

$$w|_{t=0} = w_0, \quad w \xrightarrow[|\xi|\to\infty]{} 0, \quad s \xrightarrow[|\xi|\to\infty]{} 0, \quad (10)$$

$$[w]|_{G_T} = 0, \quad [\Pi_0 T(w,s)n_0]|_{G_T} = \Pi_0 a + l_3(w),$$

$$[n_0 \cdot T(w,s)n_0]|_{G_T} - \sigma n_0 \cdot \Delta(0)\int_0^t w dt'|_{G_T} = b + \sigma\int_0^t B dt' + l_4(w,s) + \sigma\int_0^t l_5(w)dt'.$$

Here we use the notation:

$$l_1(w,s) = \nu^\pm(\nabla_u^2 - \nabla^2)w - (\nabla_u - \nabla)s/\rho^\pm =$$
$$= \nu^\pm\partial\{(A_{ij}A_{im} - \delta_j^m)\partial w/\partial\xi_m\}/\partial\xi_j - \mathcal{B}\nabla s/\rho^\pm,$$
$$l_2(w) = (\nabla_u - \nabla)\cdot w = -\mathcal{B}\nabla\cdot w,$$
$$l_3(w) = [\Pi_0 T(w,s)n_0 - \Pi_0\Pi T_u(w,s)n]|_\Gamma,$$
$$l_4(w,s) = [n_0 \cdot (T(w,s)n_0 - T_u(w,s)n)]|_\Gamma,$$
$$l_5(w) = n_0 \cdot D_t(\Delta(t) - \Delta(0))\int_0^t w|_\Gamma dt' = n_0 \cdot \{(\Delta(t) - \Delta(0))w|_\Gamma + \dot\Delta(t)\int_0^t w|_\Gamma dt'\},$$

where $\mathcal{B} = \mathcal{A} - \mathcal{I}, \mathcal{I}$ is the identity matrix, operator $\dot{\Delta}(t)$ is recieved from $\Delta(t)$ by differentiating its coefficients in t.

We observe that operators $\boldsymbol{l}_1$ and l_2 have the divergent form:

$$\boldsymbol{l}_1(\boldsymbol{w}, s) = \partial \boldsymbol{L}_{1j}(\boldsymbol{w}, s)/\partial \xi_j, \qquad \boldsymbol{L}_{1j}(\boldsymbol{w}, s) = \nu^{\pm}(A_{ij}A_{im} + B_{ij})\partial \boldsymbol{w}/\partial \xi_m - \mathcal{B}\boldsymbol{e}_j s/\rho^{\pm},$$

$$l_2(\boldsymbol{w}) = \nabla \cdot L_2(\boldsymbol{w}), \qquad L_2(\boldsymbol{w}) = -\mathcal{B}^* \boldsymbol{w}$$

($\boldsymbol{e}_j$ is ξ_j-basis vector, $\mathcal{B}^*$ is the transpose of $\mathcal{B}$).

Moreover, the expression $D_t r_1 - \nabla \cdot \boldsymbol{f}_1$ is also represented in the divergence form:

$$D_t r_1 - \nabla \cdot \boldsymbol{f}_1 = \nabla \cdot (\boldsymbol{h} + \boldsymbol{l}_6(\boldsymbol{w}, s))$$

with

$$\boldsymbol{l}_6(\boldsymbol{w}, s) = -\mathcal{B}^*(\nu^{\pm}\nabla^2 \boldsymbol{w} - \nabla s/\rho^{\pm}) - D_t \mathcal{B}^* \boldsymbol{w} - \mathcal{A}^* \boldsymbol{l}_1(\boldsymbol{w}, s) \equiv \partial \boldsymbol{L}_{6j}(\boldsymbol{w}, s)/\partial \xi_j,$$

and

$$\boldsymbol{L}_{6j}(\boldsymbol{w}, s) = -\nu^{\pm} \mathcal{B}^* \partial \boldsymbol{w}/\partial \xi_j + \mathcal{B}^* \boldsymbol{e}_j s/\rho^{\pm} - \mathcal{A}^* \boldsymbol{L}_{1j} + \partial \boldsymbol{V}/\partial \xi_j$$

where newtonian potential $\boldsymbol{V}$ is equal to

$$\boldsymbol{V}(\xi, t) = -\int_{\Omega^- \cup \Omega^+} \frac{1}{4\pi|\xi - \eta|} \left\{ \frac{\partial \mathcal{B}^*}{\partial \eta_m} (\nu^{\pm} \frac{\partial \boldsymbol{w}}{\partial \eta_m} - \frac{\boldsymbol{e}_m s}{\rho^{\pm}} + \boldsymbol{L}_{1m}) - D_t \mathcal{B}^* \boldsymbol{w} \right\} d\eta.$$

As elements $B_{ij}, i, j = 1, 2, 3$, of matrix $\mathcal{B}$ are polynomials of the second order with respect to $\int_1^t \partial u_i(\xi, t')/\partial \xi_j dt'$ the following lemma holds.

Lemma 3.1. *Let $\boldsymbol{u}$ satisfy inequality* (7) *with $\delta < 1$ then*

$$\|B_{ij}\|_{1+\gamma, D_T}^{(\alpha)} + \|\nabla B_{ij}\|_{1+\gamma, D_T}^{(\alpha)} + |B_{ij}|_{1+\gamma, D_T}^{(1+\alpha, \gamma)} +$$

$$+ \ll \nabla B_{ij} \gg_{t, 1+\gamma, D_T}^{(\frac{1+\alpha-\gamma}{2})} + \ll D_t B_{ij} \gg_{t, 1+\gamma, D_T}^{(\frac{1+\alpha-\gamma}{2})} \leq c_3 \delta,$$

$$\|D_t B_{ij}\|_{1+\gamma, D_T}^{(\alpha)} \leq c_4 \|\nabla \boldsymbol{u}\|_{1+\gamma, D_T}^{(\alpha)},$$

$$\ll B_{ij} \gg_{t, 1+\gamma, D_T}^{(\frac{1+\alpha}{2})} \leq c_5 T^{\frac{1-\alpha}{2}} \|\nabla \boldsymbol{u}\|_{1+\gamma, D_T},$$

$$\langle D_t B_{ij} \rangle_{1+\gamma, D_T}^{(1+\alpha, \gamma)} \leq c_6(|\nabla \boldsymbol{u}|_{1+\gamma, D_T}^{(1+\alpha, \gamma)} + \|\nabla \boldsymbol{u}\|_{x, 1+\gamma, D_T}^{(\gamma)}), \quad i, j = 1, 2, 3.$$

Lemma 3.2. *If $\Gamma \in C^{2+\alpha}$ then under the assumptions of Lemma 3.1 we have the estimate*

$$\|\boldsymbol{l}_1\|_{1+\gamma, D_T}^{(\alpha)} + \|l_2\|_{1+\gamma, D_T}^{(1+\alpha)} + \|l_3\|_{G_T}^{(1+\alpha)} + \|l_4\|_{G_T} + |l_4|_{G_T}^{(1+\alpha, \gamma)} + \|\Pi_0 \nabla l_4\|_{G_T}^{(\alpha)} + \|l_5\|_{G_T}^{(\alpha)} +$$

$$+ \|D_t L_2\|_{1+\gamma, D_T}^{(\alpha),} + \langle L_2 \rangle_{1+\gamma, D_T}^{(1+\alpha, \gamma)} + \|\boldsymbol{l}_6\|_{1+\gamma, D_T}^{(\alpha)} + \max_{j=1,2,3}(|\boldsymbol{L}_{6j}|_{1, \gamma, D_T}^{(1+\alpha, \gamma)} + \|\nabla \boldsymbol{L}_{6j}\|_{1, \gamma, D_T}^{(\gamma, 0)}) \leq$$

$$\leq c_7 \delta(\|\boldsymbol{w}\|_{1+\gamma, D_T}^{(2+\alpha)} + \|\nabla s\|_{1+\gamma, D_T}^{(\alpha)} + |s|_{1, \gamma, D_T}^{(1+\alpha, \gamma)} + \|s\|_{1, \gamma, D_T}^{(\gamma, 0)}) +$$

$$+ c_8(T^{\frac{1-\alpha}{2}} \|\nabla \boldsymbol{u}\|_{1+\gamma, D_T} + \|\nabla \boldsymbol{u}\|_{1+\gamma, D_T}^{(\alpha)}) \|\boldsymbol{w}(., 0)\|_{1+\gamma, \Omega^{\pm}}^{(1)} \qquad (11)$$

Proof: This result follows from Lemma 3.1 and from evaluations for newtonian potential

$$V(x,t) = \int_{\Omega^- \cup \Omega^+} \frac{F(y,t)}{|x-y|} dy,$$

namely,

$$\|\nabla V\|_{1,\gamma,D_T}^{(\gamma,0)} \leq c_9 \|F\|_{2,D_T}, \qquad |\nabla V|_{1,\gamma,D_T}^{(1+\alpha,\gamma)} \leq c_{10} \|F\|_{t,2,D_T}^{(\frac{1+\alpha-\gamma}{2})}. \qquad \square$$

Now we can demonstrate the solvability of problem (6). As zero approximation, we take $(\boldsymbol{w}^{(0)}, s^{(0)})$ where $s^{(0)}(\xi,t) = s_0(\xi)$ is a solution of (8) and $\boldsymbol{w}^{(0)}$ solves the problem

$$D_t \boldsymbol{w}^{(0)} - \nu^{\pm} \nabla^2 \boldsymbol{w} = \boldsymbol{f} - \nabla s^{(0)}/\rho^{\pm},$$

$$\boldsymbol{w}^{(0)}|_{t=0} = \boldsymbol{w}_0, \quad [\boldsymbol{w}^{(0)}]|_{\Gamma} = 0,$$

For $(\boldsymbol{w}^{(0)}, s^{(0)})$, we have

$$\|\boldsymbol{w}^{(0)}\|_{1+\gamma,D_T}^{(2+\alpha)} \leq c_{11}(\|\boldsymbol{f}\|_{1+\gamma,D_T}^{(\alpha)} + \|\nabla s_0\|_{1+\gamma,\Omega^{\pm}}^{(\alpha)} + \|\boldsymbol{w}_0\|_{1+\gamma,\Omega^{\pm}}^{(2+\alpha)}), \tag{12}$$

$$\|\nabla s^{(0)}\|_{1+\gamma,\Omega^{\pm}}^{(\alpha)} \leq c_{12} Q[0,T]. \tag{13}$$

One defines next approximations $(\boldsymbol{w}^{(m+1)}, s^{(m+1)})$, $m \geq 0$, as solutions of the initial-boundary value problems:

$$D_t \boldsymbol{w}^{(m+1)} - \nu^{\pm} \nabla^2 \boldsymbol{w}^{(m+1)} + \frac{1}{\rho^{\pm}} \nabla s^{(m+1)} = \boldsymbol{f} + \boldsymbol{l}_1(\boldsymbol{w}^{(m)}, s^{(m)}),$$

$$\nabla \cdot \boldsymbol{w}^{(m+1)} = r + l_2(\boldsymbol{w}^{(m)}) \quad in \quad D_T,$$

$$\boldsymbol{w}^{(m+1)}|_{t=0} = \boldsymbol{w}_0, \quad \boldsymbol{w}^{(m+1)} \xrightarrow[|\xi|\to\infty]{} 0, \quad s^{(m+1)} \xrightarrow[|\xi|\to\infty]{} 0, \tag{14}$$

$$[\boldsymbol{w}^{(m+1)}]|_{G_T} = 0, \quad [\Pi_0 T(\boldsymbol{w}^{(m+1)}, s^{(m+1)})\boldsymbol{n}_0]|_{G_T} = \Pi_0 \boldsymbol{a} + \boldsymbol{l}_3(\boldsymbol{w}^{(m)}),$$

$$[\boldsymbol{n}_0 \cdot T(\boldsymbol{w}^{(m+1)}, s^{(m+1)})\boldsymbol{n}_0]|_{G_T} - \sigma \boldsymbol{n}_0 \cdot \Delta(0) \int_0^t \boldsymbol{w}^{(m+1)} dt'|_{G_T} =$$

$$= b + \sigma \int_0^t B dt' + l_4(\boldsymbol{w}^{(m)}, s^{(m)}) + \sigma \int_0^t l_5(\boldsymbol{w}^{(m)}) dt'.$$

Estimates (11)-(13) give us a possibility to apply the above theorem concerning the linear problem to system (14). The convergence of the approximations $(\boldsymbol{w}^{(m+1)}, s^{(m+1)})$ to the solution $(\boldsymbol{w}, s)$ of (6) and the proof of Theorem 3.1 are recieved by the procedure of I.Sh.Mogilevskii and V.A.Solonnikov (1992) for a single liquid.

REFERENCES

DENISOVA I.V. (1990), A priori estimates of the solution of a linear time- dependent problem connected with the motion of a drop in a fluid medium, *Trudy Mat. Inst. Steklov* **188**, p. 3-21 (English transl. in *Proc. Steklov Inst. Math.*, 1991, no. 3, p. 1-24.).

DENISOVA I.V. (1993), Solvability in Hölder spaces of a linear problem concerning the motion of two fluids separated by a closed surface, *Algebra Anal.* **5**(4), p. 122-148 (English transl. in *St.-Petersburg Math. J.* **5**(4), 1994).

DENISOVA I.V. (1994), Problem of the motion of two viscous incompressible fluids separated by a closed free interface, *Acta Applicandae Math.* **37**, p. 31-40.

DENISOVA I.V. and V.A. SOLONNIKOV (1989), Solvability of the linearized problem on the motion of a drop in a fluid flow, *Zap. Nauchn. Semin. Leningrad. Otdel. Mat. Inst. Steklov.(LOMI)* **171**, p. 53-65 (English transl. in *J. Soviet Math.*, **56**, 1991, no. 2, 2309-2316).

DENISOVA I.V. and V.A. SOLONNIKOV (1991), Hölder spaces solvability of a model initial-boundary value problem generated by a problem on the motion of two fluids, *Zap. Nauchn. Semin. Leningrad. Otdel. Mat. Inst. Steklov. (LOMI)* **188**, p. 5-44 (English translaton in: *J. Math. Sciences*, **70** (3), 1994, p. 1717-1746).

MOGILEVSKII I.SH and V.A. SOLONNIKOV (1989), Solvability of a noncoercive initial-boundary value problem for the Stokes system in Hölder classes of functions (the case of half-space) (in Russian), *Zeit. Anal. Anwendungen* **8**(4), p. 329-347.

MOGILEVSKII I.SH. and V.A. SOLONNIKOV (1992), On the solvability of an evolution free boundary problem for the Navier-Stokes equations in Hölder spaces of functions, in: GALDI G.P.(ed.) *Mathematical Problems Relating to the Navier-Stokes Equations*, Ser. Adv. in Math. for Appl.Sci.,**11** (World Sci.Publ.), p. 105-181.

SOLONNIKOV V.A. (1976), Estimates for the solution of an initial-boundary value problem for a linear time-dependent system of Navier-Stokes equations, *Zap. Nauchn. Semin. Leningrad. Otdel. Mat. Inst. Steklov. (LOMI)* **59**, p. 178-254 (English transl. in *J. Soviet Math.* **10**, 1978, no. 2).

ANALYSIS OF A NONISOTHERMAL VISCOUS FLOW PROBLEM WITH FREE BOUNDARIES AND A DYNAMIC CONTACT LINE

Jürgen Socolowsky

FB Technik, FG Mathematik der
Fachhochschule Brandenburg, Magdeburger Str. 53
D-14770 Brandenburg/Havel, Germany

1. THE MATHEMATICAL PROBLEM AND GOVERNING EQUATIONS

In the present paper, we consider the plane steady-state nonisothermal flow problem with two free boundaries describing a coating process with a heavy viscous incompressible fluid onto a horizontally moving rigid wall (cf. Figure 1). One of the a priori unknown free surfaces is noncompact and the other one ends at an a priori unknown contact line on the rigid wall. The flow domain is unbounded and thermocapillary convection takes place because the surface tension depends on temperature. This type of convection is important in many technological and scientific applications; interesting examples can be found in the field of materials processing, particularly in coating and solidification processes or in crystal growth processes (see Cuvelier and Driessen, 1986; Shen, Neitzel, Jankowski and Mittelmann, 1990; Ehrhard and Davis, 1991; Puknachov, 1985). We assume that the viscosity of the liquid is constant, although in general case it may depend on temperature. The Boussinesq approximation for similar problems was used by several authors (see the references cited above, and also Cuvelier and Schulkes, 1990; Lagunova, 1986; Rivkind and Il'in, 1988; Socolowsky, 1994a).

Let us consider the mathematical model for this curtain coating process. A heavy viscous incompressible liquid fills the infinite region $V \in \mathbb{R}^2$ bounded by the straight line $\Sigma_1 = \{x \in \mathbb{R}^2; x_2 = 0\}$ and the half-lines

$$\Sigma_2 = \{x \in \mathbb{R}^2; x_1 \leq 0, x_2 = h_1 - x_1 \tan \alpha\}$$

$$\Sigma_3 = \{x \in \mathbb{R}^2; x_1 \leq 0, x_2 = h_2 - x_1 \tan \alpha, h_2 > h_1 > 0\},$$

where α is a real number with $0 \leq \alpha < \pi/2$ (cf. Fig.1). The region V is then the union of the first quadrant of $\mathbb{R}^2$ with the half-strip between Σ_2 and Σ_3.

Navier-Stokes Equations and Related Nonlinear Problems
Edited by A. Sequeira, Plenum Press, New York, 1995

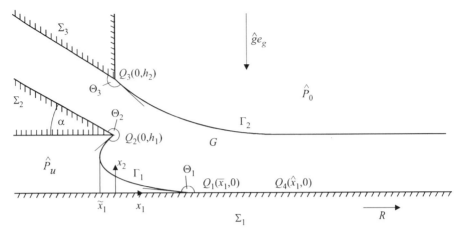

Figure 1.

The force of gravity is directed along the vector $e_g = (0, -1)^T$. We suppose that the flow of the liquid is generated by the stream $F^0(\alpha)$ and by the motion of the lower rigid wall Σ_1 with constant velocity R in x_1-direction. The values R and $F^0(\alpha)$ are assumed to be positive. In agreement with many experimental studies we further suppose that the free surfaces separate from the rigid walls at the "sharp" corner points Q_2 and Q_3. The lower free surface Γ_1 "touches" the *moving* rigid wall Σ_1 at the a priori unknown point $Q_1(\bar{x}_1, 0)$ - the so called dynamic contact point (or line). We assume that Γ_1 is described as the graph of a function ψ_1 with respect to $x_2 \in [0, h_1]$. This assumption makes sense physically and it is a key to handling the lower free surface.

The following notations are used: $\Gamma_m (m = 1, 2)$- free surfaces with the representations $x_1 = \psi_1(x_2)$ and $x_2 = \varphi_2(x_1)$; $G = \{x \in V; x_2 < \varphi_2(x_1) \text{ for } x_1 > 0 \text{ and } x_1 > \psi_1(x_2) \text{ for } x_2 < h_1\}$ - flow domain of the fluid; $\delta(t) = \{x \in G; x_1 = t\}$; n and τ are unit vectors directed along the exterior normal and along the tangent to ∂G, respectively; $\nabla = (\partial/\partial x_1, \partial/\partial x_2)^T$, $\nabla p = \text{grad } p$, $\nabla \cdot v = \text{div } v$ and $\nabla^2 = \Delta$ is the Laplace operator; $S(v)$ is the tensor of deformation velocities with elements $S_{ij} = \partial v_i/\partial x_j + \partial v_j/\partial x_i$. Finally, $a \cdot b$ is the inner product of a, b.

The dimensionless viscosity and density are assumed to be equal to 1. The symbols $\gamma, \lambda, \hat{g}, \sigma(\theta)$ denote the dimensionless thermal expansion coefficient, the thermal conductivity, the acceleration of gravity and the surface tension function, respectively.

Thus, the mathematical problem consists of the determination of the unbounded domain G occupied by the fluid, i.e. the determination of the functions ψ_1 and φ_2 inclusive of the dynamic contact point $Q_1(\bar{x}_1, 0)$, the velocity $v(x) = (v_1(x), v_2(x))^T$, the pressure $\hat{p}(x)$ and the temperature $\theta(x)$ which satisfy in G the Boussinesq approximation of the coupled heat and mass transfer

$$-\nabla^2 v + (v \cdot \nabla)v + \nabla p = -\gamma\,\theta\,e_g, \qquad \nabla \cdot v = 0, \tag{1.1}$$

$$-\lambda\nabla^2\theta + (v \cdot \nabla)\,\theta = 0 \tag{1.2}$$

and the following *boundary conditions* (=BCs)

$$v\,|_{\Sigma_j} = 0 \,(j = 2, 3), \quad v \cdot n\,|_{\Gamma_m} = 0, \ \tau \cdot S(v)n\,|_{\Gamma_m} = -b\frac{\partial\theta}{\partial\tau}|_{\Gamma_m} \quad (m = 1, 2),$$

$$\tilde{B}\tau \cdot S(v)n + [v - (R,0)^T] \cdot \tau = 0, \quad v \cdot n = 0 \quad (x \in \Sigma_1; \ \bar{x}_1 \le x_1 \le \hat{x}_1),$$

$$v \mid_{\Sigma_1} = (R,0)^T \quad (x \in \Sigma_1; \ x_1 \ge \hat{x}_1), \tag{1.3}$$

$$\theta \mid_{\Sigma_j} = \bar{\theta}_j \quad (j = 1, \ldots, 3), \qquad \frac{\partial \theta}{\partial n}\Big|_{\Gamma_m} = 0 \quad (m = 1, 2), \tag{1.4}$$

$$\frac{d}{dx_2} \frac{\psi_1'(x_2)}{[1 + (\psi_1'(x_2))^2]^{1/2}} + \beta(\theta)x_2 = \frac{1}{\sigma(\theta)} [\bar{p} + p - n \cdot S(v)n],$$

$$\frac{d}{dx_1} \frac{\varphi_2'(x_1)}{[1 + (\varphi_2'(x_1))^2]^{1/2}} - \beta(\theta)\varphi_2(x_1) = \frac{1}{\sigma(\theta)}(-p + n \cdot S(v)n),$$

$$\varphi_2(0) = h_2, \qquad \psi_1(h_1) = 0, \qquad \psi_1'(0) = \cot \theta_1 = -A. \tag{1.5}$$

The fact that the motion is caused by nonzero flux is mathematically formulated in the form

$$\int_{\delta(t)} v_1(t, x_2) \, dx_2 = F^0(\alpha). \qquad (t \in \mathbb{R}) \tag{1.6}$$

In (1.1),(1.5) the transformation $p(x) = \hat{p}(x) + \hat{g}\, x_2 - \hat{p}_0$, where $\hat{p}$ denotes the original physical pressure, is realized and the symbols $\beta(\theta) = \hat{g}/\sigma(\theta)$, $\bar{p} = \hat{p}_0 - \hat{p}_u$ are also used. In the last *equation* (=eq.) $\hat{p}_u$, $\hat{p}_0$ denote the (constant) atmospheric pressures outside Γ_1 and Γ_2, respectively.

It was shown (see the references already cited, and also Wolff, 1988) for a large number of liquids that the surface tension can be regarded as a linear function of the temperature, i.e.

$$\sigma(\theta) = a - b\,\theta, \tag{1.7}$$

with constant positive coefficients a, b. The restriction

$$\mid p(x) \mid \le \text{ const. as } x_1 \to +\infty, \tag{1.8}$$

resulting from physical considerations completes the free *boundary value problem* (=BVP) (1.1)-(1.6). The last equation in the first line of (1.3) was obtained from the original condition

$$\tau \cdot S(v)n \mid_{\Gamma_m} = \frac{\partial \sigma(\theta)}{\partial \tau}\Big|_{\Gamma_m} \quad (m = 1, 2)$$

with the help of (1.7). For a detailed discussion of BVP (1.1)-(1.6) we refer to the literature: Cuvelier and Driessen (1986), Ehrhard and Davis (1991), Socolowsky (1989). The nature of dynamic contact lines and angles as well as the necessity of slip boundary conditions like (1.3) [cf. the second line] were explained in Socolowsky (1993). Here we only assume that $\theta_1, \hat{x}_1, \tilde{B}$ are prescribed with $\pi/2 < \theta_1 \le \pi$, (i.e. $0 < A \le +\infty$) and $\tilde{B} = \text{const.} > 0$.

The present study aims to prove the unique solvability of the free BVP (1.1)-(1.6) in weighted Hölder spaces for small data $F^0(\alpha), R, |\bar{\theta}_j|$ using functional-analytic methods. Similar stationary problems in bounded domains have been analytically investigated by many authors, cf. Lagunova (1986), Rivkind and Il'in (1988), Wolff (1988), Il'in (1986). The corresponding isothermal problem (i.e. BVP (1.1),(1.3),(1.5),(1.6) with $\gamma = 0$, $b = 0$) was solved analytically and numerically a in former paper Socolowsky (1993). There is a great number of studies presenting numerical procedures for nonisothermal free BVPs, which are mentioned in the references.

Numerical results of such problems including temperature-dependent viscosities and dissipation are also given by the author (1994b). The solvability of the full free BVP (1.1)-(1.6) will be shown in several steps. Because the present study is an extension of former results (Socolowsky, 1993) to the thermal case we concentrate our attention on the equations and BCs for the temperature.

2. FUNCTION SPACES

Let B be an arbitrary domain in $\mathbb{R}^2$ and $N \subset \bar{B}$ a manifold of dimension $\bar{n} < 2$. The symbol $\rho_N(x)$ denotes the distance $\mathrm{dist}(x, N) := \inf_{y \in N} |x - y|$. Let $\beta = (\beta_1, \beta_2)$ be a multi-index in this section with

$$| \beta | = \beta_1 + \beta_2 \text{ and } D^\beta u = \frac{\partial^{|\beta|} u}{\partial x_1^{\beta_1} \partial x_2^{\beta_2}} \, (\beta_i \in \mathbb{N} \cup \{0\}) .$$

The symbol $[r]$ will denote the integer part of r.

$C^r(B)(r > 0,$ non-integer) denotes the Hölder space of functions defined in a domain $B \subset \mathbb{R}^2$ with a finite norm

$$| u |_B^{(r)} = \sum_{|\beta| < r} \sup_{x \in B} | D^\beta u | + \sum_{|\beta| = [r]} \sup_{x,y \in B} \frac{| D^\beta u(x) - D^\beta u(y) |}{| x - y |^{r - [r]}} .$$

Let $\dot{C}_s^r(B, N)$ be the weighted Hölder space of functions defined in $B \backslash N$ and having a finite norm

$$| u |_{\dot{C}_s^r(B,N)} = \sum_{|\beta| < r} \sup_{x \in B \backslash N} \rho_N^{|\beta| - s}(x) | D^\beta u(x) | +$$

$$+ \sum_{|\beta| = [r]} \sup_{x \in B \backslash N} \rho_N^{r - s}(x) \sup_{|x-y| < \frac{1}{2} \rho_N(x)} \frac{| D^\beta u(x) - D^\beta u(y) |}{| x - y |^{r - [r]}} .$$

$C_s^r(B, N)(r > s > 0; r, s$ non-integer) denotes the space of functions with a finite norm

$$| u |_{C_s^r(B,N)} := | u |_B^{(s)} + \sum_{s < |\beta| < r} \sup_{x \in B \backslash N} \rho_N^{|\beta| - s}(x) | D^\beta u(x) | +$$

$$+ \sum_{|\beta| = [r]} \sup_{x \in B \backslash N} \rho_N^{r - s}(x) \sup_{|x-y| < \frac{1}{2} \rho_N(x)} \frac{| D^\beta u(x) - D^\beta u(y) |}{| x - y |^{r - [r]}} .$$

Clearly, $\dot{C}_s^r(B, N)$ is a subspace of $C_s^r(B, N)$ consisting of functions vanishing on N together with their derivatives of order up to $[s]$. For $s < 0$ assume $C_s^r(B, N) := \dot{C}_s^r(B, N)$.

For the definition of a generalized solution to the linear auxiliary problem we need also the following function spaces. Note that the domain G is unbounded.

By $C_0^\infty(G, \Gamma)$ we mean the set of all infinitely differentiable vector functions v vanishing for $|x| \gg 1$ and satisfying the following BCs

$$v |_{\Sigma_j} = 0 \, (j = 2, 3), \quad v |_{\Sigma_{12}} = 0, \quad v \cdot n = 0 \, (x \in \Gamma_1 \cup \Gamma_2 \cup \Sigma_{12}).$$

Further we have:

$$J_0^\infty(G, \Gamma) := \{v \in C_0^\infty(G, \Gamma), \, \mathrm{div} \, v = 0\}.$$

The Sobolev spaces $D(G)$ and $H(G)$ are the completions of the sets $C_0^\infty(G, \Gamma)$ and $J_0^\infty(G, \Gamma)$ with respect to the Dirichlet norm

$$\| u_x \|_G^2 := \| u_x \|_{L_2(G)}^2 = \int_G \sum_{i,j=1}^{2} \left(\frac{\partial u_i}{\partial x_j} \right)^2 dx.$$

$C_{\theta,0}^\infty(G,\Gamma)$ is the set of all infinitely differentiable scalar fields $\theta(x)$ vanishing for $|x| \gg 1$ and satisfying the BCs

$$\theta\,|_{\Sigma_j} = 0 \, (j = 1, \dots, 3).$$

The Sobolev space $D_\theta(G)$ is then the completion of the set $C_{\theta,0}^\infty(G,\Gamma)$ with respect to the Dirichlet norm

$$\| \theta_x \|_G^2 := \| \theta_x \|_{L_2(G)}^2 = \int_G \sum_{i=1}^2 \left(\frac{\partial\theta}{\partial x_i} \right)^2 \mathrm{d}x.$$

Finally we define the weighted Hölder spaces to which the generalized solutions of the problem (1.1)-(1.6) belong. For this purpose we use the following notations: $Q^* := Q_1 \cup Q_2 \cup Q_3 \cup Q_4$, $J_1 :=]0, h_1[$, $Y_1 := \{0, h_1\}$, $G^0 := \{x \in G; \tilde{x}_1 - 2 < x_1 < \hat{x}_1 + 2\}$, $G^+ := \{x \in G; x_1 > \hat{x}_1 + 1\}$, $G^- := \{x \in G; x_1 < \tilde{x}_1 - 1\}$, $\hat{Q} = Q_1 \cup Q_2$, $J_2^0 := \{x_1 \in \mathbb{R}, 0 < x_1 < \hat{x}_1 + 2\}$, $J_2^+ := \{x_1 \in \mathbb{R}, x_1 > \hat{x}_1 + 1\}$. Here $\tilde{x}_1$ denotes the value $\tilde{x}_1 := \min_{x_2 \in [0, h_1]} \psi_1(x_2)$.

For an arbitrary real number $z > 0$ define the space

$$C_{s,z}^r(G) = \{u(x), u\,|_{G^0} \in C_s^r(G^0, Q^*), \exp(zx_1)u(x)\,|_{G^+} \in C^r(G^+),$$

$$\exp(-zx_1)u(x)\,|_{G^-} \in C^r(G^-)\}$$

with the norm:

$$\| u \|_{G,s}^{r,z} := | u\,|_{C_s^r(G^0,Q^*)} + | \exp(zx_1)u\,|_{G^+}^{(r)} + | \exp(-zx_1)u\,|_{G^-}^{(r)}.$$

For functions $f(x_1)$ defined in $\Delta := \mathbb{R}_+^1$ we introduce the space $C_{s,z}^r(\Delta)$ with the norm

$$\| f \|_{\Delta,s}^{r,z} = | f\,|_{C_s^r(J_2^0,0)} + | f(x_1)\exp(zx_1)\,|_{J_2^+}^{(r)}.$$

The norm $| \tilde{g}\,|_{C_s^r(J_1,Y_1)}$ is defined analogously.

3. LINEAR AUXILIARY PROBLEM

The linear auxiliary problem in the domain G with *fixed* boundaries consists of the linearized Eqs.(1.1),(1.2) with given right-hand sides f, r and g, respectively, and the homogeneous BCs (1.3),(1.4), where $\partial\sigma/\partial\tau$ was set at zero. Thus, we obtain

$$-\nabla^2 v + \nabla p = f, \qquad \nabla \cdot v = r, \qquad (x \in G), \tag{3.1}$$

$$-\nabla^2\theta = g, \qquad (x \in G), \tag{3.2}$$

$$v = 0 \quad (x \in \Sigma_{12} \cup \Sigma_2 \cup \Sigma_3), \qquad v \cdot n = 0 \quad (x \in \Gamma_1 \cup \Gamma_2 \cup \Sigma_{11}),$$

$$\tau \cdot S(v)n\,|_{\Gamma_m} = 0 \; (m = 1, 2), \qquad \tilde{B}\tau \cdot S(v)n + v \cdot \tau = 0 \; (x \in \Sigma_{11}), \tag{3.3}$$

$$\theta\,|_{\Sigma_j} = 0 \quad (j = 1, \dots, 3), \qquad \frac{\partial\theta}{\partial n}\,|_{\Gamma_m} = 0. \quad (m = 1, 2) \tag{3.4}$$

This problem can be decomposed into a BVP (3.1),(3.3) for v, p and a second BVP (3.2),(3.4) for θ. The solution to BVP (3.1),(3.3) was completely presented by the author (1993). Let us give the concept of a weak solution to BVP (3.2),(3.4).

Definition 3.1. *By a weak solution to BVP (3.2),(3.4) we understand a scalar field $\theta \in D_\theta(G)$ satisfying the integral identity*

$$E_\theta(\theta, \xi) := \int_G (\nabla\theta \cdot \nabla\xi)\,\mathrm{d}x = \int_G g\xi\,\mathrm{d}x \tag{3.5}$$

205

for all scalar fields $\xi \in D_\theta(G)$.

The existence of such a solution can be shown in the same manner as in Socolowsky (1994a), where a modified unbounded domain for two liquids was considered. This procedure is well known for the Laplace equation in bounded domains. Let ψ_1, φ_2 be the representation functions described in Section 1. We define the dynamic contact angle θ_1 and both static contact angles θ_2, θ_3 (cf. Fig.1) by the relations

$$\theta_1 = \pi/2 + \arctan[-\psi_1'(0)], \theta_2 = 3\pi/2 - \alpha - \arctan[-\psi_1'(h_1)], \theta_3 = \pi + \alpha + \arctan[-\varphi_2'(0)].$$

Analyzing some model problems (Socolowsky, 1992; Socolowsky and Wolff, 1994) for θ in a neighbourhood of Q_1, Q_2, Q_3, Q_4, it could be shown that the weak solution θ belongs to $C_s^{s+2}(G^0)$ with s satisfying the condition $0 < s < s_t := \min\limits_{j \in \{1,2,3\}} [1/3, \pi/(2\theta_j)]$. Assume now that the fixed boundaries Γ_m $(m = 1, 2)$ are of class C_s^{3+s}. Finally, we need the set $G(\mu, t) := \{x \in G, \mu - t < x_1 < \mu + t\}$.

Theorem 3.1. *There is a positive real number z_t such that for any $g \in C_{s-2,z}^s(G)$ with $s \in]0, s_t[$ and $z \in]0, z_t[$ the generalized solution $\theta \in D_\theta(G)$ to BVP (3.2),(3.4) satisfies the inequality*

$$\int_{G(\mu,1)} |\nabla\theta|^2 \, dx \le c_0 e^{-2z|\mu|} \left(\|g\|_{G,s-2}^{s,z} \right)^2, \tag{3.6}$$

where $|\mu| > |\tilde{x}_2| + 2$ holds and c_0 does not depend on μ.

This theorem is proved in the same way as the analogous theorem in Socolowsky (1994a). There one can find a detailed proof.

Theorem 3.2. *For arbitrary $g \in C_{s-2,z}^s(G)$ with $s \in]0, s_t[$ and $z \in [0, z_t[$ the BVP (3.2),(3.4) has a unique solution $\theta \in C_{s,z}^{s+2}(G)$ and the estimate*

$$\|\theta\|_{G,s}^{s+2,z} \le c_1 \|g\|_{G,s-2}^{s,z}, \tag{3.7}$$

holds where c_1 does not depend on g.

Note that Eq.(3.2) is elliptic in the sense of Douglis-Nirenberg (Agmon, Douglis and Nirenberg, 1964; Solonnikov, 1964), and that the BCs (3.4) fulfill the complementarity conditions (Solonnikov, 1964). Therefore, one is able to prove Theorem 3.2. in a well-known manner, too. Joining the solution θ to BVP (3.2),(3.4) given here with the solution v, p to BVP (3.1),(3.3), we obtain the solvability of the full linear problem (3.1)-(3.4).

4. NONLINEAR AUXILIARY PROBLEM AND FREE BVP

Consider the BVP (1.1)-(1.4),(1.6) in G with *fixed* boundary. Condition (1.7) is also included. Let $h_2^* > 0$ be a given constant and $(\varphi_2 - h_2^*) \in C_{1+s,z}^{3+s}(\Delta) \psi_1 \in C_{1+s}^{3+s}(J_1, Y_1)$. The space parameters s and z fulfill the conditions

$$0 < s < s_0 := \min[s_*, s_t], \qquad 0 < z < z_0 := \min[z_*, z_t], \tag{4.1}$$

where s_t, z_t are taken from Theorem 3.1. and s_*, z_* denote some constants resulting from analogous model problems (Socolowsky, 1993) for the isothermal BVP (3.1),(3.3). Furthermore, let $\bar{\theta}_j$ $(j = 1, 2)$ be constant and $\bar{\theta}_3 = \bar{\theta}_2$.

By $(u^{(-)}, p^{(-)}, \theta^{(-)})$ we understand a solution to BVP (1.1)-(1.4),(1.6) in the channel G^- (here the BCs on $\Sigma_1, \Gamma_1, \Gamma_2$ are neglected). Finally, $(u^{(+)}, p^{(+)}, \theta^{(+)})$ denotes a solution to BVP (1.1)-(1.4),(1.6) in $\hat{G}^+$, where $\hat{G}^+$ is a domain approximating the curve Γ_2 by the half-line $x_2 = h_2^*$ as $x_1 > 1$ (here the BCs on $\Sigma_j (j = 2, 3)$ are neglected). These solutions can explicitly be calculated. They have the form

$$u_1^{(-)}(x_1, x_2) = \frac{6F^0(\alpha)}{(h_2 - h_1)^3}(x_2 - h_1 + x_1 \tan \alpha)(h_2 - x_2 - x_1 \tan \alpha),$$

$$u_2^{(-)}(x_1, x_2) = -\frac{6F^0(\alpha) \tan \alpha}{(h_2 - h_1)^3}(x_2 - h_1 + x_1 \tan \alpha)(h_2 - x_2 - x_1 \tan \alpha),$$

$$p^{(-)}(x_1, x_2) = -\frac{12F^0(\alpha)}{(h_2 - h_1)^3 \cos^2 \alpha}[x_1 - (x_2 - h_1) \tan \alpha] + \gamma \theta_2 (x_2 - h_1),$$

$$\theta^{(-)}(x_1, x_2) \equiv \theta_2 = \text{ const.} \tag{4.2}$$

$$u_1^{(+)}(x_1, x_2) = R + \frac{3}{(h_2^*)^2}\left(R - \frac{F^0(\alpha)}{h_2^*}\right) x_2 \left(\frac{x_2}{2} - h_2^*\right), \qquad (x \in \bar{G}^+)$$

$$u_2^{(+)} \equiv 0, \ \theta^{(+)} \equiv \theta_1, \ p^{(+)} = \frac{3}{(h_2^*)^2}\left(R - \frac{F^0(\alpha)}{h_2^*}\right) x_1 + \gamma \theta_1 x_2, \qquad (x \in \bar{G}^+). \tag{4.3}$$

Now the solution $(u^{(+)}, p^{(+)}, \theta^{(+)})$ hitherto only defined on $\hat{G}^+$ can be extended or restricted to $G^+ \neq \hat{G}^+$ with the help of formula (4.3). In this case the notation is preserved. Let $M(F, R, \bar{\theta}) := \max(F^0(\alpha), R, |\bar{\theta}_1|, |\bar{\theta}_2|)$ and g be a smooth real function vanishing for $t \leq 1$ and being equal to 1 for $t \geq 2$.

Theorem 4.1. *For a sufficiently small number $M(F, R, \theta)$ and for s, z satisfying condition (4.1) the BVP (1.1)-(1.6) has a unique solution (v, p, θ) permitting the representation*

$$v = g(-x_1)u^{(-)} + g(x_1)u^{(+)} + w, \qquad p = g(-x_1)p^{(-)} + g(x_1)p^{(+)} + q,$$

$$\theta = g(-x_1)\theta^{(-)} + g(x_1)\theta^{(+)} + \theta_0, \tag{4.4}$$

where $u^{(+)}, u^{(-)}, p^{(+)}, p^{(-)}, \theta^{(+)}$ and $\theta^{(-)}$ are given by formulae (4.2),(4.3) and $w, \theta \in C_{s,z}^{s+2}(G), \nabla p \in C_{s-2,z}^s(G)$. Moreover,

$$\| w \|_{G,s}^{s+2,z} + \| \nabla q \|_{G,s-2}^{s,z} + \| \theta_0 \|_{G,s}^{s+2,z} \leq c_2(F^0(\alpha), R, |\bar{\theta}_1|, |\bar{\theta}_2|)$$

holds and $c_2 \to 0$ for $M(F, R, \theta) \to 0$.

This principal result can be proved in the same way as in the isothermal case using Banach's fixed point theorem and results for the linear problem considered in Section 3 as well as results for the related *nonhomogeneous* linear problem which can also be handled as in Socolowsky (1993).

Now we study the solvability of the complete BVP (1.1)-(1.6) under the conditions (1.7) and (1.8). We consider v, p and θ in Eqs.(1.5) as the solution of the nonlinear auxiliary problem (1.1)-(1.4),(1.6) depending on the functions $\psi_1(x_2), \varphi_2(x_1)$, and we show that for sufficiently small $M(F, R, \bar{\theta})$ the functions ψ_1, φ_2 are determined from (1.5) uniquely. Firstly, it follows from the representation of the solution to BVP (1.1)-(1.6) that it is necessary to choose $h_2^* = F^0(\alpha)/R$ in order to satisfy the condition (1.8). This implies

$$\varphi_2(x_1) \to \frac{F^0(\alpha)}{R} \text{ as } x_1 \to +\infty. \tag{4.5}$$

In this case the pressure $p(x)$ is bounded for $x_1 > 0$. Note that $\theta \in C_{s,z}^{s+2}(G)$ yields the continuity of θ up to the boundaries of G. Thus the inverse surface tension $\sigma^{-1}(\theta) = (a - b\theta)^{-1}$ is also a continuous function provided that $\theta < a/b$ is sufficiently small. For this reason we can conclude that

$$p(x) \to p_* = \frac{F^0(\alpha)}{R} \quad \text{as } x_1 \to +\infty \tag{4.6}$$

hold applying the same operations as in the isothermal case.

Let be $\omega_2(x_1) := \varphi_2(x_1) - \bar{\varphi}_2(x_1)$, where $\bar{\varphi}_2(x_1)$ denotes a solution to the following BVP

$$\frac{\mathrm{d}}{\mathrm{d}x_1} \frac{\bar{\varphi}_2'(x_1)}{[1 + (\bar{\varphi}_2'(x_1))^2]^{1/2}} - \beta(0)\bar{\varphi}_2(x_1) = -\beta(0)\frac{F^0(\alpha)}{R}, \quad (x_2 \in]0, h_1[)$$

$$\bar{\varphi}_2(0) = h_2, \qquad \varphi_2(x_1) \to \frac{F^0(\alpha)}{R} \text{ as } x_1 \to +\infty. \tag{4.7}$$

Analogously, let be $\omega_1(x_2) := \psi_1(x_2) - \bar{\psi}_1(x_2)$, where $\bar{\psi}_1(x_2)$ denotes a solution to the following BVP

$$\frac{\mathrm{d}}{\mathrm{d}x_2} \frac{\bar{\psi}_1'(x_2)}{[1 + (\bar{\psi}_1'(x_2))^2]^{1/2}} + \beta(0)x_2 = \beta(0)\frac{F^0(\alpha)}{R} + \frac{\beta\bar{p}}{\hat{g}}, \quad (x_1 \in \Delta)$$

$$\bar{\psi}_1(h_1) = 0, \qquad \bar{\psi}_1'(0) = -A = \cot\theta_1. \tag{4.8}$$

These BVPs result from (1.5) in the case $v \equiv 0, p = p_* = \text{const.}, \theta = 0$ being a solution to the auxiliary BVP (1.1)-(1.4),(1.6) for the parameters $R = F^0(\alpha) = \bar{\theta}_1 = \bar{\theta}_2 = 0$. It was easy to show that the problem (4.7) has a unique solution, if the condition $| \bar{h}_2 - h_2^* |< [2/\beta(0)]^{1/2}$ is satisfied. Furthermore, the difference $(\varphi_2(x_1) - h_2^*)$ is equivalent to $\exp(-\sqrt{\beta(0)}x_1)$ as $x_1 \to +\infty$.

The BVP (4.8) was studied by the author (1993) in detail and the corresponding results will be included in Theorem 4.2. For the unknown functions $\omega_m(x_1)$ we obtain a two-point BVP like BVP (8.8) from Socolowsky (1992), subtracting Eqs.(4.7) and (4.8) from Eq.(1.5). A difference to BVP (8.8) consists in the following. We have to substitute β by $\beta(0)$ everywhere and, additionally, we introduce the operators $T_m^{(3)}$ defined as

$$T_m^{(3)}\omega_m := \frac{b\theta}{\sigma(\theta)}\omega_m = \frac{\sigma(0) - \sigma(\theta)}{\sigma(\theta)}\omega_m. \tag{4.9}$$

The remaining part of the proof of the main theorem can be realized as a modified repetition of the proof of Theorem 8.1 in Socolowsky (1992). Instead of the operator Eq.(8.10) therein we have to study the following one:

$$\omega_m = L_m(T_m^{(1)}\omega_m + T_m^{(2)}\omega_m + T_m^{(3)}\omega_m) =: B_m\omega_m \tag{4.10}$$

with $T_m^{(3)}$ given in (4.9). Since $T_m^{(3)}$ is a contraction operator for small θ we can conclude in the same manner that B_m is a contraction operator too in the ball $\| \omega_m \|_{\Delta, 1+s}^{3+s,z} < \epsilon$. Consequently, we have proved the main result of the present paper.

Theorem 4.2. *There exist positive real numbers $\bar{s} < s_0$, $\bar{z} \le \min[z_0, \sqrt{\beta(0)}]$ and M_0 such that for arbitrary $s \in]0, \bar{s}[, z \in]0, \bar{z}[, M(F, R, \theta) < M_0$ and for positive $h_1, h_2, F_0(\alpha)$, R satisfying the conditions*

$$| h_2 - \frac{F_0(\alpha)}{R} |< \sqrt{\frac{2}{\beta(0)}}, \qquad 0 < h_1 < \sqrt{\frac{2(\tilde{A} + 1)}{\beta(0)}},$$

$$h_1 \leq \frac{F_0(\alpha)}{R} + \frac{(\hat{p}_o - \hat{p}_u)}{\hat{g}} < \frac{h_1}{2} + \frac{\tilde{A} + 1}{\beta(0)h_1} \qquad (4.11)$$

the free BVP (1.1)-(1.6) has a unique solution $\{v, p, \theta, \varphi_2, \psi_1\}$ which can be represented in the form

$$v = g(-x_1)u^{(-)} + g(x_1)(R, 0)^T + w, \qquad \varphi_2(x_1) = \bar{\varphi}_2(x_1) + \omega_2(x_1),$$

$$p = g(-x_1)p^{(-)} + p_0(x) + p_*, \qquad \psi_1(x_2) = \bar{\psi}_1(x_2) + \omega_1(x_2),$$

$$\theta = g(-x_1)\,\theta^{(-)} + g(x_1)\,\theta_1 + \theta_0(x), \qquad (4.12)$$

where $g, (u^{(-)}, p^{(-)}, \theta^{(-)})$ are taken from Theorem 4.1 and $\tilde{A} := A(1 + A^2)^{-1/2}$ was set. Moreover, $\theta_0, w \in C_{s,z}^{s+2}(G), p_0 \in C_{s-1,z}^{s+1}(G^0 \cup G^+), \nabla p_0 \in C_{s-2,z}^s(G)$ and $\omega_2 \in C_{1+s,z}^{3+s}(\Delta), \omega_1 \in C_{1+s}^{3+s}(J_1, Y_1)$ hold.

REFERENCES

AGMON, S., DOUGLIS, A. and L. NIRENBERG (1964), Estimates near the boundary for solutions of elliptic partial differential equations satisfying general boundary conditions. I., *Comm. Pure Appl. Math.* 12: 623 (1959), and II., *ibid.* **17**: 35.

ALLAIN, G. (1990), Role de la tension superficielle dans la convection de Benard, *M²AN- Math. Modelling and Numerical Analysis* **24**: 153.

CUVELIER, C. and J.M. DRIESSEN (1986), Thermocapillary free boundaries in crystal growth, *J. Fluid Mech.* **169**, 1.

CUVELIER, C. and R.M.S.M. SCHULKES (1990), Some Numerical Methods for Computation of Capillary Free Boundaries Governed by the Navier-Stokes Equations, Report 90-11, TU Delft.

EHRHARD, P. and S.H. DAVIS (1991), Nonisothermal spreading of liquid drops on horizontal plates, *J. Fluid Mech.* **229**: 365.

IL'IN, A.V. (1986), Numerical study of flow problems for thin viscous liquid layers, PhD thesis, University of St.Petersburg, (in Russian).

KARAGIANNIS, A., HRYMAK, A.N. and J. VLACHOPOULOS (1989), Three-dimensional nonisothermal extrusion flows, *Rheol. Acta* **28**: 121.

LAGUNOVA, M.V. (1986), On the solvability of the plane problem of thermocapillary convection, *in*: Probl. Matemat. Anal. **10** Lin. and nonlin. BVPs. Spectr. theory, Izdat. Univ., St. Petersburg, (in Russian).

SHEN, Y., NEITZEL, G.P., JANKOWSKI, D.F. and H.D. MITTELMANN (1990), Energy stability of thermocapillary convection in a model of the float-zone crystal-growth process, *J. Fluid Mech.* **217**: 639.

PUKNACHOV, V.V. (1985), Free boundary problems in the theory of thermocapillary convection, in: Research Notes in Mathematics **121** - Free Boundary Problems: Applications and Theory, Vol.IV, A. BOSSAVIT, A. DAMLAMIAN and M. FREMOND (eds.), Pitman Advanced Publishing Program, Boston-London-Melbourne.

RIVKIND, V.J. and A.V. IL'IN (1988), On a numerical method for solving the non-isothermal flow problem of a liquid layer, in: Numerical methods **15**, Izdat. Univ., St. Petersburg, (in Russian).

SACKINGER, P.A., BROWN, R.A. and J.J. DERBY (1989), A finite element method for analysis of fluid flow, heat transfer and free interfaces in Czochralski crystal growth, *Int. J. Num. Meth. Fluids* **9**: 453.

SOCOLOWSKY, J. (1992), Solvability of a stationary problem on the plane motion of two viscous incompressible liquids with non-compact free boundaries, *Z. Angew. Math. Mech. (ZAMM)* **72**: 251.

SOCOLOWSKY, J. (1993), The solvability of a free boundary problem for the stationary Navier-Stokes equations with a dynamic contact line, *Nonlinear Analysis, Theory, Methods & Applications (JNA-TMA)* **21**: 763.

SOCOLOWSKY, J. (1994a), Existence and uniqueness of the solution to a free boundary-value problem with thermocapillary convection in an unbounded domain, *Acta Applicandae Mathematicae* **37**: 181.

SOCOLOWSKY, J. (1994b), The numerical analysis of nonisothermal coating flows, (submitted for publication).

SOCOLOWSKY, J. and M. WOLFF (1994), On some model problems in flow problems with contact angles, *Z. Angew. Math. Mech.(ZAMM)* **74**: 557.

SOLONNIKOV, V.A. (1964), General boundary value problems for Douglis-Nirenberg elliptic systems. I, *Izv. Akad. Nauk SSSR-Ser.Math.* **28**: 665, (in Russian).

SUGENG, F., PHAN-THIEN, N. and R.I. TANNER (1987), A study of nonisothermal non-Newtonian extrudate swell by a mixed boundary element and finite element method, *J. Rheology* **31**: 37.

WOLFF, M. (1988), Stationary flow of heat-conducting fluids with free surface between concentric cylinders, *Seminarbericht* No.**99**, Humb.-Univ. Berlin.

ON THE FREE BOUNDARY PROBLEM FOR THE NAVIER-STOKES EQUATIONS GOVERNING THE MOTION OF A VISCOUS INCOMPRESSIBLE FLUID IN A SLOWLY ROTATING CONTAINER

V.A. Solonnikov

St. Petersburg Branch of V.A.Steklov Mathematical Institute
of the Russian Academy of Sciences
Fontanka 27, 191011 St. Petersburg, Russia

1. INTRODUCTION

This article is a continuation of the series of papers (cf. Pukhnachov and Solonnikov, 1982; Solonnikov, 1993, 1994a, 1994b, 1995) devoted to stationary free boundary problems for the Navier-Stokes equations with moving contact points. There were investigated problems governing a viscous flow in a capillary, a coating flow, and a piston problem. Here one more problem of this type is studied.

Let a heavy viscous incompressible liquid partially fill a circular container $V \subset \mathbb{R}^2$ of the radius R_0 rotating about its center with a small angular velocity ω (see Fig.1). We suppose that the force of gravity is directed along the vector $-e_2 = (0, -1)$, and we denote by Ω a subdomain of V occupied with the liquid. The boundary of Ω consists of two parts: $\Sigma = \partial\Omega \cap \partial V$ (a part of a rigid wall $\partial\Omega$) and $\Gamma = \partial\Omega \setminus \partial V$ (a free boundary). The set $M = \bar\Sigma \cap \bar\Gamma$ is a union of two contact points: x_- and x_+. We are concerned with the following free boundary problem: find $\Omega \subset V$ (or, what is the same, a free boundary Γ), the velocity vector field $\mathbf{v}(x) = (v_1, v_2)$ and the pressure $p(x)$ satisfying in Ω the Navier-Stokes equations

$$-\nu\nabla^2\mathbf{v} + (\mathbf{v} \cdot \nabla)\mathbf{v} + \nabla p = 0, \quad x \in \Omega \tag{1.1}$$

and the boundary conditions

$$\mathbf{v}|_\Sigma = \mathbf{a}, \quad \mathbf{v} \cdot \mathbf{n}|_\Gamma = 0, \quad \tau \cdot \mathbf{S}(\mathbf{v})\mathbf{n}|_\Gamma = 0, \tag{1.2}$$

$$\sigma H - g x_2 - \mathbf{n} \cdot \mathbf{T}(\mathbf{v}, p)\mathbf{n}|_\Gamma = -p_1 = Const. \tag{1.3}$$

Here $\mathbf{a} = \omega R_0 \tau_0$, τ_0 is a tangential vector to Σ, τ and $\mathbf{n}$ are a tangential and an exterior normal vectors to Γ, respectively, $\mathbf{T}$ and $\mathbf{S}$ are the stress and the deformation tensors, i.e.

$$\mathbf{T}(\mathbf{v}, p) = -pI + \nu\mathbf{S}(\mathbf{v}), \quad S_{ij} = \frac{\partial v_i}{\partial x_j} + \frac{\partial v_j}{\partial x_i},$$

Navier-Stokes Equations and Related Nonlinear Problems
Edited by A. Sequeira, Plenum Press, New York, 1995

and ν, σ, g are constant positive coefficients of viscosity, of the surface tension, and the acceleration of gravity, respectively. In addition, we fix the volume of the liquid, i.e. the area of Ω is

$$|\Omega| = Q < \pi R_0^2$$

and we assume that the contact angle θ, i.e. the angle between Γ and Σ at the contact points, equals π. This means that Γ is tangential to $\partial\Omega$ at these points. For $\theta \in (0, \pi]$ our problem can not be solved in the class of vector fields $\mathbf{v}$ with a finite Dirichlet integral (see Pukhnachov and Solonnikov, 1980; Solonnikov, 1994a, 1995).

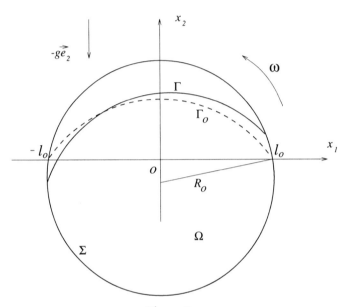

Figure 1.

Problem governing the motion of a viscous fluid in a rotating container was formulated and considered by a different approach in the paper by Baiocchi and Pukhnachov (1990) who were able to reduce it to a certain variational inequality. However, it has required some modifications of the formulation of the problem, in particular, the prescription of Γ.

Let us recall the definition of weighted Hölder spaces in which we are going to work. For arbitrary non-integral $l, s > 0$, arbitrary domain $G \subset \mathbb{R}^n$ and a closed set $F \subset \partial G$ we define the space $C_s^l(G, F)$ as the set of scalar- or vector-valued functions $u(x)$, $x \in G$, with the norm

$$|u|_{C_s^l(G,F)} = |u|_{C^s(G)} + \sum_{s < |j| < l} \sup_{x \in G} \rho^{|j|-s}(x)|D^j u(x)| + [u]_{C_s^l(G,F)},$$

where

$$[u]_{C_s^l(G,F)} = \sum_{|j|=[l]} \sup_{x \in G} \rho^{l-s}(x) \sup_{y \in K(x)} \frac{|D^j u(x) - D^j u(y)|}{|x - y|^{l-[l]}},$$

$$\rho(x) = \mathrm{dist}(x, F), \quad K(x) = \{y \in G : |x - y| \le \rho(x)/2\}$$

and

$$|u|_{C^s(G)} = \sum_{|j|<l} \sup_{x \in G} |D^j u(x)| + \sum_{|j|=[s]} \sup_{x,y \in G} \frac{|D^j u(x) - D^j u(y)|}{|x - y|^{s-[s]}}$$

212

is a usual Hölder norm in G.

The spaces $C_s^l(G, F)$ can be also introduced for $s < 0$, in which case the norm is given by

$$|u|_{C_s^l(G,F)} = \sum_{|j|<l} \sup_{x \in G} \rho^{|j|-s}(x)|D^j u(x)| + [u]_{C_s^l(G,F)}. \tag{1.4}$$

They can be defined for functions given on manifolds, in particular, on Γ. Finally, we say that $\Gamma \in C_s^l$ if this line may be given by the equations $\mathbf{x} = \mathbf{x}(t)$ where $t \in (0, d)$ is a parameter and $\mathbf{x} \in C_s^l(I, \partial I)$.

We prove the following theorem.

Theorem 1.1. *Suppose that*

$$|\Omega| \in (\pi R_0^2/2 + b_1, \pi R_0^2 - b_2), \quad b_1, b_2 > 0, \tag{1.5}$$

and $g/\sigma > B_0 > 0$ (see Proposition 3.1). For arbitrary sufficiently small ω problem (1.1)-(1.3) has a unique solution with the following properties:

1. *$\Gamma \in V$ is a curve of the class $C_{1+\gamma}^{3+\alpha}$ ($\alpha \in (0,1)$, $\gamma \in (1/2, 1)$), which is close to the curve Γ_0 corresponding to the rest state*

2. *$\mathbf{v} \in C_\beta^{2+\alpha}(\Omega, M)$, $p \in C_{\beta-1}^{1+\alpha}(\Omega, M)$, with a positive $\beta < 1/2$, and*

$$|\mathbf{v}|_{C_\beta^{2+\alpha}(\Omega,M)} + |p|_{C_{\beta-1}^{1+\alpha}(\Omega,M)} \le c_1|\omega|. \tag{1.6}$$

We shall construct the solution of (1.1)-(1.3) according to the scheme applied in Solonnikov (1993, 1994a, 1994b, 1995) to other free boundary problems with moving contact points. We consider at first the rest state, then we construct a formal solution of (1.1)-(1.3) without paying attention to the property $\Gamma \subset V$ which can be established on the basis of the local analysis of the solution carried out in Solonnikov (1993, 1994a). The main difficulties in this problem are connected with the formal construction of the solution, and it is on this point that we concentrate our attention. As for the asymptotics of the solution near the contact points, all the necessary information (i.e. the study of the behaviour of the solution both for receding and for advancing contact line with a contact angle π at the smooth rigid wall of arbitrary shape) is contained in the paper of Solonnikov (1994a).

2. THE REST STATE

In the rest state, when $\omega = 0$, $\mathbf{v} = 0$ and $p = p_0$ is a constant, the free boundary Γ_0 is defined by the equation

$$\sigma H - gx_2 = -p_0. \tag{2.1}$$

We recall that the force of gravity is directed opposite to x_2-axis and we choose the origin in such a way that the contact points $x_\pm$ have coordinates $(\pm l_0, 0)$, $l_0 < R_0$. Under the condition (1.5) (which is purely technical) the curve Γ_0 can be given by the equation

$$x_2 = \varphi_0(x_1), \quad x_1 \in (-l_0, l_0),$$

where φ_0 is an even function and

$$\varphi_0(\pm l_0) = 0. \tag{2.2}$$

Equation (2.1) can be written in the form

$$\frac{d}{dx_1} \frac{\varphi_0'}{\sqrt{1 + \varphi_0'^2}} - B\varphi_0 = -\frac{p_0}{\sigma}, \quad x_1 \in (-l_0, l_0), \tag{2.3}$$

or

$$\frac{d}{dx_1} \sin \alpha - B\varphi_0 = -\frac{p_0}{\sigma}, \tag{2.4}$$

where $B = g/\sigma$ and α is the angle between the tangential vector to Γ_0 and x_1-axis ($\tan \alpha(x_1) = \varphi_0'(x_1)$).

Let us consider φ_0 as a solution of equation (2.3) satisfying the boundary conditions

$$\varphi_0'(-l_0) = \tan \alpha_0, \quad \varphi_0'(l_0) = -\tan \alpha_0, \tag{2.5}$$

where $\alpha_0 = \alpha(-l_0)$. Choosing p_0 in an appropriate way, we can satisfy also the conditions (2.2). It is well known that for arbitrary $\alpha_0 \in (0, \pi/2)$ the problem (2.3), (2.5) has a unique infinitely differentiable solution φ_0, which is an even function of x_1 and satisfies the inequality $\varphi_0(x) > \varphi(\pm l_0) = 0$. Let us verify that $\Gamma \subset V$. Differentiation of (2.4) gives

$$\frac{d^2}{dx_1^2} \sin \alpha = B \tan \alpha > 0 \quad (x_1 \in (-l_0, 0)).$$

In addition

$$\sin \alpha(-l_0) = \sin \alpha^{(0)}(-l_0), \quad \sin \alpha(0) = \sin \alpha^{(0)}(0) = 0,$$

where $\alpha^{(0)}(x_1)$ is the angle between x_1-axis and the tangential vector to the part of the circle ∂V located above this axis. Since $\sin \alpha^{(0)}(x_1)$ is a linear function of x_1, the above relations imply

$$\sin \alpha(x_1) < \sin \alpha^{(0)}(x_1), \quad x_1 \in (-l_0, 0),$$

which shows that Γ_0 lies between x_1-axis and the upper part of ∂V.

Next, we prove that the curves Γ_0 corresponding to different values of α_0 do not intersect each other, more exactly, the curve corresponding to the greater value of α_0 is located under the curve corresponding to the smaller value of this angle. We write equation (2.4) in the form

$$\frac{d}{dx_1} \sin \alpha(x_1) = By(x_1) \quad (y = \varphi - p_0/g)$$

and suppose that there are given two functions $y_1(x_1)$ and $y_2(x_1)$ satisfying this equation and the conditions

$$y_i'(0) = 0 \quad (i.e. \ \alpha_i(0) = 0, \ i = 1, 2)$$

and

$$y_1(0) < y_2(0).$$

It follows that

$$\frac{d}{dx_1} \sin \alpha_1 \big|_{x_1=0} < \frac{d}{dx_1} \sin \alpha_2 \big|_{x_1=0},$$

hence, $\alpha_1(x_1) > \alpha_2(x_1)$ and $y_1(x_1) < y_2(x_2)$ for negative x_1.

Consider two curves Γ_{0i}, $i = 1, 2$, defined by

$$x_2 = \varphi_{0i}(x_1), \quad x_1 \in (-l_{0i}, l_{0i}), \quad i = 1, 2,$$

with $l_{0i} = R_0 \sin \alpha_{0i}$, $\alpha_{01} > \alpha_{02}$. The function α_1 is less than the corresponding function for ∂V at the point $-l_2$, hence, $\alpha_1(-l_2) < \alpha_2(-l_2)$. As we have seen, this implies $\alpha_1(x_1) < \alpha_2(x_1)$ or

$$\varphi'_{01}(x_1) = \tan \alpha_1(x_1) < \tan \alpha_2(x_1) = \varphi'_{02}(x_1),$$

for $x_1 \in (-l_2, 0)$. Consequently,

$$\varphi_{01}(x_1) = \varphi_{01}(-l_2) + \int_{-l_2}^{x_1} \varphi'_{01}(\xi) d\xi < \varphi_{02}(-l_2) + \int_{-l_2}^{x_1} \varphi'_{02}(\xi) d\xi = \varphi_{02}(x_1).$$

This shows that the area of Ω is a monotone decreasing function of the angle α_0. For every value of $|\Omega| \in (\pi R_0^2/2 + b_1, \pi R_0^2 - b_2)$ there exists exactly one corresponding value of $\alpha_0 \in (d_1, \pi/2 - d_2)$, $d_i > 0$, and Γ_0 can be found from (2.3),(2.5).

At the conclusion we compute the constant p_0. Integration of (2.3) over the interval $(-l_0, l_0)$ gives

$$-2 \sin \alpha_0 - BA = -\frac{2l_0 p_0}{\sigma}, \tag{2.6}$$

where

$$A = \int_{-l_0}^{l_0} \varphi_0(x_1) dx_1 > 0$$

is the area of the domain between Γ_0 and x_1-axis. Hence,

$$p_0 = \frac{\sigma}{R_0} + \frac{BA\sigma}{2l_0} > \frac{\sigma}{R_0}. \tag{2.7}$$

3. AUXILIARY PROPOSITIONS

Let us turn our attention to problem (1.1)-(1.3). The free boundary Γ will be found as a perturbation of Γ_0, and it will be given by the equation

$$x_2 = \varphi(x_1), \quad x_1 \in (-l_1, l_2), \tag{3.1}$$

where l_i are some unknown numbers close to l_0. The points $x_- = (-l_1, \varphi(-l_1))$ and $x_+ = (l_2, \varphi(l_2))$ should be located on ∂V and the line Γ should be tangential to ∂V at these points. Let $(0, y_c)$, $y_c = -R_0 \sin \alpha_0$, be coordinates of the center of V. The equations of the semi-circles $\{x \in \partial V, \ x_2 > y_c\}$ and $\{x \in \partial V, \ x_1 > 0\}$ have the form

$$x_2 = k(x_1) \equiv y_c + \sqrt{R_0^2 - x_1^2}, \quad x_1 \in (-R_0, R_0)$$

and

$$x_1 = h(x_2) \equiv \sqrt{R_0^2 - (x_2 - y_c)^2}, \quad x_2 \in (y_c - R_0, y_c + R_0),$$

respectively, hence, the above conditions reduce to

$$-l_1 = -h(\varphi(-l_1)), \qquad l_2 = h(\varphi(l_2)),$$

$$\varphi'(-l_1) = k'(-l_0), \qquad \varphi'(l_2) = k'(l_2). \tag{3.2}$$

Equation (1.3) may be written in the form

$$\frac{d}{dx_1} \frac{\varphi'}{\sqrt{1 + \varphi'^2}} - B\varphi = t(x_1) - q, \quad x_1 \in (-l_1, l_2), \tag{3.3}$$

with $q = p_1/\sigma$, $t(x_1) = \sigma^{-1}\mathbf{n} \cdot \mathbf{Tn}|_{x_2=\varphi(x_1)}$. It is convenient to map the interval $(-l_0, l_0)$ onto $(-l_1.l_2)$ by means of a linear transformation

$$x_1 = \mu(\xi - \bar{\xi})$$

with

$$\mu = \frac{l_1 + l_2}{2l_0}, \qquad \bar{\xi} = l_0\frac{l_1 - l_2}{l_1 + l_2},$$

and to introduce the function

$$\check{\varphi}(\xi) = \varphi(\mu(\xi - \bar{\xi})).$$

Then relations (3.3),(3.2) are transformed into

$$\frac{1}{\mu}\frac{d}{d\xi}\frac{\check{\varphi}'(\xi)}{\sqrt{\mu^2 + \check{\varphi}'(\xi)}} - B\check{\varphi}(\xi) = \tilde{t}(\xi) - q, \tag{3.4}$$

$$\check{\varphi}'(l_0) = \mu k'(l_2), \qquad \check{\varphi}'(-l_0) = \mu k'(-l_1), \tag{3.5}$$

$$-l_1 = -h(\check{\varphi}(-l_0)), \qquad l_2 = h(\check{\varphi}(l_0)),$$

with $\tilde{t}(\xi) = t(\mu(\xi - \bar{\xi}))$. The constant q may be found by the integration of (3.3) with respect to x_1 which gives

$$\frac{\varphi'(l_2)}{\sqrt{1 + \varphi'^2(l_2)}} - \frac{\varphi'(-l_1)}{\sqrt{1 + \varphi'^2(-l_1)}} - B\int_{-l_1}^{l_2}\varphi(x_1)dx_1 = \frac{1}{\sigma}\int_{-l_1}^{l_2}\mathbf{n} \cdot \mathbf{Tn}dx_1 - (l_2 + l_1)q. \tag{3.6}$$

Since

$$\int_{-l_1}^{l_2}\varphi dx_1 = \int_\Gamma x_2 n_2 dS = |\Omega| - \int_\Sigma x_2 n_2 dS,$$

the last relation is equivalent to

$$\frac{\check{\varphi}'(l_0)}{\sqrt{\mu^2 + \check{\varphi}'^2(l_0)}} - \frac{\check{\varphi}'(-l_0)}{\sqrt{\mu^2 + \check{\varphi}^2(-l_0)}} - B|\Omega| + B\int_\Sigma x_2 n_2 dS = \frac{1}{\sigma}I_t - 2l_0\mu q \tag{3.7}$$

where $I_t = \int_{-l_1}^{l_2}\mathbf{n} \cdot \mathbf{Tn}dx_1$. Similar equation holds for $q_0 = p_0/\sigma$ (see (2.6)):

$$\frac{\varphi_0'(l_0)}{\sqrt{1 + \varphi_0'^2(l_0)}} - \frac{\varphi_0'(-l_0)}{\sqrt{1 + \varphi_0'^2(-l_0)}} - B|\Omega_0| + B\int_{\Sigma_0} x_2 n_2 dS = -2l_0 q_0. \tag{3.8}$$

Let us write (3.4),(3.5) as a boundary value problem for the function

$$\psi(\xi) = \check{\varphi}(\xi) - \varphi_0(\xi).$$

We need to compute $l_2 - l_0$, $l_1 - l_0$, $\mu - 1$, $\psi'(\pm l_0)$, $q - q_0$. Taking the conditions $\varphi_0(\pm l_0) = 0$ into account, we obtain

$$\begin{aligned}
l_2 - l_0 &= h(\check{\varphi}(l_0)) - h(\varphi_0(l_0)) = \psi(l_0)\int_0^1 h'(s\psi(l_0))ds = \\
&= h'(0)\psi(l_0) + \psi(l_0)\int_0^1[h'(s\psi(l_0)) - h'(0)]ds \equiv h'(0)\psi(l_0) + L_2, \\
l_1 - l_0 &= h(\check{\varphi}(-l_0)) - h(\varphi_0(-l_0)) = \psi(-l_0)\int_0^1 h'(s\psi(-l_0))ds = \\
&= h'(0)\psi(-l_0) + \psi(-l_0)\int_0^1[h'(s\psi(-l_0)) - h'(0)]ds \equiv h'(0)\psi(-l_0) + L_1. \quad (3.9)
\end{aligned}$$

These equations imply

$$\mu - 1 = \frac{1}{2l_0}(l_2 + l_1 - 2l_0) \equiv \delta\mu + M, \tag{3.10}$$

where

$$\delta\mu = \frac{h'(0)}{2l_0}[\psi(l_0) + \psi(-l_0)]$$

is a linear part of the right-hand side with respect to ψ and

$$M = \frac{L_1 + L_2}{2l_0} = \frac{\psi(l_0)}{2l_0}\int_0^1 [h'(s\psi(l_0)) - h'(0)]ds + \frac{\psi(-l_0)}{2l_0}\int_0^1 [h'(s\psi(-l_0)) - h'(0)]ds$$

is the remainder consisting of higher order terms.

Further we have

$$\psi'(l_0) = \mu k'(l_2) - k'(l_0) = (\mu - 1)k'(l_0) + k'(l_2) - k'(l_0)+$$

$$+(\mu - 1)(k'(l_2) - k'(l_0)) = \delta\mu k'(l_0) + k''(l_0)h'(0)\psi(l_0) + M_+, \tag{3.11}$$

$$\psi'(-l_0) = \mu k'(-l_1) - k'(-l_0) = \delta\mu k'(-l_0) - k''(-l_0)h'(0)\psi(-l_0) + M_- \tag{3.12}$$

where

$$M_+ = Mk'(l_0) + (l_2 - l_0)\int_0^1 [k''(l_0 + s(l_2 - l_0)) - k''(l_0)]ds+$$

$$+k''(l_0)L_2 + (\mu - 1)(k'(l_2) - k'(l_0)),$$

$$M_- = Mk'(-l_0) - (l_1 - l_0)\int_0^1 [k''(-l_0 - s(l_1 - l_0)) - k''(-l_0)]ds-$$

$$-k''(-l_0)L_1 + (\mu - 1)(k'(-l_1) - k'(-l_0)).$$

For the computation of $q - q_0$ we subtract (3.8) from (3.7) which leads to

$$\mu q - q_0 = -\frac{1}{2l_0}\left[\left(\frac{\tilde{\varphi}'(l_0)}{\sqrt{\mu^2 + \tilde{\varphi}'^2(l_0)}} - \frac{\varphi_0'(l_0)}{\sqrt{1 + \varphi'^2(l_0)}}\right) - \right.$$

$$\left. - \left(\frac{\tilde{\varphi}'(-l_0)}{\sqrt{\mu^2 + \tilde{\varphi}'^2(-l_0)}} - \frac{\varphi_0'(-l_0)}{\sqrt{1 + \varphi_0'^2(-l_0)}}\right)\right] - \frac{B}{2l_0}\left(\int_\Sigma x_2 n_2 dS - \right.$$

$$\left. - \int_{\Sigma_0} x_2 n_2 dS\right) + \frac{I_t}{2l_0} + \frac{B}{2l_0}(|\Omega| - |\Omega_0|). \tag{3.13}$$

We transform the right-hand side making use of the formula

$$\frac{\tilde{\varphi}'(\xi)}{\sqrt{\mu^2 + \tilde{\varphi}'^2(\xi)}} - \frac{\varphi_0'(\xi)}{\sqrt{1 + \varphi_0'^2(\xi)}} = \left(\frac{\tilde{\varphi}'(\xi)}{\sqrt{\mu^2 + \tilde{\varphi}'^2(\xi)}} - \frac{\varphi_0'(\xi)}{\sqrt{\mu^2 + \varphi_0'^2(\xi)}}\right)+$$

$$+\left(\frac{\varphi_0'(\xi)}{\sqrt{\mu^2 + \varphi_0'^2(\xi)}} - \frac{\varphi_0'(\xi)}{\sqrt{1 + \varphi_0'^2(\xi)}}\right) = \psi'(\xi)\int_0^1 \frac{\mu^2 ds}{[\mu^2 + (\varphi_0' + s\psi')^2]^{3/2}}-$$

$$-\int_0^1 \frac{(\mu - 1)\varphi_0'(\xi)ds}{[(1 + s(\mu - 1))^2 + \varphi_0'^2(\xi)]^{3/2}} = \frac{\psi'(\xi) - \delta\mu\varphi_0'(\xi)}{(1 + \varphi'^2(\xi))^{3/2}} + \Phi(\xi) \tag{3.14}$$

where

$$\Phi(\xi) = \psi'(\xi) \int_0^1 \left[\frac{\mu^2}{[\mu^2 + (\varphi_0' + s\psi')]^{3/2}} - \frac{1}{(1 + \varphi_0'^2)^{3/2}} \right] ds - \frac{M\varphi_0'(\xi)}{(1 + \varphi_0'(\xi))^{3/2}} -$$

$$-(\mu - 1)\varphi_0'(\xi) \int_0^1 \left\{ \frac{1}{[(1 + s(\mu - 1))^2 + \varphi_0'^2(\xi)]^{3/2}} - \frac{1}{[1 + \varphi_0'^2(\xi)]^{3/2}} \right\} ds$$

is the sum of all the terms in (3.14) which are at least quadratic with respect to ψ. If $|\Omega| = |\Omega_0|$, then the last term in (3.13) vanishes, and (3.14) implies

$$\mu q - q_0 = -\frac{1}{2l_0} \left[\frac{\psi'(l_0) - \delta\mu\varphi_0'(l_0)}{(1 + \varphi_0'^2(l_0))^{3/2}} - \frac{\psi'(-l_0) - \delta\mu\varphi_0'(-l_0))}{(1 + \varphi_0'^2(-l_0))^{3/2}} + \right.$$

$$\left. + \Phi(l_0) - \Phi(-l_0) \right] - \frac{B}{2l_0} \left(\int_\Sigma x_2 n_2 dS - \int_{\Sigma_0} x_2 n_2 dS \right) + \frac{I_t}{2l_0}.$$

Next, we apply formulas (3.11), (3.12) and take account of the fact that φ_0 is an odd function of ξ. We obtain

$$\frac{\psi'(l_0) - \delta\mu\varphi_0'(l_0)}{(1 + \varphi_0'^2(l_0))^{3/2}} - \frac{\psi'(-l_0) - \delta\mu\varphi_0'(-l_0)}{(1 + \varphi_0'^2(-l_0))^{3/2}} = \frac{\psi'(l_0) - \psi'(-l_0) - 2k'(l_0)\delta\mu}{(1 + \varphi_0'^2(l_0))^{3/2}} =$$

$$= \frac{k''(l_0)h'(0)[\psi(l_0) + \psi(-l_0)] + M_+ - M_-}{(1 + k'^2(l_0))^{3/2}} = -\frac{h'(0)}{R_0}[\psi(l_0) + \psi(-l_0)] +$$

$$+ \frac{M_+ - M_-}{(1 + k'^2(l_0))^{3/2}},$$

hence,

$$\mu q - q_0 = \frac{\delta\mu}{R_0} - \frac{1}{2l_0} \left[\frac{M_+ - M_-}{(1 + k'^2(l_0))^{3/2}} + \Phi(l_0) - \Phi(-l_0) \right] -$$

$$- \frac{B}{2l_0} \left(\int_\Sigma x_2 n_2 dS - \int_{\Sigma_0} x_2 n_2 dS \right) + \frac{I_t}{2l_0},$$

$$q - q_0 = -\frac{\mu - 1}{\mu} q_0 + \frac{\delta\mu}{\mu R_0} - \frac{1}{2l_0\mu} \left[\frac{M_+ - M_-}{(1 + k'^2(l_0))^{3/2}} + \Phi(l_0) - \Phi(-l_0) \right] -$$

$$- \frac{B}{2l_0\mu} \left(\int_\Sigma x_2 n_2 dS - \int_{\Sigma_0} x_2 n_2 dS \right) + \frac{I_t}{2l_0\mu}.$$

or

$$q - q_0 = \delta q + Q + \frac{I_t}{2l_0\mu}$$

where

$$\delta q = -\delta\mu q_0 + \frac{\delta\mu}{R_0}$$

is a linear part of $q - q_0$ with respect to ψ, and

$$Q = \frac{\delta\mu(\mu - 1) - M}{\mu} q_0 + \frac{\delta\mu(1 - \mu)}{R_0\mu} - \frac{1}{2l_0\mu} \left[\frac{M_+ - M_-}{(1 + k'^2(l_0))^{3/2}} + \right.$$

$$\left. + \Phi(l_0) - \Phi(-l_0) \right] - \frac{B}{2l_0\mu} \left(\int_\Sigma x_2 n_2 dS - \int_{\Sigma_0} x_2 n_2 dS \right)$$

is the sum of higher order terms.

It remains to write the differential equation for ψ. We subtract (2.3) from (3.4) and take account of (3.14) which leads to

$$\frac{d}{d\xi} \frac{\psi'(\xi) - \delta\mu\varphi_0'(\xi)}{(1 + \varphi_0'^2(\xi))^{3/2}} - \delta\mu \frac{d}{d\xi} \frac{\varphi_0'}{(1 + \varphi_0'^2)^{1/2}} - B\psi = \tilde{t}(\xi) - (q - q_0) + F_1(\xi),$$

or

$$\frac{d}{d\xi} \frac{\psi'(\xi) - \delta\mu\varphi_0'(\xi)}{(1 + \varphi_0'^2(\xi))^{3/2}} - B(\psi + \delta\mu\varphi_0) + \frac{\delta\mu}{R_0} = \tilde{t}(\xi) - \frac{I_t}{2l_0\mu} + F_1(\xi) - Q \qquad (3.15)$$

with

$$F_1(\xi) = \frac{\mu - 1}{\mu} \frac{d}{d\xi} \frac{\psi'(\xi) - \delta\mu\varphi_0'(\xi)}{(1 + \varphi_0'^2(\xi))^{3/2}} - \frac{1}{\mu} \frac{d\Phi}{d\xi} - \frac{\delta\mu(\mu - 1) - M}{\mu} \frac{d}{d\xi} \frac{\varphi_0'(\xi)}{(1 + \varphi_0'^2(\xi))^{1/2}}.$$

For given $\tilde{t}(\xi)$, we consider (3.15), (3.11), (3.12) as a boundary value problem for ψ. Let us study a linearized problem

$$L[\psi] \equiv \frac{d}{d\xi} \frac{\psi' - \delta\mu\varphi_0'}{(1 + \varphi_0'^2)^{3/2}} - B(\psi + \delta\mu\varphi_0) + \frac{\delta\mu}{R_0} = f(\xi),$$

$$\psi'(l_0) - \delta\mu k'(l_0) - k''(l_0)h'(0)\psi(l_0) = a_+, \qquad (3.16)$$

$$\psi'(-l_0) - \delta\mu k'(-l_0) + k''(-l_0)h'(0)\psi(-l_0) = a_-.$$

Proposition 3.1. *There exists such $B_0 > 0$ that for $B > B_0$ problem (3.16) has a unique solution $\psi \in C_{1+\beta}^{3+\alpha}(I, \partial I)$ $(\alpha, \beta \in (0,1),\ I = (-l_0, l_0))$ for arbitrary $a_+, a_- \in R,\ f \in C_{\beta-1}^{1+\alpha}(I, \partial I)$. The solution satisfies the inequality*

$$|\psi|_{C_{1+\beta}^{3+\alpha}(I,\partial I)} \leq c_1(|f|_{C_{\beta-1}^{1+\alpha}(I,\partial I)} + |a_+| + |a_-|). \qquad (3.17)$$

Proof: It is convenient to introduce a new unknown function

$$\tilde{\psi}(\xi) = \psi(\xi) + \delta\mu\varphi_0(\xi).$$

Since $\varphi(\pm l_0) = 0$, we have

$$\delta\mu = \frac{h'(0)}{2l_0}[\tilde{\psi}(l_0) + \tilde{\psi}(-l_0)]$$

and we may express ψ in terms of $\tilde{\psi}$ by the formula

$$\psi(\xi) = \tilde{\psi}(\xi) - \delta\mu\varphi_0(\xi).$$

Problem (3.16) takes the form

$$\frac{d}{d\xi} \frac{\tilde{\psi}' - 2\delta\mu\varphi_0'}{(1 + \varphi_0'^2)^{3/2}} - B\tilde{\psi} + \frac{\delta\mu}{R_0} = f(\xi),$$

$$\tilde{\psi}'(l_0) - \frac{\tilde{\psi}(l_0)}{l_0 \cos^2\alpha_0} + 2\tan\alpha_0\delta\mu = a_+,$$

$$\tilde{\psi}'(-l_0) + \frac{\tilde{\psi}(-l_0)}{l_0 \cos^2\alpha_0} - 2\tan\alpha_0\delta\mu = a_-.$$

219

A weak solution of this problem can be defined as a function $\tilde{\psi} \in W_2^1(I)$ satisfying the integral identity

$$L[\tilde{\psi}, \eta] \equiv \int_{-l_0}^{l_0} \left(\frac{\tilde{\psi}' - 2\delta\mu\varphi_0'}{(1 + \varphi_0'^2)^{3/2}} \eta' + B\tilde{\psi}\eta - \frac{\delta\mu}{R_0}\eta \right) d\xi -$$

$$- \frac{1}{(1 + k'^2(l_0))^{3/2}} \left[\left(\frac{\tilde{\psi}(l_0)}{l_0 \cos^2 \alpha_0} - 2\tan\alpha_0\delta\mu \right) \eta(l_0) + \right.$$

$$\left. + \left(\frac{\tilde{\psi}(-l_0)}{l_0 \cos^2 \alpha_0} - 2\tan\alpha_0\delta\mu \right) \eta(-l_0) \right] +$$

$$+ \frac{2\delta\mu k'(l_0)}{(1 + k'^2(l_0))^{3/2}} [\eta(l_0) + \eta(-l_0)] = \frac{a_+\eta(l_0) - a_-\eta(-l_0)}{(1 + k'^2(l_0))^{3/2}} - \int_{-l_0}^{l_0} f\eta d\xi \qquad (3.18)$$

for arbitrary $\eta \in W_2^1(I)$.

If the coefficient B is large enough, then the quadratic form $L[\tilde{\psi}, \tilde{\psi}]$ is positive definite

$$L[\tilde{\psi}, \tilde{\psi}] \geq c_2 \int_{-l_0}^{l_0} (|\tilde{\psi}'|^2 + |\tilde{\psi}|^2) d\xi. \qquad (3.19)$$

Indeed, it is easy to see that

$$L[\tilde{\psi}, \tilde{\psi}] \geq c_3 \int_{-l_0}^{l_0} \tilde{\psi}'^2 d\xi + B \int_{-l_0}^{l_0} \tilde{\psi}^2 d\xi - \frac{|\delta\mu|}{R_0} \int_{-l_0}^{l_0} |\tilde{\psi}| d\xi - c_4(|\tilde{\psi}(l_0)|^2 + |\tilde{\psi}(-l_0)|^2), \qquad (3.20)$$

with c_3, c_4 independent of B. For the estimate of $\tilde{\psi}$ we use the identity

$$\tilde{\psi}^2(l_0) + \tilde{\psi}^2(-l_0) = \frac{1}{l_0} \int_{-l_0}^{l_0} \tilde{\psi}^2(t) dt + \frac{2}{l_0} \int_{-l_0}^{l_0} t\tilde{\psi}(t)\tilde{\psi}'(t) dt,$$

which implies

$$\tilde{\psi}^2(l_0) + \tilde{\psi}^2(-l_0) \leq \epsilon \int_{-l_0}^{l_0} \tilde{\psi}'^2(t) dt + \left(\frac{1}{\epsilon} + \frac{1}{l_0} \right) \int_{-l_0}^{l_0} \tilde{\psi}^2(t) dt, \qquad \forall \epsilon > 0.$$

Similar estimate holds for $|\delta\mu|^2$. It is clear now that (3.19) follows from (3.20) in the case of large B.

For arbitrary $f \in C_{\beta-1}^{1+\alpha}(I, \partial I)$, $\eta \in W_2^1(I)$ we have

$$\left| \int_{-l_0}^{l_0} f\eta d\xi \right| \leq \sup_I \rho^{1-\beta}(\xi)|f(\xi)| \int_{-l_0}^{l_0} \rho^{\beta-1}(\xi)|\eta(\xi)| d\xi \leq$$

$$\leq c_5 \sup_I |\eta(\xi)| \sup_I \rho^{1-\beta}(\xi)|f(\xi)| \leq c_6 \|\eta\|_{W_2^1(I)} \|f\|_{C_{\beta-1}^{1+\alpha}(I,\partial I)}.$$

Hence, the existence of a unique weak solution follows from the theorem of Lax-Milgram. Setting $\eta = \tilde{\psi}$ in (3.18) we easily obtain

$$\sup_I |\tilde{\psi}(\xi)| \leq c_7 \|\tilde{\psi}\|_{W_2^1(I)} \leq c_8 (\sup_I \rho^{1-\beta}(\xi)|f(\xi)| + |a_+| + |a_-|), \qquad (3.21)$$

and $\delta\mu$ can also be evaluated by the expression in the right-hand side. Now, we consider $\tilde{\psi}$ as a solution to the problem

$$\frac{d}{d\xi} \frac{\tilde{\psi}'}{(1 + \varphi_0'^2)^{3/2}} - B\tilde{\psi} = f_1,$$

$$\tilde{\psi}'(l_0) = b_+, \qquad \tilde{\psi}'(-l_0) = b_-$$

where

$$f_1 = f + 2\delta\mu \frac{d}{d\xi} \frac{\varphi_0'}{(1+\varphi_0'^2)^{1/2}} - \frac{\delta\mu}{R_0},$$

$$b_+ = a_+ - 2\delta\mu \tan\alpha_0 + \frac{\tilde{\psi}(l_0)}{l_0 \cos^2\alpha_0}, \qquad b_- = a_- + 2\delta\mu \tan\alpha_0 - \frac{\tilde{\psi}(-l_0)}{l_0 \cos\alpha_0}.$$

This problem was studied in Solonnikov (1979) where, in particular, the following estimate for the solution was established:

$$|\tilde{\psi}|_{C^{3+\alpha}_{1+\beta}(I,\partial I)} \le c_9 (|f_1|_{C^{1+\alpha}_{\beta-1}(I,\partial I)} + |b_+| + |b_-|) \tag{3.22}$$

(the assumption $\beta = \alpha$ made in Solonnikov (1979) is not essential). Estimate (3.17) is a consequence of (3.21) and (3.22). The proposition is proved. $\quad\square$

Our second auxiliary proposition concerns the construction of a special mapping $Y : \Omega_0 \to \Omega$.

Proposition 3.2. *Suppose that the line Γ is given by equations (3.1) on the interval $(-l_0, l_0)$ with φ satisfying conditions (3.2). Moreover, assume that $\tilde{\varphi}(\xi) = \varphi(\mu(\xi-\dot{\xi}))$ belongs to $C^{3+\alpha}_{1+\beta}(I,\partial I)$ and that*

$$|\tilde{\varphi} - \varphi_0|_{C^{3+\alpha}_{1+\beta}(I,\partial I)} \le \delta_1,$$

with a small positive δ_1. Let $\Omega \subset V$ be a domain with $\partial\Omega = \Sigma \cup \Gamma \cup M, \quad M = \{x_+, x_-\}$. There exists a mapping $Y : \Omega \to \Omega_0$ with the following properties

1. *Y is invertible, continuous in $\bar{\Omega}$ and has bounded derivatives. Moreover, $Y|_{\Sigma_0} \in C^{3+\alpha}_{1+\beta}(\Sigma_0, M_0)$ and $Y|_{\Gamma_0} \in C^{3+\alpha}_{1+\beta}(\Gamma_0, M_0)$ where $M_0 = \bar{\Sigma}_0 \cap \bar{\Gamma}_0 = \{x_+^{(0)}, x_-^{(0)}\}, \quad x_\pm^{(0)} = (\pm l_0, 0)$. For $\xi \in \Gamma_0$*

$$Y(z) = (\mu(z_1 - \bar{\xi}), \tilde{\varphi}(z_1)) \tag{3.23}$$

where

$$\mu = \frac{l_2 + l_1}{2l_0}, \qquad \bar{\xi} = l_0 \frac{l_1 - l_2}{l_1 + l_2}.$$

2. *The Jacobian matrix J of the inverse transformation Y^{-1} satisfies the inequalities*

$$|J - I|_{C^{2+\alpha}_\beta(\Sigma_0 M)} + |J - I|_{C^{2+\alpha}_\beta(\Gamma_0 . M)} \le c_{10}|\tilde{\varphi} - \varphi_0|_{C^{3+\alpha}_{1+\beta}(I,\partial I)}, \tag{3.24}$$

$$\sup_{\Omega_0} |J(z) - I| + \sum_{|j|=1,2} \sup_{\Omega_0} \rho^{|j|}(z)|D^j J_0(z)| +$$

$$+ \sum_{|j|=2} \sup_{z\in\Omega_0} \rho^{2+\alpha}(z) \sup_{y\in K(z)} |y - z|^{-\alpha} |D^j J_0(z) - D^j J_0(y)| \le c_{11}|\tilde{\varphi} - \varphi_0|_{C^{3+\alpha}_{1+\beta}(I,\partial I)} \tag{3.25}$$

where $\rho(x) = \text{dist}(x, M_0)$.

Proof: We construct the mapping Y in the form

$$Y(z) = z + \Phi(z) = (z_1 + \Phi_1(z), z_2 + \Phi_2(z)). \tag{3.26}$$

Equation (3.23) determines Φ_i on Γ_0

$$\Phi_1(z_1, \varphi_0(z_1)) = (\mu - 1)z_1 - \mu\bar{\xi},$$

$$\Phi_2(z_1, \varphi_0(z_1)) = \tilde{\varphi}(z_1) - \varphi_0(z_1) \equiv \psi(z_1).$$

In particular,

$$\Phi_1(-l_0, 0) = -(\mu - 1)l_0 - \mu l_0 \frac{l_1 - l_2}{l_1 + l_2} = -l_1 + l_0, \quad \Phi_1(l_0, 0) = l_2 - l_0,$$

$$\Phi_2(-l_0, 0) = \tilde{\varphi}(-l_0) = \varphi(-l_1), \quad \Phi_2(l_0, 0) = \varphi(l_2), \tag{3.27}$$

which implies

$$Y(x_{\pm}^{(0)}) = x_{\pm}^{(0)}. \tag{3.28}$$

Next, we extend $Y(z)$ onto Σ_0 in such a way that $Y\Sigma_0 = \Sigma$. Let (r, φ) be the standard polar coordinates in $\mathbb{R}^2$ with the center in $(0, y_c)$. A general form of automorphisms of ∂V is

$$x_1 = R_0 \cos(\varphi + h(\varphi)), \quad x_2 = R_0 \sin(\varphi + h(\varphi)).$$

Clearly, this transformation can be written in the form (3.26) with $z = (R_0 \cos \varphi, R_0 \sin \varphi)$ and

$$\Phi_1(z) = R_0[\cos(\varphi + h(\varphi)) - \cos \varphi] =$$
$$= R_0[\cos \varphi(\cos h(\varphi) - 1) - \sin \varphi \sinh(\varphi)],$$

$$\Phi_2(z) = R_0[\sin(\varphi + h(\varphi)) - \sin \varphi] =$$
$$= R_0[\sin \varphi(\cos h(\varphi) - 1) + \cos \varphi \sin h(\varphi)]. \tag{3.29}$$

These equations imply

$$R_0 \sin h(\varphi) = \Phi_2(z) \cos \varphi - \Phi_1(z) \sin \varphi. \tag{3.30}$$

We make the extension of Y by the construction of an appropriate function $h(\varphi)$ on Σ_0. We find the values of h at the points $x_{\pm}$ using relation (3.28). Because of this relation, the functions Φ_i computed at $x_{\pm}$ (they are given by (3.27)) satisfy (3.29) with certain $h_{\pm}$ which are determined by (3.30). It is elementary to construct a smooth (at least $C^{3+\alpha}$-smooth) function h on Σ_0 satisfying the conditions $h(x_{\pm}) = h_{\pm}$ and the inequality

$$|h|_{C^{3+\alpha}(\Sigma_0)} \le c_{12}(|h_+| + |h_-|) \le c_{13}(|\tilde{\varphi}(l_0) - \varphi_0(l_0)| + |\tilde{\varphi}(-l_0) - \varphi_0(-l_0)|). \tag{3.31}$$

The extensions of Φ_i are defined now by (3.29), and it is clear that their $C^{3+\alpha}(\Sigma_0)$-norms can also be evaluated by the right-hand side of (3.31). Now Φ_i are defined on $\partial\Omega_0$, and they can be extended farther into Ω_0. A special care should be taken in the neighbourhoods of $x_{\pm}$ (see also Solonnikov, 1994b). Let Φ_{i1} and Φ_{i2} be extensions of $\Phi_i|_{\Sigma_0}$ and $\Phi_i|_{\Gamma_0}$ made in such a way that

$$|\Phi_{i1}|_{C_{1+\beta}^{3+\alpha}(\Omega_0, M_0)} \le c_{15} |\Phi_i|_{C_{1+\beta}^{3+\alpha}(\Sigma_0, M_0)},$$

$$|\Phi_{i2}|_{C_{1+\beta}^{3+\alpha}(\Omega_0, M_0)} \le c_{16} |\Phi_i|_{C_{1+\beta}^{3+\alpha}(\Gamma_0, M_0)}.$$

We can define $\Phi_i(z)$ in the neighbourhood of x_-, for example, by the formula

$$\Phi_i(z) = \chi_1(z)(\Phi_{i1}(z) - \Phi_i(x_-)) + \chi_2(z)(\Phi_{i2}(z) - \Phi_i(x_-)) + \Phi_i(x_-),$$

where χ_i are functions defined near x_- and possessing the following properties

$$\chi_1|_{\Sigma_0} = 1, \quad \chi_1|_{\Gamma_0} = 0, \quad \frac{\partial \chi_1}{\partial n}\Big|_{\partial\Omega_0} = 0,$$

$$\chi_2|_{\Sigma_0} = 0, \quad \chi_2|_{\Gamma_0} = 1, \quad \frac{\partial \chi_2}{\partial n}|_{\partial \Omega_0} = 0,$$

χ_i are smooth everywhere except at the point x_-, $0 \le \chi_i \le 1$ and

$$|D^j \chi_i(z)| \le c_{17}|z - x_-|^{-|j|}.$$

Inequalities (3.24) and (3.25) are easily verified. Away from $x_\pm$, the construction of extensions is quite standard. The proposition is proved. $\qquad\square$

Remark 3.1. Let φ_1 and φ_2 be two functions satisfying the hypotheses of the preceding proposition, and let Y_i be the corresponding transformations. Since all the extension operators used in the proposition are linear, it is easily verified that the differences $J_1 - J_2$ satisfy the estimates (3.24) and (3.25) with the norms of the differences $\tilde{\varphi}_1 - \tilde{\varphi}_2$ in the right-hand sides.

4. PROOF OF THEOREM 1.1

The proof of Theorem 1.1 is based on the investigation of two auxiliary problems, the problem (1.1),(1.2) in a given domain Ω and the problem (3.15),(3.11),(3.12).

Theorem 4.1. *Suppose that Γ is given by equation (3.1) with the function φ satisfying the hypotheses of Proposition 3.2 and that ω is sufficiently small:*

$$|\omega| < \epsilon. \tag{4.1}$$

Then the problem (1.1),(1.2) possesses a unique solution $\mathbf{v} \in C_\beta^{2+\alpha}(\Omega, M)$, $p \in C_{\beta-1}^{1+\alpha}(\Omega, M)$, and

$$|\mathbf{v}|_{C_\beta^{2+\alpha}(\Omega,M)} + |p|_{C_{\beta-1}^{1+\alpha}(\Omega,M)} \le c_1|\omega|. \tag{4.2}$$

Let φ_1 and φ_2 be two functions satisfying hypotheses of Proposition 3.2 and defining the lines Γ_1 and Γ_2, and let $\mathbf{v}_1, p_1$ and $\mathbf{v}_2, p_2$ be solutions of (1.1),(1.2) in Ω_1 and Ω_2, respectively. The functions $\tilde{t}_i(\xi) = \mathbf{n} \cdot \mathbf{T}(\mathbf{v}_i, p_i)\mathbf{n}|_{x_2 = \tilde{\varphi}_i(\xi)}$ satisfy the inequality

$$|\tilde{t}_1 - \tilde{t}_2|_{C_{\beta-1}^{1+\alpha}(I,\partial I)} \le c_2|\omega||\tilde{\varphi}_1 - \tilde{\varphi}_2|_{C_{1+\beta}^{3+\alpha}(I,\partial I)}. \tag{4.3}$$

Proof: The first part of the theorem is known. The linearized problem was studied in Solonnikov (1981). For small ω, the nonlinear problem can be solved by application of the contraction mapping principle, since the nonlinear term satisfies the inequality

$$|(\mathbf{v} \cdot \nabla)\mathbf{w}|_{C_{\beta-2}^\alpha(\Omega,M)} \le c_2|\mathbf{v}|_{C_\beta^{2+\alpha}(\Omega,M)}|\mathbf{w}|_{C_\beta^{2+\alpha}(\Omega,M)}.$$

Inequality (4.3) is also established by a well known procedure. We map the domain Ω_1 onto Ω_2 by means of the transformation $y = Z(x)$, where $Z = Y_2 \circ Y_1^{-1}$ and $Y_i : \Omega_0 \to \Omega_i$ are mappings constructed in Proposition 3.2, and we write problem (1.1),(1.2) for $\mathbf{v}_1, p_1$ in new coordinates. It is easy to see that $\mathbf{v}_1 - \mathbf{v}_2 = \mathbf{w}, p_1 - p_2 = s$ can be considered as a solution of the linear problem

$$-\nabla^2 \mathbf{w} + \nabla s = \mathbf{f}, \qquad \nabla \cdot \mathbf{w} = r,$$

$$\mathbf{w}|_{\Sigma_2} = \mathbf{w}_0, \quad \mathbf{w} \cdot \mathbf{n}|_{\Gamma_2} = b, \quad \tau \cdot \mathbf{S}(\mathbf{w})\mathbf{n}|_{\Gamma_2} = d, \tag{4.4}$$

223

where $\mathbf{f}, r, \mathbf{v}_0, b, d$ are functions satisfying the inequality

$$|\mathbf{f}|_{C^\alpha_{\beta-2}(\Omega_2, M_2)} + |r|_{C^{1+\alpha}_{\beta-1}(\Omega_2, M_2)} + |\mathbf{w}_0|_{C^{2+\alpha}_\beta(\Sigma_2, M_2)} +$$

$$+|d|_{C^{1+\alpha}_{\beta-1}(\Gamma_2, M_2)} + |b|_{C^{2+\alpha}_\beta(\Gamma_2, M_2)} \le c_3 |\omega| |\tilde\varphi_1 - \tilde\varphi_2|_{C^{3+\alpha}_{1+\beta}(I, \partial I)}. \tag{4.5}$$

This inequality follows from (4.2) and from Remark 3.1 (see some details in section 5 of Solonnikov, 1995). The inequality (4.3) is a consequence of (4.5) and of a coercive estimate of the solution of (4.5) in weighted Hölder norms (see Solonnikov, 1981). □

Let us consider the problem (3.15),(3.11),(3.12).

Theorem 4.2. *Suppose that condition $B > B_0$ is satisfied. For arbitrary $\tilde t \in C^{1+\alpha}_{\beta-1}(I, \partial I)$, with a small norm*

$$|\tilde t|_{C^{1+\alpha}_{\beta-1}(I, \partial I)} \le \epsilon_1, \tag{4.6}$$

the problem (3.15),(3.11),(3.12) has a unique solution $\psi \in C^{3+\alpha}_{\beta-1}(I, \partial I)$, and for this solution the estimate

$$|\psi|_{C^{3+\alpha}_{1+\beta}(I, \partial I)} \le c_4 |\tilde t|_{C^{1+\alpha}_{\beta-1}(I, \partial I)} \tag{4.7}$$

holds.

Proof: Consider $L_1, L_2, M, M_\pm$ etc. as nonlinear functionals depending on ψ. It is clear that

$$|L_1| + |L_2| + |M| + |M_+| + |M_-| \le c_5(|\psi(l_0)|^2 + |\psi(-l_0)|^2), \tag{4.8}$$

provided that $\psi(\xi)$ is small enough, for instance,

$$|\psi|_{C^{3+\alpha}_{1+\beta}(I, \partial I)} \le \delta. \tag{4.9}$$

Moreover,

$$|\Phi|_{C^{2+\alpha}_\beta(I, \partial I)} \le c_6 |\mu - 1|(|\psi|_{C^{3+\alpha}_{1+\beta}(I, \partial I)} + |\mu - 1|) \le c_7 |\psi|^2_{C^{3+\alpha}_{1+\beta}(I, \partial I)}, \tag{4.10}$$

$$|\Phi(l_0)| + |\Phi(-l_0)| \le c_8(|\psi(l_0)|^2 + |\psi(-l_0)|^2),$$

and, finally, since the endpoints of Σ_0 are located on x_1-axis, we have

$$\left| \int_\Sigma x_2 n_2 dS - \int_{\Sigma_0} x_2 n_2 dS \right| \le c_9(|\psi(l_0)|^2 + |\psi(-l_0)|^2),$$

hence,

$$|Q| \le c_{10}(|\psi(l_0)|^2 + |\psi(-l_0)|^2).$$

Let ψ_1 and ψ_2 be two functions from the ball (4.9) and let $L_i[\psi_j], M[\psi_j]$ etc. be the corresponding functionals. It is easy to see that

$$\sum_{i=1}^{i=2} |L_i[\psi_1] - L_i[\psi_2]| \le c_{11}\delta(|\psi_1(l_0) - \psi_2(l_0)| + |\psi_1(-l_0) - \psi_2(-l_0)|),$$

$$|\Phi[\psi_1] - \Phi[\psi_2]|_{C^{2+\alpha}_\beta(I, \partial I)} \le c_{12}\delta|\psi_1 - \psi_2|_{C^{3+\alpha}_{1+\beta}(I, \partial I)}, \tag{4.11}$$

$$|Q[\psi_1] - Q[\psi_2]| \le c_{13}\delta(|\psi_1(l_0) - \psi_2(l_0)| + |\psi_1(-l_0) - \psi_2(-l_0)|).$$

These inequalities make it possible to deduce the solvability of the problem (3.15),(3.11),(3.12) from the contraction mapping principle. We write it in the form of equation

$$\psi = \mathcal{A}[F_1 - Q, M_+, M_-] + \mathcal{A}[\tilde{t} - \frac{1}{2l_0\mu}I_t, 0, 0] \equiv \mathcal{B}\psi \qquad (4.12)$$

where $\mathcal{A}$ is a linear operator which makes correspond a solution of problem (3.16) to the data $[f, a_+, a_-]$. For arbitrary $\tilde{t}(\xi)$ satisfying condition (4.6) and arbitrary ψ, ψ_1, ψ_2 from the ball (4.9) we have

$$|\mathcal{B}\psi|_{C^{3+\alpha}_{1+\beta}(I,\partial I)} \leq c_{14}(\delta^2 + \epsilon_1),$$

$$|\mathcal{B}\psi_1 - \mathcal{B}\psi_2|_{C^{3+\alpha}_{1+\beta}(I,\partial I)} \leq c_{15}(\delta + \epsilon_1)|\psi_1 - \psi_2|_{C^{3+\alpha}_{1+\beta}(I,\partial I)}.$$

Hence, $\mathcal{B}$ is a nonlinear contraction operator in the ball (4.9), if

$$c_{14}(\delta^2 + \epsilon_1) < \delta, \quad c_{15}(\delta + \epsilon_1) < 1.$$

These inequalities are satisfied, if

$$\delta < \min(c_{14}^{-1}, c_{15}^{-1})/2, \quad \epsilon_1 < \min(\delta c_{14}^{-1}, c_{15}^{-1})/2.$$

Then the solvability of equation (4.12) follows from the contraction mapping principle. The theorem is proved. $\qquad\square$

The solution of the problem (3.15),(3.11),(3.12) determines a curve Γ with $\partial\Gamma \in \partial V$ which is tangential to ∂V at the endpoints, and the area of the corresponding domain Ω equals Q. Indeed, if we set

$$\tilde{\varphi}(\xi) = \varphi_0(\xi) + \psi(\xi),$$

$$l_1 = h(\tilde{\varphi}(-l_0)), \quad l_2 = h(\tilde{\varphi}(l_0)), \quad \mu = \frac{l_1 + l_2}{2l_0}$$

and define $\varphi(x_1)$ as in Section 3, i.e. by equation

$$\varphi(\mu(\xi - \bar{\xi})) = \tilde{\varphi}(\xi),$$

then (3.11) is equivalent to

$$\psi'(l_0) = \mu k'(l_2) - k'(l_0),$$

or to

$$\tilde{\varphi}(l_0) = \mu k(l_2),$$

which immediately gives $\varphi'(l_2) = k'(l_2)$. Exactly in the same way the condition $\varphi'(-l_1) = k'(-l_1)$ can be verified. Finally, the addition of (3.15) and (2.3) leads to (3.4) (i.e. to (3.3)) with a constant q satisfying (3.13) without the last term. Integration of (3.3) gives

$$\frac{B}{2l_0}(|\Omega| - |\Omega_0|) = 0,$$

q.e.d.

We are ready now to carry out a formal construction of the solution of the free boundary problem (1.1)-(1.3). We use the following iterative procedure. Let $\mathbf{v}^{(0)} = 0$,

$p^{(0)} = p_0$, $\Omega^{(0)} = \Omega_0$ and let $\mathbf{v}^{(1)}, p^{(1)}$ be a solution of the first auxiliary problem in $\Omega^{(0)}$. Further, we solve the second auxiliary problem with the function

$$\tilde{t}^{(1)}(\xi) = \frac{1}{\sigma}\mathbf{n} \cdot \mathbf{T}(\mathbf{v}^{(1)}, p^{(1)})\mathbf{n}|_{x_2 = \varphi_0(\xi)}$$

in the right-hand side. This determines the curve $\Gamma^{(1)}$ and the domain $\Omega^{(1)}$. This procedure is repeated: we define $\psi^{(m+1)}$ as a solution of the second auxiliary problem with $\tilde{t}^{(m+1)}$ in the right-hand side, and

$$t^{(m+1)} = \sigma^{-1}\mathbf{n} \cdot \mathbf{T}(\mathbf{v}^{(m+1)}, p^{(m+1)})\mathbf{n}$$

where $\mathbf{v}^{(m+1)}, p^{(m+1)}$ is the solution of the first auxiliary problem in $\Omega^{(m)}$. Let us show that the sequence $\psi^{(m)}$ is convergent in $C^{3+\alpha}_{1+\beta}(I, \partial I)$. According to (4.12), we have

$$\psi^{(m+1)} = \mathcal{A}[F_1^{(m+1)} - Q^{(m+1)}, M_+^{(m+1)}, M_-^{(m+1)}] + \mathcal{A}[\tilde{t}^{(m+1)} - \frac{I_t^{(m+1)}}{2l_0\mu^{(m+1)}}, 0, 0],$$

hence,

$$
\begin{aligned}
\psi^{(m+1)} \quad - \quad & \psi^{(m)} = \\
= \quad & \mathcal{A}[F_1^{(m+1)} - F_1^{(m)} - Q^{(m+1)} + Q^{(m)}, M_+^{(m+1)} - M_+^{(m)}, M_-^{(m+1)} - M_-^{(m)}] + \\
+ \quad & \mathcal{A}[\tilde{t}^{(m+1)} - \tilde{t}^{(m)} - \frac{I_t^{(m+1)} - I_t^{(m)}}{2l_0\mu^{(m+1)}} + I_t^{(m)}[(2l_0\mu^{(m+1)})^{-1} - (2l_0\mu^{(m)})^{-1}], 0, 0]
\end{aligned}
$$

Suppose that $\psi^{(m)}$ satisfies the condition (4.6). In virtue of (4.7) and (4.2),

$$|\psi^{(m+1)}|_{C^{3+\alpha}_{1+\beta}(I, \partial I)} \leq c_{16}|\tilde{t}^{(m+1)}|_{C^{1+\alpha}_{\beta-1}(I, \partial I)} \leq c_{17}|\omega|. \tag{4.13}$$

For small ω, the right-hand side does not exceed δ. Hence, we see that all the approximations ψ^m satisfy (4.6).

Further, estimates (4.11) and (4.3) imply

$$|\psi^{(m+1)} - \psi^{(m)}|_{C^{3+\alpha}_{1+\beta}/I, \partial I} \leq c_{19}[(\delta + |\omega|)|\psi^{(m+1)} - \psi^{(m)}|_{C^{3+\alpha}_{1+\beta}(I, \partial I)} +$$

$$+ |\omega||\psi^{(m)} - \psi^{(m-1)}|_{C^{3+\alpha}_{1+\beta}(I, \partial I)}]$$

We see that if

$$c_{19}(\delta + |\omega|) < 1/2,$$

then

$$|\psi^{(m+1)} - \psi^{(m)}|_{C^{3+\alpha}_{1+\beta}(I, \partial I)} \leq 2c_{19}|\omega||\psi^{(m)} - \psi^{(m-1)}|_{C^{3+\alpha}_{1+\beta}(I, \partial I)}$$

which guarantees the convergence of $\{\psi^{(m)}\}$, since $2c_{19}|\omega| < 1$. It is evident that all the smallness conditions can be satisfied by the choice of small ω.

By virtue of (4.13), the limiting function ψ satisfies the inequality

$$|\psi|_{C^{3+\alpha}_{1+\beta}(I, \partial I)} \leq c_{17}|\omega|. \tag{4.14}$$

This function defines the domain Ω, and $\mathbf{v}$ and p can be obtained as a solution of (1.1),(1,2).

Inequality (4.14) does not guarantee that $\Gamma \subset V$, since the space $C^{3+\alpha}_{1+\beta}(I, \partial I)$ is too wide and its elements may have singular second derivatives at the points $x_\pm$. As a consequence, the curves Γ corresponding to such elements may leave V. To show that this can not happen, we should study the asymptotics of solution near the contact points. We are not able to make it here and we refer the reader to the papers Puknachov and Solonnikov (1982), Solonnikov (1992), Solonnikov (1994a). In particular, it is shown in Solonnikov (1994a) that the free boundary is more regular and it belongs, as a minimum, to the class $C_{1+\gamma}^{3+\alpha}$ with $\gamma \in (1/2, 1)$, and that it is contained in V, provided that

$$p_1 > \frac{\sigma}{R_0}.$$

This condition is guaranteed by (2.7) and by the smallness of $p_1 - p_0$, hence, the solution we have obtained is physically reasonable. The proof of Theorem 1.1 is now complete.

REFERENCES

BAIOCCHI, C. and V.V. PUKNACHOV (1990), Unilateral problems for the Navier-Stokes equations and problem of dynamical contact angle, *J. Appl. Mech. Techn. Physics* Nr. **2**, 27-42.

PUKNACHOV, V.V. and V.A. SOLONNIKOV (1982), On the problem of dynamic contact angle, *J. Appl. Math. Mech.* **46**, 771-779.

SOLONNIKOV, V.A. (1979), Solvability of a problem on the plane motion of a heavy viscous incompressible capillary liquid partially filling a container, *Izv. Akad. Nauk SSSR* **43**, 203-236.

SOLONNIKOV, V.A. (1981), On the Stokes equations in domains with non-smooth boundaries ond on viscous incompressible flow with a free surface, in: Brezis, H. and J.-L. Lions (eds.) *Nonlinear partial differential equations and their applications*, College de France Seminars, Vol. 111, , Research Notes in Mathematics **70**, 340-423.

SOLONNIKOV, V.A. (1993), On the problem of a moving contact angle, Preprint University of Paderborn.

SOLONNIKOV, V.A. (1994a), On free boundary problems for the Navier-Stokes equations with moving contact points, *Zapiski Nauchn. Semin. POMI* **213**, 179-205.

SOLONNIKOV, V.A. (1994b), Some free boundary problems for the Navier-Stokes equations with moving contact points and lines, in: *Partial Differential Equations: Models in Physics and Biology*, Math. Res., **82**.

SOLONNIKOV, V.A. (1995), On some free boundary problems for the Navier-Stokes equations with moving contact points and lines, *Math. Annalen*, (to appear).

Non-Newtonian Fluids

HEAT-CONDUCTING INCOMPRESSIBLE VISCOUS FLUIDS

Herbert Amann

Institut für Mathematik, Universität Zürich
Winterthurerstr. 190, CH–8057 Zürich, Switzerland

1. INTRODUCTION

The motion of a heat-conducting incompressible viscous fluid is governed by the following set of equations (e.g., Serrin, 1959; Truesdell and Noll, 1965):

$$\nabla \cdot \boldsymbol{v} = 0 \ ,$$
$$\rho\big(\partial_t \boldsymbol{v} + (\boldsymbol{v} \cdot \nabla)\boldsymbol{v}\big) = -\nabla p + \nabla \cdot \boldsymbol{S} + \rho \boldsymbol{b} \ , \qquad (1.1)$$
$$\rho c(\partial_t \theta + \boldsymbol{v} \cdot \nabla \theta) = \mathrm{tr}(\boldsymbol{SD}) - \nabla \cdot \boldsymbol{q} + \rho r \ ,$$

where the velocity $\boldsymbol{v}$, the pressure p, and the (absolute) temperature $\theta > 0$ are the unknown variables. Moreover, $\rho > 0$ is the constant density, $\boldsymbol{S}$ is the viscous part of the stress tensor (the extra stress), $\boldsymbol{b} := \boldsymbol{b}(x,t,\theta)$ is the external body force, $c := c(\theta) > 0$ is the heat capacity, $\boldsymbol{q}$ is the heat flux vector, and $r := r(x,t,\theta)$ is the heat production density (radiant heating). Lastly,

$$\boldsymbol{D}(\boldsymbol{v}) := \frac{1}{2}\Big[\nabla \boldsymbol{v} + (\nabla \boldsymbol{v})^\mathsf{T}\Big]$$

is the rate of strain tensor and $\mathrm{tr}(\cdot)$ denotes the trace.

The equations in (1.1) are the basic laws of conservation of mass, momentum, and energy, respectively.

We consider (1.1) in a bounded smooth domain Ω of $\mathbb{R}^3$ and suppose that $\boldsymbol{b}$, c, and r depend smoothly upon their respective variables. Then we have to augment (1.1) by boundary conditions. For the velocity field we impose no-slip conditions:

$$\boldsymbol{v} = 0 \qquad \text{on } \partial\Omega \ . \qquad (1.2)$$

As for the temperature, we prescribe θ on one part of $\partial\Omega$ and the value of the heat flux through the remaining part. Thus we put $\partial\Omega = \Gamma_0 \cup \Gamma_1$, where Γ_0 and Γ_1 are disjoint

Navier-Stokes Equations and Related Nonlinear Problems
Edited by A. Sequeira, Plenum Press, New York, 1995

and both open and closed in $\partial\Omega$. Of course, either Γ_0 or Γ_1 can be empty. In such a case we use obvious modifications. Then, denoting by $\boldsymbol{\nu}$ the outer unit normal vector field on $\partial\Omega$, we require that

$$
\begin{aligned}
\theta &= \theta_\Gamma &&\text{on } \Gamma_0 \ , \\
-\boldsymbol{q}\cdot\boldsymbol{\nu} &= h &&\text{on } \Gamma_1 \ ,
\end{aligned}
\tag{1.3}
$$

where $\theta_\Gamma := \theta_\Gamma(x,t)$ and $h := h(x,t,\theta)$ are given smooth functions.

In addition, we impose the initial conditions

$$
\boldsymbol{v} = \boldsymbol{v}^0 \ , \qquad \theta = \theta^0 \quad \text{on } \Omega \quad \text{at } t=0 \ ,
\tag{1.4}
$$

where $\boldsymbol{v}^0 := \boldsymbol{v}^0(x)$ and $\theta^0 := \theta^0(x)$ are also smooth and given.

Lastly, we have to specify constitutive laws for the stress tensor $\boldsymbol{S}$ and the heat flux vector $\boldsymbol{q}$. As for $\boldsymbol{S}$, we suppose that it is a function of $\boldsymbol{D}$ and the thermodynamic variable θ only, that is,

$$
\boldsymbol{S} = \boldsymbol{S}(\boldsymbol{D},\theta) \ .
\tag{1.5}
$$

This means that we consider so-called **Stokesian fluids**. (Recall that, thanks to the incompressibility of the fluid, the pressure p is not a thermodynamic variable.) Then the principle of material frame indifference requires $\boldsymbol{S}$ to be an isotropic function of $\boldsymbol{D}$ (cf. Serrin, 1959; Truesdell and Noll, 1965). Thus, thanks to a well-known representation theorem (cf. Truesdell and Noll, 1965, Section 12),

$$
\boldsymbol{S} = \beta_1\boldsymbol{D} + \beta_2\boldsymbol{D}^2 \ ,
\tag{1.6}
$$

where

$$
\beta_j = \beta_j\!\left(\operatorname{tr}(\boldsymbol{D}^2), \operatorname{tr}(\boldsymbol{D}^3), \theta\right) \ , \qquad j=1,2 \ .
\tag{1.7}
$$

(Here we used the fact that $\operatorname{tr}(\boldsymbol{D}) = 0$ and, consequently, $3\det(\boldsymbol{D}) = \operatorname{tr}(\boldsymbol{D}^3)$ by the Cayley-Hamilton theorem.)

As for the flux vector $\boldsymbol{q}$, we assume that it is a function of $\boldsymbol{D}$, θ, and $\nabla\theta$, that is,

$$
\boldsymbol{q} = \boldsymbol{q}(\boldsymbol{D},\theta,\nabla\theta) \ .
\tag{1.8}
$$

The principle of material frame indifference implies in this case that $\boldsymbol{q}$ is an isotropic function of $\nabla\theta$ and $\boldsymbol{D}$. Thus, by the representation theorem (13.8) in Truesdell and Noll (1965),

$$
\boldsymbol{q} = -(\gamma_0\boldsymbol{1} + \gamma_1\boldsymbol{D} + \gamma_2\boldsymbol{D}^2)\nabla\theta \ ,
\tag{1.9}
$$

where

$$
\gamma_k := \gamma_k\!\left(\operatorname{tr}(\boldsymbol{D}^2), \operatorname{tr}(\boldsymbol{D}^3), \theta, |\nabla\theta|^2, \nabla\theta\cdot\boldsymbol{D}\nabla\theta, \nabla\theta\cdot\boldsymbol{D}^2\nabla\theta\right)
\tag{1.10}
$$

for $k=0,1,2$. Henceforth, the functions β_j and γ_k are supposed to be smooth.

The initial-boundary value problem (1.1)–(1.4) under the constitutive laws (1.5) and (1.8) represents a very general model of non-Newtonian fluids and comprises many special cases occurring in application. First, suppose that

$$
\beta_1 =: 2\rho\nu > 0 \ , \quad \beta_2 = 0 \ ,
\tag{1.11}
$$

where ν, the kinematic viscosity, is constant. Then the first two equations in (1.1) reduce to the incompressible *Navier-Stokes equations*:

$$
\begin{aligned}
\nabla\cdot\boldsymbol{v} &= 0 \ , \\
\partial_t\boldsymbol{v} + (\boldsymbol{v}\cdot\nabla)\boldsymbol{v} - \nu\Delta\boldsymbol{v} &= -\rho^{-1}\nabla p + \boldsymbol{b}(x,t,\theta) \ .
\end{aligned}
\tag{1.12}
$$

Next, assume that

$$c \text{ and } \gamma_0 \text{ are positive constants} , \quad \gamma_1 = \gamma_2 = 0 , \qquad (1.13)$$

so that γ_0 is the thermal conductivity, and define κ, the thermal diffusivity, by

$$\kappa := (\rho c)^{-1} \gamma_0 .$$

Then the energy equation reduces to the convective heat equation

$$\partial_t \theta + \boldsymbol{v} \cdot \nabla \theta - \kappa \Delta \theta = 2\nu c^{-1} \operatorname{tr}\left(\boldsymbol{D}^2(\boldsymbol{v})\right) + c^{-1} r(x, t, \theta) , \qquad (1.14)$$

and boundary conditions (1.3) take the form

$$\begin{aligned} \theta &= \theta_{\Gamma_0} & &\text{on } \Gamma_0 , \\ \kappa \partial_\nu \theta &= (\rho c)^{-1} h(y, t, \theta) & &\text{on } \Gamma_1 , \end{aligned} \qquad (1.15)$$

where $\partial_\nu := \boldsymbol{\nu} \cdot \nabla$ is the outer normal derivative. In particular, the second equation in (1.15) is the no-flux boundary condition $\partial_\nu \theta = 0$ on Γ_1, provided $h = 0$. If, in addition, the 'dissipation term' $2\nu c^{-1} \operatorname{tr}(\boldsymbol{D}^2)$ is omitted in (1.14) then (1.12), together with

$$\partial_t \theta + \boldsymbol{v} \cdot \nabla \theta - \kappa \Delta \theta = c^{-1} r(x, t, \theta) \qquad (1.16)$$

and the boundary conditions

$$v = 0 , \qquad \theta = \theta_\Gamma \quad \text{on } \Gamma := \partial \Omega , \qquad (1.17)$$

constitute the so-called *Boussinesq approximation* for the motion of a viscous compressible fluid driven by buoyancy forces. More precisely, in the Boussinesq approximation it is usually assumed that $\boldsymbol{b}$ is an affine function of θ (e.g., Joseph, 1976).

If assumption (1.11) is satisfied, we are dealing with Newtonian fluids. In all other cases the fluid is said to be **non-Newtonian**.

In spite of the fact that the Stokesian flow model has been criticized for its limitations, mainly for its incapability of describing normal stress differences in simple shear flows (e.g., Truesdell and Noll, 1965, Section 119), it has been found to describe adequately many flow problems for colloids and suspensions in a variety of polymeric liquids, flows in glaciology, geology, and blood-rheology. This is true, in particular, for the submodel of **generalized Newtonian flows** in which it is assumed that β_1 is independent of $\operatorname{tr}(\boldsymbol{D}^3)$, and $\beta_2 = 0$, that is,

$$\beta_1 = \beta_1\left(\operatorname{tr}(\boldsymbol{D}^2)\right) , \quad \beta_2 = 0 .$$

Thus in this case the first two equations in (1.1) differ from the Navier-Stokes equation by a nonconstant shear-rate dependent viscosity coefficient only. We refer to Bird, Armstrong and Hassager (1987), Chapter 4, and Schowalter (1978), Chapter 10, for many concrete examples, as well as for arguments in favour of these models, and for their limitations (also see Málek, Rajagopal and Růžička, 1994, for an extensive list of references). It should be mentioned that rather often it is assumed that β_1 is a 'power nonlinearity' of the form $\beta_1(\tau) = \alpha_1 + \alpha_2 \tau^\delta$ for $\tau \geq 0$, where α_1 is a nonnegative and α_2 and δ are positive constants.

Generalized Newtonian flow models have already attracted considerable mathematical interest (cf. Cioranescu, 1992; Ladyženskaja, 1967, 1969; Málek, Nečas and

Růžička, 1993; Málek, Rajagopal and Růžička, 1994). In Kaniel (1970) a wider subclass of the general class satisfying (1.6), (1.7) has been investigated. The only result treating the full class of isothermal Stokesian fluids (that is, the case where $\boldsymbol{b}$ and β_j are independent of θ), under the sole assumption that $\beta_1(0,0) > 0$, is due to the author (Amann, 1994).

All mathematical investigations of non-Newtonian fluids mentioned above are concerned with the isothermal case. The only paper dealing with temperature effects in non-Newtonian fluids is Málek, Růžička and Thäter (1993), where a Boussinesq approximation for a particular class of Stokesian fluids is considered, in which $\boldsymbol{S}$ is the gradient of a scalar potential for $\boldsymbol{D}$ satisfying suitable coercivity assumptions. Lastly, we refer to Goncharova, 1990, where a Boussinesq approximation with a temperature dependent viscosity coefficient is studied.

It is the purpose of this paper to extend the results of Amann (1994) to the full system (1.1)–(1.5), (1.8) of heat-conducting Stokesian fluids.

2. ASSUMPTIONS AND CONVENTIONS

Now we impose the simplifying assumption that the γ_k are independent of $\nabla\theta$, that is,

$$\gamma_k = \gamma_k\big(\mathrm{tr}(\boldsymbol{D}^2), \mathrm{tr}(\boldsymbol{D}^3), \theta\big) , \qquad k = 0, 1, 2 . \tag{2.1}$$

We also define smooth symmetric tensor fields $\boldsymbol{M} := \boldsymbol{M}(\boldsymbol{v},\theta)$, $\boldsymbol{K} := \boldsymbol{K}(\boldsymbol{v},\theta)$, and $\boldsymbol{R} := \boldsymbol{R}(\boldsymbol{v},\theta)$ by

$$\boldsymbol{M} := \rho^{-1}(\beta_1\boldsymbol{1} + \beta_2\boldsymbol{D}) , \quad \boldsymbol{K} := (\rho c)^{-1}(\gamma_0\boldsymbol{1} + \gamma_1\boldsymbol{D} + \gamma_2\boldsymbol{D}^2) , \quad \boldsymbol{R} := (\rho c)^{-1}\boldsymbol{S} ,$$

respectively, a smooth vector field $\boldsymbol{k} = \boldsymbol{k}(\boldsymbol{v},\theta)$ by

$$\boldsymbol{k}(\boldsymbol{v},\theta) := \boldsymbol{v} + \rho c(\theta)\boldsymbol{K}(\boldsymbol{v},\theta)\nabla\big([\rho c(\theta)]^{-1}\big) ,$$

and smooth functions $k = k(x,t,\theta)$ and $g = g(y,t,\theta)$ for $x \in \overline{\Omega}$ and $y \in \Gamma_1$ by

$$k := c^{-1}r , \quad g := (\rho c)^{-1}h ,$$

respectively. Using these notations, equations (1.1) take the form

$$\nabla \cdot \boldsymbol{v} = 0 ,$$
$$\partial_t\boldsymbol{v} - \nabla\cdot\big(\boldsymbol{M}(\boldsymbol{v},\theta)\boldsymbol{D}(\boldsymbol{v})\big) + \boldsymbol{v}\cdot\nabla\boldsymbol{v} = -\rho^{-1}\nabla p + \boldsymbol{b}(x,t,\theta) , \tag{2.2}$$
$$\partial_t\theta - \nabla\cdot\big(\boldsymbol{K}(\boldsymbol{v},\theta)\nabla\theta\big) + \boldsymbol{k}(\boldsymbol{v},\theta)\cdot\nabla\theta = \mathrm{tr}\big(\boldsymbol{R}(\boldsymbol{v},\theta)\boldsymbol{D}(\boldsymbol{v})\big) + k(x,t,\theta) ,$$

and the boundary conditions read:

$$\boldsymbol{v} = 0 \qquad\qquad \text{on } \partial\Omega ,$$
$$\theta = \theta_\Gamma \qquad\qquad \text{on } \Gamma_0 , \tag{2.3}$$
$$\boldsymbol{\nu} \cdot \boldsymbol{K}(\boldsymbol{v},\theta)\nabla\theta = g(y,t,\theta) \qquad \text{on } \Gamma_1 .$$

Of course, the initial conditions,

$$\boldsymbol{v} = \boldsymbol{v}^0 , \qquad \theta = \theta^0 \quad \text{on } \Omega , \tag{2.4}$$

remain unchanged.

For a smooth manifold X, $s \in \mathbb{R}$, and $1 \leq q \leq \infty$, we let $W_q^s(X) := W_q^s(X, \mathbb{R})$ be the usual Sobolev-Slobodeckii spaces, and $\boldsymbol{W}_q^s(X) := W_q^s(X, \mathbb{R}^3)$. Moreover, if $X = \Omega$, we omit X in these notations. Similarly, $\boldsymbol{C}^\alpha := C^\alpha(\overline{\Omega}, \mathbb{R}^3)$ for $\alpha \in \mathbb{R}^+ := [0, \infty)$, etc. We write τ_j for the trace operator on Γ_j, $j = 0, 1$. Recall that

$$\tau_j \in \mathcal{L}\left(W_q^s, W_q^{s-1/q}(\Gamma_j)\right) \tag{2.5}$$

for $1 < q < \infty$ and $s > 1/q$. We use the same symbol in the vector-valued case. Of course, $\mathcal{L}(E, F)$ is the Banach space of all bounded linear operators from the Banach space E to the Banach space F, and $\mathcal{L}(E) := \mathcal{L}(E, E)$. We also fix an extension operator

$$\mathcal{R} \in \mathcal{L}\left(W_q^{s-1/q}(\Gamma_0), W_q^s\right), \qquad s > 1/q, \tag{2.6}$$

satisfying

$$\tau_0 \mathcal{R} = \mathrm{id} \quad \text{and} \quad \mathcal{R}\mathbf{1} = \mathbf{1}, \tag{2.7}$$

where $\mathbf{1}$ is the function (whose domain may vary from occurrence to occurrence) which is constantly equal to 1 (cf. Amann, 1990, Theorem B.3). We write τ for the trace operator on $\partial\Omega$ and recall that (2.5) holds for τ if Γ_j is replaced by $\partial\Omega$.

We put

$$\langle u, v \rangle := \int_\Omega u \cdot v \, dx, \quad \langle w, z \rangle_\partial := \int_{\Gamma_1} w \cdot z \, d\sigma \tag{2.8}$$

for $u, v : \overline{\Omega} \to \mathbb{R}^N$ and $w, z : \Gamma_1 \to \mathbb{R}^N$, respectively, where it will be clear from the context which value of N has to be taken (usually $N = 1$ or $N = 3$), and where $d\sigma$ denotes the Riemann-Lebesgue volume measure of $\partial\Omega$.

Throughout the remainder of this paper we fix $q \in (3, \infty)$ arbitrarily and use overdots to denote time-derivatives. We also put, letting $\tilde{\tau} := \tau$ or $\tilde{\tau} := \tau_0$,

$$W_{q,\tilde{\tau}}^s := \begin{cases} \{u \in W_q^s \; ; \; \tilde{\tau}u = 0\}, & 1/q < s < \infty, \\ W_q^s, & -2 + 1/q < s < 1/q, \\ (W_{q',\tilde{\tau}}^{-s})', & -\infty < s < -2 + 1/q, \end{cases} \tag{2.9}$$

where $q' := q/(q-1)$ and the dual space is taken with respect to the standard L_q-duality pairing $\langle \cdot, \cdot \rangle$, defined in (2.8). Of course, W in (2.9) can be replaced by $\boldsymbol{W}$.

By a global $\boldsymbol{W}_q^2 \times W_q^1 \times W_q^1$-solution $(\boldsymbol{v}, p, \theta)$ of problem (2.2)–(2.4) we mean a triple of functions

$$\boldsymbol{v} \in C(\mathbb{R}^+, \boldsymbol{W}_{q,\tau}^2) \cap C^1(\mathbb{R}^+, \boldsymbol{L}_q), \quad p \in C(\mathbb{R}^+, W_q^1), \quad \theta \in C(\mathbb{R}^+, W_q^1)$$

with

$$\theta - \mathcal{R}\theta_\Gamma \in C(\mathbb{R}^+, W_{q,\tau_0}^1) \cap C^1(\mathbb{R}^+, W_{q,\tau_0}^{-1}),$$

satisfying the first two equations in (2.2) in the strong sense, the third equation in the **weak L_q-sense** [that is, in the sense that

$$\langle \psi, \dot{\theta} \rangle + \langle \nabla\psi, \boldsymbol{K}(\boldsymbol{v}, \theta)\nabla\theta \rangle + \langle \psi, \boldsymbol{k}(\boldsymbol{v}, \theta) \cdot \nabla\theta \rangle$$
$$= \langle \psi, \mathrm{tr}\big(\boldsymbol{R}(\boldsymbol{v}, \theta)\boldsymbol{D}(\boldsymbol{v})\big) + k(\cdot, t, \theta) \rangle + \langle \tau_1\psi, g(\cdot, t, \tau_1\theta) \rangle_\partial \tag{2.10}$$

holds for $\psi \in W_{q',\tau_0}^1$ and $t \geq 0$], the first two equations in (2.3) in the sense of traces, and the initial conditions (2.4).

In the following, we suppose that there exists $\theta_0 \in \mathbb{R}^+$ such that the partial derivatives of $\boldsymbol{b}$, k, and g with respect to θ are independent of t at $\theta = \theta_0$. Thus these functions can be written in the form

$$
\begin{aligned}
\boldsymbol{b}(x,t,\theta) &= \boldsymbol{b}_1(x,\theta)(\theta - \theta_0) + \boldsymbol{b}_0(x,t) , \\
k(x,t,\theta) &= k_1(x,\theta)(\theta - \theta_0) + k_0(x,t) , \\
g(y,t,\theta) &= g_1(y,\theta)(\theta - \theta_0) + g_0(y,t)
\end{aligned}
\tag{2.11}
$$

for $x \in \overline{\Omega}$, $y \in \Gamma_1$, and $t,\theta \in \mathbb{R}^+$, where $\boldsymbol{b}_1$, k_1, and g_1 are smooth.

We also suppose that

$$
\nu_0 := (2\rho)^{-1}\beta_1(0,0,\theta_0) > 0 , \quad \kappa_0 := \bigl(\rho c(\theta_0)\bigr)^{-1}\gamma_0(0,0,\theta_0) > 0 .
\tag{2.12}
$$

Remark 2.1. We recall that the second law of thermodynamics in the form of the Clausius-Duhem inequality implies the general relations

$$
\mathrm{tr}\bigl(\boldsymbol{S}(\boldsymbol{A},\theta)\boldsymbol{A}\bigr) \geq 0 , \quad \boldsymbol{q}(0,\theta,\boldsymbol{a}) \cdot \boldsymbol{a} \leq 0
\tag{2.13}
$$

for all traceless symmetric tensors $\boldsymbol{A}$ and all vectors $\boldsymbol{a}$ (cf. Dunn and Fosdick, 1974, Section 3, Corollary 2; Rajagopal, 1993). Hence we see that (2.12) is consistent with (2.13) if we restrict ourselves to small values of $\boldsymbol{D}(\boldsymbol{v})$, which we shall do. $\qquad\square$

We let $\lambda_0 := \lambda_0(\theta_0)$ be the least eigenvalue of the linear elliptic eigenvalue problem

$$
\begin{aligned}
-\kappa_0 \Delta u - k_1(\cdot,\theta_0)u &= \lambda u && \text{in } \Omega , \\
u &= 0 && \text{on } \Gamma_0 , \\
\kappa_0 \partial_\nu u - g_1(\cdot,\theta_0)u &= 0 && \text{on } \Gamma_1 .
\end{aligned}
\tag{2.14}
$$

We write $\boldsymbol{L}_{q,\sigma}$ for the closure in $\boldsymbol{L}_q$ of the subspace of all smooth solenoidal vector fields with compact supports in Ω. Then the Helmholtz projector P of $\boldsymbol{L}_q$ onto $\boldsymbol{L}_{q,\sigma}$, parallel to the subspace of gradient vectorfields, is a well-defined continuous linear operator. We denote by $S := -\nu_0 P\Delta|\boldsymbol{W}^2_{q,\tau,\sigma}$, where $\boldsymbol{W}^2_{q,\tau,\sigma} := \boldsymbol{W}^2_{q,\tau} \cap \boldsymbol{L}_{q,\sigma}$, the Stokes operator and by μ_0 the least eigenvalue of S. Then we assume that

$$
\omega_0 := \min\{\lambda_0, \mu_0\} > 0 .
\tag{2.15}
$$

Remark 2.2. It is well-known that $\mu_0 > 0$. Hence, thanks to (2.12), assumption (2.15) reduces to the hypothesis: $\lambda_0 > 0$. It is an easy consequence of the variational formulation of eigenvalue problem (2.14) and of Poincaré's inequality that the following conditions:

$$
k_1(\cdot,\theta_0) \leq 0 , \quad g_1(\cdot,\theta_0) \leq 0
$$

and

$$
\mathrm{vol}(\Gamma_0) + \|k_1(\cdot,\theta_0)\|_C + \|g_1(\cdot,\theta_0)\|_{C(\Gamma_1)} > 0
$$

are sufficient for guaranteeing $\lambda_0 > 0$.

3. THE THEOREM

Given a Banach space E and $\alpha \in (0,1)$, we denote by $BUC^\alpha(\mathbb{R}^+, E)$ the Banach space of all bounded and uniformly α-Hölder continuous functions from $\mathbb{R}^+$ to E, endowed with the usual norm, which we denote by $\|\cdot\|_{C^\alpha(\mathbb{R}^+,E)}$, or even by $\|\cdot\|_{C^\alpha}$, if no confusion seems possible. Moreover, we write $e^{\omega t}f$ for the function $t \mapsto e^{\omega t}f(t)$.

Now we can formulate the main result of this paper, namely, the following global existence and uniqueness result:

Theorem 3.1. *Let assumptions* (2.1), (2.11), (2.12), *and* (2.15) *be satisfied. Then, given* $\omega \in [0, \omega_0)$, *there exist positive constants* a_0 *and* c_0 *such that, given* $\alpha \in (0,1)$ *and*

$$(\boldsymbol{v}^0, \theta^0) \in \boldsymbol{W}_{q,\tau}^2 \times W_{q,\tau_0}^1 , \tag{3.1}$$

satisfying

$$\|\boldsymbol{v}^0\|_{\boldsymbol{W}_q^2} + \|\theta^0 - \theta_0\|_{W_q^1} \leq a_0 \tag{3.2}$$

and the compatibility conditions

$$\nabla \cdot \boldsymbol{v}^0 = 0 , \quad \boldsymbol{v}|\partial\Omega = 0 , \quad \theta^0|\Gamma_0 = \theta_0 , \tag{3.3}$$

problem (2.2)–(2.4) *has a unique global* $\boldsymbol{W}_q^2 \times W_q^1 \times W_q^1$-*solution* $(\boldsymbol{v}, p, \theta)$ *with*

$$\int_\Omega p(t)\,dx = 0 , \qquad t \geq 0 , \tag{3.4}$$

provided

$$(\boldsymbol{b}_0, k_0, \theta_\Gamma, g_0) \in BUC^\alpha \left(\mathbb{R}^+, \boldsymbol{L}_q \times L_q \times W_q^{1-1/q}(\Gamma_0) \times L_q(\Gamma_1) \right) \tag{3.5}$$

and satisfies

$$\|e^{\omega t}(\boldsymbol{b}_0, k_0, \theta_\Gamma - \theta_0, g_0)\|_{C^\alpha} \leq a_0 . \tag{3.6}$$

Moreover,

$$\|(\boldsymbol{v}, p, \theta - \theta_0)(t)\|_{\boldsymbol{W}_q^2 \times W_q^1 \times W_q^1} + \|(\dot{\boldsymbol{v}}, \dot{\theta})(t)\|_{\boldsymbol{L}_q \times W_{q,\tau_0}^{-1}} \leq c_0 e^{-\omega t} \tag{3.7}$$

for $t \geq 0$.

In the special case $\boldsymbol{b}_0 = 0$, $k_0 = 0$, $g_0 = 0$, and $\theta_\Gamma = \theta_0$, problem (2.2), (2.3) possesses the rest state $\boldsymbol{v} = 0$, $p = 0$, $\theta = \theta_0$. Thus Theorem 3.1 shows that the rest state $\boldsymbol{v} = 0$, $p = 0$ of the fluid and the uniform temperature distribution $\theta = \theta_0$ are stable under small perturbations. In fact, the rest state $(0, 0, \theta_0)$ of (2.2), (2.3) is even exponentially stable, provided $\boldsymbol{b}_0$, k_0, $\theta_\Gamma - \theta_0$, and g_0 decay exponentially with time. In particular, Theorem 3.1 implies that, given our assumptions, the *principle of linearized stability is valid for non-Newtonian Stokesian fluids.* This seems to be the first rigorous mathematical justification of this principle which is usually applied formally without proof (e.g., Joseph, 1976).

Observe that Theorem 3.1 covers — as a very special case — the Boussinesq approximation (1.12)–(1.15), even taking into account heat-production by viscous heating, that is, the nonlinear term $2\nu c^{-1}\operatorname{tr}(\boldsymbol{D}^2(\boldsymbol{v}))$ in (1.14). It also covers the standard Boussinesq approximation since in the proof below the term $2\nu c^{-1}\operatorname{tr}(\boldsymbol{D}^2(\boldsymbol{v}))$ can formally be put equal to zero.

Remark 3.2. The smoothness assumptions have been imposed for convenience and can be considerably weakened.

Remark 3.3. Assumption (2.1) can be omitted, provided we restrict ourselves to the no-flux condition

$$\boldsymbol{q} \cdot \boldsymbol{\nu} = 0 \qquad \text{on } \Gamma_1 \,,$$

that is, if we suppose that $h = 0$. In this case we have to assume that $\theta^0 \in W_q^2$ and satisfies the additional compatibility condition $\partial_\nu \theta^0 = 0$ on Γ_1, that

$$\theta_\Gamma - \theta_0 \in BUC^\alpha(\mathbb{R}^+, W_q^{2-1/q}(\Gamma_0)) \,,$$

and that in the smallness conditions (3.2) and (3.6) the W_q^1- and $W_q^{1-1/q}(\Gamma_0)$-norms are replaced by the W_q^2- and $W_q^{2-1/q}(\Gamma_0)$-norms, respectively. Then we obtain that $(\boldsymbol{v}, p, \theta)$ is a global $\boldsymbol{W}_q^2 \times W_q^1 \times W_q^2$-solution, which means that

$$\theta \in C(\mathbb{R}^+, W_q^2) \cap C^1(\mathbb{R}^+, L_q) \,,$$

and that $(\boldsymbol{v}, p, \theta)$ is a strong solution to (2.2)–(2.4). Moreover, estimate (3.7) takes the form

$$\|(\boldsymbol{v}, p, \theta - \theta_0)(t)\|_{\boldsymbol{W}_q^2 \times W_q^1 \times W_q^2} + \|(\dot{\boldsymbol{v}}, \dot{\theta})(t)\|_{\boldsymbol{L}_q \times L_q} \leq c_0 e^{-\omega t}$$

for $t \geq 0$.

Remark 3.4. Theorem 3.1 contains the main result of Amann (1994), where the isothermal case has been studied, as a special case.

4. THE PROOF

Since the proof of Theorem 3.1 is a modification of the one of the main result in Amann (1994), we will be rather brief and refer the reader to that paper for more details and to Amann (1993, 1995) for background information on abstract evolution equations, the theory of function spaces, and elliptic differential operators.

Given a Banach space E, we denote by $BUC(\mathbb{R}^+, E)$ the Banach space of all bounded and uniformly continuous functions $u \colon \mathbb{R}^+ \to E$, endowed with the supremum norm $\|\cdot\|_\infty$. Moreover, given $\alpha \in (0, 1)$ and $\varepsilon > 0$,

$$[u]^*_{\alpha, [\varepsilon, 2\varepsilon]} := \sup_{\substack{\varepsilon < s < t < 2\varepsilon \\ t - s < 1}} \frac{\|u(s) - u(t)\|}{|s - t|^\alpha}$$

and

$$[\![u]\!]_\alpha := \sup_{\varepsilon > 0} (1 \wedge \varepsilon)^\alpha [u]^*_{\alpha, [\varepsilon, 2\varepsilon]} \,, \qquad u \in BUC(\mathbb{R}^+, E) \,.$$

Then $BUC^\alpha_\alpha(\mathbb{R}^+, E)$ is the Banach space of all $u \in BUC(\mathbb{R}^+, E)$ satisfying $[\![u]\!]_\alpha < \infty$ and

$$\lim_{\varepsilon \to 0} \varepsilon^\alpha [u]^*_{\alpha, [\varepsilon, 2\varepsilon]} = 0 \,,$$

equipped with the norm

$$\|\cdot\|_{C^\alpha} := \|\cdot\|_\infty + [\![\cdot]\!]_\alpha \,.$$

We also define the Banach space $BUC^{1+\alpha}_\alpha(\mathbb{R}^+, E)$ to consist of all $u \in BUC^\alpha_\alpha(\mathbb{R}^+, E)$ such that $\partial u := \dot{u} \in BUC^\alpha_\alpha(\mathbb{R}^+, E)$, endowed with the norm

$$u \mapsto \|u\|_{C^{1+\alpha}} := \|u\|_{C^\alpha} + \|\dot{u}\|_{C^\alpha} \,.$$

Lastly, given $\omega \in \mathbb{R}^+$ and $\beta \in \{\alpha, 1 + \alpha\}$,

$$e^{-\omega} BUC_\alpha^\beta(\mathbb{R}^+, E) := \left\{ u \in BUC(\mathbb{R}^+, E) \ ; \ e^{\omega t} u \in BUC_\alpha^\beta(\mathbb{R}^+, E) \right\} .$$

These are Banach spaces too with the norms

$$u \mapsto \|u\|_{e^{-\omega} C_\alpha^\beta} := \left\| e^{\omega t} u \right\|_{C_\alpha^\beta} .$$

It is obvious that

$$BUC^\alpha(\mathbb{R}^+, E) \hookrightarrow BUC_\alpha^\alpha(\mathbb{R}^+, E) . \tag{4.1}$$

Let (E_0, E_1) be a densely injected Banach couple, that is, E_0 and E_1 are Banach spaces such that $E_1 \hookrightarrow E_0$ and E_1 is dense in E_0. We denote by $\mathcal{H}(E_1, E_0)$ the set of all operators $B \in \mathcal{L}(E_1, E_0)$ such that $-B$, considered as a linear operator in E_0 with domain E_1, is the infinitesimal generator of a strongly continuous analytic semigroup on E_0. We write $s(B)$ for the lower spectral bound of $B \in \mathcal{H}(E_1, E_0)$, defined by

$$s(B) := \inf\{ \operatorname{Re} \lambda \ ; \ \lambda \in \sigma(B) \} ,$$

where $\sigma(B)$ is the spectrum of B.

We put

$$E_0 := \boldsymbol{L}_{q,\sigma} \times W_{q,\tau_0}^{-1} , \quad E_1 := \boldsymbol{W}_{q,\tau,\sigma}^2 \times W_{q,\tau_0}^1 .$$

Then (E_0, E_1) is a densely injected Banach couple and

$$E_1 \hookrightarrow \boldsymbol{C}^1 \times C \tag{4.2}$$

since $q > 3$, thanks to Sobolev's imbedding theorem.

Given $u := (\boldsymbol{v}, w) \in E_1$ and $\xi \in W_q^{1-1/q}(\Gamma_0)$, we define $A(u, \xi) \in \mathcal{L}(E_1, E_0)$ by

$$A(u, \xi) := \begin{bmatrix} A_1(u, \xi) & 0 \\ 0 & A_2(u, \xi) \end{bmatrix} ,$$

where, letting $\ell := \ell(w, \xi) := w + \mathcal{R}\xi + \theta_0 \boldsymbol{1}$,

$$A_1(u, \xi)\tilde{\boldsymbol{v}} := P\left[-\nabla \cdot \left(\boldsymbol{M}(\boldsymbol{v}, \ell) \boldsymbol{D}(\tilde{\boldsymbol{v}}) \right) + (\boldsymbol{v} \cdot \nabla)\tilde{\boldsymbol{v}} - \boldsymbol{b}_1(\cdot, \ell)(\tilde{w} + \mathcal{R}\xi) \right]$$

and

$$\begin{aligned} \left\langle \psi, A_2(u, \xi)\tilde{w} \right\rangle := &\left\langle \nabla\psi, \boldsymbol{K}(\boldsymbol{v}, \ell)\nabla\tilde{w} \right\rangle \\ &+ \left\langle \psi, \boldsymbol{k}(\boldsymbol{v}, \ell)\nabla\tilde{w} - \operatorname{tr}\left(\boldsymbol{R}(\boldsymbol{v}, \ell)\boldsymbol{D}(\tilde{\boldsymbol{v}}) \right) - k_1(\cdot, \ell)(\tilde{w} + \mathcal{R}\xi) \right\rangle \\ &- \left\langle \tau_1\psi, g_1(\cdot, \tau_1\ell)\tau_1(\tilde{w} + \mathcal{R}\xi) \right\rangle_\partial \end{aligned}$$

for $\psi \in W_{q',\tau_0}^1$. By (2.5), (4.2), and well-known properties of Nemytskii operators on spaces of continuous functions it is not difficult to see that

$$\left[(u, \xi) \mapsto A(u, \xi) \right] \in C^\infty\left(E_1 \times W_q^{1-1/q}(\Gamma_0), \mathcal{L}(E_1, E_0) \right) . \tag{4.3}$$

We put

$$\tau_{1,0} := \tau_1 | W_{q,\tau_0}^1 \in \mathcal{L}\left(W_{q,\tau_0}^1, W_q^{1-1/q}(\Gamma_1) \right)$$

and denote by $\tau_{1,0}'$ the dual of $\tau_{1,0}$. Then we verify that

$$B := A(0,0) := \begin{bmatrix} S & P\boldsymbol{b}_1(\cdot, \theta_0) \\ 0 & -\kappa_0\Delta - k_1(\cdot, \theta_0) - \tau_{1,0}'g_1(\cdot, \theta_0)\tau_{1,0} \end{bmatrix} ,$$

where the Laplace operator in the second row has to be interpreted in the appropriate weak sense. It is well-known that $S \in \mathcal{H}(\boldsymbol{W}_{q,\tau,\sigma}^2, \boldsymbol{L}_q)$ (cf. Giga, 1981; Miyakawa,

1981; Solonnikov, 1977; von Wahl, 1985). From Amann (1993), we know that the operator in the lower right corner of the above matrix belongs to $\mathcal{H}(W_{q,\tau_0}^1, W_{q,\tau_0}^{-1})$. Since $P\boldsymbol{b}_1(\cdot, \theta_0) \in \mathcal{L}(W_{q,\tau_0}^1, \boldsymbol{L}_{q,\sigma})$, we infer from Amann (1995), Theorem 1.6.1, that

$$B \in \mathcal{H}(E_1, E_0) \ . \tag{4.4}$$

It is also easily verified that $s(B) = \omega_0$.

We fix $\omega \in [0, \omega_0)$ and $\alpha \in (0, 1)$ and put

$$\mathbb{E}_0 := e^{-\omega} BUC_\alpha^\alpha(\mathbb{R}^+, E_0)$$

and

$$\mathbb{E}_1 := e^{-\omega} BUC_\alpha^\alpha(\mathbb{R}^+, E_1) \cap e^{-\omega} BUC_\alpha^{1+\alpha}(\mathbb{R}^+, E_0) \ .$$

We also let

$$\mathbb{F}_0 := e^{-\omega} BUC_\alpha^\alpha\big(\mathbb{R}^+, \boldsymbol{L}_q \times \boldsymbol{L}_q \times W_q^{1-1/q}(\Gamma_0) \times L_q(\Gamma_1)\big)$$

and $\mathbb{E} := \mathbb{E}_0 \times E_1$, $\mathbb{F} := \mathbb{F}_0 \times E_1$.

Then we define $T \in \mathcal{L}(\mathbb{F}, \mathbb{E}_0)$ by

$$T\varphi := (T_1\varphi, T_2\varphi) \ , \quad \varphi := \big((\boldsymbol{b}_0, k_0, \eta, g_0), u^o\big) \in \mathbb{F}_0 \times E_1 \ , \tag{4.5}$$

where

$$T_1\varphi := P\boldsymbol{b}_1$$

and

$$\langle \psi, T_2\varphi \rangle := \langle \psi, k_0 \rangle + \langle \tau_1, \psi, g_0 \rangle_\partial \ , \quad \psi \in W_{q',\tau_0}^1 \ .$$

Finally, we put

$$\Psi(u, \varphi)(t) := \big(\partial u(t) + A(u(t), \eta(t))u(t) - T\varphi(t), \gamma u - u^o\big)$$

for $(u, \varphi) \in \mathbb{E}_1 \times \mathbb{F}$ and $t \geq 0$, where the generic representation of φ, given in (4.5), is being used. Thanks to (4.3), it is not difficult to see that

$$\Psi \in C^\infty(\mathbb{E}_1 \times \mathbb{F}, \mathbb{E}) \ . \tag{4.6}$$

Note that $\Psi(0, 0) = 0$ and $\partial_1 \Psi(0, 0) = (\partial + B, \gamma)$. Hence it follows from (4.4) and Amann (1994), Proposition 1 that $\partial_1 \Psi(0, 0)$ is an isomorphism from $\mathbb{E}_1$ onto $\mathbb{E}$. Thus the implicit function theorem guarantees the existence of a neighborhood $U \times V$ of $(0, 0)$ in $\mathbb{E}_1 \times \mathbb{F}$ and of a map

$$\overline{u} \in C^\infty(U, V) \tag{4.7}$$

such that, given any $\varphi \in V$, the equation $\Psi(\cdot, \varphi) = 0$ has a unique solution in U, namely $\overline{u}(\varphi)$, that is,

$$\Psi\big(\overline{u}(\varphi), \varphi\big) = 0 \ , \quad \varphi \in V \ . \tag{4.8}$$

Let $(u, \varphi) \in \mathbb{E}_1 \times \mathbb{F}$ be given such that $\Psi(u, \varphi) = 0$. Write

$$u := (\boldsymbol{v}, w) \ , \quad \varphi := ((\boldsymbol{b}_0, k_0, \eta, g_0), u^o) \ , \quad u^o := (\boldsymbol{v}^o, w^o) \ ,$$

where these decompositions are naturally induced by the product structures of E_1 (and E_0) and of $\mathbb{F}_0$. Put

$$\theta := w + \theta_0 \ , \quad \theta_\Gamma := \eta + \theta_0 \ , \quad \theta^0 := w^0 + \theta_0 \ .$$

Then it is not difficult to see that $(\boldsymbol{v}, \theta)$ is a global $\boldsymbol{W}^2_{q,\tau,\sigma} \times W^1_q$-solution (in the obvious sense) of the system

$$\partial_t \boldsymbol{v} - P\nabla \cdot \left(\boldsymbol{M}(\boldsymbol{v}, \theta)\boldsymbol{D}(\boldsymbol{v}) \right) + P(\boldsymbol{v} \cdot \nabla)\boldsymbol{v} = P\boldsymbol{b}_1(\cdot, \theta)(\theta - \theta_0) + P\boldsymbol{b}_0(\cdot, t)$$
$$\partial_t \theta - \nabla \cdot \left(\boldsymbol{K}(\boldsymbol{v}, \theta)\nabla\theta \right) + \boldsymbol{k}(\boldsymbol{v}, \theta) \cdot \nabla\theta = \operatorname{tr}\left(\boldsymbol{R}(\boldsymbol{v}, \theta)\boldsymbol{D}(\boldsymbol{v}) \right) + k_1(\cdot, \theta)(\theta - \theta_0) + k_0(\cdot, t)$$

in Ω, subject to the boundary conditions (2.3) and the initial conditions (2.4). By means of standard techniques (cf. the corresponding considerations in Amann, 1994, Section 3) we can now determine a unique

$$p \in e^{-\omega}BUC^\alpha_\alpha(\mathbb{R}^+, W^1_q)$$

satisfying the normalization condition (3.4) in such a way that $(\boldsymbol{v}, p, \theta)$ is a global $\boldsymbol{W}^2_q \times W^1_q \times W^1_q$-solution of (2.2)–(2.4), subject to condition (2.12). Conversely, given a $\boldsymbol{W}^2_q \times W^1_q \times W^1_q$-solution of (2.2)–(2.4), subject to (2.12), we apply the projector P to the first equation in (2.2) to eliminate the pressure p, as well as the gradient part of $\boldsymbol{b}_0$. Then it is not difficult to see that, letting $w := \theta - \theta_0$, $\eta := \theta_\Gamma - \theta_0$, and $w^0 := \theta^0 - \theta_0$, the functions $u := (\boldsymbol{v}, w)$ and φ, defined by the obvious component representation, satisfy $\Psi(u, \varphi) = 0$, provided (u, φ) is regular enough to belong to $\mathbb{E}_1 \times \mathbb{F}$. This shows that, given $\varphi \in \mathbb{F}$, (2.2)–(2.4), subject to (2.12), is equivalent to the equation $\Psi(v, \varphi) = 0$, provided $v \in \mathbb{E}_1$. Hence the above considerations prove Theorem 3.1 if we stay in the 'regularity class' $\mathbb{E}_1 \times \mathbb{F}$.

In order to give a complete proof of Theorem 3.1, it remains to show, thanks to (4.1), that problem (2.2)–(2.4), subject to (2.12), has at most one solution in the larger class of $\boldsymbol{W}^2_q \times W^1_q \times W^1_q$-solutions. This can be done by adapting the interpolation-extrapolation technique of Amann (1994), Section 3, that is, the uniqueness part of the proof given there, to the present situation. We leave the details to the reader. □

Remark 4.1. The only place where the boundedness hypothesis for Ω has been significantly used is to guarantee the validity of (4.4) and that the lower spectral bound $s(B)$ is strictly positive. It is known that this is also the case for certain unbounded domains. Thus Theorem 3.1 extends to domains of this type.

Remark 4.2. The assumption that θ_0 be constant can be replaced by the more general hypothesis that $\boldsymbol{v} = \boldsymbol{0}$, $p = 0$, $\theta = \theta_0$ is a stationary solution of problem (2.2)–(2.4), provided $\beta_1(0, 0, \theta_0)$ is a positive constant. In contrast to this, the function

$$\kappa_0 := (\rho c(\theta_0))^{-1}\gamma_0(0, 0, \theta_0)$$

need only be strictly positive everywhere, but not constant. It is then clear how assumptions (2.11) and (2.15), as well as the statement of Theorem 3.1, together with this proof, have to be modified in this case. Details are left to the reader.

Remark 4.3. As mentioned after Theorem 3.1, this theorem establishes the validity of the principle of linearized stability for Stokesian fluids. Given suitable hypotheses, it is also possible, using the theory of abstract parabolic evolution equations (cf. Amann, 1995; Lunardi, 1995) along the lines of this paper, to give a rigorous proof of the 'principle of linearized instability' and to establish bifurcation results.

Remark 4.4. The above proof shows that the solution depends smoothly on all data (see, in particular, (4.7)).

Supported in part by Schweizerischer Nationalfonds and European Community (Science Plan, Project "Evolutionary Systems").

REFERENCES

AMANN, H. (1990), Dynamic theory of quasilinear parabolic equations – II. Reaction-diffusion systems, *Diff. Int. Equ.* **3**, p. 13–75.

AMANN, H. (1993), Nonhomogeneous linear and quasilinear elliptic and parabolic boundary value problems, in: H.J. SCHMEISSER and H. TRIEBEL (eds.), *Function Spaces, Differential Operators and Nonlinear Analysis* , p. 9–126 (Teubner, Stuttgart, Leipzig).

AMANN, H. (1994), Stability of the rest state of a viscous incompressible fluid, *Arch. Rat. Mech. Anal.* **126**, p. 231–242.

AMANN, H. (1995), *Linear and Quasilinear Parabolic Problems*, (Birkhäuser, Basel).

BIRD, R.B., ARMSTRONG, R.C. and O. HASSAGER (1987), *Dynamics of Polymeric Liquids*, Vol. 1: *Fluid Mechanics*, (Wiley, New York).

CIORANESCU, D. (1992), Quelques examples de fluides Newtonien generalisés, in: J.F. RODRIGUES and A. SEQUEIRA (eds.), Mathematical Topics in Fluid Mechanics, *Pitman Research Notes in Math.* **274**, p. 1–31.

DUNN, J.E. and R.L. FOSDICK (1974), Thermodynamics, stability, and boundedness of fluids of complexity 2 and fluids of second grade, *Arch. Rat. Mech. Anal.* **56**, p. 191–252.

GIGA, Y. (1981), Analyticity of the semigroup generated by the Stokes operator in L_r spaces, *Math. Z.* **178**, p. 297–329.

GONCHAROVA, O.N. (1990), Solvability of the nonstationary problem for the free convection equation with temperature dependent viscosity (in Russian), *Dinamika sploshnoi sredy, Novosibirsk* **96**, p. 35–58.

JOSEPH, D.D. (1976), *Stability of Fluid Motions*, (Springer Verlag, Berlin).

KANIEL, S. (1970), On the initial value problem for an incompressible fluid with nonlinear viscosity, *J. Math. Mech.* **19**, p. 681–707.

LADYŽENSKAYA, O.A. (1967), New equations for the description of motion of viscous incompressible fluids and solvability in the large of boundary value problems for them, *Proc. Steklov Inst. Math.* **102**, p. 95–118.

LADYŽENSKAYA, O.A. (1969), *Mathematical Theory of Viscous Incompressible Flow*, (Gordon & Breach, New York).

LUNARDI, A. (1995), *Analytic Semigroups and Optimal Regularity in Parabolic Equations*, (Birkhäuser, Basel).

MÁLEK, J., NEČAS, J. and M. RŮŽIČKA (1993), On the non-Newtonian incompressible fluids, *Math. Meth. & Models in Appl. Sciences* (M^3AS) **3**, p. 35–63.

MÁLEK, J., RŮŽIČKA, M. and G. THÄTER (1993), *Fractal dimension, attractors and Boussinesq approximation in three dimensions*, (preprint).

MÁLEK, J., RAJAGOPAL, K.R. and M. RŮŽIČKA (1994), *Existence and regularity of solutions and the stability of the rest state for fluids with shear dependent viscosity*, (preprint).

MIYAKAWA, T. (1981), On the initial value problem for the Navier-Stokes equations in L^p spaces, *Hiroshima Math. J.* **11**, p. 9–20.

RAJAGOPAL, K.R. (1993), Mechanics of non-Newtonian fluids, in: G.P. GALDI and J. NEČAS (eds.), Recent Developments in Theoretical Fluid Dynamics, p. 129–162, *Pitman Research Notes in Math* **291**.

SCHOWALTER, W.R. (1978), *Mechanics of Non-Newtonian Fluids*, (Pergamon Press, Oxford).

SERRIN, J. (1959), Mathematical principles of classical fluid dynamics, *Handbuch der Physik* **VIII/1**, p. 125–263 (Springer Verlag).

SOLONNIKOV, V.A. (1977), Estimates for solutions of nonstationary Navier-Stokes equations, *J. Soviet Math.* **8**, p. 467–529.

TRUESDELL, C. and W. NOLL (1965), The Non-Linear Field Theories of Mechanics, *Handbuch der Physik* **III/3**, (Springer Verlag).

VON WAHL, W. (1985), *The Equations of Navier-Stokes and Abstract Parabolic Equations*, (Vieweg & Sohn, Braunschweig).

ON A CLASS OF ABSTRACT EVOLUTION EQUATIONS RELATED TO THE MOTION OF COMPLEXITY 2 FLUIDS

Vincenzo Coscia[1] and Juha H. Videman[2]

[1] Università di Ferrara, Dipartimento di Matematica
Via Machiavelli, 35, 44100 - Ferrara, Italy

[2] Instituto Superior Técnico, Departamento de Matemática
Av. Rovisco Pais, 1, 1096 - Lisboa Codex, Portugal

1. INTRODUCTION

This paper is concerned with the existence of classical solutions to the following nonlinear system of evolution equations

$$
\begin{cases}
\dfrac{\partial}{\partial t}\mathbf{A}(\mathbf{v}) + \lambda\mathbf{B}(\mathbf{v}) + \mathbf{v}\cdot\nabla\mathbf{A}(\mathbf{v}) + \mathbf{N}(\mathbf{v}) + \nabla p = \mathbf{f} \\
\nabla\cdot\mathbf{v} = 0
\end{cases}
\tag{1.1}
$$

in a space-time region $\Omega \times (0, T)$, where Ω denotes a bounded two- or three-dimensional domain. By $\mathbf{A}$ and $\mathbf{B}$ above, we denote linear differential operators with constant coefficients, $\mathbf{N}$ is a nonlinear differential operator and $\lambda \in \mathbf{R}$. For the problem to be well set, we need, of course, to supplement the equations (1.1) with appropriate initial and boundary conditions. Equations of this type arise naturally, *e.g.* from the balance of linear momentum and from the equation of continuity governing the motion of non-Newtonian fluids, with p denoting the pressure, $\mathbf{v}$ the velocity field and $\mathbf{f}$ the given body force.

As is well-known, for a large class of fluids of rheological interest, the rate of deformation is not linearly proportional to the stress, leading to partial differential equations containing highly nonlinear terms in addition to the usual inertial one. For this kind of fluids, generically called non-Newtonian (see *e.g.* Huilgol, 1975; Schowalter, 1978), the mathematical problem regarding existence and uniqueness of flows is still largely an open question. Such a problem is of fundamental interest, not only from a purely theoretical point of view, but since it serves as a crucial test for the underlying physical model.

Over the past years there has been a notable progress in studying the issues concerning the well-posedness of nonlinear problems in hydrodynamics. This breakthrough is largely based on a simple idea of decoupling the nonlinear problem into two (or more) linear ones, and using, consequently, a fixed point argument to prove the equivalence between the original and the coupled problem, see *e.g.* Beirão da Veiga (1980), Galdi, Grobbelaar and Sauer (1993), Novotny and Padula (1994).

Here we shall adopt and generalize this way of treating systems of nonlinear equations. In fact, what we do is express, in view of the Helmholtz decomposition, the vector $\mathbf{A}(\mathbf{v})$ in $(1.1)_1$ as the $\mathbf{L}^2(\Omega)$-orthogonal sum of a divergent-free vector $\mathbf{w}$ and of the gradient of a scalar function π. This allows us to split the original nonlinear system into two (linear) auxiliary problems, one being an elliptic system, in our applications a (resolvent) Stokes system, and the other one a system of transport equations. By making certain assumptions on the differential operators $\mathbf{A}$, $\mathbf{B}$ and $\mathbf{N}$, we are able to show that the linear problems are well posed, and that we can, moreover, by joining the problems back together through the fixed point theory, find a solution to the original nonlinear problem.

The applications considered in this paper arise in the study of the equations of motion of non-Newtonian fluids, called fluids of differential type of complexity n, also commonly referred to as Rivlin-Ericksen fluids (cf. Rivlin and Ericksen, 1955; Truesdell and Noll, 1965). In an incompressible Rivlin-Ericksen fluid of complexity n the Cauchy stress tensor $\mathbf{T}$ is given by

$$\mathbf{T} = -p\mathbf{I} + \mathbf{f}(\mathbf{A}_1, \ldots, \mathbf{A}_n),$$

where $-p\mathbf{I}$ is the sphere symmetric stress field due to the incompressibility condition, i.e. p denotes the pressure, and $\mathbf{f}$ is a tensor-valued isotropic function of the first n Rivlin-Ericksen tensors, $\mathbf{A}_1, \ldots, \mathbf{A}_n$ defined recursively by

$$\mathbf{A}_1 = \nabla\mathbf{v} + (\nabla\mathbf{v})^T$$

$$\mathbf{A}_n = \frac{d}{dt}\mathbf{A}_{n-1} + (\nabla\mathbf{v})^T\mathbf{A}_{n-1} + \mathbf{A}_{n-1}\nabla\mathbf{v},$$

where $\frac{d}{dt}$ stands for the material derivative and $\mathbf{v}$ is the velocity field. See Dunn and Rajagopal (1995), for an exhaustive discussion about the thermomechanics of fluids of complexity n.

Fluids of grade n form an important subclass of fluids of complexity n (Coleman and Noll, 1960; Truesdell and Noll, 1965). For instance, in a fluid of grade 2 the Cauchy stress tensor $\mathbf{T}$ is given by

$$\mathbf{T} = -p\mathbf{I} + \mu\mathbf{A}_1 + \alpha_1\mathbf{A}_2 + \alpha_2\mathbf{A}_1^2 \tag{1.2}$$

and in a *thermodynamically compatible* fluid of grade 3 (see Fosdick and Rajagopal, 1980) by

$$\mathbf{T} = -p\mathbf{I} + \mu\mathbf{A}_1 + \alpha_1\mathbf{A}_2 + \alpha_2\mathbf{A}_1^2 + \beta_3(\mathrm{tr}\mathbf{A}_1^2)\mathbf{A}_1, \tag{1.3}$$

where μ, α_1, α_2 and β_3 are material coefficients (μ is the coefficient of viscosity). These are both fluids of complexity 2.[1]

The second and third grade fluids have been extensively studied during the last years, cf. Ting (1963), Dunn and Fosdick (1974), Fosdick and Rajagopal (1980), Cioranescu and Ouazar (1984), Galdi, Grobbelaar and Sauer (1993), Coscia and Galdi (1994), Sequeira and Videman (1995). However, concerning the question of existence of classical solutions, the answer was shown to be affirmative only very recently. First,

---

[1]To be precise, a third grade fluid, with its constitutive law $\mathbf{T} = -p\mathbf{I} + \mu\mathbf{A}_1 + \alpha_1\mathbf{A}_2 + \alpha_2\mathbf{A}_1^2 + \beta_1\mathbf{A}_3 + \beta_2[\mathbf{A}_1\mathbf{A}_2 + \mathbf{A}_2\mathbf{A}_1] + \beta_3(\mathrm{tr}\mathbf{A}_1^2)\mathbf{A}_1$ is a complexity 3 fluid. However, if we assume the thermodynamical conditions of Fosdick and Rajagopal (1980), then, in particular, $\beta_1 = \beta_2 = 0$.

Galdi, Grobbelaar and Sauer (1993), using an approach that gave the idea to this present generalization, studied the second grade fluids and showed existence and uniqueness of classical solutions for short time. Moreover, their results hold globally (in time) if α_1 is positive and sufficiently large, and if the data is suitably restricted in size. Later, Galdi and Sequeira (1994) managed to remove this rather artificial condition on α_1 and proved, using a somewhat different method of decomposing the problem, existence of a global unique classical solution, for any $\alpha_1 > 0$. The corresponding steady problem for second grade fluids was investigated by Coscia and Galdi (1994). Finally, in their very recent work, Sequeira and Videman (1995), considered the system of equations of a third grade fluid and showed that also this problem admits a unique global classical solution for small data.

The particular class of non-Newtonian fluids that we shall discuss here is defined by a constitutive equation of the form

$$\mathbf{T} = -p\mathbf{I} + \mu_\infty \mathbf{A}_1 + \mu_0 |\mathbf{A}_1|^{q-2}\mathbf{A}_1 + \alpha_1 \mathbf{A}_2 + \alpha_2 \mathbf{A}_1^2 \tag{1.4}$$

where $|\mathbf{A}_1| = (\mathrm{tr}\ \mathbf{A}_1^T\mathbf{A}_1)^{1/2}$ denotes the usual tensor (trace) norm, $\mu_\infty, \mu_0, \alpha_1$ and α_2 are material coefficients and $q \in (1, +\infty)$. These fluids belong also to the class of differential type fluids of complexity 2. Notice that setting in (1.4) $q = 2$ and $q = 4$ we find, respectively, the constitutive laws of a second and a third grade fluid, (1.2) and (1.3).

Amann (1994) studied a class of complexity 1 fluids including as a special case the fluids arising from (1.4), if we set $\alpha_1 = 0$. Also Málek, Rajagopal and Růžička (1995) have recently considered a class of complexity 1 fluids. Their study contains, in particular, all the commonly used power-law models, for instance, the model found from (1.4) by putting $\alpha_1 = \alpha_2 = 0$. None of these results cannot, however, be recovered nor complemented here, since the class of fluids of complexity 1 does not fit into our general setting, i.e. we must assume that $\alpha_1 \neq 0$. [2]

After this introduction, we go, in Section 2, briefly through our notations and formulate a system of evolution equations. Then, we give an outline of our results, namely, that this system admits a unique classical solution, global in time, for sufficiently smooth and small data. In Section 3 we consider a particular set of nonlinear equations, governing the non-steady motion of an incompressible non-Newtonian fluid defined by (1.4). We conclude, from our results for the general problem, that, if $q \geq 2$ and $\alpha_1 > 0$, this set of equations has a unique, global classical solution for sufficiently smooth and small data. It is worth of noticing here, that the global (in time) existence is established, for the first time for any complexity 2 fluid, in a (not necessarily simply) connected flow domain. Moreover, the rest state of the solution is asymptotically stable. Furthermore, we show that the solution exists also for sufficiently large negative values of α_1, for a short period of time, where the lenght of this time period depends only on the size of the data. Finally, we point out that for the uniqueness we do not need the adherence boundary condition, but only that the normal component of $\mathbf{v}$ vanishes at the boundary. Similar results hold also for a system of equations modeling the steady flow of a fluid given by (1.4). For a more extensive treatment, with complete proofs, including also the steady case, we refer to Coscia, Sequeira and Videman (1995).

---

[2]As a consequence, neither can we study the linearly viscous Navier-Stokes fluid here.

2. ABSTRACT PROBLEM

First, let us introduce our notations. In all that follows Ω denotes a bounded domain in $\mathbf{R}^n$, $n=2$ or 3, having a sufficiently smooth boundary $\partial\Omega$. For scalar-valued measurable functions having distributional derivatives up to order m, $m \geq 0$ integer, we define the usual Hilbert spaces $H^m(\Omega)$. For vector-valued functions the corresponding spaces are denoted by the boldface letter, e.g. $\mathbf{H}^m(\Omega) = [H^m(\Omega)]^n$, a convention that will be used repeatedly in the sequel. The inner product and the norm in $H^m(\Omega)$ are denoted, respectively, by $(\cdot,\cdot)_m$ and $\|\cdot\|_m$. We will also need the following two closed subspaces of $\mathbf{H}^m(\Omega)$

$$\mathbf{V}_m = \{\mathbf{v} \in \mathbf{H}^m(\Omega) \mid \nabla \cdot \mathbf{v} = 0\}, \qquad \mathbf{X}_m = \{\mathbf{v} \in \mathbf{V}_m \mid \mathbf{v} \cdot \mathbf{n} = 0 \text{ on } \partial\Omega\},$$

where $\mathbf{n}$ is the exterior unit normal to $\partial\Omega$. They are both Hilbert spaces equipped with the inner product (and the norm) of $\mathbf{H}^m(\Omega)$. Let then $(Y, \|\cdot\|_Y)$ be a Banach space of functions defined on Ω and let $I = (t_0, t_0 + T)$, $t_0 \geq 0$, $T > 0$ be a given time interval. For $1 \leq q < \infty$ we define the Banach space $L^q(I;Y)$ as being the space of all measurable functions $v : t \to v(t) \in Y$ for which the norm $\int_I \|v(t)\|_Y^q dt$ is finite. For $q = \infty$, $L^\infty(I;Y)$ is the Banach space of all measurable, essentially bounded functions defined on I with values in Y. Furthermore, for $k \geq 0$, we denote by $\|\cdot\|_{k,m,I}$ the usual norm in the Sobolev space $W^{k,\infty}(I; H^m(\Omega))$ and, in particular, for $k = 0$, i.e. in $L^\infty(I; H^m(\Omega))$, the norm will be written as $\|\cdot\|_{m,I}$. If $I = (0,T)$, $T > 0$, we simplify the notation and write $\Omega_T = \Omega \times (0,T)$, $\partial\Omega_T = \partial\Omega \times (0,T)$ and denote by $\|\cdot\|_{k,m,T}$ the norm in $W^{k,\infty}(0,T; H^m(\Omega))$. Finally, $C^m(I;Y)$ denotes the space of m times continuously differentiable functions on the closed interval $I = [t_0, t_0 + T]$ with values in Y.

Let us consider the problem

$$
\begin{cases}
\dfrac{\partial}{\partial t}\mathbf{A}(\mathbf{v}) + \lambda\mathbf{B}(\mathbf{v}) + \mathbf{v} \cdot \nabla\mathbf{A}(\mathbf{v}) + \mathbf{N}(\mathbf{v}) + \nabla p = \mathbf{f} & \text{in } \Omega_T \\[2mm]
\nabla \cdot \mathbf{v} = 0 & \\[2mm]
\mathbf{v} = 0 & \\
& \text{on } \partial\Omega_T \\
BC & \\[2mm]
\mathbf{v}(\mathbf{x},0) = \mathbf{v}_0(\mathbf{x}) & \text{in } \Omega,
\end{cases}
\tag{2.1}
$$

where $\mathbf{A}$ and $\mathbf{B}$ are bounded linear differential operators [3] with constant coefficients of order l, $l \geq 2$, mapping, for fixed $t \in (0,T)$ and for any $m \geq 0$, $\mathbf{H}^{m+l}(\Omega)$ into $\mathbf{H}^m(\Omega)$, and $\mathbf{N}$ is a nonlinear differential operator of order l_1, $l_1 \leq l$, that maps, for fixed $t \in (0,T)$ and for any $m \geq 0$, $\mathbf{H}^{m+l_1}(\Omega)$ into $\mathbf{H}^m(\Omega)$. Moreover, $\mathbf{f}$ and $\mathbf{v}_0$ are given vector fields, λ is a real-valued constant and by BC we denote the set of boundary conditions that may eventually be needed when the order l of the operators $\mathbf{A}$ and $\mathbf{B}$ is strictly greater than 2.

We are concerned with the existence of a unique classical solution for the problem (2.1). Towards this aim, we recall that, in view of Helmholtz decomposition, it is possible to express any vector in $\mathbf{H}^m(\Omega)$ as the $\mathbf{L}^2(\Omega)$-orthogonal sum of a vector in $\mathbf{X}_m$ and of the gradient of a scalar function in $H^{m+1}(\Omega)$. Bearing this in mind, we may formally write, for any $\mathbf{v} \in \mathbf{H}^{m+l}(\Omega)$

$$\mathbf{A}(\mathbf{v}) = \mathbf{w} + \nabla\pi, \tag{2.2}$$

---

[3]Here, and in the sequel, whenever we deal with differential operators we solely mean differentiation with respect to the spatial variable.

where $\mathbf{w} \in \mathbf{X}_m$ and $\nabla\pi \in \mathbf{H}^m(\Omega)$. Using the substitution (2.2) in (2.1), the original nonlinear problem can be reformulated as a coupled system of the following two auxiliary linear problems

$$\left\{ \begin{array}{ll} \mathbf{A}(\mathbf{v}) = \nabla\pi + \mathbf{w} & \\ & \text{in } \Omega_T \\ \nabla \cdot \mathbf{v} = 0 & \\ \mathbf{v} = 0 & \\ & \text{on } \partial\Omega_T \\ BC & \end{array} \right. \tag{2.3}$$

$$\left\{ \begin{array}{ll} \dfrac{\partial}{\partial t}\mathbf{w} + \mathbf{v} \cdot \nabla\mathbf{w} + \lambda\mathbf{w} + \nabla\tau & \\ \qquad = \lambda(\mathbf{A}(\mathbf{v}) - \mathbf{B}(\mathbf{v})) - \mathbf{v} \cdot \nabla\nabla\pi - \mathbf{N}(\mathbf{v}) + \mathbf{f} & \text{in } \Omega_T \\ \nabla \cdot \mathbf{w} = 0 & \\ \mathbf{w} \cdot \mathbf{n} = 0 & \text{on } \partial\Omega_T \\ \mathbf{w}(\mathbf{x}, 0) = \mathbf{w}_0(\mathbf{x}) := P(\mathbf{A}(\mathbf{v}))|_{t=0} & \text{in } \Omega, \end{array} \right. \tag{2.4}$$

where $\tau = p + \lambda\pi + \partial_t\pi$, and where P is the orthogonal projection of $\mathbf{H}^m(\Omega)$ onto $\mathbf{X}_m$.

Now, let us suppose, for a moment, that the problems (2.3) and (2.4) are independent of each other. Considering first the solvability of the problem (2.3), with $\mathbf{w}$ replaced by a prescribed function φ, i.e.

$$\left\{ \begin{array}{ll} \mathbf{A}(\mathbf{v}) = \nabla\pi + \varphi & \\ & \text{in } \Omega_T \\ \nabla \cdot \mathbf{v} = 0 & \\ \mathbf{v} = 0 & \\ & \text{on } \partial\Omega_T, \\ BC & \end{array} \right. \tag{2.5}$$

and assuming then $(\mathbf{v}, \nabla\pi)$, the solution to the system (2.5), along with $\mathbf{f}$ and $\mathbf{w}_0$, as given data for the problem (2.4), we observe that, formally, we may introduce the mapping

$$\mathcal{F} : \varphi \to \mathbf{w} \tag{2.6}$$

as being the composite map $\varphi \to \mathbf{v} \to \mathbf{w}$, defined by the auxiliary problems. Now, the solution to the coupled system (2.5)-(2.4) satisfies the original problem (2.1) if we can show that the map $\mathcal{F}$ admits a fixed point in some function class G of sufficiently smooth functions.

To our purpose, it is sufficient to assume that the problem (2.5) is uniquely solvable, along with a strong regularity estimate. Therefore, we make our first

Assumption 2.1. *Let k and m be non-negative integers. The problem (2.5), with φ prescribed in $W^{k,\infty}(0, T; \mathbf{X}_m)$, admits, for all $T > 0$, a unique solution*

$$(\mathbf{v}, \nabla\pi) \in W^{k,\infty}(0, T; \mathbf{X}_{m+l}) \times W^{k,\infty}(0, T; \mathbf{H}^m(\Omega)).$$

Moreover there exists a constant $C > 0$ such that

$$\|\mathbf{v}\|_{k,m+l,T} + \|\nabla\pi\|_{k,m,T} \leq C\|\varphi\|_{k,m,T}. \tag{2.7}$$

Next, let us turn our attention to the second auxiliary problem, formulated, for reasons that become obvious in the sequel, in a space-time region $\Omega \times (t_0, t_0 + T)$

$$
\begin{cases}
\dfrac{\partial \mathbf{w}}{\partial t} + \mathbf{v} \cdot \nabla \mathbf{w} + \lambda \mathbf{w} + \nabla \tau \\[2mm]
\qquad = \lambda(\mathbf{A}(\mathbf{v}) - \mathbf{B}(\mathbf{v})) - \mathbf{v} \cdot \nabla \nabla \pi - \mathbf{N}(\mathbf{v}) + \mathbf{f} & \text{in } \Omega \times I \\[2mm]
\nabla \cdot \mathbf{w} = 0 \\[2mm]
\mathbf{w} \cdot \mathbf{n} = 0 & \text{on } \partial \Omega \times I \\[2mm]
\mathbf{w}(\mathbf{x}, t_0) = \mathbf{w}_{t_0}(\mathbf{x}) & \text{in } \Omega.
\end{cases}
\tag{2.8}
$$

In order to show that the problem (2.8) is well-posed it is enough to assume that the nonlinear operator $\mathbf{N}$ is bounded. However, since later on we are forced to make further restrictions on $\mathbf{N}$ we shall give the assumption on $\mathbf{N}$ already in its final form.

Assumption 2.2. *Let $m \geq 1$ be an integer and let $\mathbf{v} \in L^{\infty}(0, T; \mathbf{H}^{m+l}(\Omega))$. There exist constants $C > 0$ and $p > 1$ such that for all $t \in (0, T)$ one has*

$$
\|\mathbf{N}(\mathbf{v})\|_m \leq C \|\mathbf{v}\|_{m+l}^p.
\tag{2.9}
$$

Moreover, for any $\mathbf{v}_1, \mathbf{v}_2 \in L^{\infty}(0, T; \mathbf{H}^{l+1}(\Omega))$, satisfying $\|\mathbf{v}_i\|_{l+1,T} \leq D$, $i = 1, 2$, for some $D > 0$, there exists a constant $L > 0$, depending on D, such that for all $t \in (0, T)$ it holds

$$
\|\mathbf{N}(\mathbf{v}_1) - \mathbf{N}(\mathbf{v}_2)\|_0 \leq L \|\mathbf{v}_1 - \mathbf{v}_2\|_l.
\tag{2.10}
$$

Now, making use of the estimates proved in Galdi, Grobbelaar and Sauer (1993), see also Coscia and Galdi (1994), and employing the Galerkin method (see *e.g.* Lions, 1969), we can prove the following

Theorem 2.1. *Let $t_0 \geq 0$ and $T > 0$ be arbitrary. Suppose that, for $m \geq 1$, it holds*

$\mathbf{w}_{t_0} \in \mathbf{X}_m,$

$\mathbf{f} \in L^{\infty}(I; \mathbf{H}^m(\Omega)),$ with $\|\mathbf{f}\|_{m,I} \leq M_1,$

$(\mathbf{v}, \nabla \pi) \in L^{\infty}(I; \mathbf{X}_{m+l}) \times L^{\infty}(I; \mathbf{H}^m(\Omega)),$ with $\|\mathbf{v}\|_{m+l,I} + \|\nabla \pi\|_{m,I} \leq M_2,$

where $M_1, M_2 > 0$. Assume further that $\mathbf{v} = 0$ at $\partial \Omega_T$. Then the problem (2.8) has a unique solution $(\mathbf{w}, \nabla \tau)$ satisfying

$$\mathbf{w} \in L^{\infty}(I; \mathbf{X}_m) \cap W^{1,\infty}(I; \mathbf{X}_{m-1})$$

$$\nabla \tau \in L^{\infty}(I; \mathbf{H}^m(\Omega))$$

$$\|\mathbf{w}\|_{m,I} + \left\|\frac{\partial \mathbf{w}}{\partial t}\right\|_{m-1,I} + \|\nabla \tau\|_{m,I} \leq C,$$

where the constant $C = C(m, \Omega, M_1, M_2, \lambda, \|\mathbf{w}_{t_0}\|_m, T) > 0$. $\qquad \square$

Now, we are ready to turn back and consider the original initial-boundary value problem (2.1). Applying Schauder's fixed point theorem to the composite mapping (2.6), we shall prove that there exists, at any time interval $I = (t_0, t_0 + T)$ and for any smooth enough data, a local regular solution for the problem (2.1). Moreover, the regularity of the solution depends only on the smoothness of the data, and T is independent of t_0.

First, let us recall the following version of Schauder's fixed point theorem

Theorem 2.2. *Let G be a closed convex set in a Banach space Y, and let $\mathcal{F}$ be a continuous operator from G into Y such that $\mathcal{F}G \subset G$ and such that the closure of $\mathcal{F}G$ is compact. Then $\mathcal{F}$ has a fixed point in G.* $\square$

Hence, in order to apply Schauder's fixed point method we must choose an appropriate Banach space to work with. To this end, for all $m \geq 1$, let us introduce

$$Y_{m-1} = C(I; \mathbf{X}_{m-1})$$

and define G as a following subset of Y

$$G = \{\boldsymbol{\varphi} \in Y_{m-1} \mid \boldsymbol{\varphi} \in L^\infty(I; \mathbf{H}^m(\Omega)), \ \|\boldsymbol{\varphi}\|_{m,I} \leq D, \ \boldsymbol{\varphi}(\mathbf{x}, t_0) = \mathbf{w}(\mathbf{x}, t_0) \in \mathbf{X}_m\},$$

where $D > 0$ is an arbitrary constant.

Next, we show that the composite map $\mathcal{F}$ defined by the auxiliary problems (2.5) and (2.8) satisfies the hypotheses of the Schauder's fixed point theorem and thus admits a fixed point in G. We point out that, besides Assumption 2.1, we now need to use the Lipschitzian property (2.10) of the nonlinear operator $\mathbf{N}$. For the proof, see Coscia, Sequeira and Videman (1995).

Theorem 2.3. (Local Existence) *Let $\lambda \in \mathbf{R}$ and $t_0 \geq 0$ and $D > 0$ be arbitrary. Let further $m \geq 1$ and suppose that $\mathbf{w}(\mathbf{x}, t_0) = \mathbf{w}_{t_0} \in \mathbf{X}_m$ and $\mathbf{f} \in L^\infty(I; \mathbf{H}^m(\Omega))$. Then provided $\|\mathbf{w}_{t_0}\|_m \leq D_1 < D$ and $\|\mathbf{f}\|_{m,I} \leq D$, there exists $T > 0$ such that problem (2.1) admits, on the time interval $I = (t_0, t_0 + T)$, a unique solution $(\mathbf{v}, \nabla p)$ satisfying*

$$\mathbf{v} \in C(I; \mathbf{X}_{m+l-1}) \bigcap L^\infty(I; \mathbf{H}^{m+l}(\Omega))$$

$$\frac{\partial \mathbf{v}}{\partial t} \in L^\infty(I; \mathbf{H}^{m+l-1}(\Omega)) \qquad (2.11)$$

$$\nabla p \in L^\infty(I; \mathbf{H}^m(\Omega)).$$

$\square$

Remark 2.1. We wish to emphasize that the uniqueness result in the previous theorem remains valid also with a more relaxed boundary condition on $\mathbf{v}$, namely $\mathbf{v}_* \cdot \mathbf{n} = 0$ at $\partial\Omega_T$, where $\mathbf{v}_*$ is a prescribed function at $\partial\Omega_T$. Moreover, we observe that the local existence theorem is, in fact, valid for any operator $\mathbf{B}$, of order $\leq l$, satisfying (only) the Lipschitz condition (2.10). See Coscia, Sequeira and Videman (1995), for more details.

The preceding theorem states that, for any $t_0 \geq 0$, there exists a positive period of time T, depending only on the upper bound of the $\mathbf{H}^m(\Omega)$-norm of $\mathbf{w}_{t_0}$, such that problem (2.1) admits a unique solution on the time interval $I = (t_0, t_0 + T)$. In order to establish an existence theorem, valid for all time T, we need to make further assumptions on the linear differential operators $\mathbf{A}$ and $\mathbf{B}$. Hence, we give the following

Assumption 2.3. *Let $l \geq 2$ be an even integer.*
Suppose that $\mathbf{u}, \mathbf{v} \in L^\infty(0, T; \mathbf{H}^l(\Omega))$ and that $\mathbf{u}$ and $\mathbf{v}$ satisfy the boundary conditions in (2.1). Then, for all $t \in (0, T)$, $\mathbf{u}, \mathbf{v} \mapsto (\mathbf{A}(\mathbf{u}), \mathbf{v})_0$ is a symmetric bilinear form. Moreover, there exist positive constants k_1 and k_2 such that the operators $\mathbf{A}$ and $\mathbf{B}$ satisfy, for all $t \in (0, T)$, the following conditions

$$(\mathbf{A}(\mathbf{v}), \mathbf{v})_0 \geq k_1 \|\mathbf{v}\|_{\frac{l}{2}}^2 \qquad (2.12)$$

$$(\mathbf{B}(\mathbf{v}), \mathbf{v})_0 \geq k_2 \|\mathbf{v}\|_{\frac{1}{2}}^2. \qquad (2.13)$$

Finally, let $m \geq 1$ be an integer and let $\mathbf{v} \in L^\infty(0, T; \mathbf{H}^{m+l}(\Omega))$. Then it holds, for all $t \in (0, T)$

$$\|\mathbf{A}(\mathbf{v}) - \mathbf{B}(\mathbf{v})\|_m \leq C \|\mathbf{v}\|_{m+\frac{l}{2}-1}, \qquad (2.14)$$

where $C > 0$ is a constant.

Now, assuming that $\lambda > 0$, we may prove, see Coscia, Sequeira and Videman (1995), the following global a *priori* estimate for the solution $(\mathbf{v}, \nabla p)$ of (2.1).

Theorem 2.4. *Let $m \geq 1$ be an integer. Suppose that $\lambda > 0$ and that $\mathbf{v}_0 \in \mathbf{X}_{m+l}$ and $\mathbf{f} \in L^\infty(0, T; \mathbf{H}^m(\Omega))$. Assume that $(\mathbf{v}, \nabla p, \mathbf{w}, \nabla\tau)$ solves the problem (2.3)-(2.4) with*

$$\mathbf{v} \in L^\infty(0, T; \mathbf{X}_{m+l}) \cap W^{1,\infty}(0, T; \mathbf{H}^{m+l-1}(\Omega)) \cap C(0, T; \mathbf{X}_{m+l-1})$$

$$\nabla p \in L^\infty(0, T; \mathbf{H}^m(\Omega))$$

$$\mathbf{w} \in L^\infty(0, T; \mathbf{X}_m) \cap W^{1,\infty}(0, T; \mathbf{H}^{m-1}(\Omega)) \cap C(0, T; \mathbf{X}_{m-1})$$

$$\nabla\tau \in L^\infty(0, T; \mathbf{H}^m(\Omega)).$$

There exist constants $\varepsilon_i = \varepsilon_i(m, \Omega, \lambda) > 0$, $i = 1, 2$, such that, if $\|\mathbf{v}_0\|_{m+l} < \varepsilon_1$, and $\int_0^T \|\mathbf{f}(s)\|_m^2 ds < \varepsilon_2$, then one has the estimates

$$\|\mathbf{v}\|_{m+l,T}^2 + \int_0^T \|\mathbf{v}(s)\|_{m+l}^2 ds \leq \delta_1 \int_0^T \|\mathbf{f}(s)\|_m^2 ds + \delta_2 \|\mathbf{v}_0\|_{m+l}^2,$$

$$\|\frac{\partial \mathbf{v}}{\partial t}\|_{m+l-1,T} \leq \delta_3, \qquad (2.15)$$

where $\delta_i = \delta_i(m, \Omega, \lambda)$, $i = 1, 2$, and $\delta_3 \to 0$ when $\|\mathbf{v}_0\|_{m+l} \to 0$ and $\|\mathbf{f}\|_{m,T} \to 0$. $\qquad \square$

The global existence theorem is now a simple consequence of the two preceding ones.

Theorem 2.5. (Global Existence) *Let $T > 0$ be arbitrary, let $\lambda > 0$ and $l \geq 2$ and assume that $\mathbf{v}_0 \in \mathbf{X}_{m+l}$ and $\mathbf{f} \in L^\infty(0, T; \mathbf{H}^m(\Omega))$, for some $m \geq 1$. Then, there exists a $\varepsilon = \varepsilon(m, \Omega, \lambda) > 0$ such that, provided $\|\mathbf{v}_0\|_{m+l} < \varepsilon$ and $\int_0^T \|\mathbf{f}(s)\|_m^2 ds < \varepsilon$, the problem (2.1) admits a unique global solution $(\mathbf{v}, \nabla p)$ satisfying*

$$\mathbf{v} \in C(0, T; \mathbf{X}_{m+l-1}) \cap L^\infty(0, T; \mathbf{H}^{m+l}(\Omega))$$

$$\frac{\partial \mathbf{v}}{\partial t} \in L^\infty(0, T; \mathbf{H}^{m+l-1}(\Omega))$$

$$\nabla p \in L^\infty(0, T; \mathbf{H}^m(\Omega)).$$

Proof: Letting $t_0 \geq 0$ be arbitrary, we recall from Theorem 2.3 that there exists a $T_* > 0$, independent of t_0, such that the problem (2.1) admits a unique solution $\mathbf{v} \in L^\infty(I; \mathbf{X}_{m+l}) \cap C(I; \mathbf{X}_{m+l-1})$, for all $t \in I = (t_0, t_0 + T_*)$. Since this implies, moreover, that $\|\mathbf{v}(t)\|_{m+l}$ is finite for all $t \in I = [t_0, t_0 + T_*]$, we may extend the solution progressively to any time interval $[0, T]$, starting from $[0, T_*]$ and applying the a *priori* estimate (2.15). $\qquad \square$

Finally, let us show that, provided the data is regular enough, the solution obtained in the previous theorem is indeed a classical one (see Coscia, Sequeira and Videman, 1995).

Theorem 2.6. *Let $\lambda > 0$ and let $l \geq 2$ be an even integer. Suppose that $\mathbf{v}_0 \in \mathbf{X}_{l+5}$ and $\mathbf{f} \in L^\infty(0,T;\mathbf{H}^5(\Omega)) \cap W^{1,\infty}(0,T;\mathbf{H}^3(\Omega))$ and that the Assumptions 2.1-2.3 are valid, for $m = 5$. Assume further that $\mathbf{N}(\mathbf{v}) \in L^\infty(0,T;\mathbf{H}^5(\Omega)) \cap W^{1,\infty}(0,T;\mathbf{H}^3(\Omega))$. Then, there exists a $\varepsilon = \varepsilon(\Omega,\lambda) > 0$ such that if*

$$\|\mathbf{v}_0\|_{l+5} < \varepsilon \qquad and \qquad \int_0^T \|\mathbf{f}(s)\|_5^2 ds < \varepsilon,$$

then the initial-boundary value problem (2.1) has a unique classical solution

$$\mathbf{v} \in C^1(0,T;\mathbf{C}^{l+1}(\Omega))$$

$$\nabla p \in C(0,T;\mathbf{C}(\Omega)),$$

for all $T > 0$. □

Remark 2.2. If we assume that the vector field $\mathbf{f}$ is conservative, we easily deduce that there exists a positive constant k such that

$$\frac{d}{dt}E(t) + kE(t) \leq 0,$$

where $E(t) = \|\mathbf{v}\|_{\frac{l}{2}}^2 + \|\mathbf{w}\|_1^2$. This implies that the solution $\mathbf{v}$ is exponentially stable in the sense that there exist positive constants $\sigma_1, \sigma_2 > 0$ such that, for all $t \geq 0$

$$\|\mathbf{v}\|_{l+1} \leq \sigma_1 \|\mathbf{v}_0\|_{l+1} e^{-\sigma_2 t}.$$

See Coscia, Sequeira and Videman (1995) for details.

3. APPLICATIONS

There exist numerous substances which exhibit flow behaviour that cannot be properly described by the linearly viscous (Newtonian) fluid model. Amongst these materials are e.g. biological fluids, polymeric liquids, liquid crystals, geological materials and many food products. During the past decades several constitutive relations have been suggested in order to capture the non-Newtonian characteristics of fluids (cf. e.g. Huilgol, 1975; Schowalter, 1978). A well accepted, and widely applied, class of constitutive assumptions are fluids of differential type of complexity n, cf. Rivlin and Ericksen (1955), Truesdell and Noll (1965).

In this section we shall study the issues concerning the well-posedness of the equations of motion of some differential type fluids of complexity 2. We recall that in an incompressible, homogeneous differential type fluid of complexity 2, the Cauchy stress tensor $\mathbf{T}$ is given by

$$\mathbf{T} = -p\mathbf{I} + \mathbf{S}(\mathbf{A}_1, \mathbf{A}_2),$$

where $-p\mathbf{I}$ is the spherical part of the stress due to the constraint of incompressibility and $\mathbf{S}$ is a tensor-valued function of the first two Rivlin-Ericksen tensors $\mathbf{A}_1$ and $\mathbf{A}_2$ defined by[4]

$$\mathbf{A}_1 = \nabla\mathbf{v} + (\nabla\mathbf{v})^T$$

$$\mathbf{A}_2 = \frac{d}{dt}\mathbf{A}_1 + (\nabla\mathbf{v})^T\mathbf{A}_1 + \mathbf{A}_1\nabla\mathbf{v}.$$

---

[4]Generally, $\mathbf{S}$ may also depend on temperature and its gradient. Nevertheless, in the sequel, we will only consider the isothermal case, i.e. we suppose that the temperature field is uniform over Ω and constant in time for all $t \geq 0$.

The particular complexity 2 fluids that we shall be discussing here have constitutive equations of the form

$$\mathbf{T} = -p\mathbf{I} + \mu_\infty\mathbf{A}_1 + \mu_0|\mathbf{A}_1|^{q-2}\mathbf{A}_1 + \alpha_1\mathbf{A}_2 + \alpha_2\mathbf{A}_1^2, \tag{3.1}$$

where $|\mathbf{A}_1| = (\operatorname{tr}\mathbf{A}_1^2)^{1/2}$ and $\mu_\infty, \mu_0, \alpha_1, \alpha_2$ are (constant) material coefficients. We shall call these fluids *second grade fluids with shear-dependent viscosity*. With $\mu_\infty = 0$, they have sometimes been referred to as modified second grade fluids or power-law fluids of grade 2, and have been used, for instance, in modeling the flow of glaciers (cf. Man, 1992; Man and Sun, 1987). We recall that, if in (3.1) we choose $q = 2$ (or alternatively $\mu_0 = 0$) we end up with the constitutive law of a second grade fluid (1.2) and if we set $q = 4$ we find the third grade fluid model (1.3).

The equations of motion for an incompressible, homogeneous second grade fluid with shear-dependent viscosity, transform into

$$\begin{cases} \dfrac{\partial}{\partial t}(\rho\mathbf{v} - \alpha_1\Delta\mathbf{v}) - \mu_\infty\Delta\mathbf{v} + \mathbf{v}\cdot\nabla(\rho\mathbf{v} - \alpha_1\Delta\mathbf{v}) - \\[2mm] \qquad -\alpha_1(\nabla\mathbf{v})^T : \nabla\mathbf{A}_1 - \alpha_1\nabla\cdot[(\nabla\mathbf{v})^T\mathbf{A}_1 + \mathbf{A}_1\nabla\mathbf{v}] - \quad \text{in } \Omega_T \\[2mm] \qquad -\alpha_2\nabla\cdot\mathbf{A}_1^2 - \mu_0\nabla\cdot[|\mathbf{A}_1|^{q-2}\mathbf{A}_1] + \nabla p = \rho\mathbf{f} \\[2mm] \nabla\cdot\mathbf{v} = 0 \\[2mm] \mathbf{v} = 0 \qquad\qquad\qquad\qquad\qquad\qquad\qquad\qquad\quad \text{on } \partial\Omega_T \\[2mm] \mathbf{v}(\mathbf{x},0) = \mathbf{v}_0(\mathbf{x}) \qquad\qquad\qquad\qquad\qquad\qquad\quad \text{in } \Omega, \end{cases} \tag{3.2}$$

where $\rho > 0$ is the density of the fluid, $\mathbf{f}$ a given body force per unit mass of fluid and $\mathbf{v}_0$ a prescribed initial velocity distribution. Dividing the equations $(3.2)_1$ by the constant density ρ, denoting by ν_∞ and ν_0 the kinematic viscosity coefficients and defining their second-grade analogues by

$$\lambda_1 = \frac{\alpha_1}{\rho} \qquad\qquad \lambda_2 = \frac{\alpha_2}{\rho},$$

the system of equations (3.2) becomes

$$\begin{cases} \dfrac{\partial}{\partial t}(\mathbf{v} - \lambda_1\Delta\mathbf{v}) - \nu_\infty\Delta\mathbf{v} + \mathbf{v}\cdot\nabla(\mathbf{v} - \lambda_1\Delta\mathbf{v}) + \\[2mm] \qquad\qquad\qquad +\mathbf{N}_1(\mathbf{v}) + \mathbf{N}_2(\mathbf{v}) + \nabla p = \mathbf{f} \quad \text{in } \Omega_T \\[2mm] \nabla\cdot\mathbf{v} = 0 \\[2mm] \mathbf{v} = 0 \qquad\qquad\qquad\qquad\qquad\qquad\qquad\quad \text{on } \partial\Omega_T \\[2mm] \mathbf{v}(\mathbf{x},0) = \mathbf{v}_0(\mathbf{x}) \qquad\qquad\qquad\qquad\qquad\quad \text{in } \Omega, \end{cases} \tag{3.3}$$

where the density has been absorbed into the pressure and where we have defined the nonlinear terms $\mathbf{N}_i(\mathbf{v}), i = 1, 2$, by

$$\begin{aligned} \mathbf{N}_1(\mathbf{v}) &= -\lambda_1(\nabla\mathbf{v})^T : \nabla\mathbf{A}_1 - \lambda_1\nabla\cdot[\mathbf{A}_1\mathbf{W} - \mathbf{W}\mathbf{A}_1] - (\lambda_1 + \lambda_2)\nabla\cdot\mathbf{A}_1^2 \\ \mathbf{N}_2(\mathbf{v}) &= -\nu_0\nabla\cdot[|\mathbf{A}_1|^{q-2}\mathbf{A}_1], \end{aligned} \tag{3.4}$$

with $\mathbf{W}$ denoting the skew-symmetric part of $\nabla\mathbf{v}$.

Dunn (1982) discussed the thermodynamics and mechanical stability of the rest state for a class of complexity 2 fluids for which the stress $\mathbf{T}$ is given by

$$\mathbf{T} = -p\mathbf{I} + \tilde{\mu}(\theta, \mathbf{A}_1)\mathbf{A}_1 + \tilde{\alpha_1}(\theta, \mathbf{A}_1)\mathbf{A}_2 + \tilde{\alpha_2}(\theta, \mathbf{A}_1)\mathbf{A}_1^2, \tag{3.5}$$

where θ denotes the temperature field. Ignoring the temperature dependence, we observe that for the fluids given by (3.1), the material coefficients $\tilde{\alpha}_1 = \alpha_1$ and $\tilde{\alpha}_2 = \alpha_2$ are absolute constants and

$$\tilde{\mu} = \tilde{\mu}(|\mathbf{A}_1|) = \mu_\infty + \mu_0|\mathbf{A}_1|^{q-2}.$$

Hence, we conclude from the results of Dunn (1982) that, as a consequence of the Clausius-Duhem inequality and of the assumption that the Helmholtz free energy is a minimum in equilibrium, the second grade fluids with shear-dependent viscosity are compatible with the thermodynamics if

$$\tilde{\mu}(|\mathbf{A}_1|) \geq 0$$

$$\alpha_1 \geq 0 \tag{3.6}$$

$$|\alpha_1 + \alpha_2| \leq \sqrt{6}\, \frac{\tilde{\mu}(|\mathbf{A}_1|)}{|\mathbf{A}_1|}, \quad \text{for all} \quad |\mathbf{A}_1| \neq 0.$$

Remark 3.1. If, in (3.1), we set $q = 2$, then $\tilde{\mu} = \mu_\infty + \mu_0$ is a constant and the conditions (3.6) reduce to the well-known thermodynamical restrictions for second-grade fluids, i.e. $\mu \geq 0$, $\alpha_1 \geq 0$ and $\alpha_1 + \alpha_2 = 0$ (cf. Dunn and Fosdick, 1974). On the other hand, if we choose $q = 4$ in (3.1), then the condition $(3.6)_3$ can be written as

$$\mu_0|\mathbf{A}_1|^2 + \frac{|\alpha_1 + \alpha_2|}{\sqrt{6}}|\mathbf{A}_1| + \mu_\infty \geq 0,$$

from which it follows that $|\alpha_1 + \alpha_2| \leq \sqrt{24\mu_\infty\mu_0}$ (cf. Fosdick and Rajagopal, 1980, inequality (4.15c)).

Now, we shall show that the problem (3.3) has a unique global classical solution for any sufficiently small and regular data, provided $\nu_\infty > 0$ and $\lambda_1 > 0$. First, it is easy to see that system (3.3) is of the general form (2.1) with

$$\mathbf{A}(\mathbf{v}) = \mathbf{v} - \lambda_1\Delta\mathbf{v},$$

$$\mathbf{B}(\mathbf{v}) = -\lambda_1\Delta\mathbf{v},$$

$$\lambda = \frac{\nu_\infty}{\lambda_1}, \quad \lambda_1 \neq 0$$

and with $\mathbf{N}_i(\mathbf{v})$, i=1,2, given by (3.4).

Next, let us check the validity of Assumptions 2.1-2.3. If $\lambda_1 > 0$ then Assumption 2.1 follows directly from classical results for the Stokes problem (cf. Cattabriga, 1961; Agmon, Douglis and Nirenberg, 1964). Moreover, we observe that the Stokes (resolvent) system

$$\begin{cases} \mathbf{v} - \lambda_1\Delta\mathbf{v} = \boldsymbol{\varphi} + \nabla\pi & \text{in } \Omega_T \\ \nabla \cdot \mathbf{v} = 0 & \tag{3.7} \\ \mathbf{v} = 0 & \text{on } \partial\Omega_T \end{cases}$$

is solvable also for negative values of λ_1, i.e for negative α_1. To see this, take the $\mathbf{L}^2(\Omega)$-inner product of $(3.7)_1$ with $\mathbf{v}$ and use the Poincaré inequality to obtain the estimate

$$(\frac{|\lambda_1|}{C_P^2} - 1)\|\mathbf{v}\|_0 \leq \|\boldsymbol{\varphi}\|_0, \tag{3.8}$$

with C_P denoting the Poincaré constant of $\mathbf{v}$ in Ω. This estimate is sufficient to guarantee that problem (3.7) admits a unique solution for all negative λ_1 such that $|\lambda_1| > C_P^2$.

Regarding Assumption 2.2, it can be shown without difficulties that, for any $m \geq 1$, it holds

$$\|\mathbf{N}_1(\mathbf{v})\|_m \leq C\|\mathbf{v}\|_{m+2}^2$$

$$\|\mathbf{N}_2(\mathbf{v})\|_m \leq C\|\mathbf{v}\|_{m+2}^{q-1},$$

provided $q > 2$. (If $q = 2$ then $\mathbf{N}_2(\mathbf{v}) = -\nu_0\Delta\mathbf{v}$ is linear and it may be added to the other dissipative term $-\nu_\infty\Delta\mathbf{v}$.) The Lipschitz condition (2.10) follows also directly, as soon as we assume that $q \geq 2$, with the constant L depending on the $\mathbf{H}^3(\Omega)$-norm of $\mathbf{v}_1$ and $\mathbf{v}_2$.

Therefore, we conclude from Theorem 2.3. that the initial-boundary value problem (3.3) admits, at least for short time, a unique regular solution, for any (regular enough) data, provided

$$\lambda_1 > 0 \qquad \text{or} \qquad (\lambda_1 < 0 \text{ and } |\lambda_1| > C_P^2)$$

and $q \geq 2$, independently of the values of the other material coefficients ν_∞, ν_0 and λ_2.

In order to attain global existence, we need to fulfill the requirements of Assumption 2.3 concerning the linear operators $\mathbf{A}$ and $\mathbf{B}$. After a simple integration by parts we obtain

$$(\mathbf{A}(\mathbf{v}), \mathbf{v})_0 = \|\mathbf{v}\|_0^2 + \lambda_1\|\nabla\mathbf{v}\|_0^2,$$

$$(\mathbf{B}(\mathbf{v}), \mathbf{v})_0 = \lambda_1\|\nabla\mathbf{v}\|_0^2.$$

Therefore, in view of the Poincaré inequality, it follows that the coercivity conditions (2.12) and (2.13) are valid, as soon as $\lambda_1 > 0$. Estimate (2.14) is trivially satisfied since $\mathbf{A}(\mathbf{v}) - \mathbf{B}(\mathbf{v}) = \mathbf{v}$.

Taking into account that, if $m \geq 4$, then

$$\frac{\partial\mathbf{v}}{\partial t} \in L^\infty(0, T; \mathbf{X}_{m+1})$$

implies that

$$\frac{\partial\mathbf{N}_i(\mathbf{v})}{\partial t} \in L^\infty(0, T; \mathbf{X}_{m-2}), \qquad i = 1, 2,$$

we may state the following main result

Theorem 3.1. *Let $\lambda_1 > 0, \nu_\infty > 0$ and $q \geq 2$. Furthermore, assume that*

$$\mathbf{v}_0 \in \mathbf{X}_7$$

$$\mathbf{f} \in L^\infty(0, T; \mathbf{H}^5(\Omega)) \cap W^{1,\infty}(0, T; \mathbf{H}^3(\Omega)).$$

There exists $\varepsilon = \varepsilon(\Omega, \nu_\infty, \nu_0, \lambda_1, \lambda_2, q) > 0$ such that if

$$\|\mathbf{v}_0\|_7 < \varepsilon \qquad and \qquad \int_0^T \|\mathbf{f}(s)\|_5^2 ds < \varepsilon,$$

then the initial-boundary value problem (3.3) has a unique classical solution

$$\mathbf{v} \in C^1(0, T; \mathbf{C}^3(\Omega))$$

$$\nabla p \in C(0, T; \mathbf{C}(\Omega)),$$

for all $T > 0$. $\qquad\qquad\qquad\qquad\qquad\qquad\qquad\qquad\qquad\qquad\qquad\qquad\qquad$ $\square$

Remark 3.2. If the body force field is conservative, *e.g.* if the external forces are purely gravitational, we conclude from Remark 2.2. that the rest state of the solution **v** is asymptotically stable, i.e.

$$\|\mathbf{v}\|_3 \leq \sigma_1 \|\mathbf{v}_0\|_3 e^{-\sigma_2 t},$$

where $\sigma_1 > 0$ and $\sigma_2 > 0$.

Remark 3.3. Since the fluid model (3.1) contains the second and third grade fluids as particular cases, and since these fluids have attracted a lot of attention during the past decades, it is worth of emphasizing again that all the previous results are valid for the fluids of grade 2 and 3.

REFERENCES

AGMON, S., A. DOUGLIS and L. NIRENBERG (1964), Estimates near the boundary for solutions of elliptic partial differential equations satisfying general boundary conditions. II, *Comm. Pure Appl. Math.* **17**, 35-92.

AMANN, H. (1994), Stability of the Rest State of a Viscous Incompressible Fluid, *Arch. Rational Mech. Anal.* **126**, 231-242.

BEIRÃO DA VEIGA, H. (1980), On an Euler type equation in hydrodynamics, *Ann. Mat. Pura App.* **125**, 279-324.

CATTABRIGA, L. (1961), Su un problema al contorno relativo al sistema di equationi di Stokes, *Rend. Sem. Mat. Univ. Padova* **31**, 308-340.

CIORANESCU, D. and E.H. OUAZAR (1984), Existence and uniqueness for fluids of second grade, in: *Research Notes in Mathematics*, Vol. **109**, 178-197 (Pitman, Boston).

COLEMAN, B.D. and W. NOLL (1960), An approximation theorem for functionals with applications in continuum mechanics, *Arch. Rational Mech. Anal.* **6**, 355-370.

COSCIA, V. and G.P. GALDI (1994), Existence, Uniqueness and Stability of Regular Steady Motions of a Second Grade Fluid, *Int. J. Non-Linear Mechanics* **29**, 493-506.

COSCIA, V., A. SEQUEIRA and J.H. VIDEMAN (1995), Existence and Uniqueness of Classical Solutions for a Class of Complexity 2 Fluids, *Int. J. Non-Linear Mechanics*, (to appear).

DUNN, J.E. (1982), On the Free Energy and Stability of Nonlinear Fluids, *J. Rheology* **26**, 43-68.

DUNN J.E. and R.L. FOSDICK (1974), Thermodynamics, stability and boundedness of fluids of complexity 2 and fluids of second grade, *Arch. Rational Mech. Anal.* **56**, 191-252.

DUNN, J.E. and K.R. RAJAGOPAL (1995), Fluids of differential type: Critical review and thermodynamic analysis, *Intl. J. Engng. Science* **33**, 689-729.

FOSDICK, R.L. and K.R. RAJAGOPAL (1980), Thermodynamics and stability of fluids of third grade, *Proc. Royal Soc. London* **339**, 351-377.

GALDI, G.P., M. GROBBELAAR and N. SAUER (1993), Existence and Uniqueness of Classical Solutions of the Equations of Motion for Second-Grade Fluids, *Arch. Rational Mech. Anal.* **124**, 221-237.

GALDI, G.P. and A. SEQUEIRA (1994), Further existence results for classical solutions of the equations of a second grade fluid, *Arch. Rational Mech. Anal.* **128**, 297-312.

HUILGOL, R.R. (1975), *Continuum Mechanics of Viscoelastic Liquids*, (Wiley, New York).

LIONS, J.L. (1969), *Quelques méthodes de résolution des problèmes aux limites non linéaires*, (Dunod, Paris).

MÁLEK, J., K.R. RAJAGOPAL and M. RŮŽIČKA (1995), Existence and Regularity of Solutions and the Stability of the Rest State for Fluids with Shear Dependent Viscosity, M^3AS, (to appear).

MAN, C.-S. (1992), Nonsteady Channel Flow of Ice as a Modified Second-Order Fluid with Power-Law Viscosity, *Arch. Rational Mech. Anal.* **119**, 35-57.

MAN, C.-S. and Q.-X. SUN (1987), On the significance of normal stress effects in the flow of glaciers, *J. Glaciology.* **33**, 268-273.

NOVOTNY, A. and M. PADULA (1994), L^p-Approach to Steady Flows of Viscous Compressible Fluids in Exterior Domains, *Arch. Rational Mech. Anal.* **126**, 243-297.

RIVLIN, R.S. and J.L. ERICKSEN (1955), Stress-deformation relations for isotropic materials, *J. Rational Mech. Anal.* **4**, 323-425.

SCHOWALTER, W.R. (1978), *Mechanics of Non-Newtonian Fluids*, (Pergamon Press, New York).

SEQUEIRA, A. and J.H. VIDEMAN (1995), Global existence of classical solutions for the equations of third grade fluids. *J. Math. Phys. Sciences*, (to appear).

TING, T.W. (1963), Certain non-steady flows of second-order fluids, *Arch. Rational Mech. Anal.* **14**, 1-26.

TRUESDELL, C. and W. NOLL (1965), *The Nonlinear Field Theories of Mechanics*, Flügge's Handbuch der Physik, **III/3**, (Springer-Verlag, Heidelberg).

ASYMPTOTIC IN TIME DECAY OF SOLUTIONS TO THE EQUATIONS OF A SECOND-GRADE FLUID FILLING THE WHOLE SPACE

Giovanni P. Galdi and Arianna Passerini

Instituto di Ingegneria, Università di Ferrara
Via Scandiana 21, Ferrara, I-44100, Italy

1. INTRODUCTION

Second grade fluids have been introduced to describe certain nonlinear effects that can not be explained by the classical theory of Navier-Stokes for Newtonian fluids (see *e.g.* Truesdell and Noll, 1965; Dunn and Fosdick, 1974; Rajagopal, 1992). In this regard, the Cauchy stress tensor $\mathbf{T}$ for an incompressible fluid of this type, is related to the kinematical variables by

$$\mathbf{T} = -\tilde{p}\mathbf{I} + \nu\mathbf{A}_1 + \alpha_1\mathbf{A}_2 + \alpha_2\mathbf{A}_1^2,$$

where $\tilde{p}$ is the pressure, ν is the viscosity, α_1 and α_2 the normal stress moduli, and the Rivlin-Ericksen tensors $\mathbf{A}_1$ and $\mathbf{A}_2$ (Rivlin and Ericksen, 1955), are defined by

$$\mathbf{A}_1 = (\operatorname{grad}\mathbf{v}) + (\operatorname{grad}\mathbf{v})^T$$

$$\mathbf{A}_2 = \frac{d\mathbf{A}}{dt} + \mathbf{A}_1(\operatorname{grad}\mathbf{v}) + (\operatorname{grad}\mathbf{v})^T\mathbf{A}_1$$

($\mathbf{v}$ denoting the velocity and $\dfrac{d}{dt}$ the material time derivative).

If a fluid modelled by these relations has to be compatible with thermodynamics in the sense that all flows of the fluid meet the Clausius-Duhem inequality and the assumption that the specific Helmholtz free energy is a minimum when the fluid is in equilibrium, then the viscosity ν and the normal stress moduli α_1, α_2, satisfy

$$\nu \geq 0, \quad \alpha_1 \geq 0, \quad \alpha_1 + \alpha_2 = 0.$$

Under the latter conditions, conservation of linear momentum and of total mass lead

Navier-Stokes Equations and Related Nonlinear Problems
Edited by A. Sequeira, Plenum Press, New York, 1995

to the following initial-boundary value problem:

$$\begin{cases} \dfrac{\partial}{\partial t}(\mathbf{v} - \alpha\Delta\mathbf{v}) - \mu\Delta\mathbf{v} = \text{grad}\, p - \text{curl}\,(\mathbf{v} - \alpha\Delta\mathbf{v}) \times \mathbf{v} \\ \qquad\qquad\qquad\qquad\qquad\qquad\qquad\qquad\qquad \text{in } \Omega \times (0, T) \\ \text{div}\,\mathbf{v} = 0. \\ \\ \mathbf{v}(x, 0) = \mathbf{v}_0(x), \quad x \in \Omega \\ \mathbf{v}(y, t) = \mathbf{v}_*(y, t), \quad (y, t) \in \partial\Omega \times (0, T). \end{cases} \qquad (1.1)$$

Here $\alpha = \alpha_1/\rho$, $\mu = \nu/\rho$ (ρ the constant density of the fluid), Ω is the region of flow, $T > 0$, p is the (modified) pressure field obtained by division by ρ, and $\boldsymbol{v}_0$ and $\boldsymbol{v}_*$ are prescribed fields.

During the last two years, extensive research has been done with the aim of describing mathematical properties of classical solutions to Eqs. (1.1), such as existence, uniqueness and stability of steady solutions under different conditions, see e.g. Galdi, Grobbelaar-Van Dalsen and Sauer (1993), Galdi and Sequeira (1994), Coscia and Galdi (1994), Coscia, Sequeira and Videman (1995). However, the cited papers treat only the case of Ω bounded. The aim of the present paper is to study some asymptotic in time properties of the solutions in the case when Ω is the whole space. In particular, we show that the kinetic energy of the fluid decays to zero logarithmically fast, provided the size of initial data is sufficiently restricted in suitable norms. Thus, our results imply that the rest solution is asymptotically energy stable. In this respect, we wish to remark that the same problem was only recently solved in the (simpler) case of the Navier-Stokes model, the main difficulty relying in the fact that the region of flow is unbounded in all directions see, e.g. Schonbek (1985), Wiegner (1987), Maremonti (1985).

Here, to solve the problem, we employ a "weighted method" introduced in Galdi and Rionero (1986). This method leads to the afore-said decay of the kinetic energy provided we show that for some number $M > 0$ the following estimates hold

$$\int_\Omega \ln(|x| + 1)v^2(x, t)dx \leq M$$

$$\int_\Omega |\text{grad}\,\mathbf{v}(x, t)|^2 dx \leq Mt^{-1}.$$

After giving some preliminary estimates concerning an elliptic problem in $\mathbb{R}^3$, in Section 3 we prove some "energy inequalities" which allow us to show the estimates above.

For the sake of simplicity, we have considered the case $\Omega = \mathbb{R}^3$. The more general case where Ω is an exterior domain will be the object of a forthcoming paper.

2. NOTATIONS AND PRELIMINARY RESULTS

In what follows, we shall consider functions depending on a space variable $x \in \Omega = \mathbb{R}^3$ and on a time variable $t \in (0, T)$, where, possibly, $T = \infty$.

The functions can be either scalar, vector or tensor-valued, according to the context. Along with the usual Sobolev spaces

$$W^{m,q}(\Omega) := \{\mathbf{u} \in L^q(\Omega) : D^\alpha\mathbf{u} \in L^q(\Omega) \; \forall|\alpha| = 1, m \quad m \geq 0, q \in (1, \infty)\},$$

where D^α is the partial differential operator with multi-index α, endowed with the norm

$$\|\mathbf{u}\|_{m,q}^q = \sum_{|\alpha|=0}^m \int_\Omega |D^\alpha\mathbf{u}(x, t)|^q dx \quad,$$

we introduce the homogeneous Sobolev space (see Galdi, 1994)

$$H(\Omega) := \{\mathbf{u} \in L^6(\Omega) : \operatorname{grad} \mathbf{u} \in L^2(\Omega), \quad \operatorname{div} \mathbf{u} = 0\} .$$

For m an integer, we put

$$\|D^m \mathbf{u}\|_q = \left(\sum_{|\alpha|=m} \int_\Omega |D^\alpha \mathbf{u}|^q dx \right)^{1/q} .$$

Also, we shall say that $\mathbf{u} \in L^q(0,T;X)$ if, almost everywhere in $(0,T)$, $\mathbf{u}(t) \in X(\Omega)$ and, moreover,

$$\int_0^T \|\mathbf{u}(t)\|_X^q \, dt < \infty .$$

We shall use the following standard notations: for $R > 0$ and Ω an arbitrary domain, we set

$$B_R(x) := \{y \in \mathbb{R}^3 : |x - y| < R\}$$

and

$$\Omega_R = \Omega \cap B_R(0), \quad \Omega^R = \Omega - \overline{\Omega}_R .$$

Finally, by D_j, $j = 1,3$, we denote the partial derivative with respect to x^j.

We begin to show the following result, which generalizes that given by Galdi and Rionero (1986), and which holds in an arbitrary unbounded domain Ω (not necessarily coinciding with the whole space).

Lemma 2.1. *Let $\rho(|x|)$ be a positive function, which is non-decreasing for $|x| > \overline{R} > 0$ and such that*

$$\lim_{|x| \to \infty} \rho(|x|) = \infty .$$

Let $\mathbf{u}(x,t)$ be a function for which there exists $c > 0$ such that, $\forall t > \overline{t} > 0$ and $\forall R > \overline{R}$,

$$\int_{\Omega_R} u^2(x,t) \, dx \leq cR^\gamma t^{-\beta} \qquad \gamma, \beta > 0 . \tag{2.1}$$

If, for some $c_1 > 0$ and $\forall t > \overline{t}$,

$$\int_\Omega \rho(|x|) u^2(x,t) \, dx \leq c_1 , \tag{2.2}$$

then, there exists a constant $c_2 > 0$ such that, $\forall \alpha \in (0, \beta/\gamma)$ and $\forall t > \overline{t}$,

$$\int_\Omega u^2(x,t) \, dx \leq c_2 \left(t^{\gamma\alpha - \beta} + \frac{1}{\rho(t^\alpha)} \right) . \tag{2.3}$$

Proof: The estimate in Eq. (2.3) is a direct consequence of the inequality

$$\int_{\Omega_R} u^2(x,t) \, dx + \int_{\Omega^R} u^2(x,t) \, dx \leq cR^\gamma t^{-\beta} + \frac{1}{\rho(R)} \int_{\Omega^R} \rho(|x|) u^2(x,t) \, dx ,$$

provided we set $R = t^\alpha$.

Remark 2.1. If

$$\int_{\Omega_R} u^2(x,t) \, dx = o(1) \text{ as } t \to \infty,$$

by the proof just given, we can nevertheless deduce, in the same limit,

$$\int_\Omega u^2(x,t)\,dx \,=\, o(1) \,.$$

Remark 2.2. Lemma 2.1 can be generalized to the following situation

$$\int_{\Omega_R} u^2(x,t)\,dx \,\leq\, c(R)g(t) \,,$$

where c and g are respectively increasing and decreasing functions, such that

$$\lim_{t\to\infty} c(t^\alpha)g(t) \,=\, 0.$$

Corollary 2.1. *Assume*

$$\int_\Omega |\operatorname{grad}\mathbf{u}(x,t)|^2\,dx \,\leq\, ct^{-\beta} \quad \forall t > \bar{t} \,, \tag{2.4}$$

then the estimate in Eq.(2.1) holds with $\gamma = 2$.

Proof: By the Hölder and Sobolev inequalities we get

$$\int_{\Omega_R} u^2(x,t)\,dx \,\leq\, cR^2\|\mathbf{u}\|_6^2 \,\leq\, cR^2\|\operatorname{grad}\mathbf{u}\|_2^2 \,.$$

Thus, the corollary follows. $\qquad\square$

In the present paper, Lemma 2.1 will be applied to the function

$$\mathbf{u} = \mathbf{v} + \sum_{j=1}^3 D_j\mathbf{v} \,,$$

with $\mathbf{v}$ solution of the system given in Eqs. (1.1). Indeed, the estimate in Eq. (2.4) will be proved with $\beta = 1$. Moreover, we choose as weight function

$$\rho(|x|) \,=\, \ln(1 + |x|) \,, \tag{2.5}$$

so that, ultimately, we shall find

$$\|\mathbf{v}\|_2^2 \,=\, O\left(\frac{1}{\ln(t)}\right) \,, \quad t \to \infty \,.$$

We need a result on an elliptic problem in $\mathbb{R}^3$.

Lemma 2.2. *Given $\mathbf{w} \in H(\Omega)$, there exists one and only one $\mathbf{v} \in H(\Omega)$ such that*

$$-\Delta\mathbf{v} + \mathbf{v} = \mathbf{w} \tag{2.6}$$

Moreover, the following estimates hold

$$\|\operatorname{grad}\mathbf{v}\|_2 \,\leq\, \|\operatorname{grad}\mathbf{w}\|_2 \tag{2.7}$$

$$\|D^3\mathbf{v}\|_2 \,\leq\, c\|\operatorname{grad}\mathbf{w}\|_2 \,, \tag{2.8}$$

with c a positive constant.

Proof: Let us formally differentiate both sides of Eq. (2.6). Then, multiplying by grad $\mathbf{v}$ and integrating, we obtain

$$(\mathrm{grad}\,\mathbf{v}, \mathrm{grad}\,\mathbf{v}) + (D^2\mathbf{v}, D^2\mathbf{v}) = (\mathrm{grad}\,\mathbf{w}, \mathrm{grad}\,\mathbf{w})\ .$$

From this equation, using the Schwarz and Cauchy inequalities, Eq. (2.7) follows. Let us next consider Eq. (2.6) as a Dirichlet problem in Ω, corresponding to the "force" $\mathbf{f} = \mathbf{v} - \mathbf{w}$. We can apply the homogeneus estimates for the Dirichlet problem in $\mathbb{R}^3$ (see Simader and Sohr, 1995), which ensure that if, for some integer m, $D^m\mathbf{f} \in L^q(\Omega)$, then there exists $c > 0$ such that

$$\|D^{m+2}\mathbf{v}\|_q \le c\|D^m\mathbf{f}\|_q\ .$$

By choosing $m = 1$ and $q = 2$, we get

$$\|D^3\mathbf{v}\|_2 \le c(\|\mathrm{grad}\,\mathbf{v}\|_2 + \|\mathrm{grad}\,\mathbf{w}\|_2)\ .$$

From this inequality and Eq. (2.7), we find the estimate given in Eq. (2.8). The existence can easily be deduced from these *a priori* estimates. On the other hand, by taking the divergence of both sides of Eq. (2.6) and setting $\psi = \mathrm{div}\,\mathbf{v}$, we immediately see that $\|\psi\|_{1,2}^2$ must be zero. $\qquad\qquad \square$

3. ENERGY INEQUALITIES AND PROOF OF THE MAIN RESULT

In what follows, we assume that integration by parts is allowed in Ω by tacitly understanding summability properties of the functions we are dealing with.

Lemma 3.1. *Let* $(\mathbf{v}, p) \in L^2(0, T; H) \times L^2(0, T; L^2)$ *be a solution of Eqs.(1.1), and let* ρ *be the function defined in Eq.(2.5).*
Assume that

$$\sqrt{\rho}\,\mathbf{v}_0 \in L^2(\Omega), \qquad \sqrt{\rho}\,(\sum_{j=1}^{3} D_j\mathbf{v}_0) \in L^2(\Omega)\ .$$

Then, setting

$$\mathbf{u} = \mathbf{v} + \sum_{j=1}^{3} D_j\mathbf{v}\ ,$$

for all $t \in (0, T)$, *the following estimate holds true*

$$\|\sqrt{\rho}\,\mathbf{u}(t)\|_2^2 \le 2\int_\Omega \rho(v_0^2 + |\mathrm{grad}\,\mathbf{v}_0|^2)\,dx$$

$$+c\int_0^T \left(\|\mathrm{grad}\,\mathbf{v}(t)\|_2^2 + \|p(t)\|_2^2 + \|\frac{\partial}{\partial t}\mathrm{grad}\,\mathbf{v}(t)\|_2^2\right) dt, \tag{3.1}$$

with c *a positive constant.*

Proof: If we multiply Eq. $(1.1)_1$ by $\rho(|x|)\mathbf{v}(x, t)$ and integrate over Ω, we obtain

$$\frac{1}{2}\frac{d}{dt}\int_\Omega \rho(|x|)(v^2(x, t) + |\mathrm{grad}\,\mathbf{v}(x, t)|^2)\,dx = -\int_\Omega \rho(|x|)|\mathrm{grad}\,\mathbf{v}(x, t)|^2\,dx$$

$$+\frac{1}{2}\int_\Omega \Delta\rho(|x|)v^2(x, t)\,dx - \int_\Omega p(x, t)\mathbf{v}(x, t)\cdot\mathrm{grad}\,\rho(|x|)\,dx$$

$$- \int_\Omega \operatorname{grad} \rho(|x|) \cdot \frac{\partial}{\partial t} (\operatorname{grad} \mathbf{v}(x,t)) \cdot \mathbf{v}(x,t) \, dx \ .$$

Thus, choosing $\rho(|x|)$ as in Eq. (2.5), so that

$$|\operatorname{grad} \rho(|x|)| = \frac{1}{|x|+1}$$

$$|\Delta \rho(|x|)| \leq \frac{c}{|x|^2}$$

and using the well-known inequality

$$\int_\Omega \frac{v^2(x,t)}{|x|^2} \, dx \leq c \int_\Omega |\operatorname{grad} \mathbf{v}(x,t)|^2 \, dx \ ,$$

we finally get, again by means of Schwarz and Cauchy inequalities,

$$\frac{d}{dt} \int_\Omega \rho(v^2 + |\operatorname{grad} \mathbf{v}|^2) \, dx \leq c \left(\|\operatorname{grad} \mathbf{v}\|_2^2 + (\|p\|_2 + \|\frac{\partial}{\partial t}\operatorname{grad} \mathbf{v}\|_2)\|\operatorname{grad} \mathbf{v}\|_2 \right) \leq$$

$$\leq c' \left(\|\operatorname{grad} \mathbf{v}\|_2^2 + \|p\|_2^2 + \|\frac{\partial}{\partial t}\operatorname{grad} \mathbf{v}\|_2^2 \right) \ .$$

Eq. (3.1) comes out by integrating this last inequality over $(0,t)$, increasing up to T the bound of integration and taking into account that

$$\|\sqrt{\rho}\,\mathbf{u}(t)\|_2^2 \leq 2 \int_\Omega \rho(v^2 + |\operatorname{grad} \mathbf{v}|^2) \, dx \ .$$

The proof of the main result will be achieved by showing that the right hand side of Eq. (3.1) is bounded uniformly in T, that is, with suitable hypotheses on the initial datum, any solution of Eqs. (1.1) verifies the condition $(\mathbf{v},p) \in L^2(0,\infty;H) \times L^2(0,\infty;L^2)$, and also $\frac{\partial}{\partial t}\operatorname{grad} \mathbf{v} \in L^2(0,\infty;L^2)$.

On the other hand, the condition given in Eq. (2.4) is also needed. It will be obtained, with $\beta = 1$, through further differential inequalities.

Let us begin by proving

Lemma 3.2. *Let* $\mathbf{v}$ *be a solution of Eqs.(1.1) with* $\mathbf{v}_0 \in W^{1,2}(\Omega)$, *then the following "energy equality" holds true*

$$\frac{d}{dt}\|\mathbf{v}\|_{1,2}^2 = -2\|\operatorname{grad} \mathbf{v}\|_2^2 \ . \tag{3.2}$$

Hence, in particular,

$$\|\mathbf{v}(t)\|_{1,2} \leq \|\mathbf{v}_0\|_{1,2} \quad \forall t > 0 \ , \tag{3.3}$$

and $\mathbf{v} \in L^2(0,\infty;H)$.

Proof: Eq. (3.2) is obtained by multiplying Eq. $(1.1)_1$ by $\mathbf{v}(x,t)$ and then by integrating over Ω. The property given in Eq. (3.3) simply follows from Eq. (3.2). Moreover, from Eq. (3.2) and Eq. (3.3) we infer, by integration,

$$\int_0^\infty \|\operatorname{grad} \mathbf{v}\|_2^2 \, dt \leq \|\mathbf{v}_0\|_{1,2}^2 - \lim_{t \to \infty} \|\mathbf{v}(t)\|_{1,2}^2 \ . \tag{3.4}$$

We can rewrite Eqs. (1.1) in the following equivalent form

$$\begin{cases} \dfrac{\partial \mathbf{w}}{\partial t} + (\operatorname{curl} \mathbf{w}) \times \mathbf{v} = \operatorname{grad} p + (\mathbf{v} - \mathbf{w}) \\[2mm] \operatorname{div} \mathbf{w} = 0 \\[2mm] \mathbf{w}(x,0) = \mathbf{v}_0 - \Delta \mathbf{v}_0, \end{cases} \qquad (3.5)$$

where $\mathbf{w}$ is defined by Eq. (2.6) and $\operatorname{div} \mathbf{v} = 0$. If $\mathbf{v}_0 \in W^{3,2}(\Omega)$, then $\mathbf{w} \in W^{1,2}(\Omega)$. Hence, we can apply Lemma 2.2 to estimate the derivatives of $\mathbf{v}$. To this end, we will look for a differential inequality involving the norm of $\operatorname{grad} \mathbf{w}$ only.

From Eqs. (3.5), we can deduce the following result.

Lemma 3.3. *Let $\mathbf{w}$ be the solution of Eqs.(3.5), with $\mathbf{v}$ solution of Eqs.(1.1), corresponding to $\mathbf{v}_0 \in W^{3,2}(\Omega)$. Then, there exists a constant $c > 0$ such that*

$$\frac{d}{dt}\|\operatorname{grad} \mathbf{w}\|_2^2 \leq (c\|\operatorname{grad} \mathbf{w}\|_2 - 1)\|\operatorname{grad} \mathbf{w}\|_2^2 + \|\operatorname{grad} \mathbf{v}\|_2^2 \qquad (3.6)$$

$$\frac{d}{dt}\|\operatorname{grad} \mathbf{w}\|_2^2 \leq c\|\operatorname{grad} \mathbf{w}\|_2^3 \qquad (3.7)$$

Proof: Let us take the derivative with respect to x_j of Eq. (3.5)$_1$, multiply by $D_j\mathbf{w}$ and integrate over Ω. If we use the identity

$$(\operatorname{curl} \mathbf{w}) \times \mathbf{v} = -\operatorname{grad}(\mathbf{w} \cdot \mathbf{v}) + \mathbf{v} \cdot \operatorname{grad} \mathbf{w} + \mathbf{w} \cdot \operatorname{grad} \mathbf{v} + \mathbf{w} \times (\operatorname{curl} \mathbf{v}),$$

we find

$$\frac{1}{2}\frac{d}{dt}\|D_j\mathbf{w}\|_2^2 = -(D_j\mathbf{w}, D_j\mathbf{v} \cdot \operatorname{grad} \mathbf{w}) - (D_j\mathbf{w}, D_j\mathbf{w} \cdot \operatorname{grad} \mathbf{v}) -$$

$$(D_j\mathbf{w}, D_j\mathbf{w} \times (\operatorname{curl} \mathbf{v})) - (D_j\mathbf{w}, \mathbf{w} \times \operatorname{curl}(D_j\mathbf{v})) -$$

$$(D_j\mathbf{w}, \mathbf{w} \cdot \operatorname{grad}(D_j\mathbf{v})) - (D_j\mathbf{w}, \mathbf{v} \cdot \operatorname{grad}(D_j\mathbf{w})) -$$

$$\|\operatorname{grad} \mathbf{w}\|_2^2 + (\operatorname{grad} \mathbf{v}, \operatorname{grad} \mathbf{w}) \qquad (3.8)$$

The first three terms on the right hand side of Eq. (3.8) are bounded by $\|\operatorname{grad} \mathbf{v}\|_\infty \|\operatorname{grad} \mathbf{w}\|_2^2$. The fourth and the fifth are bounded by

$$\|D^2\mathbf{v}\|_3 \|\mathbf{w}\|_6 \|\operatorname{grad} \mathbf{w}\|_2.$$

The sixth vanishes identically.

Thus, using Sobolev, Cauchy and Schwarz inequalities, we obtain

$$\frac{1}{2}\frac{d}{dt}\|\operatorname{grad} \mathbf{w}\|_2^2 \leq c'\left(\|\operatorname{grad} \mathbf{v}\|_\infty + \|D^2\mathbf{v}\|_3 - \frac{1}{c'}\right)\|\operatorname{grad} \mathbf{w}\|_2^2 + \|\operatorname{grad} \mathbf{v}\|_2^2 \qquad (3.9)$$

On the other hand, by the Sobolev embedding theorems we have

$$\|\operatorname{grad} \mathbf{v}\|_\infty \leq c\|\operatorname{grad} \mathbf{v}\|_{2,2},$$

$$\|D^2\mathbf{v}\|_3 \leq c\|D^2\mathbf{v}\|_{1,2}. \qquad (3.10)$$

In order to estimate the right hand sides of Eqs. (3.10) in terms of $\|\operatorname{grad} \mathbf{w}\|_2$ we need to use, in addition to Eq. (2.7) and Eq. (2.8), also Ehrling's inequality, which states that

$$\|\operatorname{grad} \mathbf{u}\|_q \leq c(\epsilon)\|\mathbf{u}\|_q + \epsilon\|D^2\mathbf{u}\|_q \quad \forall q \in (1, \infty), \quad \forall \epsilon > 0,$$

where $c(\epsilon) \to \infty$ as $\epsilon \to 0$.

In this way, we find

$$\|\operatorname{grad} \mathbf{v}\|_2 + \|D^2\mathbf{v}\|_2 + \|D^3\mathbf{v}\|_2 \leq$$

$$\leq \|\operatorname{grad} \mathbf{w}\|_2 + c(\epsilon)\|\operatorname{grad} \mathbf{v}\|_2 + \epsilon\|D^3\mathbf{v}\|_2 + c\|\operatorname{grad} \mathbf{w}\|_2 \leq c'\|\operatorname{grad} \mathbf{w}\|_2,$$

$$\|D^2\mathbf{v}\|_2 + \|D^3\mathbf{v}\|_2 \leq$$

$$\leq c(\epsilon)\|\operatorname{grad} \mathbf{v}\|_2 + \epsilon\|D^3\mathbf{v}\|_2 + c\|\operatorname{grad} \mathbf{w}\|_2 \leq c'\|\operatorname{grad} \mathbf{w}\|_2.$$

Finally, the inequalities

$$\|\operatorname{grad} \mathbf{v}\|_\infty \leq c\|\operatorname{grad} \mathbf{w}\|_2,$$

$$\|D^2\mathbf{v}\|_3 \leq c\|\operatorname{grad} \mathbf{w}\|_2, \tag{3.11}$$

imply the inequality of Eq. (3.6), by direct substitution in Eq. (3.9), and that of Eq. (3.7), using again Eq. (2.7) for the last term on the right hand side of Eq. (3.9). □

The next result leads, among other things, to the condition of Eq. (2.4).

Lemma 3.4. *Let* $\mathbf{v}$ *be a solution of Eqs.(1.1), with* $\mathbf{v}_0 \in W^{3,2}(\Omega)$. *Let* $\mathbf{w}$ *be the corresponding solution of Eqs.(3.5). There exists a positive constant* k, *such that, if*

$$\|\mathbf{v}_0\|_{3,2} \leq k, \tag{3.12}$$

then

$$\|\mathbf{v}(t)\|_{1,2} + \|\operatorname{grad} \mathbf{w}(t)\|_2 \leq \|\mathbf{v}_0\|_{3,2} \quad \forall t > 0. \tag{3.13}$$

Moreover, $\mathbf{w} \in L^2(0, \infty; H)$ *and*

$$\|\operatorname{grad} \mathbf{v}(t)\|_2^2 + \|D^2\mathbf{v}(t)\|_2^2 = O\left(\frac{1}{t}\right) \tag{3.14}$$

as $t \to \infty$.

Proof: Let us add to Eq. (3.2) the inequality of Eq. (3.6). Then, by suitable use of the Cauchy inequality, we can get, for positive constants k_1 and k_2,

$$\frac{d}{dt}(\|\mathbf{v}\|_{1,2}^2 + \|\operatorname{grad} \mathbf{w}\|_2^2) \leq (k_1\|\operatorname{grad} \mathbf{w}\|_2^2 - k_2)\|\operatorname{grad} \mathbf{w}\|_2^2 - \|\operatorname{grad} \mathbf{v}\|_2^2 \tag{3.15}$$

Let us add on the right hand side $\|\operatorname{grad} \mathbf{v}\|_2^2 + k_1\|\mathbf{v}\|_{1,2}^2 \|\operatorname{grad} \mathbf{w}\|_2^2$ and set

$$y(t) = \|\mathbf{v}(t)\|_{1,2}^2 + \|\operatorname{grad} \mathbf{w}(t)\|_2^2 \quad z(t) = \|\operatorname{grad} \mathbf{w}(t)\|_2^2,$$

from Eq. (3.15), it follows that

$$\frac{d}{dt}y(t) \leq (k_1 y(t) - k_2)z(t). \tag{3.16}$$

By choosing

$$y(0) < \frac{k_2}{k_1} := k,$$

one has

$$y(t) \leq k \quad \forall t > 0 \,. \tag{3.17}$$

In fact, if not, there must exist at least one $t_1 > 0$ for which

$$y(t_1) = \frac{k_2}{k_1}$$

and

$$\frac{d}{dt}y(t_1) > 0 \,.$$

This contradicts the inequality of Eq. (3.16). Inserting the estimate of Eq. (3.17) in the right hand side of Eq. (3.16), it comes out that $y(t)$ is actually non-increasing and this proves Eq. (3.13). Indeed, since $y(t)$ is monotonical and bounded, integrating Eq. (3.16) over $t \in (0, \infty)$, we find

$$(k_2 - k_1 y(0)) \int_0^\infty z(t)\,dt \leq y(0) - \lim_{t \to \infty} y(t) < \infty \,, \tag{3.18}$$

that is to say $\mathbf{w} \in L^2(0, \infty; H)$, provided the initial datum is sufficiently small. It remains to prove Eq. (3.14) As a consequence of the just showed summability of $\|\mathbf{w}(t)\|_2^2$, a diverging sequence $\{t_n\}_{n \in \mathbb{N}}$ exists such that

$$\lim_{n \to \infty} z(t_n) = 0 \,.$$

On the other hand, by taking the modulus in both sides of Eq. (3.8) and recalling the estimates of Eq. (3.11) and Eq. (3.13), we get

$$\left| \frac{d}{dt} z(t) \right| \leq c(z(t) + z(t)^{3/2}) \leq c(1 + y(0)^{1/2}) z(t) \,.$$

Thus, $\frac{d}{dt} z(t) \in L^1(0, \infty)$ and from

$$z(t) = z(t_n) + \int_{t_n}^t \frac{d}{ds} z(s)\,ds \,,$$

it follows that

$$\lim_{t \to \infty} z(t) = 0 \,. \tag{3.19}$$

We next multiply Eq. (1.1)$_1$ by $\Delta \mathbf{v}$ and integrate over Ω, to obtain

$$\frac{d}{dt}(\|\operatorname{grad} \mathbf{v}\|_2^2 + \|\Delta \mathbf{v}\|_2^2) = -\|\Delta \mathbf{v}\|_2^2 + ((\mathbf{v} - \Delta \mathbf{v}) \cdot \operatorname{grad} \mathbf{v} \,, \Delta \mathbf{v}) \tag{3.20}$$

Since, by Sobolev imbedding theorems, one has

$$\sup_{x \in \Omega} |\mathbf{v}(x, t)| \leq c(\|\mathbf{v}\|_6 + \|\operatorname{grad} \mathbf{v}\|_6) \leq c'(\|\operatorname{grad} \mathbf{v}\|_2 + \|D^2 \mathbf{v}\|_2)$$

we find

$$|(\mathbf{v} \cdot \operatorname{grad} \mathbf{v} \,, \Delta \mathbf{v})| \leq \sup_{x \in \Omega} |\mathbf{v}(x, t)| \, \|\operatorname{grad} \mathbf{v}\|_2 \, \|\Delta \mathbf{v}\|_2$$

$$\leq c(\|\operatorname{grad} \mathbf{v}\|_2^2 \|D^2 \mathbf{v}\|_2 + \|\operatorname{grad} \mathbf{v}\|_2 \, \|D^2 \mathbf{v}\|_2^2)$$

$$|(\Delta \mathbf{v} \cdot \operatorname{grad} \mathbf{v} \,, \Delta \mathbf{v})| \leq \sup_{x \in \Omega} |\operatorname{grad} \mathbf{v}(x, t)| \, \|\Delta \mathbf{v}\|_2^2$$

$$\leq c(\|D^2 \mathbf{v}\|_2 + \|D^3 \mathbf{v}\|_2) \|\Delta \mathbf{v}\|_2^2 \tag{3.21}$$

As is known (see Ladyzhenskaya, 1969), in $\Omega = \mathbb{R}^3$

$$\|D^2\mathbf{v}\|_2 = \|\Delta\mathbf{v}\|_2 \,.$$

Thus, by means of Cauchy and Ehrling inequalities, together with the estimates of Eq. (2.7) and Eq. (2.8), the inequalities of Eq. (3.21) imply

$$|((\mathbf{v} - \Delta\mathbf{v}) \cdot \operatorname{grad}\mathbf{v}, \Delta\mathbf{v})| \leq c\|\operatorname{grad}\mathbf{v}\|_2^4 + (\epsilon + \|\operatorname{grad}\mathbf{w}\|_2)\|\Delta\mathbf{v}\|_2^2, \qquad (3.22)$$

where ϵ can be taken as small as we want and, for t sufficiently large, $\|\operatorname{grad}\mathbf{w}\|_2$ too, because of Eq. (3.19).

So, using Eq. (3.22) to estimate the right hand side of Eq. (3.20), we obtain, for large t,

$$\frac{d}{dt}(\|\operatorname{grad}\mathbf{v}\|_2^2 + \|\Delta\mathbf{v}\|_2^2) \leq c\left(\|\operatorname{grad}\mathbf{v}\|_2^2 + \|\Delta\mathbf{v}\|_2^2\right)^2 \,.$$

We can rewrite this last inequality setting

$$f(t) := \|\operatorname{grad}\mathbf{v}\|_2^2 + \|\Delta\mathbf{v}\|_2^2 \,,$$

so that it looks like

$$\frac{d}{dt}f(t) \leq c\,f^2(t)\,.$$

As a consequence of Eq. (2.7), Eq. (2.8) and Ehrling's inequality we have also

$$f(t) \leq c\,z(t)\,.$$

Hence, $f(t)$ belongs to $L^1(0,\infty)$. In order to see that all this implies Eq. (3.14), we set $\Phi(t) = 1 + ct f(t)$ and we see that

$$\frac{d}{dt}\Phi(t) \leq cf(t) + ct\frac{d}{dt}f(t) \leq cf(t)\Phi(t)\,.$$

If we divide by Φ and recall that $f \in L^1(0,\infty)$, we obtain that $\ln(\Phi)$ is bounded and then $f < c'/t$. $\qquad\square$

Coming back to the estimate of Eq. (3.1), what is still to be proved, for the application of Lemma 2.1, is that $\|p(t)\|_2^2$ and $\|\frac{\partial}{\partial t}\operatorname{grad}\mathbf{v}(t)\|_2^2$ are in $L^1(0,T)$. The following Lemma contains the right estimate for $\|p\|_2$

Lemma 3.5. *Let $(\mathbf{v}, p)$ be a solution of Eqs.(1.1), with $\mathbf{v}_0 \in W^{3,2}(\Omega)$. Assume that the condition of Eq.(3.12) is satisfied.*

Then, there exists $c > 0$ such that

$$\|p\|_2 \leq c\|\operatorname{grad}\mathbf{w}\|_2\,, \qquad (3.23)$$

where $\mathbf{w}$ is the solution of Eqs.(3.5) corresponding to $\mathbf{v}$.

Proof: By taking the divergence of both sides of Eq. (1.1)$_1$, we see that $p(x,t)$ solves

$$\Delta p = \operatorname{div}((\operatorname{curl}\mathbf{w}) \times \mathbf{v}) \qquad (3.24)$$

Since, by Sobolev's inequality,

$$\|p\|_2 \leq \|\operatorname{grad} p\|_{6/5}\,,$$

we multiply Eq. (3.24) by a test function ϕ such that $\operatorname{grad}\phi \in L^6(\Omega)$ and, after integration on the space variable, we deduce

$$\|\operatorname{grad} p\|_{6/5} \leq \|(\operatorname{curl}\mathbf{w})\times\mathbf{v}\|_{6/5} \leq \|\operatorname{grad}\mathbf{w}\|_2\|\mathbf{v}\|_3 \leq$$

$$\leq \|\operatorname{grad}\mathbf{w}\|_2\|\mathbf{v}\|_2^{1/2}\|\mathbf{v}\|_6^{1/2} \leq \|\operatorname{grad}\mathbf{w}\|_2\|\mathbf{v}\|_2^{1/2}\|\operatorname{grad}\mathbf{v}\|_2^{1/2}\,.$$

Here, use has been made of Hölder and interpolation inequalities.

Finally, the estimate of Eq. (3.23), which implies that for all T, $\|p(t)\|_2^2 \in L^1(0,T)$, becomes a consequence of Eq. (3.13). $\qquad\square$

The next Lemma gives the searched estimate for $\|\frac{\partial}{\partial t}\operatorname{grad}\mathbf{v}(t)\|_2^2 \in L^1(0,T)$.

Lemma 3.6. *Let $\mathbf{v}$ be a solution of Eqs.(1.1), with $\mathbf{v}_0 \in W^{3,2}(\Omega)$. Assume that the condition of Eq.(3.12) is satisfied. Then, there exists $c > 0$ such that*

$$\|\frac{\partial}{\partial t}\mathbf{v}\|_2^2 + \|\frac{\partial}{\partial t}\operatorname{grad}\mathbf{v}\|_2^2 \leq c\|\operatorname{grad}\mathbf{w}\|_2^2 - \frac{d}{dt}\|\operatorname{grad}\mathbf{v}\|_2^2\,, \qquad (3.25)$$

where $\mathbf{w}$ is the solution of Eqs.(3.5) corresponding to $\mathbf{v}$.

Proof: Let us multiply both sides of Eq. $(1.1)_1$ by $\frac{\partial}{\partial t}\mathbf{v}$ and then integrate over Ω. We get

$$\|\frac{\partial}{\partial t}\mathbf{v}\|_2^2 + \|\frac{\partial}{\partial t}\operatorname{grad}\mathbf{v}\|_2^2 = -\frac{1}{2}\frac{d}{dt}\|\operatorname{grad}\mathbf{v}\|_2^2 + \int_\Omega (\operatorname{curl}\mathbf{w})\times\mathbf{v}\cdot\frac{\partial}{\partial t}\mathbf{v}\,dx\,. \qquad (3.26)$$

The integral in Eq. (3.26) can be increased by $\|\operatorname{grad}\mathbf{w}\|_2\|\frac{\partial}{\partial t}\mathbf{v}\|_2\|\mathbf{v}\|_\infty$. Recalling that

$$\sup_{x\in\Omega}|\mathbf{v}(x,t)| \leq c(\|\operatorname{grad}\mathbf{v}\|_2 + \|D^2\mathbf{v}\|_2)$$

$$\leq c'\|\operatorname{grad}\mathbf{w}\|_2 \leq c'\|\mathbf{v}_0\|_{3,2}\,,$$

we see that the non linear term is finally bounded by $\frac{1}{2}\|\frac{\partial}{\partial t}\mathbf{v}\|_2^2 + c\|\operatorname{grad}\mathbf{w}\|_2^2$. From this consideration, the estimate of Eq. (3.25) immediately follows, as a consequence of Eq. (3.26). $\qquad\square$

We can now state the main result

Theorem 3.1. *Let $\mathbf{v}$ be a solution of Eqs.(1.1), with $\mathbf{v}_0 \in W^{3,2}(\Omega)$ verifying*

$$\sqrt{\rho}\,\mathbf{v}_0 \in L^2(\Omega)\,, \qquad \sqrt{\rho}(\sum_{j=1}^3 D_j\mathbf{v}_0) \in L^2(\Omega)\,,$$

and

$$\|\mathbf{v}_0\|_{3,2} \leq k\,,$$

where $k > 0$ is the constant found in Lemma 3.3.
Then,

$$\|\mathbf{v}(t)\|_2^2 = O\left(\frac{1}{\ln(t)}\right) \qquad t\to\infty\,. \qquad (3.27)$$

Proof: The result is a consequence of Lemmas 2.1 and 3.1. In particular, Lemmas 3.2, 3.5 and 3.6 ensure that the right-hand side of Eq. (3.1) is bounded, so that Lemma 3.1

allows us to recover the condition in Eq. (2.2). On the other hand, Lemma 3.4, with the estimate of Eq. (3.14), and the obvious inequality

$$\|\operatorname{grad} \mathbf{u}\|_2^2 \leq 2(\|\operatorname{grad} \mathbf{v}\|_2^2 + \|D^2\mathbf{v}\|_2^2),$$

leads to the condition of Eq. (2.1) and the proof is a consequence of Corollary 2.1. □

REFERENCES

COSCIA, V. and G.P. GALDI (1994), Existence, Uniqueness and Stability of Regular Steady Motions of a Second-Grade Fluid, *Int. J. Non-Linear Mechanics* **29**, 493.

COSCIA, V., SEQUEIRA, A. and J. VIDEMAN (1995), Existence and Uniqueness of Classical Solutions for a Class of Complexity 2 Fluids, *Int. J. Non-Linear Mechanics*, (to appear).

DUNN, J.E. and R.L. FOSDICK (1974), Thermodynamics, Stability and Boundedness of Fluids of Complexity 2 and Fluids of Second Grade, *Arch. Rational Mech. Anal.* **56**, 191.

GALDI, G.P. (1994), *An Introduction to the Mathematical Theory of the Navier-Stokes Equations: Linearized Steady Problems*, Springer Tracts in Natural Philosophy, Vol. **38**, (Springer-Verlag).

GALDI, G.P., GROBBELAAR-VAN DALSEN, M. and N. SAUER (1993), Existence and Uniqueness of Classical Solutions of the Equations of Motion for Second-Grade Fluids, *Arch. Rational Mech. Anal.* **124**, 221.

GALDI, G.P. and S. RIONERO (1986), *Weighted Energy Methods in Fluid Dynamics and Elasticity*, Springer Lecture Notes in Mathematics, Vol. 1134.

GALDI, G.P. and A. SEQUEIRA (1994), Further Existence Results for Classical Solutions of the Equations of a Second Grade Fluid, *Arch. Rational Mech. Anal.* **128**, 297.

LADYZHENSKAYA, O.A. (1969), *The Mathematical Theory of Viscous Incompressible Flow*, (Gordon & Breach, New York).

MAREMONTI, P. (1985), Stabilità Asintotica in Media per Moti Fluidi Viscosi in Domini Esterni, *Ann. Mat. Pura Appl.* (IV) **142**, 57.

RAJAGOPAL, K.R. (1992), Flow of Viscoelastic Fluids between Rotating Plates, *Theor. and Comput. Fluid Dyn.* **3**, 185.

RIVLIN, R.S. and J.L. ERICKSEN (1955), Stress-Deformation Relations for Isotropic Materials, *J. Rational Mech. Anal.* **3**, 323.

SCHONBEK, M.E. (1985), L^2-Decay for Weak Solutions of the Navier-Stokes Equations, *Arch. Rational Mech. Anal.* **88**, 209.

SIMADER, C.G. and H. SOHR (1995), *The Weak and Strong Dirichlet Problem for Δ in L^q in Bounded and Exterior Domains*, Pitman Research Notes in Mathematics, (in press).

TRUESDELL, C. and W. NOLL (1965), *The Nonlinear Field Theories of Mechanics*, Handbuch der Physik, **III**, **3**, (Springer-Verlag, Heidelberg).

WIEGNER, M. (1987), Decay Results for Weak Solutions of the Navier-Stokes Equations in $\mathbb{R}^n$, *J. London Math. Soc.* **35**, 303.

ON BOUNDARY CONDITIONS FOR FLUIDS OF THE DIFFERENTIAL TYPE

K.R. Rajagopal

Department of Mechanical Engineering
University of Pittsburgh
Pittsburgh, PA 15261, USA

1. INTRODUCTION

There have been several, recent investigations devoted to the study of the existence and uniqueness of solutions to the equations governing the flows of fluids of the differential type. As these equations are usually higher order partial differential equations than the Navier-Stokes equations, we need to address the issue of whether the "no slip" boundary condition is sufficient to have a well-posed problem. While this issue has been raised earlier (see Rajagopal, 1984; Rajagopal and Kaloni, 1989; Rajagopal, 1992 and Rajagopal and Gupta, 1984), given the critical role it plays in shaping the theory, it has not been accorded the importance it deserves. This question cannot be answered in any generality for fluids of the differential type of complexity n, for arbitrary n. However, if attention is confined to fluids of grade 2 or grade 3, we can indeed provide some definitive answers, while some partial answers are possible for fluids of grade n (Rajagopal, 1984).

We recall that in the case of Euler fluids, we prescribe impenetrability at a solid impervious boundary, that is, that there is no normal component of the velocity at the impervious boundary, while in the case of the Navier-Stokes fluid, we also presume that there is no slip between the fluid and the solid impervious boundary, that is, the tangential velocity of the fluid is the same as that of the boundary. With fluids that lead to equations that are of higher order than Navier-Stokes, the question is whether we need additional conditions at the boundary. The immediate response, based on previous experience with partial differential equations may be in the affirmative. However, the recent work of Galdi, Grobbelaar and Sauer (1993), Coscia, Sequeira and Videman (1995), Galdi and Sequeira (1994) as well as the much earlier work of Dunn and Fosdick (1974), Fosdick and Rajagopal (1980) and Rajagopal (1982) imply that for a class of flows, the usual boundary condition of no slip between

Navier-Stokes Equations and Related Nonlinear Problems
Edited by A. Sequeira, Plenum Press, New York, 1995

the fluid and the solid boundary is sufficient to prove that the solution is unique. However, the works of Rajagopal and Gupta (1984) and Rajagopal and Kaloni (1989) show that the "no–slip" boundary conditions are in general quite inadequate. We shall now proceed to discuss the issues connected with the specification of boundary conditions in some detail.

2. EQUATIONS OF MOTION

For the purpose of discussion, we shall consider a fluid of second grade (see Truesdell and Noll, 1965) rather than a general fluid of complexity n. The Cauchy stress $\mathbf{T}$ in an incompressible homogeneous fluid of second grade is given by (Truesdell and Noll, 1965)

$$\mathbf{T} = -p\mathbf{1} + \mu\mathbf{A}_1 + \alpha_1\mathbf{A}_2 + \alpha_2\mathbf{A}_1^2 \tag{2.1}$$

where $-p\mathbf{1}$ is the indeterminate part of the stress due to the constraint of incompressibility, μ is the viscosity and α_1 and α_2 are the normal stress moduli.

Also,

$$\mathbf{A}_1 = (grad\,\mathbf{v}) + (grad\,\mathbf{v})^T, \tag{2.2}$$

and

$$\mathbf{A}_2 = \frac{d\mathbf{A}_1}{dt} + \mathbf{A}_1(grad\,\mathbf{v}) + (grad\,\mathbf{v})^T\mathbf{A}_1, \tag{2.3}$$

where $\dfrac{d}{dt}$ is the usual material time derivative and $\mathbf{v}$ denotes the velocity.

Since the fluid is incompressible, it can undergo only isochoric motion, i.e.,

$$div\,\mathbf{v} = 0. \tag{2.4}$$

It has been shown that if the fluid model (2.1) is to be compatible with thermodynamics (see Dunn and Fosdick, 1974), then

$$\mu \geq 0, \quad \alpha_1 \geq 0, \quad \alpha_1 + \alpha_2 = 0. \tag{2.5}$$

The restrictions (2.5) are the subject of much debate and we refer the reader to the recent review by Dunn and Rajagopal (1995) for a comprehensive discussion of the same.

Substituting (2.1) into the balance of linear momentum

$$div\,\mathbf{T} + \rho\mathbf{b} = \rho\frac{d\mathbf{v}}{dt}, \tag{2.6}$$

we obtain

$$\mu\,\Delta\mathbf{v} + \alpha_1(\Delta\mathbf{w} \times \mathbf{v}) + \alpha_1\Delta\mathbf{v}_t + (\alpha_1 + \alpha_2)\{\mathbf{A}_1\Delta\mathbf{v} + 2\,div\,[(grad\,\mathbf{v})(grad\,\mathbf{v})^T]\} - \\ - \rho\,(\mathbf{w} \times \mathbf{v}) - \rho\,\mathbf{v}_t = grad\,\mathbf{P}, \tag{2.7}$$

where

$$\mathbf{w} = curl\,\mathbf{v}, \tag{2.8}$$

$$\mathbf{P} = p - \alpha_1(\mathbf{v}\cdot\Delta\mathbf{v}) - \frac{(2\alpha_1 + \alpha_2)}{4}|\mathbf{A}_1|^2 + \frac{\rho}{2}|\mathbf{v}|^2 + \rho\phi, \tag{2.9}$$

where it has been assumed that $\mathbf{b} = -grad\,\phi$, and the suffix t denotes partial derivative with respect to time.

274

We shall consider steady flows, and in this case the above equation simplifies to

$$\mu\Delta\mathbf{v} + \alpha_1(\Delta\mathbf{w} \times \mathbf{v}) + (\alpha_1 + \alpha_2)\{\mathbf{A}_1\Delta\mathbf{v} + 2\,div[(grad\,\mathbf{v})(grad\,\mathbf{v})^T]\}$$
$$- \rho(\mathbf{w} \times \mathbf{v}) = grad\,\mathbf{P}. \tag{2.10}$$

We notice that (2.7) is one order higher spatially than the Navier-Stokes equation. Let us suppose that the flow domain Ω is compact and that

$$\mathbf{v} = 0 \quad \text{on } \partial\Omega. \tag{2.11}$$

We first recognize that with respect to existence of solution, proving that a solution exists which satisfies fewer boundary conditions is by no means surprising notwithstanding the mathematical complexity and elegance of such proofs. However, the uniqueness results require some explanation. The uniqueness results for the steady-state case have all been established for small data. This is to be expected from our experience of the Navier-Stokes theory. However, a marked difference between the case of the Navier-Stokes theory and that for fluids of second grade bears note. While ignoring the non-linearity in the Navier-Stokes equation does not lower the order of the equation, ignoring the higher order non-linearities in the case of the second grade fluid reduces the order of the equation. This is the crux of the matter, as the uniqueness results have been established for small data where the non-linear terms can be neglected (or controlled), and the equation is essentially of the same order as the Navier-Stokes equation, and hence the condition of "no slip" at the boundary suffices. This issue in fact becomes evident when one considers the examples due to Rajagopal and Gupta (1984), and Rajagopal and Kaloni (1989), where the uniqueness results break down. In the examples considered by Rajagopal and Gupta (1984), and Rajagopal and Kaloni (1989), the non-linear terms reduce to a linear term and the term cannot be ignored as being small with respect to the linear part of the equation. In the case of the initial-boundary value problem, the conditon (2.11) suffices, for arbitrary data in marked similarity to the Navier-Stokes case. But this result is counter-intuitive and requires some rumination.

In order to clarify the above points, let us briefly take a look at the example considered by Rajagopal and Gupta (1984) of the flow of a fluid of second grade past an infinite porous flat plate, with velocity component along the x-direction tending to U as y tends to infinity. Let us seek a solution of the form

$$\mathbf{v} = u(y)\mathbf{i} + v(y)\mathbf{j}. \tag{2.12}$$

It follows from (2.4) that

$$v(y) = v_0 = \text{constant}, \tag{2.13}$$

and (2.6) reduces to (see Rajagopal and Gupta, 1984)

$$\mu\,u'' - \alpha_1 v_0 u''' + \rho\,v_0 u' = \frac{\partial p}{\partial x}. \tag{2.14}$$

We shall not document the balance of linear momentum along the other co-ordinate directions or discuss the nature of the solution. The point of interest is the fact that the (2.14) is a linear equation which is of higher order than the Navier-Stokes equation. From the fact that the fluid has an asymptotic value U for its x-component of the velocity as $y \to \infty$, we conclude that the pressure gradient along the x-direction is zero.

The appropriate boundary condition at the porous plate is

$$u(0) = 0, \tag{2.15}$$

while

$$u \to U \quad \text{as} \quad y \to \infty, \tag{2.16}$$

and we thus have one boundary condition less than that necessary to solve (2.14) completely. While it is possible to augment the boundary conditions based on asymptotic structures for the velocity field or the stress, Rajagopal and Gupta (1984) show that it can yet lead to ill-posed problems.

A similar situation is obtained when there is an impervious plate at $y = h$, with suction at the plate at $y = 0$. In this case the condition (2.16) is replaced by

$$u(h) = U. \tag{2.17}$$

It is trivial to verify that (2.14) cannot be solved completely with the aid of (2.15) and (2.17).

Rajagopal and Kaloni (1989) have provided an example of non–uniqueness due to the inadequacy of the "no slip" boundary condition in a bounded domain by considering the flow in the annular region between two porous rotating cylinders. Fosdick and Bernstein (1969) had originally studied the same problem by setting one of the constants that appear in the solution to be zero. However, there is no apparent physical reason that can be advanced for such a choice. The non–uniqueness of solutions is due to the inadequacy of boundary conditions and is manifest for even the smallest of data. In other words, if the equation were to have a higher order term that was linear, then the proof would not go through. This point cannot be over-emphasized enough. For instance, if the representation for the stress has terms like $\Delta \mathbf{A}_1$ which leads to additional higher order terms that are linear, then we would definitely need boundary conditions in addition to (2.11).

It is also worth pointing out that for the problem of the flow of the fluid of second grade due to a stretching sheet (see Rajagopal, Na and Gupta, 1984), on using the similarity transformation:

$$u = x f'(y) + g(y), \quad v = -f(y), \quad w = 0, \tag{2.18}$$

where u, v and w denote the components of the velocity in the x, y and z directions respectively, the equations of motion reduce to

$$f''' + f f'' - (f')^2 + k\{2 f''' f' - (f'')^2 - f f^{IV}\} = 0, \tag{2.19}$$

and

$$g''' + f g'' - f'g' + k\{g''' f' + f''' g' - f'' g'' - f g^{IV}\} = 0, \tag{2.20}$$

where k is a non-Newtonian parameter; $k = 0$ corresponds to the Navier-Stokes case. Here f and g are appropriate non-dimensional functions. Let us consider the case $g \equiv 0$.

The appropriate boundary conditions for $f(y)$ are

$$f(0) = 0, \quad f'(0) = 1, \tag{2.21}$$

and

$$f' \to 0 \quad \text{as} \quad y \to \infty. \tag{2.22}$$

The conditions (2.21) are due to the assumption of "no-slip" at the stretching sheet, and (2.22) is a consequence of the fluid being quiescent at infinity.

We notice that (2.19) is a fourth order equation, while (2.21) and (2.22) provide only three boundary conditions for f. It is well-known that (2.19) subject to (2.21) has non-unique solutions (see Chang, Kazarinoff and Lu, 1991). Of course, for small data, we notice that the higher order non-linear term would disappear. It is important to recognize for the above problem, the domain is a half-space. Nonetheless, it serves to point out that even for problems in which there is no normal component to the velocity field on the boundary, non-unique solutions are possible if the flow domain is not compact.

Similarly, in the problem of the flow between two rotating parallel plates, the usual von Karman similarity transformation leads to a coupled system of equations that is one order higher than the corresponding equations for the Navier–Stokes fluid and the no–slip boundary condition does not lead to an adequate number of boundary conditions for solving the equations. Of course, the velocity fields become unbounded as the radius of the plates go to infinity in this problem, but such is the case also for a Navier–Stokes fluid wherein the "no–slip" condition is sufficient to determine a solution to the problem.

Using k as a perturbation parameter in (2.19) lowers the order of the equation (singular perturbation), and there have been many studies which resort to such an approach. However, if it is known apriori that the problem has a unique solution with the "no–slip" boundary conditions, then we can carry out such a perturbance, provided that the series converges. Mansutti, Pontrelli and Rajagopal (1993) studied the steady flow of three non–linear fluids of the differential type flowing past a porous plate subject to suction or injection at the plate. They solved the resulting equation numerically by augmenting the boundary conditions and also by a singular perturbation without augmenting the boundary conditions. Of course, they carry out their perturbation only upto first order and do not prove that the perturbation converges. Even so, they find excellent agreement between the two solutions suggesting that the singular perturbation provides the solution that comes from appropriately augmenting the boundary conditions.

REFERENCES

CHANG, W.D., KAZARINOFF, N.D. and C. LU (1991), *Arch. Rational Mech. Analysis* **113**, 191.

COSCIA, V., SEQUEIRA, A. and J. VIDEMAN (1995), Existence and uniqueness of classical solutions for a class of complexity 2 fluids, *Intl. J. Non-Linear Mechanics*, (to appear).

DUNN, J.E. and R.L. FOSDICK (1974), *Arch. Rational Mech. Anal.* **56**, 191.

DUNN, J.E. and K.R. RAJAGOPAL (1995), Fluids of differential type: critical review and thermodynamic analysis, *Intl. J. Engng. Science* (to appear).

FOSDICK, R.L. and B. BERNSTEIN (1969), *Int. J. Engng. Science* **7**, 555.

FOSDICK, R.L. and K.R. RAJAGOPAL (1980), *Proc. Roy. Soc. London* A**339**, 351.

GALDI, G.P., GROBBELAAR, M. and N. SAUER (1993), *Arch. Rational Mech. Anal.* **124**, 221.

GALDI, G.P. and A. SEQUEIRA (1994), *Arch. Rational Mech. Anal.* **128**, 297.

MANSUTTI, D. PONTRELLI, G. and K.R. RAJAGOPAL (1993), *Int. J. Num. Methods in Fluids* **17**, 927.

RAJAGOPAL, K.R. (1982), *Acta Ciencia Indica* **18**, 1.

RAJAGOPAL, K.R. (1984), *J. Non-Newtonian Fluid Mechanics* **15**, 239.

RAJAGOPAL, K.R. (1992), *Theoretical and Computational Fluid Dynamics* **3**, 185.

RAJAGOPAL, K.R. and A.S. GUPTA (1984), *Meccanica* **19**, 1948.

RAJAGOPAL, K.R. and P.N. KALONI (1989), Some remarks on boundary conditions for flows of fluids of the differential type, in: *Cont. Mech. & its Applications*, (Hemisphere Press, New York).

RAJAGOPAL, K.R., NA, T.Y. and A.S. GUPTA (1984), *Rheological Acta* **23**, 213. Chang, Kazarinoff and Lu (1991)

TRUESDELL, C. and W. NOLL (1965), *The Non-Linear Field Theories of Mechanics, Handbuch der Physik* (ed. W. Flugge), III/3, (Springer, Berlin).

THERMOCONVECTION WITH DISSIPATION
OF QUASI-NEWTONIAN FLUIDS IN TUBES

José Francisco Rodrigues

C. M. A. F. / University of Lisbon
Av. Prof Gama Pinto, 2 1699 - Lisboa Codex, Portugal

1. INTRODUCTION

The flow of an incompressible viscous fluid, as it is well-known, may be described by the equations of conservation of mass, momentum and energy

$$\nabla \cdot \mathbf{v} = 0 \tag{1.1}$$

$$\mathbf{v}_t + (\mathbf{v} \cdot \nabla)\mathbf{v} = \nabla \cdot \mathbf{S} + \mathbf{f} \tag{1.2}$$

$$e_t + \mathbf{v} \cdot \nabla e = \mathbf{S} : \mathbf{D} - \nabla \cdot \mathbf{q} + g \tag{1.3}$$

where we take the constant density $\rho \equiv 1$. Hence we denote by $\mathbf{v} = (v_i)$ the velocity, $\mathbf{S} = (S_{ij})$ the stress tensor, $\mathbf{f}$ the body forces, e the internal energy, $e_t = \dfrac{\partial e}{\partial t}$,

$$\mathbf{D} = (D_{ij}) = \frac{1}{2}(v_{i,j} + v_{j,i}), \quad (v_{i,j} = \frac{\partial v_i}{\partial x_j})$$

the strain tensor, $\mathbf{q}$ the heat flux, g the heat source and the term $\mathbf{S} : \mathbf{D} = S_{ij}D_{ij}$ represents the energy dissipation, with the sum convention.

We shall be concerned here with a class of non Newtonian fluids governed by the constitutive law

$$\mathbf{S} = -p\,\mathbf{I} + 2\,\nu(\vartheta, 2\sqrt{D_{II}})\,\mathbf{D} \tag{1.4}$$

where p denotes the pressure, $\mathbf{I} = (\delta_{ij})$ the identity matrix, and $\nu = \nu(\vartheta, 2\sqrt{D_{II}}) > 0$ represents the viscosity, which is supposed to depend on the temperature ϑ and on the second invariant $D_{II} = \frac{1}{2}\mathbf{D} : \mathbf{D} = \frac{1}{2}D_{ij}D_{ij}$. If $\nu \equiv$ constant > 0 we have the classical Newtonian fluids.

In tubes, it is often possible to assume the flow parallel to the axis $\mathbf{x}_3$ of the cylinder, i.e. , to restrict the velocity field to be of the form

$$\mathbf{v} = (0, 0, u(x_1, x_2, t)), \tag{1.5}$$

Navier-Stokes Equations and Related Nonlinear Problems
Edited by A. Sequeira. Plenum Press, New York, 1995

since, by the incompressibility condition, u must be independent of the variable x_3. In particular, the strain tensor will be of the form

$$\mathbf{D} = \frac{1}{2} \begin{pmatrix} 0 & 0 & u_{,1} \\ 0 & 0 & u_{,2} \\ u_{,1} & u_{,2} & 0 \end{pmatrix}$$

and, consequently, $D_{II} = \frac{1}{4}|\nabla u|^2$ and $\mathbf{S}:\mathbf{D} = \nu |\nabla u|^2$, due to (1.4).

It is then natural to assume also the temperature $\vartheta = \vartheta(x_1, x_2, t)$ and the body forces in the form $\mathbf{f} = (0, 0, \varphi(x_1, x_2, t))$. Therefore, if $\Omega \in \mathbb{R}^2$ denotes the arbitrary cross - section of the tube, from (1.2) it is easy to see that convection terms vanish and we are reduced to a single equation

$$u_t = \nabla \cdot (\nu(\vartheta, |\nabla u|)\nabla u) + f \qquad \text{in } \Omega, \ t > 0, \tag{1.6}$$

where $f = \varphi(x_1, x_2, t) - p_{,3}(t)$ is supposed to be given. Note that, our restrictions imply that the pressure must be independent of the variables $(x_1, x_2) \in \Omega$ and, hence, $p_{,3}$ is a function of t only, which is assumed to be known. On the boundary of the tube we assume the usual no slip condition, i.e.,

$$u = 0 \qquad \text{on } \partial\Omega, \ t > 0, \tag{1.7}$$

and we suppose also known the inicial velocity

$$u(x, 0) = u_0(x), \qquad x = (x_1, x_2) \in \Omega. \tag{1.8}$$

However, in the velocity equation (1.6) we keep the temperature as a parameter in the viscosity coefficient. Hence we need to consider also the temperature equation, obtained from (1.3) with the following linearized constitutive relations

$$e = c\vartheta \qquad \text{and} \qquad q = -k\nabla\vartheta \tag{1.9}$$

where $c > 0$ is the specific heat and $k > 0$ the thermal conductivity, both assumed here constants for simplicity. By (1.5) the convective term also vanishes and, due to (1.4), we obtain the following heat equation

$$c\vartheta_t - k\Delta\vartheta = \nu(\vartheta, |\nabla u|)|\nabla u|^2 + g \text{ in } \Omega, \ t > 0. \tag{1.10}$$

In this equation the coupling with the velocity is given through the dissipation term, which is at least quadratic in its gradient.

On the lateral boundary we prescribe a Neumann type boundary condition

$$k\frac{\partial\vartheta}{\partial n} + \lambda\vartheta = h \qquad \text{on } \partial\Omega, \ t > 0 \tag{1.11}$$

where $\lambda = \lambda(x, t) \geq 0$ and $h = h(x, t)$, $x \in \partial\Omega$, $t > 0$ are known functions. Finally we also give the initial temperature

$$\vartheta(x, 0) = \vartheta_0(x), \qquad x \in \Omega. \tag{1.12}$$

As examples of non-newtonian fluids we are particularly interested in generalized fluids including the asymptotically Newtonian class, in which $\mu(\sigma) \xrightarrow{\sigma\to\infty} \mu_\infty > 0$ (see, for instance, Cioranescu, 1992, and its references), and $\nu(\tau, \sigma) = \alpha(\tau)\mu(\sigma)$, as in the
i) Prandtl - Eyring model: $\mu(\sigma) = \mu_\infty + a\,log(\sigma\lambda + \sqrt{\sigma^2\lambda^2 + 1})/\sigma\lambda$, $\qquad a, \lambda > 0$;

ii) Cross model: $\qquad \mu(\sigma) = \mu_\infty + a/(1 + \sigma^{2/3})$, $\qquad a > 0$;

iii) Williamson model: $\qquad \mu(\sigma) = \mu_\infty + a/(\lambda + \sigma)$, $\qquad a, \lambda > 0$;

iv) Carreau model: $\qquad \mu(\sigma) = \mu_\infty + a(1 + \lambda\sigma^2)^{(r-2)/2}$, $\qquad a, \lambda > 0, 1 < r < 2$.

In all the cases we shall always assume here that both nonlinear functions α and μ are continous, bounded and strictly positive.

In this work we show how the linear estimates for parabolic equations can be used to prove the existence of at least a solution to the strongly nonlinear system obtained for the couple velocity-temperature. This is done in the next section, where the results of Rodrigues (1992) for Newtonian fluids are extended here to a class of quasi-Newtonian fluids, so called in analogy with the name of the corresponding partial differencial equations (1.6). Finally in Section 3 we give a stability result, which implies the uniqueness of solution but only holds for a class of more regular solutions. In the special Newtonian case in which the total oscilation of the viscosity is not too high we obtain enough regularity and hence also uniqueness of solutions.

2. EXISTENCE OF A SOLUTION $\{u, \vartheta\}$

We assume $\Omega \in \mathbb{R}^2$ is a bounded domain with smooth boundary $\partial\Omega$ (say $\partial\Omega \in \mathcal{C}^2$) and we denote $(x, t) \in Q_T = \Omega \times]0, T[, 0 < T < \infty$, and $\Sigma_T = \partial\Omega \times]0, T[$. We suppose the viscosity is given by a continous function $\nu = \nu(\tau, \sigma) : \mathbb{R} \times \mathbb{R}_+ \to \mathbb{R}_+$, which is assumed to be continuously differenciable in the second variable and satisfying

$$0 < \alpha \le \frac{\partial}{\partial\sigma}(\sigma\nu(\tau, \sigma)) \le \gamma, \qquad \forall\tau \in \mathbb{R}, \forall\sigma \in \mathbb{R}_+, \tag{2.1}$$

for given constants $\gamma > \alpha > 0$.

For the nonhomogeneous data we shall assume

$$f \in L^p(0, T; W^{-1,p}(\Omega)) \text{ and } u_0 \in W_0^{1,p}, \text{ for some } p > 2, \tag{2.2}$$

$$g \in L^q(Q_T), \ h \in W_q^{1-1/q, 1/2-1/2q}(\Sigma_T), \ \vartheta_0 \in W^{2-2/q, q}(\Omega), \text{ for some } 1 < q < 3 \tag{2.3}$$

and we suppose that $\lambda = \lambda(x, t)$ in (1.1) is a nonnegative Lipschitz function.

Here we use the standard notations for Sobolev spaces $W^{s,p}(\Omega)$ (for $s \in \mathbb{R}$), $W_0^{1,p}(\Omega) = \mathring{W}_p^1(\Omega)$, $W^{1,p}(Q_T) = W_P^{1,1}(Q_T)$, $W_p^{0,1}(Q_T) = L^p(0, T; W^{1,p}(\Omega))$,, and

$$W_p^{2,1}(Q_T) = W^{1,p}(0, T; L^p(\Omega)) \cap L^p(0, T; W^{2,p}(\Omega)), \quad 1 \le p \le \infty.$$

$W_p^{1-1/p, 1/2-1/2p}(\Sigma_T)$ and $W^{2-2/p,p}(\Omega)$ denote respectively the space of normal traces on Σ_T and initial traces on $\Omega \times \{0\}$ for the functions of $W_p^{2,1}(Q_T)$ (see Ladyženskaja, Solonnikov and Ural'ceva 1968). We also use $W_0^{1,2}(\Omega) = H_0^1(\Omega)$ and $W^{-1,2}(\Omega) = H^{-1}(\Omega)$, for simplicity of notations.

Theorem 2.1. *Under the assumptions (2.1)–(2.3) there exists at least one pair $\{u, \vartheta\}$ in the class*

$$u \in L^r(0, T; W_0^{1,r}(\Omega)) \cap W^{1,r}(0, T; W^{-1,r}(\Omega)) \quad \text{for some } r, \ 2 < r \le p \tag{2.4}$$

$$\vartheta \in W_s^{2,1}(Q_T) \quad \text{for } s = \min(q, r/2) > 1 \tag{2.5}$$

solving, in the generalized sense, the system

$$u_t - \nabla \cdot (\nu(\vartheta, |\nabla u|)\nabla u) = f \quad \text{in} \quad Q_T \tag{2.6}$$

$$c\,\vartheta_t - k\,\Delta\vartheta = \nu\,(\vartheta,\,|\nabla u|)\,|\nabla u|^2 + g \quad in \quad Q_T \tag{2.7}$$

$$u = 0 \quad and \quad k\frac{\partial\vartheta}{\partial n} + \lambda\vartheta = h \quad on \quad \Sigma_T \tag{2.8}$$

$$u\,(0) = u_0 \quad and \quad \vartheta\,(0) = \vartheta_0 \quad on \quad \Omega. \tag{2.9}$$

$\square$

Remark 2.1. From the regularity (2.4) and (2.5), in particular, we have that u and ϑ are, at least, in $C^0([0,T];\,L^s(\Omega))$ and the initial conditions (2.9) are satisfied in $L^s(\Omega)$, and hence also almost everywhere in Ω. Analogously, the boundary conditions (2.8) are satisfied in the trace sense and almost everywhere on Σ_T. The equation (2.7) is also satisfied almost everywhere in Q_T, but the equation (2.6) holds in the distributional sense in $L^r(0,\,T;\,W^{-1,r}(\Omega))$.

The proof of this theorem is based on a fixed point argument and on the following parabolic estimates for two auxiliary linear problems.

For the heat equation with Neumann data

$$c\,v_t - k\,\Delta v = \tilde{g} \text{ in } Q_T, \tag{2.10}$$

$$k\frac{\partial v}{\partial n} + \lambda\,v = h \text{ on } \Sigma_T \text{ and } v\,(0) = v_0 \text{ on } \Omega, \tag{2.11}$$

we can apply the well-known parabolic estimate

$$\|v\|_{W_q^{2,1}(Q_T)} \leq C_q \left\{ \|\tilde{g}\|_{L^q(Q_T)} + \|h\|_{W_q^{1-1/q,\,1/2-1/2q}(\Sigma_T)} + \|v_0\|_{W_q^{2-2/q}(\Omega)} \right\}, \tag{2.12}$$

which can be found, for instance, in Ladyženskaja, Solonnikov and Ural'ceva (1968). Notice that for $1 < q < 3$ no additional assumption is required, but for $q > 3$ we need to suppose also the compatibility condition $\dfrac{\partial v_0}{\partial n} + \lambda\,(0)v_0|_{\partial\Omega} = h\,(0)$, and for $q = 3$ an additional technical condition is necessary.

For the linear Dirichlet problem with a parabolic operator with discontinuous coefficients $a = a\,(x,\,t)$, $(x,\,t) \in Q_T$:

$$w_t - \nabla \cdot (a\,(x,\,t)\nabla w) = f \quad on \quad Q_T \tag{2.13}$$

$$w = 0 \text{ on } \Sigma_T \text{ and } w\,(0) = w_0 \text{ on } \Omega \tag{2.14}$$

we recall the following results, under the assumptions (2.2) and

$$0 < \alpha \leq a\,(x,\,t) \leq \gamma, \quad a.\,e.\ (x,\,t) \in Q_T. \tag{2.15}$$

Lemma 2.1. *(i) For each $p > 2$, there exists a constant $\Lambda_p > 0$ such that if $(1 - \alpha/\gamma) < 1/\Lambda_p$, then the unique solution w of the problem (2.13)–(2.14) satisfies the estimate:*

$$\|w\|_{L^p(0,T;\,W_0^{1,p}(\Omega))} \leq C_p^* \left\{ \|f\|_{L^p(0,T;\,W^{-1,p}(\Omega))} + \|w_0\|_{W_0^{1,p}(\Omega)} \right\} \tag{2.16}$$

for some $C_p^* > 0$.

(ii) There exists a $p^* > 2$ and a positive constant $C_p^* = C\,(\alpha,\,\gamma,\,\Omega,\,p^*)$ such that (2.16) holds for all p, $2 < p \leq p^*$.

Remark 2.2. For $p = 2$, the estimate (2.16) is straightforward by variational methods and only require the L^2-norm of w_0. For $\alpha = \gamma = 1$ (i.e. $a \equiv 1$) the estimate (2.16) just means that the linear operator $\mathcal{L} : \{f, w_0\} \rightarrow w$, w solution of (2.13)–(2.14), is an isomorphism from $L^p(0, T; W^{-1,p}(\Omega)) \times W_0^{1,p}(\Omega)$ into $L^p(0, T; W_0^{1,p}(\Omega))$. The proof of Lemma 2.1 (i) follows from this fact, as shown in Rodrigues (1992), and the constant Λ_p can be chosen as the norm of the linear operator $\mathcal{L}$. The proof of Lemma 2.1 (ii) is more delicate. It corresponds to the parabolic version of Meyer's estimate and it can be found in Bensoussan, Lions and Papanicolau (1978) (see Theorem 2.2 of page 272).

In the construction of the fixed point, we start with an arbitrary function $\tau = \tau(x, t) \in L^1(Q_T)$ and we consider the auxiliary quasilinear parabolic problem

$$w_t - \nabla \cdot (a^\tau(x, t, |\nabla w|)\nabla w) = f \quad \text{on} \quad Q_T \tag{2.17}$$

$$w = 0 \quad \text{on} \quad \Sigma_T \quad \text{and} \quad w(0) = w_0 \quad \text{on} \quad \Omega \tag{2.18}$$

where, we define the coefficient a^τ by

$$a^\tau(x, t, \sigma) = \nu(\tau(x, t), \sigma), \quad \text{a.e. } (x, t) \in Q_T, \sigma \in \mathbb{R}_+. \tag{2.19}$$

Proposition 2.1. *Let $f \in L^2(0, T; H^{-1}(\Omega))$ and $w_0 \in L^2(\Omega)$. Then for each $\tau \in L^1(Q_T)$ there exists a unique*

$$w = w^\tau \in \mathcal{W}^2 \equiv L^2(0, T; H_0^1(\Omega)) \cap C^0([0, T]; L^2(\Omega)),$$

which solves (2.17)–(2.18). Moreover the nonlinear mapping $\mathcal{S} : L^1(Q_T) \ni \tau \rightarrow w^\tau \in \mathcal{W}^2$ is continuous for the respective strong topologies. In addition, if (2.2) holds then $w^\tau \in L^r(0, T; W_0^{1,r}(\Omega))$, for some r, $2 < r \leq p$, and

$$\mathcal{S}_r : L^1(Q_T) \ni \tau \rightarrow w^\tau \in L^r(0, T; W_0^{1,r}(\Omega)), \tag{2.20}$$

is also strongly continuous.

Proof: Let us denote by $A^\tau : \mathcal{V} \rightarrow \mathcal{V}'$ the nonlinear operator in $\mathcal{V} = L^2(0, T; H_0^1(\Omega))$ defined for a.e. $t \in [0, T]$

$$\langle A^\tau(t)w, v \rangle = \int_\Omega a^\tau(x, t, |\nabla w|)\nabla w \cdot \nabla v \, dx, \qquad \forall v, w \in H_0^1(\Omega) \tag{2.21}$$

for each $\tau \in L^1(Q_T)$. In particular, by the Lemma 2.2 below, $A^\tau(t)$ is a continuous and strongly coercive operator in $H_0^1(\Omega)$ and we can apply well-known abstract existence and uniqueness results to (2.17)–(2.18) in the form (see Lions, 1969, for instance)

$$w' + A^\tau w = f \quad \text{in } \mathcal{V}', \quad w(0) = w_0 \quad \text{in } \mathcal{C} \tag{2.22}$$

where we denote also $w' = w_t$, $\mathcal{C} = C^0([0, T]; L^2(\Omega))$ and $\mathcal{V}' = L^2(0, T; H^{-1}(\Omega))$.

In order to show the continuous dependence with respect to τ, consider an arbitrary sequence $\tau_n \rightarrow \tau$ in $L^1(Q_T)$ and the corresponding unique solutions of (2.22) $w_n = \mathcal{S}\tau_n$ and $w = \mathcal{S}\tau$.

In particular, by the property (2.23) below, we may conclude, at least for some subsequence, that

$$w_n \xrightarrow[n]{} w_* \text{ in } \mathcal{V}, \quad w'_n \xrightarrow[n]{} w'_* \text{ and } A^{\tau_n} \xrightarrow[n]{} \chi \text{ in } \mathcal{V}'.$$

Then, by standard monotonicity methods (see Lions, 1969) we easily conclude that $w_* = w$ and $\chi = A^\tau w$ and it remains to show the strong convergence in the norms of $\mathcal{C}$ and $\mathcal{V}$, respectively, $\sup\limits_{0<t<T} \|v(t)\|_{L^2(\Omega)}$ and $\|\nabla v\|_{L^2(Q_T)}$. Multiplying the equation, for a. e. t,

$$(w - w_n)' + A^{\tau_n} w - A^{\tau_n} w_n = A^{\tau_n} w - A^\tau w$$

by the difference $(w - w_n)(t)$, we obtain, using again (2.23),

$$\frac{1}{2} \sup_{0 \le t \le T} \int_\Omega |(w - w_n)(t)|^2 \, dx + \alpha \int_0^T \int_\Omega |\nabla(w - w_n)|^2 \, dx dt \le \rho_n$$

where, by $\tau_n \xrightarrow{n} \tau$ in $L^1(Q_T)$, we have

$$\rho_n \equiv \langle A^{\tau_n} w - A^\tau w, w - w_n \rangle = \int_{Q_T} (a^{\tau_n} - a^\tau) \nabla w \cdot \nabla(w - w_n) \, dx dt \xrightarrow{n} 0,$$

since, by Lebesgue theorem, $|[a^{\tau_n}(|\nabla w|) - a^\tau(|\nabla w|)]\nabla w| \xrightarrow{n} 0$ in $L^2(Q_T)$.

Under the assumption (2.2), the regularity $w^\tau \in L^r(0, T; W_0^{1,r}(\Omega))$ is a direct consequence of Lemma 2.1 (ii) and the quasilinear structure of (2.17), since the coefficient $a^\tau = \nu(\tau(x, t), |\nabla w^\tau(x, t)|)$ satisfies (2.15), by the Remark 2.3 below.

The strong continuous dependence in $L^r(0, T; W_0^{1,r}(\Omega))$, for some $r > 2$, can also be obtained as a direct application of the estimate (2.16) to the difference $w_n - w$, corresponding also to an arbitrary sequence $(\tau_n - \tau) \xrightarrow{n} 0$ in $L^1(Q_T)$. In fact, it is sufficient to observe that $w_n - w$ satisfies

$$(w_n - w)_t - \nabla \cdot [\nu(\tau_n, |\nabla w_n|)\nabla(w_n - w)] = \nabla \cdot F_n \quad \text{in } Q_T$$

with

$$F_n = [\nu(\tau_n, |\nabla w_n|) - \nu(\tau, |\nabla w|)] \nabla w \xrightarrow{n} 0 \quad \text{in } L^r(Q_T)$$

since we know that $\nabla w \in L^r(Q_T)$ and $\nu(\tau_n, |\nabla w_n|) \xrightarrow{n} \nu(\tau, |\nabla w|)$ a. e. in Q_T. □

Lemma 2.2. *The nonlinear operator $A^\tau = A^\tau(t) : H_0^1(\Omega) \to H^{-1}(\Omega)$ defined, for each $\tau \in L^1(Q_T)$ by (2.19), under the assumptions (2.1) and (2.21), satisfies for a. e. $t \in [0, T]$:*

$$\alpha \|\nabla(u - v)\|_{L^2(\Omega)}^2 \le \langle A^\tau u - A^\tau v, u - v \rangle, \quad \forall u, v \in H_0^1(\Omega). \tag{2.23}$$

Proof: This is a straightforward consequence of writing

$$\langle A^\tau u - A^\tau v, u - v \rangle = \int_\Omega [\varphi(1) - \varphi(0)] \, dx$$

since, for a $\mu \in]0, 1[$ we have

$$\varphi(1) - \varphi(0) = \frac{d}{d\lambda} \varphi|_{\lambda=\mu} = [\nu(\tau, |\xi_\mu|) + |\xi_\mu| \frac{\partial \nu}{\partial \sigma}(\nu, |\xi_\mu|)]|\xi - \zeta|^2$$

with the definitions $\xi = \nabla u$, $\zeta = \nabla v$, $\xi_\lambda = \zeta + \lambda(\xi - \tau)$ and $\varphi(\lambda) = (\nu(\tau, |\xi_\lambda|)\xi_\lambda, \xi - \zeta)$ for $\lambda \in [0, 1]$. □

Remark 2.3. From the proof of Lemma 2.2 we conclude, in particular with $\zeta = 0$ and $\varphi(1) = \nu(\tau, |\xi|)|\xi|^2$, that (2.1) implies

$$0 < \alpha \le \nu(\tau, \sigma) \le \gamma, \quad \forall \tau \in \mathbb{R}, \ \forall \sigma \in \mathbb{R}_+.$$

284

Proof of Theorem 2.1: We construct the mapping $\mathcal{T} = \mathcal{R} \circ \mathcal{L}_s \circ \mathcal{S}_r$:

$$\mathcal{T} : L^1(Q_T) \xrightarrow{\mathcal{S}_r} L^r(0, T; W_0^{1,r}(\Omega)) \xrightarrow{\mathcal{L}_s} W_s^{2,1}(Q_T) \xrightarrow{\mathcal{R}} L^1(Q_T)$$

where $w^\tau = \mathcal{S}_r(\tau)$ is the solution of (2.17)–(2.18) (with $w_0 = u_0$), $v^\tau = \mathcal{L}_s(\tau, w^\tau)$ solves the linear problem (2.10)–(2.11) with

$$\tilde{g} = g + \nu\left(\tau, |\nabla w^\tau|\right)|\nabla w^\tau|^2 \in L^{r/2}(Q_T), \quad s = \min(q, r/2) > 1,$$

and $\mathcal{R}$ is the compact imbedding given by the Rellich-Kondrachov theorem. Clearly the fixed point $\vartheta = \mathcal{T}\vartheta$ together with $u = \mathcal{S}_r\vartheta$ provide a solution to (2.6)–(2.9).

In order to apply the Shauder fixed point theorem to $\mathcal{T}$ it is sufficient to show that $\mathcal{T}(L^1(Q_T))$ is a bounded subset, since $\mathcal{S}_r$ and $\mathcal{L}_s$ are continuous mappings by the Proposition 2.1 and by the estimate (2.12), respectively. The boundedness of $\mathcal{T}$ is an immediate consequence of the *a priori* estimate (2.16) for w^τ, i.e.,

$$\mathcal{S}_r(L^1(Q_T)) \subset \left\{v \in L^r(0, T; W_0^{1,r}(\Omega)) : \|\nabla v\|_{L^r(Q_T)} \le C_r\right\},$$

where the constant $C_r > 0$ depends only on f, u_0, Ω, r and the constants α and γ from (2.1). $\qquad\square$

Remark 2.4. As in the Newtonian case of Rodrigues (1992), we can extend this existence result to nonlinear viscosity function ν depending also on the gradient of the temperature, i.e., $\nu = \nu(\vartheta, \nabla\vartheta, |\nabla u|)$. In that case, under similar assumptions, we should apply the fixed point in $L^1(0, T; W^{1,1}(\Omega)) = W_1^{1,0}(Q_T)$. Analogously to Rodrigues (1992), we can consider also a Dirichlet boundary condition for the temperature, instead of (1.11).

3. STABILITY RESULTS AND FURTHER REMARKS

In this section we suppose that the viscosity function ν is uniformly Lipschitz continuous in each variable, i.e., for some constants l, $\mu > 0$ it satisfies:

$$|\nu(\tau, \sigma) - \nu(\tau^*, \sigma)| \le l\,|\tau - \tau^*|, \quad \forall \tau, \tau^* \in R, \sigma \in \mathbb{R}_+, \tag{3.1}$$

$$|\nu(\tau, \sigma) - \nu(\tau, \sigma^*)| \le \mu\,|\sigma - \sigma^*|, \quad \forall \tau, \in R, \sigma, \sigma^* \in \mathbb{R}_+. \tag{3.2}$$

We consider solutions $\{u, \vartheta\}$ of the system (2.6)–(2.9) with some additional regularity, namely in the following classes

$$\mathcal{U}_T^\infty = \{u \in L^\infty(0, T; W_0^{1,\infty}(\Omega)) : u_t \in L^2(0, T; H^{-1}(\Omega))\},$$

$$\Theta_T^2 = \{\vartheta \in L^2(0, T; H^1(\Omega)) : \vartheta_t \in L^2(0, T; (H^1(\Omega))')\}.$$

Theorem 3.2. *Let $\{u, \vartheta\}$ and $\{u^*, \vartheta^*\}$ denote two solutions of (2.6)–(2.9) in the class $\{\mathcal{U}_T^\infty, \Theta_T^2\}$, corresponding to the inicial conditions $\{u_0, \vartheta_0\}$ and $\{u^*, \vartheta^*\}$, respectively. Then their differences $\bar{u} = u - u^*$, $\bar{\vartheta} = \vartheta - \vartheta^*$ satisfy*

$$|\bar{u}(t)|^2 + |\bar{\vartheta}(t)|^2 \le e^{\beta t}(|\bar{u}_0|^2 + |\bar{\vartheta}_0|^2), \quad \forall 0 < t < T \tag{3.3}$$

for some constant β, where $|v| = \left(\int_\Omega |v|^2\,dx\right)^{1/2}$ denotes the $L^2(\Omega)$-norm. In particular, there is uniqueness of solutions in the class $\{\mathcal{U}_T^\infty, \Theta_T^2\}$.

Proof: By substracting the equation (2.6) for u and u^* we obtain

$$\bar{u}_t - \nabla \cdot [\nu(\vartheta, |\nabla u|) - \nu(\vartheta, |\nabla u^*|)\nabla u^*] = \nabla \cdot \{[\nu(\vartheta, |\nabla u^*|) - \nu(\vartheta^*, |\nabla u^*|)]\nabla u^*\},$$

which, multiplied by $\bar{u} = u - u^*$ and integrated by parts in Ω, for a.e. $t > 0$, yields

$$
\begin{aligned}
\frac{1}{2}\frac{d}{dt}\int_\Omega |\bar{u}|^2 + \alpha\int_\Omega |\nabla\bar{u}|^2 &\leq l\int_\Omega |\nabla u^*||\nabla\bar{u}||\bar{\vartheta}| \\
&\leq \frac{\alpha}{4}\int_\Omega |\nabla\bar{u}|^2 + \frac{(lM_*)^2}{\alpha}\int_\Omega |\bar{\vartheta}|^2
\end{aligned}
\tag{3.4}
$$

by (2.23), (3.1) and denoting $M_* = \|\nabla u^*\|_{L^\infty(Q_T)}$.

From equations (2.7) for ϑ and ϑ^*, by substraction and integration by parts, we obtain, for a.e. $t > 0$,

$$\frac{c}{2}\frac{d}{dt}\int_\Omega |\bar{\vartheta}|^2 + k\int_\Omega |\nabla\bar{\vartheta}|^2 + \int_{\partial\Omega}\lambda|\bar{\vartheta}|^2 = \int_\Omega (\nu|\nabla u|^2 - \nu^*|\nabla u^*|^2)\bar{\vartheta}. \tag{3.5}$$

Denoting $\nu = \nu(\vartheta, |\nabla u|)$, $\nu^* = \nu(\vartheta^*, |\nabla u^*|)$ and introducing $\nu_* = \nu(\vartheta^*, |\nabla u|)$, from (3.1) and (3.2), we obtain, respectively

$$|\nu - \nu_*| \leq l|\bar{\vartheta}| \quad \text{and} \quad |\nu_* - \nu^*| \leq \mu|\nabla\bar{u}|.$$

Letting $M = \|\nabla u\|_{L^\infty(Q_T)}$, we have

$$
\begin{aligned}
|\nu|\nabla u|^2 - \nu^*|\nabla u^*|^2| &= |\nu\nabla\bar{u}\cdot\nabla(u + u^*) + [(\nu - \nu_*) - (\nu^* - \nu_*)]|\nabla u^*|^2| \\
&\leq \gamma(M + M_*)|\nabla\bar{u}| + M_*^2[l|\bar{\vartheta}| + \mu|\nabla\bar{u}|]
\end{aligned}
$$

and, from (3.5), we obtain with $d = \frac{1}{c}(\gamma M + \gamma M_* + \mu M_*^2)$ and $\delta = 2(d^2/\alpha + lM^2/c)$,

$$
\begin{aligned}
\frac{1}{2}\frac{d}{dt}\int_\Omega |\bar{\vartheta}|^2 &\leq \frac{M^2 l}{c}\int_\Omega |\bar{\vartheta}|^2 + d\int_\Omega |\nabla\bar{u}||\bar{\vartheta}| \\
&\leq \frac{\delta}{2}\int_\Omega |\bar{\vartheta}|^2 + \frac{\alpha}{4}\int_\Omega |\nabla\bar{u}|^2.
\end{aligned}
\tag{3.6}
$$

From (3.4) and (3.6) we have, with $\beta = 2(lM_*)^2/\alpha + \delta$,

$$\frac{d}{dt}(|\bar{u}|^2 + |\bar{\vartheta}|^2) \leq \beta(|\bar{u}|^2 + |\bar{\vartheta}|^2), \quad 0 < t < T, \tag{3.7}$$

and (3.3) follows by integration between 0 and t. $\qquad\square$

From the preceeding proof it is clear that the constant β in (3.3) is positive. However, in certain cases, it is possible to find a sharper estimate for β. Assume

$$\lambda(x, t) \geq \lambda_* > 0, \ \forall(x, t) \in \Sigma_T \tag{3.8}$$

and denote by $C_0 > 0$ and $C_1 > 0$ the constants, such that,

$$C_0\int_\Omega |w|^2 \leq \int_\Omega |\nabla w|^2, \ \forall w \in H_0^1(\Omega) \tag{3.9}$$

$$C_1\int_\Omega |v|^2 \leq \int_\Omega |\nabla v|^2 + \int_{\partial\Omega}|v|^2, \ \forall v \in H^1(\Omega). \tag{3.10}$$

Corollary 3.1. *Under the condition (3.8), we may take in the estimate (3.3), with*
$$m = \frac{1}{c} \min(k, \lambda_*):$$

$$\beta = -\min\{\alpha\, C_0,\, 2\, m\, C_1 - 2\,(l\, M_*)^2/\alpha - 2\,(\gamma\, M + \gamma\, M_* + \mu\, M_*^2)^2/\alpha c^2 - 2\, M^2 l/c\}. \quad (3.11)$$

In particular, β is negative if $\alpha\, m\, C_1 > (l\, M_*)^2 + (\gamma\, M + \gamma\, M_* + \mu\, M_*^2)^2/c^2 + \alpha\, l\, M^2/c$.

Proof: From (3.5), with the assumption (3.8), instead of (3.6) we may obtain (with C_1 given by (3.10))

$$\frac{1}{2}\frac{d}{dt}\int_\Omega |\bar\vartheta|^2 + m\, C_1 \int_\Omega |\bar\vartheta|^2 \leq \frac{\delta}{2}\int_\Omega |\bar\vartheta|^2 + \frac{\alpha}{4}\int_\Omega |\nabla \bar u|^2 \qquad (3.6')$$

which, combined with (3.4) and taking (3.9) into account yields (3.7) with β given by (3.11). $\qquad\square$

Remark 3.5. Suppose that the solution $\{u, \vartheta\}$ corresponding to $f = 0$, $g = 0$ and $h = 0$ is in $\{\mathcal{U}_T^\infty, \Theta_T^2\}$ for every $T < \infty$ and satisfies $\|\nabla u\|_{L^\infty(Q_T)} \leq M$ independently of T. Then, from Corollary 3.1, if $m > \frac{m^2}{c\, C_1}(\frac{\gamma^2}{\alpha\, c} + l)$, we have $\beta < 0$ and an exponential decay holds
$$|u(t)|^2 + |\vartheta(t)|^2 \leq e^{\beta t}(|u_0|^2 + |\vartheta_0|^2) \to 0 \text{ as } t \to \infty.$$

Remark 3.6. Assume $f \in W^{-1,p}(\Omega), p > 2, g \in L^q(\Omega), q > 1$ and $h \in W^{1-1/q,q}(\partial\Omega)$ are time independent, as well as λ. Then, with a similar argument to the proof of Theorem 2.1, there exists at least a solution $\{u^*, \vartheta^*\} \in W_0^{1,r}(\Omega) \times W^{2,s}(\Omega)$ for $2 < r \leq p$, $s = \min(\frac{r}{2}, q) > 1$, to the steady-state problem

$$\begin{cases} -\nabla \cdot (\nu(\vartheta, |\nabla u|)\nabla u) = f & \text{in } \Omega, \quad u = 0 \text{ on } \partial\Omega \\ -k\Delta\vartheta = \nu(\vartheta, |\nabla u|)|\nabla u|^2 + g & \text{in } \Omega, \quad k\frac{\partial\vartheta}{\partial n} + \lambda\vartheta = h \text{ on } \partial\Omega. \end{cases}$$

If we suppose $|\nabla u^*| \in L^\infty(\Omega)$ and $|\nabla u(t)| \leq M$ for all $t < \infty$, then from (3.3) we would find $u(t) \to u_*$ and $\vartheta(t) \to \vartheta_*$ as $t \to \infty$, provided $\beta < 0$.

Remark 3.7. The uniqueness and stability results are limited to a more restrictive class of solutions than the class of the existence theorem of the Section 2. They are based on *a priori* L^∞-bounds for the gradient of the velocity, which is, in general, a delicate open question (except in certain cases of a Newtonian fluid). Moreover those L^∞-bounds, in general, may also depend on the constants α, γ, etc., and Corollary 3.1 is of limited range.

Remark 3.8. For the Newtonian fluid $\nu = \nu(\vartheta)$, i.e., when the viscosity function is independent of the velocity, as a special case of Theorem 3 of Rodrigues (1992), namely for $f \in L^p(Q_T), u_0 \in W^{2-2/p,p}(\Omega)$, with $p > 4$ and with g, h, ϑ_0 given as in (2.3) with $q > 4$, and the respective compatibility condition, as well as the assumption (2.1) satisfied with $(1 - \alpha/\gamma) < 1/\Lambda_p$ ($p > 4$), it is possible to show the existence (and uniqueness) of regular solutions $\{u, \vartheta\}$ in the class $W_p^{2,1}(Q_T) \cap C_\alpha^{1,0}(Q_T), 0 < \alpha < 1 - 4/p < 1$. This regularity is based on the estimate of Lemma 2.1 i) and on the Sobolev imbeddings in $n = 2$ (see Rodrigues, 1992).

REFERENCES

BENSOUSSAN, A., LIONS, J.L. and G. PAPANICOLAU (1978), *Asymptotic Analysis for Periodic Structures*, (North-Holland, Amsterdam).

CIORANESCU, D. (1992), *Quelques Exemples de Fluides Newtonniens Géneralisés*, Pitman Res. Notes Maths. Ser. **274**, p. 1-31.

DAUTRAY, R. and J.L. LIONS (1990), *Mathematical Analysis and Numerical Methods for Science and Technology*, Vol. 1, Phisical Origins and Classical Methods, (Springer, Heidelberg).

LIONS, J.L. (1969), *Quelques méthodes de résolution des problèmes aux limites non-linéaires*, (Dunod, Paris).

LADYŽENSKAJA, J.L., SOLONNIKOV, V.A. and N.N. URAL'CEVA (1968), Linear and Quasilinear Equations of Parabolic type, *Amer. Math. Soc. Transl.*, Vol. **23**, Providence.

RODRIGUES, J.F. (1992), A Nonlinear Parabolic System Arising in Thermodynamics and in Thermomagnetism, *Math. Mods. Meths. in Appl. Sci.*, **2**, p. 271-281.

Related Problems

REMARKS ON THE FLOW OF HOLES AND ELECTRONS IN CRYSTALLINE SEMICONDUCTORS

H. Beirão da Veiga

Dipartimento di Matematica, Università di Pisa
Via F. Buonarroti, 2, 56127 - Pisa, Italy

1. INTRODUCTION

In this paper we study the following system of nonlinear partial differential equations, that describes the transport of holes and electrons in a semiconductor device

$$
\left\{
\begin{aligned}
&\frac{\partial p}{\partial t} - \nabla \cdot (D_1 \nabla p + \mu_1 p \nabla u) = R(p, n) \ , \\
&\frac{\partial n}{\partial t} - \nabla \cdot (D_2 \nabla n - \mu_2 n \nabla u) = R(p, n) \ , \\
&-\nabla \cdot (a \nabla u) = f + p - n \qquad \text{in } \mathbb{R}_+ \times \Omega \ ,
\end{aligned}
\right.
\tag{1.1}
$$

with boundary conditions

$$
\left\{
\begin{aligned}
&p = \phi(x) \ , \ n = \psi(x) && \text{on } \mathbb{R}_+ \times D \ , \\
&(D_1 \nabla p + \mu_1 p \nabla u) \cdot \nu = (D_2 \nabla n - \mu_2 n \nabla u) \cdot \nu = 0 && \text{on } \mathbb{R}_+ \times B \ ,
\end{aligned}
\right.
\tag{1.2}
$$

$$
\left\{
\begin{aligned}
&u = U(x) \ \text{ on } \mathbb{R}_+ \times D \ , \\
&\frac{\partial u}{\partial \nu} = 0 \ \text{ on } \mathbb{R}_+ \times B \ ,
\end{aligned}
\right.
\tag{1.3}
$$

and initial conditions

$$
p(0, x) = p_0(x) \ , \ n(0, x) = n_0(x) \ \text{in } \Omega \ .
\tag{1.4}
$$

Here Ω is a bounded Lipschitzian domain in $\mathbb{R}^n$ (see Nečas, 1967). We assume that the boundary Γ of Ω is the union of two disjoint sets D and B, where D is closed. For convenience we assume that D has non vanishing $(n-1)$- dimensional measure. We denote by ν the unit outward normal to Γ. We refer to van Roosbroeck (1950), Moll (1964), Markowich, Ringhofer and Schmeiser (1990) for more detailed descriptions of

Navier-Stokes Equations and Related Nonlinear Problems
Edited by A. Sequeira, Plenum Press, New York, 1995

the model. The unknowns u, p and n denote the electrostatic potential, the free hole carrier concentration and the free electron carrier concentration. The solutions $p(x, t)$ and $n(x, t)$ are required to be nonnegative. Usually, the above equations are written in the form

$$\frac{\partial p}{\partial t} - \nabla \cdot j_p = R(p, n) \ , \quad \frac{\partial n}{\partial t} - \nabla \cdot j_n = R(p, n)$$

where the hole and electron current densities are given by

$$j_p = D_1 \nabla p + \mu_1 p \nabla u \ , \quad j_n = D_2 \nabla n - \mu_2 n \nabla u$$

respectively. We do not assume that the hole and electron diffusion coefficients D_1 and D_2 and the hole and electron mobilities μ_1 and μ_2 are (necessarily) connected by the Einstein relations $D_i = (k\vartheta_0)\mu_i$, $i = 1, 2$, where k is the Boltzmann's constant and ϑ_0 the constant temperature. We assume that D_1, D_2, μ_1, μ_2 and a (the dielectric permittivity) are positive constants. This leads to the equations (1.1).

In the sequel we assume that $\phi, \psi \in H^1(\Omega) \cap L^\infty_+(\Omega)$ and $U \in H^1(\Omega) \cap L^\infty(\Omega)$. The symbol "+" means the cone of nonnegative functions (we point out that we will not define standard notation). Finally, we assume that the net density of ionized impurities f satisfies

$$f \in L^s(0, +\infty; \ L^r(\Omega)) \tag{1.5}$$

for some fixed $s \in [4, +\infty]$ and some fixed $r \in]2N, +\infty]$. In particular, the case of an arbitrary bounded measurable $f(t, x)$ is included. The most important case in applications is that in which f is independent of time. However, it has mathematical interest considering the above more general case (a similar remark holds if $N > 3$). Concerning the initial data we assume that

$$p_0, \ n_0 \in L^\infty_+(\Omega) \ . \tag{1.6}$$

The recombination term $R(p, n)$ is assumed to be a locally Lipschitz continuous function, defined on $\mathbb{R}_+ \times \mathbb{R}_+$, such that

$$\lim_{p+n \to +\infty} \frac{R(p, n)^+}{p + n} = 0 \ , \tag{1.7}$$

where $z^+ = \max\{z, 0\}$, moreover

$$\begin{cases} R(p, 0) \geq 0 \ , \ \forall \, p \geq 0 \ , \\ \\ R(0, n) \geq 0 \ , \ \forall \, n \geq 0 \ . \end{cases} \tag{1.8}$$

We point out that (1.7), (1.8) hold for the Shockley-Read-Hall recombination term

$$R(p, n) = \frac{1 - pn}{r_0 + r_1 p + r_2 n} \ , \tag{1.9}$$

in which r_0, r_1, and r_2 are positive constants. It is worth noting that the solutions p and n must be nonnegative. Under suitable hypotheses (see below) Gajewski and Gröger (1986), show that the solution of the above problem satisfies

$$p, n \in L^\infty(0, T; L^\infty_+(\Omega)) \tag{1.10}$$

for each fixed T. However, the $L^\infty(\Omega)$-norm of $p(t)$ and $n(t)$ may blow up, at most exponentially, as t goes to $+\infty$. For different boundary conditions a similar result is proved by Seidman and Troianiello (1985). For previous, related results, see Mock (1974, 1975) and Gajewski (1985). A main open question, in order to approach the problem of the qualitative behaviour of solutions for large values of t, is to know whether the solutions are uniformly bounded in $[0, +\infty[\times \Omega$. This is our main concern here. We will prove, under no smallness or under other restrictive assumptions, that

$$p, n \in L^\infty(0, +\infty); L^\infty_+(\Omega)) . \tag{1.11}$$

A partial result in this direction was obtained by Gröger, 1986. This author exhibits a sufficient condition in order that (1.11) holds. For the Shokley-Read-Hall recombination term (1.9) Gröger's condition corresponds to the following smallness assumption on $f : -M_1 \leq f(x) \leq M_2$ a.e. in Ω, where M_1 and M_2 satisfy $M_i < a/D_i(r_1 + r_2)$, $i = 1, 2$. Here $\mu_i = D_i$.

With respect to Gajewski and Gröger (1986), we do not assume that ϕ and ψ are bounded from below by a strictly positive constant and that $\nabla (\log \phi + U)$ and $\nabla (\log \psi - U)$ are bounded. On the other hand, in Gajewski and Gröger (1986), the authors assume a more general boundary condition on the Neumann boundary B.

Before stating our main theorem we introduce some notation. We set $Q_T = (0, T) \times \Omega$, $Q = Q_\infty$. We denote by $\| \cdot \|_r$, $r \in [1, +\infty]$, the canonical norm in $L^r = L^r(\Omega)$ and by $\| \cdot \|_{r,s;T}$, $r, s \in [1, +\infty]$ and $T \in]0, +\infty]$ that in $L^s(0, T; L^r)$. For convenience, we set $\| \cdot \| = \| \cdot \|_2$ and $\| \cdot \|_{r,s} = \| \cdot \|_{r,s;+\infty}$. We denote by $|E|$ the N-dimensional Lebesgue measure of a set E.

We denote by V the Hilbert space $V = \{v \in H^1 : v = 0 \text{ on } D\}$ and by V' its dual space. In order to use here a standard notation, let us set $H = L^2(\Omega)$. By identifying H with its dual H' one has $V \hookrightarrow H \hookrightarrow V'$, where each space is dense in the next one. The spaces V, H, and V' are in a typical situation, often considered on studying weak solutions of partial differential equations. We denote by $(\cdot, \cdot)$ the scalar product in H (or in H^N). We use the same notations for scalar and for vector fields) and by $\langle \cdot, \cdot \rangle$ the duality pairing between V' and V. If v belongs to $L^2_{loc}(0, +\infty; V)$ we denote by v' the derivative of v as a distribution in $]0, +\infty[$ with values in V. Since $V \hookrightarrow V'$ it could be that $v' \in L^2_{loc}(0, +\infty; V')$. For properties connected to this (already classical) setting up we refer the reader to Lions and Magenes (1968), Dautray-Lions (1992).

We set $\mu_0 = \sqrt{\mu_1 \mu_2}$, $\mu_3 = \max \{\mu_1, \mu_2\}$, $\mu_4 = \min \{\mu_1, \mu_2\}$, $\rho = \min \{D_1/\mu_1, D_2/\mu_2\}$, $b = r_0^{-1} \mu_3$. Moreover

$$M_0 = \max \{\|p_0\|_\infty, \|n_0\|_\infty, \|\phi\|_\infty, \|\psi\|_\infty\} . \tag{1.12}$$

We denote by c_0 a positive constant such that the Poincaré's inequality

$$\int v^2 dx \leq c_0 \int |\nabla u|^2 dx, \quad \forall v \in V , \tag{1.13}$$

holds and by c_1 a positive constant such that the Sobolev's embedding theorem

$$\left(\int v^{2^*} dx \right)^{1/2^*} \leq c_1 \left(\int |\nabla v|^2 dx \right)^{1/2}, \quad \forall v \in V , \tag{1.14}$$

holds. If $N \geq 3$, we denote by 2^* the embedding Sobolev exponent $2^* = 2N/(N - 2)$ and by $\hat{2}$ its dual exponent $\hat{2} = 2N/(N + 2)$. If $N = 2$ (hence $r \in]4, +\infty]$; recall (1.5)) we set $2^* = 4r/(r - 4)$ and $\hat{2} = 4r/(4 + 3r)$. Note that (1.14) also holds for $N = 2$. Moreover $1/2^* + 1/\hat{2} = 1$.

Next we set

$$\delta_0 = \frac{\rho\mu_0^2}{4\mu_3 c_0} \tag{1.15}$$

and we define N_0 as being a positive constant such that

$$R(p,n)^+ \le \delta_0(p+n) \qquad \text{if } p+n \ge N_0 \ . \tag{1.16}$$

N_0 exists, by the assumption (1.7). We set $N_1 = 2^{-1}\delta_0 \sup\limits_{p+n \le N_0} R(p,n)^+$ and we define

$$M = \max\{M_0, 1, N_0, N_1\} \ . \tag{1.17}$$

In the particular case of the Schokley-Read-Hall recombination term (1.9), we simply set

$$M = \max\{M_0, 1\} \ . \tag{1.18}$$

Under the above hypotheses there is a weak solution (p,n,u) of problem (1.1)-(1.4) in the following class: $p - \phi$ and $n - \psi$ belong to $L^2_{\text{loc}}(0, +\infty; V)$; $u - U$ belongs to $L^r_{\text{loc}}(0, +\infty; V)$; p and n are nonnegative a.e. in Q and belong to $L^\infty_{\text{loc}}(0, +\infty; L^\infty)$. Moreover, the solution is unique in the above class. Functions p, n and u in the above class are said to be a weak solution (1.1)-(1.4) if, for each fixed $v \in V$, one has $a(\nabla u, \nabla v) = (f + p - n, v)$, and also, in the sense of $\mathcal{D}'(]0, +\infty[)$ (or equivalently, almost everywhere in $]0, +\infty[)$

$$\langle p', v\rangle + (D_1\nabla p, \nabla v) + \mu_1(p\nabla u, \nabla v) = (R(p,n), v) \ ,$$

and

$$\langle n', v\rangle + (D_2\nabla n, \nabla v) + \mu_2(-n\nabla u, \nabla v) = (R(p,n), v) \ .$$

Moreover, $p(0) = p_0$, $n(0) = n_0$. Note that p and n are continuous on $[0, +\infty[$ with values in $H = L^2(\Omega)$. We may also write the above equations in terms of $y = p - \phi$, $z = n - \psi$ and $w = u - U$.

Our main result is the following

Theorem 1.1. *The above solution (p,n) of problem (1.1-(1.4) is uniformly bounded in $Q = R_+ \times \Omega$. More precisely,*

$$\sup_Q (p(t,x) + n(t,x)) \le CM \left(1 + \|f\|_{r,s}^{2(1+\frac{1}{\chi})}\right) \tag{1.19}$$

where

$$\chi = \frac{2}{N} - \frac{4}{r} \quad \text{if} \quad N \ge 3, \quad \chi = \frac{1}{2} - \frac{2}{r} \quad \text{if} \quad N = 2 \ . \tag{1.20}$$

The constant C depends only on $N, c_0, c_1a, D_1, D_2, \mu_1, \mu_2, |\Omega|$ and r.

The reader is assumed to be well acquainted with the formulation of PDE's in weak form. We adopt here classical terminology and notations in order to bring out clearly the underlying ideas. The interpretation of some of the terminology and the justification of some of the calculations (in terms of weak solutions, distributional derivatives, duality pairing, and so on) is done by using well known standard devices. We refer the unexperienced reader to Dautray- Lions (1992), Ladyzhenskaya, Solonnikov and Ural'ceva (1968), Lions and Magenes (1968), Ladyzhenskaya (1973); see, in particular Dautray- Lions (1992) Chapt. XVIII, §§ 1 and 3.

Some words about the proof of the Theorem 1.1. are in order. The proof consists in four steps. The first one consists in giving (q, m) (see (3.2)) and in solving for (p, n, u) the linear problem (3.3)-(3.6); the definition of $\hat{q}$ and $\hat{m}$ is given in (3.1). For the time being, the value of the positive constant μ in the definition of $\hat{q}$ and $\hat{n}$ is arbitrary.

The second step consists in proving the existence of a weak solution (p, n, u) of problem (3.11)-(3.14). This is done by proving the existence of a fixed point $(p, n) = (q, m)$ for the map S which associates to each (q, m) the (unique) solution of the linear problem (3.3)-(3.6).

The third step consists in showing that the solutions p and n of problem (3.11)-(3.14) are nonnegative. In particular, from this result it follows that $\hat{p} = \max\{p, \mu\}$ and $\hat{n} = \max\{n, \mu\}$.

The proof of the above three steps is done by following Gajewski and Gröger (1986), and will be postponed to Section 3.

The fourth step is the main point in our paper. Here, we show that the solution p and n of problem (3.11)-(3.14) constructed above (solutions that depend on the particular value of the parameter μ) are bounded from above by the right hand side of equation (1.19). Hence, by choising μ larger than the above right hand side it follows that $\hat{p} = p$ and $\hat{n} = n$. Consequently, p, n (and u) are a (weak) solution of problem (1.1)-(1.4).

The above steps prove the existence part together with the main estimate (1.19). In order to give a better understanding of the underlying ideas developed in the fourth step (proof of the main estimate (1.19)) we rather prefer to present the corresponding calculations in the form of an a priori estimate. This is done in Section 2. The proof of (1.19) for the solution (p, n) of problem (3.11)-(3.14) (the above step four) is done by making (quite obvious) minor changes on the argument developed in Section 2. The few modifications to be done are indicated in Section 3.

Finally, the proof of the uniqueness of the solution follows the usual devices and is presented (at the end of Section 3) just for the reader's convenience.

Before going on let us remark that our proof can be adapted to more general situations: dependence on time of the data $\phi \geq 0$, $\psi \geq 0$, provided that they belong to $L^\infty(0, +\infty; H^1 \cap L^\infty)$ and that ϕ', ψ' belong to $L^2_{\text{loc}}(0, +\infty; V')$ (this generalization requires only a few modifications in the proofs); other boundary conditions (for instance, unilateral contraints, see Beirão da Veiga and Dias, 1972; Beirão da Veiga, 1974); dependence of the coefficients $D_i, \mu_i (i = 1, 2)$, and a on the solution itself and on (x, t), under suitable assumptions.

We also note that many regularity results follow as straightforward applications of well-known theorems or techniques. For instance, $u \in C(0, +\infty; C^{0,\alpha}(\overline{\Omega}))$ for some $\alpha > 0$, since $p, n \in C(0, +\infty; L^r)$, for arbitrarily large r; see Stampacchia (1960), and also Beirão da Veiga, 1972. We conjecture that p and n are Hölder-continuous on Q if $N = 2$ or 3 (under slight regularity assumptions on the boundary of D as a subset of $\partial\Omega$) but that (in general) this result is false if $N \geq 4$. However, we did not investigative in this direction.

It is worth noting that obvious modifications (in fact, simplifications) in our proofs (set $\partial p/\partial t = \partial n/\partial t = 0$ everywhere ...) yield the following result for *stationary solutions*.

Theorem 1.2. *Let ϕ, ψ, U be as above and let $f \in L^r(\Omega)$, $r \in\,]2N, +\infty]$. Then, the problem (1.1)-(1.3) admits a time independent solution (p, n, u) such that $p - \phi$,*

$n - \psi,\ u - U \in V$. Moreover $p \geq 0,\ n \geq 0$ and

$$\sup_{\Omega} (p(x) + n(x)) \leq C\, M(1 + \|f\|_r^{2(1+1/\chi)}) \ . \tag{1.21}$$

Here, χ and C are as in Theorem 1.1.. In the definition of M drop $\|p_0\|_\infty$ and $\|n_0\|_\infty$.

A second basic question in order to study the asymptotic behaviour of the set of solutions is that of the existence (or non existence) of a (significant) functional space X and of a bounded set $B_0 \subset X$ that atracts (uniformly) each bounded subset B of X. We prove that this property holds for $X = L^2(\Omega)$. In order to prove this result, the first step consists in showing the existence and the uniqueness of a solution

$$p, n \in C(0, +\infty; L^2(\Omega)) \tag{1.22}$$

in correspondence to each (arbitrary) pair of initial data $p_0, n_0 \in L^2(\Omega)$. We prove the existence of a weak solution (p, n) to our problem in the class $p - \phi,\ n - \psi \in L^2_{\mathrm{loc}}(0, +\infty; V),\ p_t, n_t \in L^2_{\mathrm{loc}}(0, +\infty; V_N')$ is the dual space of $V_N = \{v \in V : \nabla v \in L^N(\Omega)\}$; if $N = 2$ replace N by $q,\ q > 2$. In order to prove the uniqueness of the solution and also that $p_t, n_t \in L^2_{\mathrm{loc}}(0, +\infty; V')$ (hence that (1.22) holds) we assume the property described below. Consider the elliptic mixed boundary value problem

$$-\nabla u = g \ \text{ in } \ \Omega \ , \quad u = U \ \text{ on } \ D \ , \quad \partial u / \partial v = 0 \ \text{ on } \ B \ . \tag{1.23}$$

We assume that there is a functional space Y and a real q ($q > 2$ if $N = 2$; $q = N$ otherwise) such that if $g \in L^2(\Omega)$ and $U \in Y$ then the variational solution u of problem (1.23) satisfies

$$\|\nabla u\|_q \leq c\,(\|g\| + \|U\|_Y) \ . \tag{1.24}$$

Note that this is an assumption on $\{\Omega, B, D\}$. This assumption is out of place if $N > 4$ since $H^2(\Omega)$ is not contained in $L^N(\Omega)$. If $N = 2$, it holds if Ω is a bounded domain with a polygonal boundary (or a regular transformation of such a set). In this case $q > 2$ can be arbitrarily fixed, moreover $Y = W^{1-\frac{1}{q}, q}(B)$. This follows essentialy from results by Lorenzi (1975). Since it is sufficient to have (1.24) for some $q > 2$, it seems possible to use Gröger's results (1989).

If $N = 3$ and if Ω is bounded convex set with a polyhedral boundary (or a regular transformation of it) then the solution of problem (1.23) belongs to $H^{3/2}(\Omega)$. This follows from results of Grisvard (1992), at least if $U = 0$. Note that $\nabla u \in L^3(\Omega)$ since $H^{3/2}(\Omega) \hookrightarrow W^{1,3}(\Omega)$. It is worth noting that in Grisvard (1992), the author considers only homogeneous boundary conditions, but that looks inessential there. We also note that $W^{1,3}$-regularity holds under weaker hypotheses on the angles between faces than that needed to get $H^{3/2}$-regularity. But we do not know about precise statements in the literature.

The following result is proved in a forthcoming paper. For brevity, we assume that $f \in L^\infty(Q)$ and that $R(p, n)$ is given by (1.9).

Theorem 1.3. Let the assumption (1.24) hold and let ϕ, ψ, U, f and $R(p, n)$ be as in Theorem 1.1.; moreover $U \in Y$. Then, to each pair of initial data $(p_0, n_0) \in L^2(\Omega)$ it corresponds a unique solution (p, n) of problem (1.1)-(1.4) in the class $p - \phi, n - \psi \in L^2_{\mathrm{loc}}(0, +\infty; V);\ p_t, n_t \in L^2_{\mathrm{loc}}(0, +\infty; V')$. Moreover, $p, n \in C(0, +\infty; L^2(\Omega))$ and there is a positive constant C_0 which depends on the norms $\|\phi\|_\infty, \|\psi\|_\infty, \|\|f\|\|_\infty$ but not on p_0, n_0 and U) such that

$$\|p(t)\|^2 + \|n(t)\|^2 \leq C_0 + ce^{\nu t}\,(\|p_0\|^2 + \|n_0\|^2) \ . \tag{1.25}$$

for each $t \geq 0$. The positive constant c and ν are independent of the data ϕ, ψ, U, f, p_0 and n_0.

The above result shows that the set

$$B_0 = \{(\bar{p}, \bar{n}) \in L^2(\Omega) : \bar{p} \geq 0,\ \bar{n} \geq 0,\ \|\bar{p}\|^2 + \|\bar{n}\|^2 \leq C_0\}$$

is a global bounded atractor in the space $L^2(\Omega)$.

2. PROOF OF THE MAIN ESTIMATE

As explained in Section 1, the proof of (1.19) will be carried out here as an a priori estimate for solutions of problem (1.1)-(1.4). However, according to the above explanation, the proof should be applicable to the solution of problem (3.11)-(3.14). We assume here that the solution of problem (1.1)-(1.4) belongs to the existence class, described before the statement of Theorem 1.1., since the solution of problem (3.11)-(3.14) belongs to this class and since the justification of each single calculation is the same in both cases.

For $k \geq 0$ we set

$$\bar{w} = w^{(k)} = \max\{w - k, 0\}\ .$$

The notation $\bar{w}$ will be used when there is no danger of misunderstanding. In the sequel $k \geq M_0$. Hence $\bar{p}$ and $\bar{n}$ belong to $L^2_{loc}(0, +\infty; V)$, moreover $\bar{p}(0) = \bar{n}(0) = 0$. See, for instance, Dautray- Lions (1992) Vol. 2, Chap. IV, §7, Prop. 6. Next multiply the equation $(1.1)_1$ by $\bar{p}$, integrate over Ω and made suitable integrations by parts. This yields

$$\frac{1}{2}\frac{d}{dt}\|\bar{p}\|^2 + D_1\|\nabla\bar{p}\|^2 + \mu_1 \int \nabla u \cdot \left(\frac{1}{2}\nabla\bar{p}^2 + k\nabla\bar{p}\right)\,dx = \int R(p,n)\bar{p}\,dx \qquad (2.1)$$

where integrals are over Ω. Again by suitable integrations by parts, and also by taking into account the equation $(1.1)_3$ and the boundary conditions, one gets

$$\frac{1}{2}\frac{d}{dt}\|\bar{p}\|^2 + D_1\|\nabla\bar{p}\|^2 + \frac{\mu_1}{2a}\int(f+p-n)\bar{p}^2 dx + \frac{\mu_1}{a}k\int(f+p-n)\bar{p}\,dx =$$
$$= \int R(p,n)\bar{p}\,dx\,. \qquad (2.2)$$

In a similar way, by starting from $(1.1)_2$, one gets

$$\frac{1}{2}\frac{d}{dt}\|\bar{n}\|^2 + D_2\|\nabla\bar{n}\|^2 - \frac{\mu_2}{2a}\int(f+p-n)\bar{n}^2 dx + \frac{\mu_2}{a}k\int(f+p-n)\bar{n}\,dx =$$
$$= \int R(p,n)\bar{n}\,dx\,. \qquad (2.2')$$

Next, multiply equation (2.2) by μ_2, equation (2.2 ') by μ_1, and add both equations. This yields

$$\frac{1}{2}\frac{d}{dt}\int(\mu_2\bar{p}^2 + \mu_1\bar{n}^2)\,dx + \rho\mu_0^2\int|\nabla(\bar{p},\bar{n})|^2\,dx + \frac{\mu_0^2}{2a}\int(f+p-n)\,(\bar{p}^2 - \bar{n}^2)\,dx +$$

$$+\frac{\mu_0^2}{a}k\int(f+p-n)\,(\bar{p}-\bar{n})\,dx = \int R(p,n)\,(\mu_2\bar{p}+\mu_1\bar{n})\,dx\ , \qquad (2.3)$$

where, for convenience, we set

$$|\nabla(\overline{p}, \overline{n})|^2 = |\nabla\overline{p}|^2 + |\nabla\overline{n}|^2 \ .$$

Since p and n are nonegative, one easily proves that $(\overline{p} - \overline{n})\,(p - n) \geq 0$ and that $|\overline{p} - \overline{n}| \leq |p - n|$. In particular $(\overline{p} - \overline{n})\,(p - n) \leq (p - n)^2$. Hence

$$(\overline{p}^2 - \overline{n}^2)\,(p - n) \geq (\overline{p} - \overline{n})^2\,(\overline{p} + \overline{n}) \ .$$

On the other hand

$$|f\,(\overline{p}^2 - \overline{n}^2)| \leq \frac{1}{2}\,|f|^2\,(\overline{p} + \overline{n}) + \frac{1}{2}\,(\overline{p} - \overline{n})^2\,(\overline{p} + \overline{n}) \ .$$

It readily follows that

$$(f + p - n)\,(\overline{p}^2 - \overline{n}^2) \geq \frac{1}{2}\,(\overline{p} - \overline{n})^2\,(\overline{p} + \overline{n}) - \frac{1}{2}\,|f|^2\,(\overline{p} + \overline{n}) \ . \tag{2.4}$$

Similarly, $(f + p - n)\,(\overline{p} - \overline{n}) \geq \frac{1}{2}\,(\overline{p} - \overline{n})^2 - \frac{1}{2}\,f^2$ (however, we will estimate the corresponding term in equation (2.3) in a different way). From (2.3) and (2.4) one gets

$$\frac{1}{2}\,\frac{d}{dt}\int(\mu_2\overline{p}^2 + \mu_1\,\overline{n}^2)\,dx + \rho\mu_0^2\int|\nabla(\overline{p}, \overline{n})|^2\,dx +$$

$$\frac{\mu_0^2}{4a}\int(\overline{p} - \overline{n})^2\,(\overline{p} + \overline{n})\,dx + \frac{\mu_0^2}{a}\,k\int(\overline{p} - \overline{n})^2\,dx \tag{2.5}$$

$$\leq \frac{\mu_0^2}{4a}\int|f|^2\,(\overline{p} + \overline{n})\,dx + \frac{\mu_0^2}{a}\,k\int|f|\,|\overline{p} - \overline{n}|\,dx + \int R(p, n)\,(\mu_2\overline{p} + \mu_1\overline{n})\,dx \ .$$

Next, we estimate the last term in the above inequality. In the specific case (1.9), one has

$$R(p, n)\,(\,\mu_2\overline{p} + \mu_1\overline{n}) \leq b(\overline{p} + \overline{n}) \ . \tag{2.6}$$

In the general case (1.7), (1.8) one has $R(p, n)^+ \leq \delta_0(\overline{p} + \overline{n} + 2k)$ for $k \geq \max\{N_0, N_1\}$. Hence

$$R(p, n)\,(\mu_2\overline{p} + \mu_1\overline{n}) \leq \mu_3\delta_0(\overline{p} + \overline{n})^2 + 2\delta_0\mu_3 k(\overline{p} + \overline{n}) \ .$$

By taking into account (1.15) one gets

$$R(p, n)\,(\mu_2\overline{p} + \mu_1\overline{n}) \leq \frac{\rho\mu_0^2}{4c_0}\,(\overline{p} + \overline{n})^2 + 2\delta_0\mu_3\,k\,(\overline{p} + \overline{n}) \ . \tag{2.7}$$

Hence, by (1.13), (2.5) shows that

$$\frac{1}{2}\,\frac{d}{dt}\int(\mu_2\overline{p}^2 + \mu_1\overline{n}^2)\,dx + \frac{\rho\mu_0^2}{2}\int(|\nabla\overline{p}|^2 + |\nabla\overline{n}|^2)\,dx$$

$$\leq \frac{\mu_0^2}{4a}\int f^2(\overline{p} + \overline{n})\,dx + \frac{\mu_0^2}{a}\,k\int|f|\,(\overline{p} + \overline{n})\,dx + 2\delta_0\mu_3 k\int(\overline{p} + \overline{n})\,dx \ . \tag{2.8}$$

In the specific case (1.9) we could replace $2\delta_0\mu_3 k$ by b. However (for convenience) we rather prefer replacing $2\delta_0\mu_3$ by b and assuming that $k \geq 1$. Under this assumption (2.8) holds for each $k \geq M$.

Next, we define

$$A_k(t) = \{x \in \Omega : p(t, x) > k\} \,\cup\, \{x \in \Omega : n(t, x) > k\} \ . \tag{2.9}$$

We set

$$\alpha = \frac{(N+2)r}{4N} \quad , \quad \beta = \frac{(N+2)r}{(N+2)r - 4N} \tag{2.10}$$

if $N \geq 3$, and

$$\alpha = \frac{4+3r}{8} \quad , \quad \beta = \frac{4+3r}{3r-4} \tag{2.11}$$

if $N = 2$. Note that $\alpha^{-1} + \beta^{-1} = 1$ and that $2/(\hat{2}\beta) = 1 + \chi$.

By Hölder's inequality one gets

$$\int f^2(\overline{p} + \overline{n}) \, dx \leq \|\overline{p} + \overline{n}\|_{2^*} \|f^2\|_{\hat{2}\alpha} |A_k(t)|^{\frac{1}{2\beta}} .$$

Since $\|\overline{p} + \overline{n}\|_{2^*} \leq \sqrt{2}c_1 \|\nabla(\overline{p}, \overline{n})\|$ it follows that

$$\int f^2 (\overline{p} + \overline{n}) \, dx \leq c_1^2 \varepsilon \|\nabla(\overline{p}, \overline{n})\|^2 + \frac{1}{2\varepsilon} \|f^2\|_{\hat{2}\alpha}^2 |A_k(t)|^{1+\chi} . \tag{2.12}$$

By setting $\varepsilon = a\rho/2c_1^2$ one obtains

$$\frac{\mu_0^2}{4a} \int f^2(\overline{p} + \overline{n}) \, dx \leq \frac{\rho\mu_0^2}{8} \|\nabla(\overline{p}, \overline{n})\|^2 + \frac{c_1\mu_0^2}{4a^2\rho} \|f\|_r^4 |A_k(t)|^{1+\chi} .$$

Note that $r = 2\hat{2}\alpha$. In this section, for convenience, we denote by C positive constants that depend, at most, on the constants $N, c_0, c_1, a, D_1, D_2, \mu_1, \mu_2, r$ and $|\Omega|$. The same symbol C can be used to denote distinct constants, even in the same formula. It is worth noting that all the constants C that appear in the sequel can be easily estimated (as in (2.12)).

By replacing in equation (2.12) f^2 by kf we show that

$$k \int |f| \, (\overline{p}, \overline{n}) \, dx \leq c_1^2 \varepsilon \|\nabla(\overline{p}, \overline{n})\|^2 + \frac{k^2}{2\varepsilon} \|f\|_{r/2}^2 |A_k(t)|^{1+\chi} , \tag{2.13}$$

and by replacing in (2.13) f by $2\delta_0$ we show that

$$2\delta_0 k \int (\overline{p} + \overline{n}) \, dx \leq c_1^2 \varepsilon \|\nabla(\overline{p}, \overline{n})\|^2 + \frac{2\delta_0^2}{\varepsilon} k^2 |\Omega|^{\frac{2}{2\alpha}} |A_k(t)|^{1+\chi} . \tag{2.14}$$

By making obvious choices for ε in the above estimates, we show from (2.8) that

$$\frac{d}{dt} \int (\mu_2\overline{p}^2 + \mu_1\overline{n}^2) \, dx + \nu \int (\mu_2\overline{p}^2 + \mu_1\overline{n}^2) \, dx \leq C \left[\|f\|_r^4 + k^2(\|f\|_{r/2}^2 + 1) \right] |A_k(t)|^{1+\chi} , \tag{2.15}$$

where $\nu = \rho\mu_0^2/4c_0\mu_3$. We have used also Poincaré's inequality (1.13). Recall that $\hat{2}\alpha = r/2$. Since we are not looking for the sharpest estimates, we replace in (2.15) the term $\|f\|_r^4$ by $k^2\|f\|_r^4$. We get the simplified expression

$$y_k'(t) + \nu y_k(t) \leq Ck^2 g(t) |A_k(t)|^{1+\chi} \tag{2.16}$$

where

$$\begin{cases} y_k(t) = \int (\mu_2\overline{p}^2 + \mu_1\overline{n}^2) \, dx , \\[2mm] g(t) = 1 + \|f(t)\|_r^4 . \end{cases} \tag{2.17}$$

Note that $y_k(0) = 0$ since $k \geq M$. It readily follows from (2.16) that

$$y_k(t) \leq Ck^2 \int_0^t e^{-\nu(t-s)} g(s) |A_k(s)|^{1+\chi} \, ds . \tag{2.18}$$

In particular

$$y_k(t) \leq \frac{C}{\nu} k^2 \left(\sup_{0 \leq t < +\infty} g(t) \right) \left(\sup_{0 \leq t < +\infty} |A_k(t)|^{1+\chi} \right) . \tag{2.19}$$

Let now $h > k \geq M$. One has

$$y_k(t) \geq \mu_4 (h-k)^2 |A_h(t)| . \tag{2.20}$$

Set

$$\Phi(h) = \sup_{0 \leq t < +\infty} |A_h(t)|^{1/2} , \tag{2.21}$$

for each $h \geq M$. Note that $\Phi(h) \leq |\Omega|^{1/2}$. From (2.19) and (2.20) it follows that

$$\Phi(h) \leq \frac{\gamma k}{h-k} \Phi(k)^{1+\chi} , \quad \text{for } h > k \geq M , \tag{2.22}$$

where

$$\gamma = \left(\frac{C}{\nu \mu_4} \right)^{1/2} (1 + \|f\|_{r,\infty}^2) . \tag{2.23}$$

The proof of the following result is postponed to the end of this section.

Lemma 2.1. *Let $\Phi(\xi)$ be a function defined for $\xi \geq M$, nonegative and decreasing (not necessarily strictly decreasing) such that, for $h > k \geq M$, the estimate (2.22) holds. Then $\Phi(2d) = 0$ where*

$$d = M + 2^{\frac{2}{\chi} + \frac{1}{\chi^2}} \gamma^{1 + \frac{1}{\chi}} \Phi(M)^{1+\chi} M . \tag{2.24}$$

M and γ are nonegative constants.

Application of the above lemma shows that $|A_{2d}(t)| = 0$ for $t \in [0, +\infty[$, hence $p^{(2d)}$ and $n^{(2d)}$ vanish on Q. This shows that (1.19) holds when $s = +\infty$. Note that $\Phi(M) \leq |\Omega|^{1/2}$. □

Next, we show that (1.19) holds if

$$f \in L^s (0, +\infty; L^r) , \tag{2.25}$$

for some $s \geq 4$. In fact, from (2.18) and from Hölder's inequality one gets

$$y_k(t) \leq C k^2 \left(\frac{1}{\nu} + \frac{1}{(\nu \vartheta')^{1/\vartheta'}} \|f\|_{r,4\vartheta;t}^4 \right) \sup_{0 \leq t < +\infty} |A_k(t)|^{1+\chi} , \tag{2.26}$$

where $\vartheta \geq 1$ and $1/\vartheta + 1/\vartheta' = 1$. Hence (2.22) holds, where now γ is given by

$$\gamma = \left(\frac{C}{\nu \mu_4} \right)^{1/2} \left(1 + \frac{\nu^{1/2\vartheta}}{\vartheta'^{1/2\vartheta'}} \|f\|_{r,4\vartheta}^2 \right) . \tag{2.27}$$

Note that the right hand side of (2.27) coincides with that of (2.23) provided that $\vartheta = +\infty, \vartheta' = 1$. □

Finally, we prove the Lemma 2.1. We will prove a slightly more general result, that can be useful in other situations. This kind of results turn back to ideas of De Giorgi, and were developed by other authors in particular O. A. Ladyzhenskaya and G. Stampacchia. See Stampacchia (1963) and Ladyzhenskaya, Solonnikov and and Ural'ceva (1968).

Lemma 2.2. *Let $\phi(\xi)$ be a function defined for $\xi \geq M$, nonegative and decreasing (not necessarily strictly) such that for $h > k \geq M$ the estimate*

$$\phi(h) \leq \frac{\gamma k^{\vartheta}}{(h-k)^{\alpha}} \, \phi(k)^{1+\chi} \tag{2.28}$$

holds. Here, γ, α and χ are positive constants. Moreover $\vartheta < \alpha(1+\chi)$. Then $\phi(2d) = 0$, where $d > M$ is the root of the equation

$$d = M + \lambda M^{\vartheta/\alpha} \, d^{\frac{\vartheta-\alpha}{\chi\alpha}} \tag{2.29}$$

and

$$\lambda^{\alpha} = 2^{\frac{\alpha+\vartheta}{\chi} + \frac{\alpha}{\chi^2}} \, \gamma^{1+\frac{1}{\chi}} \phi(M)^{1+\chi} \ . \tag{2.30}$$

Proof: Set $k_j = d(2 - 2^{-j})$, $j = 0, 1, 2, \ldots$. We want to show that

$$\phi(k_j) \leq \left[\frac{d^{\alpha-\vartheta}}{2^{\alpha(j+1+1/\chi)+\vartheta} \, \gamma} \right]^{1/\chi} \ . \tag{2.31}$$

Since $\lim_{j \to +\infty} k_j = 2d$, then (2.31) implies that $\phi(2d) = 0$. Equation (2.28) for $h = k_0$ and $k = M$ shows that

$$\phi(k_0) \leq \frac{\gamma M^{\vartheta}}{(d-M)^{\alpha}} \, \phi(M)^{1+\chi} \ . \tag{2.32}$$

By replacing $(d - M)^{\alpha}$ by the value obtained from equation (2.29) it readily follows that the right hand side of (2.32) is equal to the right hand side of (2.31) for $j = 0$. Next, by supposing that (2.31) holds for some $j \geq 0$ and by using (2.28), we prove that

$$\phi(k_{j+1}) \leq \frac{2^{\vartheta+(j+1)\alpha} \, \gamma}{d^{\alpha-\vartheta}} \left[\frac{d^{\alpha-\vartheta}}{2^{\alpha(j+1+1/\chi)+\vartheta} \, \gamma} \right]^{1+1/\chi} \ . \tag{2.33}$$

Straightforward calculations show that the right hand side of (2.33) is equal to the right hand side of (2.31) if here we replace j by $j + 1$.

3. EXISTENCE

In the sequel we denote by μ a real fixed number, larger than M_0. If w is a real function we set

$$\hat{w} = \begin{cases} \mu & \text{if } w \geq \mu \ , \\ w & \text{if } 0 \leq w \leq \mu \ , \\ 0 & \text{if } w \leq 0 \ . \end{cases} \tag{3.1}$$

We assume that ϕ, ψ and U are as in Section 1. We denote by T a fixed, positive (arbitrarily large) real number and we assume that $f \in L^1(0, T; L^2)$. Let

$$q, m \in L^2(0, T; L^2) \tag{3.2}$$

and consider the auxiliary problem (see Gajewski and Gröger, 1986)

$$\begin{cases} \dfrac{\partial p}{\partial t} - \nabla \cdot D_1(\nabla p + \mu_1 \hat{q} \nabla u) = R(\hat{q}, \hat{m}) \ , \\[2mm] \dfrac{\partial n}{\partial t} - \nabla \cdot (D_2 \nabla n - \mu_2 \hat{m} \nabla u) = R(\hat{q}, \hat{m}) \ , \\[2mm] -\nabla \cdot (a \nabla u) = f + \hat{q} - \hat{m} \ , \end{cases} \tag{3.3}$$

with boundary conditions

$$\begin{cases} p = \phi(x) \ , n = \psi(x) & \text{on } (0,T) \times D \ , \\ (D_1 \nabla p + \mu_1 \hat{q} \nabla u) \cdot \nu = (D_2 \nabla n - \mu_2 \hat{m} \nabla u) \cdot \nu = 0 & \text{on } (0,T) \times N \ , \end{cases} \tag{3.4}$$

$$\begin{cases} u = U(x) \text{ on } (0,T) \times D \ , \\ \dfrac{\partial u}{\partial \nu} = 0 \quad \text{on } (0,T) \times B \ , \end{cases} \tag{3.5}$$

and initial conditions

$$p(0,x) = p_0(x) \ , \quad n(0,x) = n_0(x) \text{ in } \Omega \ . \tag{3.6}$$

The first step is to show the existence of a fixed point $(p,n) = (q,m)$. Let q and m be given. By setting $u = U + w$, the problem $(3.3)_3$, (3.5) is formulated in the following weak form. We look for $w \in V$ such that

$$\int a \, \nabla w \cdot \nabla v \, dx = -a \int \nabla U \cdot \nabla v \, dx + \int (f + \hat{q} - \hat{m}) v \, dx \ , \quad \forall \, v \in V \ . \tag{3.7}$$

The symmetric bilinear form on the left hand side of (3.7) is continuous and coercive over the Hilbert space V, moreover the right hand side of (3.7) defines a bounded linear functional on V. By Riesz-Fréchet representation theorem, the problem (3.7) admits a unique solution $w \in V$. One easily shows that the weak solution $u = U + w$ of $(3.3)_3$, (3.5) satisfies

$$\|\nabla u\| \leq 2\|\nabla U\| + c_0^{1/2} a^{-1} \left(\|f\| + \|\hat{q}\| + \|\hat{m}\| \right) \ . \tag{3.8}$$

Next, we study the problems $(3.3)_1$ and $(3.3)_2$ with boundary and initial conditions (3.4) - (3.6), in the following weak form. We set $p = y + \phi$, $n = z + \psi$, and we look for y and z in $L^2(0,T;V)$ with y' and z' in $L^2(0,T;V')$ (see, for instance Dautray- Lions, 1992, chap. XVIII, §1, specially sections 1 and 2) such that $y(0) = p_0 - \phi$, $z(0) = n_0 - \psi$ and

$$\begin{aligned} \langle y'(t), v \rangle + D_1(\nabla y, \nabla v) &= -D_1(\nabla \phi + \mu_1 \hat{q} \nabla u, \nabla v) + (R(\hat{q}, \hat{m}), v) \ , \ \forall \, v \in V \ , \\ \langle z'(t), v \rangle + (D_2(\nabla z, \nabla v) &= -(D_2 \nabla \psi - \mu_2 \hat{m} \nabla u, \nabla v) + (R(\hat{q}, \hat{m}), v), \ \forall \, v \in V. \end{aligned} \tag{3.9}$$

Since $D_1 \nabla \phi + \mu_1 \hat{q} \nabla u$ and $R(\hat{q}, \hat{m})$ belong to $L^2(0,T;L^2)$, in $(3.9)_1$ they act on V as elements of $L^2(0,T;V')$. It follows (see, for instance Dautray- Lions, 1992, Chap. XVIII §3, Theorem 2) that the above problems admit unique solutions y and z. Hence, the problems $(3.3)_1$ and $(3.3)_2$ with boundary and initial conditions (3.4) and (3.6) have unique solutions p and n in $L^2(0,T;H^1)$ with p' and n' in $L^2(0,T;V')$. □

Next we show the existence of a fixed point. By setting $v = y(t)$ in equation $(3.9)_1$ it follows that

$$\frac{1}{2} \frac{d}{dt} \|y\|^2 + D_1 \|\nabla y\|^2 \leq D_1 \left(\|\nabla \phi\| + \mu_1 \|\hat{q}\|_\infty \|\nabla u\| \right) \|\nabla y\| + c_0 \|R(\hat{q}, \hat{m})\| \|\nabla y\| \ .$$

By taking (3.8) into account, it readily follows that the right hand side of the above inequality is bounded by $C \left(1 + \|\nabla \phi\| + \|\nabla U\| + \|f\| \right) \|\nabla y\|$, hence is bounded also by

$\frac{1}{2} D_1 \|\nabla y\|^2 + (C/\mu_1)(1 + \|\nabla \phi\| + \|\nabla U\| + \|f\|)^2$, where the constant C may depend on a, D_1, μ_1 and μ. Hence

$$\frac{d\|y(t)\|^2}{dt} + D_1 \|\nabla y\|^2 \leq \frac{2C}{\mu_1}(1 + \|\nabla \phi\| + \|\nabla U\| + \|f\|)^2 . \qquad (3.10)$$

It readily follows that

$$\sup_{0 \leq t \leq T} \|y(t)\|^2 + D_1 \int_0^T \|\nabla y(t)\|^2 \, dt \leq 2\|y(0)\|^2 + \frac{4C}{\mu_1} \int_0^T (1 + \|\nabla \phi\| + \|\nabla U\| + \|f\|)^2 \, dt .$$

Hence, there is a constant B_0, independent of the pair $(q, m) \in L^2(0, T; L^2)$, such that the norms of the solution y of $(3.9)_1$ in the spaces $C(0, T; L^2)$, $L^2(0, T; V)$ and $L^2(0, T; L^2)$ are bounded by B_0. A similar result holds for z. From these bounds and from the equations $(3.9)_1$ and $(3.9)_2$ it follows that the norms of y' and z' in $L^2(0, T; V')$ are uniformly bounded. Since $p = y + \phi$ and $n = z + \psi$, a similar result holds for p and n. Denote by B_1 an uniform bound for these norms and set

$$\mathbb{K}_T := \left\{ (q, m) : \|q\|_{L^2(H^1)} \leq B_1, \ \|m\|_{L^2(H^1)} \leq B_1, \ \|q'\|_{L^2(V')} \leq B_1, \ \|m'\|_{L^2(V')} \leq B_1 \right\} ,$$

where $L^2(X) = L^2(0, T; X)$. $\mathbb{K}_T$ is a closed, convex, compact set with respect to the $L^2(L^2)$ topology. The map $S : (q, m) \to (p, n)$ satisfies $S(\mathbb{K}_T) \subset \mathbb{K}_T$. Moreover, S is continuous on $\mathbb{K}_T$ with respect to the $L^2(L^2)$ norm, as follows from standard arguments. Hence, Schauder's fixed point theorem guarantees the existence of a fixed point on $\mathbb{K}_T$ for the map S, i.e. $(p, n) = (q, m)$. Clearly this fixed point is a weak solution of the problem

$$\begin{cases} \dfrac{\partial p}{\partial t} - \nabla \cdot (D_1 \nabla p + \mu_1 \hat{p} \nabla u) = R(\hat{p}, \hat{n}) \\[2mm] \dfrac{\partial n}{\partial t} - \nabla \cdot (D_2 \nabla n - \mu_2 \hat{n} \nabla u) = R(\hat{p}, \hat{n}) \\[2mm] -\nabla (a \nabla u) = f + \hat{p} - \hat{n} \end{cases} \qquad (3.11)$$

$$\begin{cases} p = \phi(x) \ , \quad n = \psi(x) \quad \text{on } (0, T) \times D \ , \\[2mm] (D_1 \nabla p + \mu_1 \hat{p} \nabla u) \cdot \nu = (D_2 \nabla n - \mu_2 \hat{n} \nabla u) \cdot \nu = 0 \quad \text{on } (0, T) \times B \ , \end{cases} \qquad (3.12)$$

$$\begin{cases} u = U(x) \quad \text{on } (0, T) \times D \ , \\[2mm] \dfrac{\partial u}{\partial \nu} = 0 \quad \text{on } (0, T) \times B \ , \end{cases} \qquad (3.13)$$

$$p(0, x) = p_0(x) \ , \quad n(0, x) = n_0(x) \quad \text{in } \Omega \qquad (3.14)$$

Next, we show (by following Gajewski and Gröger, 1986) that the solution (p, n) satisfies $p \geq 0$, $n \geq 0$. Multiply the equation $(3.11)_1$ by $p^- = \min\{p, 0\}$ and integrate on Ω. By taking into account that $(\hat{p} \nabla u, \nabla p^-) = 0$ and that $(R(\hat{p}, \hat{n}), p^-) \leq 0$ (recall (1.8)) one shows that $d\|p^-(t)\|^2/dt \leq 0$. Since $p^-(0) = 0$, it follows that $p^- \equiv 0$. A similar proof shows that $n^- \equiv 0$.

Finally, we prove that the solutions p and n of problem (3.11) - (3.14) are bounded from above by the right hand side ℓ of equation (1.19). Since $\mu > \ell$, this shows that

$\hat{p} = p$ and $\hat{n} = n$. Hence (p, n) solves (1.1)-(1.4) and satisfies (1.19), for arbitrarily large T. At this point the reader should recall the explanation about the proof of the Theorem 1.1. given just after the statement of that theorem. According to that explanation, we show here how to modify the proof of (1.19) given in Section 2 in order to adapt it to the solution (p, n) of problem (3.11)-(3.14).

Set $\hat{v} = \min\{v, \mu\}$ where μ is above, and set $\overline{\hat{v}} = \max\{\hat{v} - k, 0\}$. In the sequel $k \in [M_0, \mu]$. Next, instead of multiplying (1.1)$_1$ by $\overline{p}$ (as done in Section 2) we multiply (3.11)$_1$ by $\overline{\hat{p}}$. As in Section 2, we prove (2.1) where now (p, n) is replaced by $(\hat{p}, \hat{n})$. Hence (2.2) and (2.2 ') hold by replacing (p, n) by $(\hat{p}, \hat{n})$. From now on, all the calculations done in Section 2 hold if we replace (p, n) by $(\hat{p}, \hat{n})$ since they depend just on (2.2) and (2.2 '). This shows that $\hat{p} \leq \ell$ and $\hat{n} \leq \ell$. By chosing $\mu > \ell$, one gets $(\hat{p}, \hat{n}) = (p, n)$.
$\square$

The proof of the uniqueness of the solution follows the standard argument. If (p, n, u) and (q, m, v) are two solutions of problem (1.1)-(1.4) one easily shows that

$$\frac{1}{2}\frac{d}{dt}\|p - q\|^2 + D_1 \int |\nabla(p - q)|^2 \, dx + \frac{\mu_1}{2a} \int (f + p - n)(p - q)^2 \, dx +$$

$$+ \mu_1 \int q\nabla(u - v) \cdot \nabla(p - q) dx = \int [R(p, n) - R(q, m)](p - q) \, dx \ .$$

Since p, n, q and m are bounded and R is locally Lipschitz continuous, one has

$$\frac{1}{2}\frac{d}{dt}\|p - q\|^2 + D_1 \int |\nabla(p - q)|^2 \, dx \leq C \int (p - q)^2 \, dx + C \int |f|(p - q)^2 \, dx +$$

$$+ C\|\nabla(u - v)\| \, \|\nabla(p - q)\| + C\|n - m\|^2 \ .$$

Moreover, $\|\nabla(u - v)\| \leq C\|(p - q) + (n - m)\|$ and

$$\int |f| \, (p - q)^2 \, dx \leq \|f\|_N \, \|p - q\| \, \|p - q\|_{2*} \leq \frac{D_1}{2} \|\nabla(p - q)\|^2 + C \, \|f\|_N^2 \, \|p - q\|^2 \ .$$

(Here, and below, if $N = 2$ replace N and 2^* by 4). By using similar results for $n - m$ one shows that

$$\frac{d}{dt} \left(\|p - q\|^2 + \|n - m\|^2 \right) + C \left(\|\nabla(p - q)\|^2 + \|\nabla(n - m)\|^2 \right)$$

$$\leq C \left(1 + \|f\|_N^2 \right) \left(\|p - q\|^2 + \|n - m\|^2 \right) \ .$$

Since $f \in L^2_{\text{loc}} (0, +\infty; L^N)$ one has

$$\|p(t) - q(t)\|^2 + \|n(t) - m(t)\|^2 \leq \left(\|p(0) - q(0)\|^2 + \|n(0) - m(0)\|^2 \right) \ e^{C \int_0^t (1 + \|f\|_N^2) \, ds} \ .$$

Hence, $(p, q) = (n, m)$.

REFERENCES

BEIRÃO DA VEIGA, H. (1972), Sur la régularité des solutions de l'équation div $A(x, u, \nabla u) = B(x, u, \nabla u)$ avec des conditions aux limites unilatérales et mêlées, *Ann. Mat. Pura Appl.* **93**, 173-230.

BEIRÃO DA VEIGA, H. (1974), Un principe de maximum pour les solutions d'une classe d'inéquations paraboliques quasi-linéaires, *Arch. Rat. Mech. Anal.*, **55**, 214-224.

BEIRÃO DA VEIGA, H. and J. P. DIAS (1972), Regularité des solutions d'une équation parabolique non linéaire avec des constraintes unilatérales sur la frontière, *Ann. Inst. Fourier*, **22**, 161-192.

DAUTRAY, R. and LIONS, J. L. (1992), *Mathematical Analysis and Numerical Methods for Science and Technology*, **2** and **5**, (Springer-Verlag, Berlin).

GAJEWSKI, H. (1985), On existence, uniqueness and asymptotic behavior of solutions of the basic equations for carrier transport in semiconductors, *Z. Angew. Math. Mech*, **65**, 101-108.

GAJEWSKI, H. and K. GRÖGER (1986), On the basic equations for carrier transport in semiconductors, *J. Math. Anal. Appl.*, **113**, 12-35.

GRÖGER, K. (1986), On the boundedness of solutions to the basic equations in semiconductors theory, *J. Math. Nachr.*, **129**, 167-174.

GRÖGER, K. (1989), A $W^{1,p}$-estimate for solutions to mixed boundary value problems for second order elliptic differential equations, *Math. Ann.*, **283**, 679-687.

GRISVARD, P. (1992), Singularities in Boundary Value Problems, *Research Notes in Appl. Math.*, **22** (Masson and Springer-Verlag).

LORENZI, A. (1975), A mixed boundary value problems for the Laplace equation in an angle (1.st part), *Rend. Sem. Mat. Univ. Padova*, **54**, 147-183.

LADYZHENSKAYA, O. A. (1973), The Boundary Value Problems of Mathematical Physics, (Springer-Verlag, New York); original russian edition: Nauka (1973).

LADYZHENSKAYA, O. A.; SOLONNIKOV, V. A. and N.N. URAL'CEVA (1968), Linear and Quasilinear Equation of Parabolic Type, *A.M.S., Providence*, (original russian edition: Moscow 1967).

LIONS, J. L. and E. MAGENES (1968), *Problèmes aux Limites Non Homogènes et Applications*, **1**, (Dunod, Paris).

MARKOWICH, P. A.; RINGHOFER, C.A. and C. SCHMEISER, (1990), *Semiconductor Equations*, (Springer-Verlag, Wien).

MOCK, M. S. (1974), An initial value problem from semiconductor device theory, *SIAM J. Math. Anal.*, **5**, 597-612.

MOCK, M. S. (1975), Asymptotic behavior of solutions of transport equations for semiconductor devices, *J. Math. Anal. Appl.*, **49**, 215-225.

MOLL, J. L. (1964), *Physics of Semiconductors*, (Mc Graw-Hill, New-York).

NEČAS, J. (1967), *Les Méthodes Directes en Théorie des Equations Elliptiques*, (Masson et Cie, Prague).

SEIDMAN, T. I. and G. M. TROIANIELLO (1985), Time dependent solutions of a nonlinear system arising in semiconductor in theory, *Nonlinear Anal.-TMA*, **9**, 1137-1157.

STAMPACCHIA, G. (1963) Some limit cases of L^p-estimates for solutions of second order elliptic equations, *Comm. Pure Appl. Math.*, **16**, 501-510.

STAMPACCHIA, G. (1960) Problemi al contorno ellittici, con dati discontinui, dotati di soluzioni hölderiane, *Ann. Mat. Pura Appl.*, **51**, 1-38.

VAN ROOSBROECK, W. (1950) Theory of the flow of electrons and holes in germanium and other semiconductors, *Bell Sys. Tech. J.*, **29**, 560-607.

BIFURCATION OF SOLUTIONS TO REACTION-DIFFUSION SYSTEMS WITH UNILATERAL CONDITIONS [1]

Milan Kučera

Mathematical Institute of the Academy of Sciences
of the Czech Republic,
Žitná 25, 115 67 Prague 1, Czech Republic

1. INTRODUCTION

Let Ω be a bounded domain in R^N with a lipschitzian boundary $\partial\Omega$, let Γ_D, Γ_N, Γ_U be open (in $\partial\Omega$) disjoint subsets of $\partial\Omega$, meas$(\partial\Omega \setminus \Gamma_D \cup \Gamma_N \cup \Gamma_U) = 0$, f, g real differentiable functions on R^2, $\bar{u}$, $\bar{v}$ positive constants such that $f(\bar{u},\bar{v}) = g(\bar{u},\bar{v}) = 0$. Consider a reaction-diffusion system

$$
\begin{aligned}
u_t &= d_1\Delta u + f(u,v), \\
v_t &= d_2\Delta v + g(u,v) \quad \text{on } [0,+\infty) \times \Omega
\end{aligned}
\tag{1.1}
$$

with unilateral boundary conditions

$$
u = \bar{u}, \ v = \bar{v} \quad \text{on } \Gamma_D,
$$

$$
\frac{\partial u}{\partial n} = 0, \ -\frac{\partial v}{\partial n} \in m(v) \quad \text{on } \Gamma_U,
\tag{1.2}
$$

$$
\frac{\partial u}{\partial n} = \frac{\partial v}{\partial n} = 0 \quad \text{on } \Gamma_N,
$$

where $m : R \to 2^{\bar{R}}$ is a multivalued function. Simultaneously we shall consider the system (1.1) with the classical boundary conditions

$$
u = \bar{u}, \ v = \bar{v} \quad \text{on } \Gamma_D,
$$

$$
\frac{\partial u}{\partial n} = \frac{\partial v}{\partial n} = 0 \quad \text{on } \Gamma_N \cup \Gamma_U.
\tag{1.3}
$$

---

[1]The research is supported by the Grant No. 11958 of the Academy of Sciences of the Czech Republic.

Hence, $\bar{u}$, $\bar{v}$ is a stationary spatially homogeneous solution of (1.1), (1.2) as well as of (1.1), (1.3) under the assumption $0 \in m(\bar{v})$. Set $b_{11} = \frac{\partial f}{\partial u}(\bar{u}, \bar{v})$, $b_{12} = \frac{\partial f}{\partial v}(\bar{u}, \bar{v})$, $b_{21} = \frac{\partial g}{\partial u}(\bar{u}, \bar{v})$, $b_{22} = \frac{\partial g}{\partial v}(\bar{u}, \bar{v})$ and suppose that

$$b_{11} > 0, \ b_{12} < 0, \ b_{21} > 0, \ b_{22} < 0, \ b_{11} + b_{22} < 0, \ \det(b_{ij}) > 0. \tag{1.4}$$

These assumptions correspond to systems of activator-inhibitor (or prey-predator) type for which the following effect (diffusion driven instability) occurs. The constant solution $\bar{u}$, $\bar{v}$ is stable as a solution of the equations $u_t = f(u, v)$, $v_t = g(u, v)$ but $\bar{u}$, $\bar{v}$ as a solution of (1.1), (1.3) is stable only for some parameters $[d_1, d_2] \in R_+^2$ (domain of stability) and it is unstable for the other $[d_1, d_2] \in R_+^2$ (domain of instability). See Fig. 1. Stationary spatially nonhomogeneous solutions (spatial patterns) bifurcate at the border between the domain of stability and instability. Precisely see Proposition 2.7.. This effect plays an essential role in some models in biochemistry and ecology, e.g. models of morphogenesis – see Gierer and Meinhardt (1972).

It was proved by Drábek, Kučera and Míková (1985), Drábek and Kučera (1988) that unilateral conditions for v (inhibitor or predator) described by variational inequalities given by a cone have a destabilizing effect: stationary spatially nonconstant solutions of the unilateral problem bifurcate already in the domain of stability of the corresponding classical problem (1.1), (1.3) and the stability of $\bar{u}$, $\bar{v}$ as a solution of the linearized unilateral problem is lost already in the domain of stability. The simplest case of unilateral conditions considered were boundary conditions

$$u = \bar{u}, \ v = \bar{v} \quad \text{on } \Gamma_D, \quad \frac{\partial u}{\partial n} = \frac{\partial v}{\partial n} = 0 \text{ on } \Gamma_N,$$

$$\frac{\partial u}{\partial n} = 0 \text{ on } \Gamma_U, \ v \geq \bar{v}, \ \frac{\partial v}{\partial n} \geq 0, \ (v - \bar{v})\frac{\partial v}{\partial n} = 0 \text{ on } \Gamma_U.$$

(These conditions can be written as (1.2) with $m(v) = \{0\}$ for $v > \bar{v}$, $m(\bar{v}) = [-\infty, 0]$, $m(v) = \{-\infty\}$ for $v < \bar{v}$.) The proof was based on a certain homotopical joining of the inequality with the corresponding equation and on the transfer of information about the existence of small solutions from the equation to the inequality. This approach was developed originally for the investigation bifurcations of variational inequalities of another nonsymmetric type by Kučera (1982). A different method for the study of eigenvalues and bifurcations for inequalities with nonsymmetric operators was found by Quittner (1986) and the results concerning the destabilizing effect of unilateral conditions given by a cone mentioned were generalized by Quittner (1987). The Quittner's method is based on a direct use of the topological degree and it is simpler than the homotopy method. However, during last years, the homotopy method proved to be usefull in some situations where it is not clear how to use the Quittner's approach. Its nontrivial modification is the basis for the proof of the existence of a bifurcation of periodic solutions to ordinary differential inequalities (Kučera, 1991; Eisner and Kučera, 1995).

The aim of this paper is to show how the homotopy method can be used also for the investigation of bifurcations for inclusions of a certain type. We shall show that the bifurcation of stationary spatially nonconstant solutions of (1.1) with conditions given by inclusions occurs already in the domain of stability of the classical problem (1.1), (1.3). Our aim is to explain main principles of the method and therefore we shall confine ourself to the case $N = 1$ which is technically simpler. Moreover, we shall suppose $\Gamma_D \neq \emptyset$ because the bifurcation point obtained can be shifted to infinity in a certain sense if $\Gamma_D = \emptyset$. (For the special case of variational inequalities cf. also Drábek, Kučera and Míková, 1985.) The general case (when the idea is hidden in technical problems) will be treated in a forthcoming paper by J. Eisner and the author. Further, we shall consider $\bar{u} = \bar{v} = 0$ without loss of generality in our mathematical considerations and we shall keep in mind that, in interpretation, $\bar{u}$, $\bar{v}$ represent e.g. positive concentrations. In fact, we shall study only stationary solutions. Hence, setting $n_1(u,v) = b_{11}u + b_{12}v - f(u,v)$, $n_2(u,v) = b_{21}u + b_{22}v - g(u,v)$, we can write our equations as

$$d_1 u_{xx} + b_{11}u + b_{12}v - n_1(u,v) = 0, \quad d_2 v_{xx} + b_{21}u + b_{22}v - n_2(u,v) = 0 \text{ on } (0,1). \quad (1.5)$$

The conditions (1.2) and (1.3) have the form

$$u(0) = v(0) = 0, \quad u_x(1) = 0, \quad -v_x(1) \in m(v(1)) \quad (1.6)$$

and

$$u(0) = v(0) = u_x(1) = v_x(1) = 0 \quad (1.7)$$

respectively. (Of course, the role of the end points 0, 1 can be exchanged.) But we can consider also other examples of unilateral conditions given by inclusions describing some regulation in $(0,1)$ (not only at the boundary) — see Examples 2.3., 2.4..

2. MAIN RESULT

2.1. Notation

κ_j, e_j $(j = 1, 2, \ldots)$ — the eigenvalues and eigenvectors of $-\Delta u = \lambda u$ with (1.3) or (1.7);

$C_j = \{d = [d_1, d_2] \in R_+^2 ; d_2 = \frac{b_{12}b_{21}/\kappa_j^2}{d_2 - b_{11}/\kappa_j} + \frac{b_{22}}{\kappa_j}\}$, $j = 1, 2, 3, \ldots$ (see Fig. 1);

C — the envelope of the hyperbolas C_j, $j = 1, 2, 3, \ldots$ (see Fig. 1);

G_S, G_I — the set of all $d \in R_+^2$ lying on the right and on the left, respectively, from C;

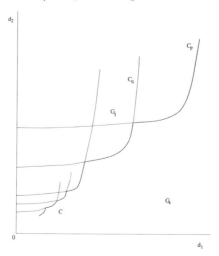

Figure 1.

R_+ — the set of all positive reals, $R_+^2 = R_+ \times R_+$, $\bar{R} = \{-\infty\} \cup R$;

$\mathbf{V}$, A, N_j — the space and the operators from Remark 2.1. or Remark 2.9.;

M, M_0 and P^τ — multivalued mappings of $\mathbf{V} \times \mathbf{V}$ into $2^{\mathbf{V} \times \mathbf{V}}$ and operators in $\mathbf{V} \times \mathbf{V}$, respectively, satisfying (2.6)–(2.15) (see also Examples 2.2., 2.3., 2.4.);

$U = [u, v]$ elements of $\mathbf{V} \times \mathbf{V}$, $AU = [Au, Av]$ for $U = [u, v] \in \mathbf{V} \times \mathbf{V}$;

$U^* = [b_{21} b_{12}^{-1} u, v]$ for $U = [u, v]$ (see also Proposition 2.7.);

$\langle \cdot, \cdot \rangle$, $\|.\|$ — the inner product and the norm in $\mathbf{V}$ (Remarks 2.1., 2.9.) or in $\mathbf{V} \times \mathbf{V}$, i.e.

$$\langle U, W \rangle = \langle u, w \rangle + \langle v, z \rangle, \quad \|U\|^2 = \|u\|^2 + \|v\|^2 \text{ for } U = [u, v], W = [w, z] \in \mathbf{V} \times \mathbf{V};$$

$K = \{U \in \mathbf{V} \times \mathbf{V}; \ 0 \in M_0(U)\}$;

∂K, K^0 — the boundary and the interior of K;

$\to$, $\rightharpoonup$ — strong convergence, weak convergence;

$$D(d) = \begin{bmatrix} d_1 & 0 \\ 0 & d_2 \end{bmatrix}, \ D^{-1}(d) = \begin{bmatrix} 1/d_1 & 0 \\ 0 & 1/d_2 \end{bmatrix}, \ B = \begin{bmatrix} b_{11} & b_{12} \\ b_{21} & b_{22} \end{bmatrix}, \ B^* = \begin{bmatrix} b_{11} & b_{21} \\ b_{12} & b_{22} \end{bmatrix}$$

$E_B(d) = \{U \in \mathbf{V} \times \mathbf{V}; \ D(d)U - BAU = 0\}$, $E_{B^*}(d) = \{U \in \mathbf{V} \times \mathbf{V}; \ D(d)U - B^* AU = 0\}$

$E_I(d) = \{U \in \mathbf{V} \times \mathbf{V}; \ D(d)U - BAU \in -M_0(U)\}$;

critical point of (2.1) or (2.20) — a parameter $d \in R_+^2$ or $s \in R$ for which $E_B(d) \neq \{0\}$ or $E_B(d(s)) \neq \{0\}$, respectively.

Remark 2.1. Set $\mathbf{V} = \{u \in W_2^1(0,1); \ u(0) = 0\}$,

$$\langle u, \varphi \rangle = \int_0^1 u_x \varphi_x \, dx \quad \text{for all } u, \varphi \in \mathbf{V}.$$

Then $\langle \cdot, \cdot \rangle$ is the inner product on $\mathbf{V}$, the corresponding norm $\|.\|$ is equivalent to the usual Sobolev norm on $\mathbf{V}$. Introduce operators $A : \mathbf{V} \to \mathbf{V}$ and $N_j : \mathbf{V} \times \mathbf{V} \to \mathbf{V}$ by

$$\langle Au, \varphi \rangle = \int_0^1 u\varphi \, dx \text{ for all } u, \varphi \in \mathbf{V},$$

$$\langle N_j(U), \varphi \rangle = \int_0^1 n_j(u, v)\varphi \, dx \text{ for all } U = [u, v] \in \mathbf{V} \times \mathbf{V}, \ \varphi \in \mathbf{V}, \ j = 1, 2.$$

Then the weak solution of the problem (1.5), (1.7) is the solution of the operator equations

$$\begin{aligned} d_1 u - b_{11} Au - b_{12} Av + N_1(u, v) &= 0, \\ d_2 u - b_{21} Au - b_{22} Av + N_2(u, v) &= 0 \end{aligned} \tag{2.1}$$

which can be written also as

$$D(d)U - BAU + N(U) = 0. \tag{2.2}$$

(Of course, weak solutions are simultaneously classical solutions in our simple model example.) It follows from well-known embedding theorems (see e.g. Nečas, 1967) that $A : \mathbf{V} \to \mathbf{V}$ is a linear, symmetric, positive and completely continuous operator, $N_j : \mathbf{V} \times \mathbf{V} \to \mathbf{V}$ are completely continuous nonlinear operators satisfying

$$\lim_{\|U\| \to 0} \frac{\|N_j(U)\|}{\|U\|} = 0 \quad (j = 1, 2). \tag{2.3}$$

Example 2.2. Consider a mapping $m : R \to 2^{\bar{R}}$ described by a real continuous function on $(-\infty, 0) \cup (0, +\infty)$ and multivalued only at $s = 0$ such that

$$m(s) = 0 \text{ for } s > 0, \; m(s) < 0 \text{ for } s < 0, \; \lim_{s \to 0_-} m(s) = m^0,$$

$$m(0) = [m^0, 0] \quad \text{with some } m^0 \in [-\infty, 0]$$

(see Fig. 2). Define a multivalued mapping $M_2 : \mathbf{V} \times \mathbf{V} \to 2^{\mathbf{V}}$ by

$$M_2(U) = M_2(v) = \{z \in \mathbf{V}; \; \langle z, \psi \rangle \in m(v(1))\psi(1) \text{ for all } \psi \in \mathbf{V}\} \quad \text{for } U = [u, v].$$

A weak solution of (1.5), (1.6) can be introduced as a solution of the inclusion

$$
\begin{aligned}
d_1 u - b_{11} A u - b_{12} A v + N_1(u, v) &= 0, \\
d_2 u - b_{21} A u - b_{22} A v + N_2(u, v) &\in -M_2(v)
\end{aligned}
\tag{2.4}
$$

which can be written also in the vector form

$$D(d)U - BAU + N(U) \in -M(U) \tag{2.5}$$

where $N(U) = [N_1(U), N_2(U)]$, $M(U) = [\{0\}, M_2(U)]$ for $U \in \mathbf{V} \times \mathbf{V}$. (From formal reasons, we write multivalued mappings with the negative sign — see also Remark 2.5..)

Further, introduce the homogeneous multivalued mapping $M_0 : \mathbf{V} \times \mathbf{V} \to 2^{\mathbf{V} \times \mathbf{V}}$ corresponding to M and defined by $M_0(U) = [\{0\}, M_{02}(U)]$ with

$$M_{02}(U) = M_{02}(v) = \{z \in \mathbf{V}; \; \langle z, \psi \rangle \in m_0(v(1))\psi(1) \text{for } \psi \in \mathbf{V}\}$$

$$\text{for all } U = [u, v] \in \mathbf{V} \times \mathbf{V},$$

where $m_0(s) = 0$ for $s > 0$, $m_0(0) = [-\infty, 0]$, $m_0(s) = -\infty$ for $s < 0$. It satisfies the following conditions which are essential for our general considerations:

$$K = \{U \in \mathbf{V} \times \mathbf{V}; \; 0 \in M_0(U)\} \text{ is a closed convex cone in } \mathbf{V} \times \mathbf{V}$$
$$\text{with the vertex at the origin, } K \neq \mathbf{V} \times \mathbf{V}; \tag{2.6}$$

$$M_0(tU) = tM_0(U) \text{ for all } t > 0, \; U \in \mathbf{V} \times \mathbf{V}; \tag{2.7}$$

$$\text{if } U_n \to 0, \; \frac{U_n}{\|U_n\|} \to W, \; Z_n \to Z, d_n \to d \in R_+^2,$$
$$\frac{D(d_n)U_n}{\|U_n\|} + Z_n \in -\frac{M(U_n)}{\|U_n\|} \text{ then } \frac{U_n}{\|U_n\|} \to W, \; D(d)W + Z \in -M_0(W); \tag{2.8}$$

$$\text{if } U \in \mathbf{V} \times \mathbf{V} \text{ then } \langle Z, W \rangle \leq 0 \text{ for all } Z \in M_0(U), \; W \in K. \tag{2.9}$$

There exists a system of continuous functions $p^\tau : R \to R$ ($\tau \in [0, +\infty)$) such that

$$p^0 \equiv 0, \; p^\tau(s) = 0 \text{ for } s \geq 0, \; p^\tau(s) \in (m(s), 0) \text{ for } s < 0$$

and satisfying the following conditions:

if $\tau_n \to \tau \in [0, +\infty)$, $s_n \to s$, then $p^{\tau^n}(s_n) \to p^\tau(s)$;

if $\tau_n \to 0_+$, $s_n \to 0_-$, then $p^{\tau_n}(s_n)/s_n \to 0$, $\liminf p^{\tau_n}(s_n)/\tau_n s_n > 0$;

if $\tau_n \to +\infty$, $s_n \to s$, $p^{\tau_n}(s_n) \to p$ then $p \in m(s)$ or $p = m(s)$ for $s = 0$ or $s \neq 0$.

Define $P^\tau : \mathbf{V} \times \mathbf{V} \to \mathbf{V} \times \mathbf{V}$, $\tau \in [0, +\infty)$, by $P^\tau(U) = [0, P_2^\tau(v)]$ for $U = [u, v]$,

$$\langle P_2^\tau(v), \psi \rangle = p^\tau(v(1))\psi(1) \text{ for all } v \in \mathbf{V}, \ \psi \in \mathbf{V}.$$

It is not hard to show that the following conditions are fulfilled:

$$\langle P^\tau(U), U \rangle \geq 0 \text{ for all } U \in \mathbf{V} \times \mathbf{V}; P^\tau(U) = 0 \text{ for all } U \in K,$$

$$\langle P^\tau(U), V \rangle \leq 0 \text{ for all } U \in \mathbf{V} \times \mathbf{V}, \ V \in K, \ \tau \in [0, +\infty); \tag{2.10}$$

$$\text{if } U_n \rightharpoonup U, \ \tau_n \geq 0 \text{ then } \liminf \langle P^{\tau_n}(U_n), U_n - U \rangle \geq 0;$$

if, moreover, $U = 0$, $\dfrac{P^{\tau_n}(U_n)}{\|U_n\|}$ are bounded and $W_n = \dfrac{U_n}{\|U_n\|} \rightharpoonup W$ $\qquad$ (2.11)

$$\text{then } \liminf \frac{\langle P^{\tau_n}(U_n), W_n - W \rangle}{\|U_n\|} \geq 0;$$

$$\text{if } U_n \to U, \ \tau_n \to \tau \in [0, +\infty) \text{ then } P^{\tau_n}(U_n) \to P^\tau(U)$$

$$\text{if } U_n \to U, \ \tau_n \to +\infty, P^{\tau_n}(U_n) \to Z \text{ then } Z \in M(U); \tag{2.12}$$

$$\text{if } U_n \to 0, \ \tau_n \to 0 \text{ then } \frac{P^{\tau_n}(U_n)}{\|U_n\|} \to 0; \tag{2.13}$$

if $U_n \to 0$, $\dfrac{U_n}{\|U_n\|} \to W \notin K$, $\tau_n \geq \tau_0 > 0$, $V \in K^0$

$$\text{then } \limsup \frac{\langle P^{\tau_n}(U_n), V \rangle}{\|U_n\|} < 0; \tag{2.14}$$

if $U_n \to 0$, $\dfrac{U_n}{\|U_n\|} \to W \notin K$, $\tau_n \to 0_+$, $V \in K^0$

$$\text{then } \limsup \frac{\langle P^{\tau_n}(U_n), V \rangle}{\tau_n \|U_n\|} < 0. \tag{2.15}$$

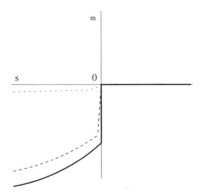

Figure 2. A possible shape of the graph of m and p^τ. (The monotonicity is not essential).

Example 2.3. Consider the space $\mathbf{V}$, the operators A, N_j and the functions m, m_0 and p^τ as in Remark 2.1. and Example 2.2.. Define $M(U) = [\{0\}, M_2(U)]$,

$$M_2(U) = M_2(v) = \left\{ z \in \mathbf{V}; \ \langle z, \psi \rangle \in \sum_{i=1}^k m(v(x_i))\psi(x_i) \text{ for all } \psi \in \mathbf{V} \right\}$$

$$\text{for } U = [u, v] \in \mathbf{V} \times \mathbf{V},$$

with $x_i \in (0,1)$ $(i = 1, ..., k)$. The operator $M_0(U) = [\{0\}, M_{02}(v)]$ is given by

$$M_{02}(v) = \left\{ z \in \mathbf{V}; \ \langle z, \psi \rangle \in \sum_{i=1}^k m_0(v(x_i))\psi(x_i) \text{ for all } \psi \in \mathbf{V} \right\}.$$

A solution of (2.4) is a weak solution of the problem

$$d_1 u_{xx} + f(u, v) = 0,$$

$$d_2 v_{xx} + g(u, v) = 0 \text{ on } (x_i, x_{i+1}), \ i = 0, \dots, k \ (x_0 = 0, \ x_{k+1} = 1),$$

$$u(0) = v(0) = u_x(1) = v_x(1) = 0, \ u, v \text{ are continuous on } [0, 1],$$

$$u_x(x_i+) - u_x(x_i-) = 0, \ v_x(x_i+) - v_x(x_i-) \in m(v(x_i))$$

$(u(x_i\pm)$ denotes the right and the left derivative at x_i.) If we define $P^\tau = [0, P_2^\tau]$,

$$\langle P_2^\tau(U), \psi \rangle = \sum_{i=1}^k p^\tau(v(x_i))\psi(x_i) \text{ for all } U = [u, v] \in \mathbf{V} \times \mathbf{V}, \ \psi \in \mathbf{V}$$

then the assumptions (2.6)–(2.15) are fulfilled.

Example 2.4. Analogously we can consider mappings of the type

$$M_2(v) = \left\{ w \in \mathbf{V}; \ \int_0^1 \underline{m}(v)\psi \, dx \le \langle w, \psi \rangle \le \int_0^1 \overline{m}(v)\psi \, dx \text{ for } \psi \in \mathbf{V}, \ \psi \ge 0 \right\}$$

where $\underline{m}(s) = \overline{m}(s) = m(s)$ for $s \ne 0$, $\underline{m}(0) = m_0$, $\overline{m}(0) = 0$.

Remark 2.5. In many reasonable examples, a multivalued mapping M can be written as a subdifferential of some proper convex functional Γ. Then the inclusion (2.5) with $M = \partial\Gamma$ (the subdifferential of Γ) is equivalent to the inequality

$$\langle D(d)U - BAU + N(U), Z - U \rangle + \Gamma(Z) - \Gamma(U) \ge 0 \quad \text{for all } Z \in \mathbf{V} \times \mathbf{V}$$

– cf. Duvaut and Lions (1972). An analogy of Theorem 2.10. for the particular case of such inequalities was proved by J. Eisner during preparation of his PhD thesis.

Remark 2.6. Recall that if $\text{Re}\,\lambda \le -\varepsilon < 0$ for all eigenvalues of the problem

$$d_1 \Delta u + b_{11}u + b_{12}v = \lambda u, \ d_2 \Delta v + b_{21}u + b_{22}v = \lambda v \tag{2.16}$$

with the boundary conditions (1.3) (with $\bar{u} = \bar{v}=0$) then the trivial solution of (1.1), (1.3) is stable and if there exists an eigenvalue of (2.16), (1.3) satisfying $\text{Re}\,\lambda > 0$ then the trivial solution of (1.1), (1.3) is unstable (see Kielhöfer 1974 for details). In the special situation of Remark 2.1., the weak formulation of (2.16), (1.7) is

$$D(d)U - BAU + \lambda AU = 0. \tag{2.17}$$

The eigenvalues and the eigenvectors of (2.16), (1.7) coincide with those of (2.17).

Proposition 2.7. *Let the assumption* (1.4) *be fulfilled. For d lying on the left from C_j or in a right neighbourhood of C_j (including C_j), there exists a real eigenvalue $\lambda_j(d)$ of* (2.16), (1.3) *(or of* (2.17) *in the special situation of* Remark 2.1.) *with the corresponding eigenvector $[\frac{d_2\kappa_j - b_{22} + \lambda_j(d)}{b_{21}} e_j, e_j]$. This eigenvalue $\lambda_j(d)$ is negative for d on the right from C_j and positive on the left from C_j, C_j being the set of all d for which $\lambda_j(d) = 0$. The problem* (2.16), (1.3) *(or* (2.17)) *has no other real eigenvalues. Particularly, $\bigcup_{j=1}^{\infty} C_j$ is the set of all critical points of* (2.1). *The problem* (2.16), (1.3) *(or* (2.17)) *has also some complex eigenvalues but their real parts are always negative for all $d \in R_+^2$. The same assertions hold for eigenvalues $\tilde{\lambda}_j(d)$ of*

$$D(d)U - BAU + \lambda U = 0,$$

where the corresponding eigenvectors are $[\frac{d_2\kappa_j - b_{22} + \kappa_j\lambda_j(d)}{b_{21}} e_j, e_j]$. Particularly, if $d^0 \in C \cap C_j$, $\dim E_B(d^0) = 1$ then there exists a neighbourhood $\mathcal{U}(d^0)$ of d^0 such that the operator $D^{-1}(d)BA - I$ has precisely one positive eigenvalue for $d \in \mathcal{U}(d^0) \cap G_I$ and this eigenvalue is simple, while all eigenvalues of $D^{-1}(d)BA - I$ have negative real parts for $d \in G_S$. Finally, $U = [u, v] \in E_B(d)$ if and only if $U^* = [\frac{b_{21}}{b_{12}}u, v] \in E_{B^*}(d)$.

Proof: For the special case $N = 1$ and the problem (2.16) with Nemann boundary conditions see e.g. Nishiura (1982), for the general case see Drábek and Kučera (1986).

Further, we shall consider a curve d given by a differentiable mapping $d : R \to R_+^2$ intersecting the border C between the domain of stability and instability and such that

$$d(s) \in G_S \text{ for all } s \in (s_0, +\infty), \quad \lim_{s \to +\infty} d_1(s) = +\infty, \quad \lim_{s \to +\infty} \inf d_2(s) > 0. \quad (2.18)$$

We shall study the bifurcation problem for the inclusion

$$D(d(s))U - BAU + N(U) \in -M(U) \quad (2.19)$$

with the bifurcation parameter s. Hence, by a bifurcation point of (2.19) we mean $s_I \in R$ such that for any neighbourhood $\mathcal{U}$ of $[s_I, 0]$ in $R \times \mathbf{V} \times \mathbf{V}$ there exists a couple $[s, U] \in \mathcal{U}$ satisfying (2.19), $\|U\| \neq 0$.

Remark 2.8. Of course, any bifurcation point of

$$D(d(s))U - BAU + N(U) = 0 \quad (2.20)$$

is simultaneously a critical point. Hence, if $d(s) \in G_S$ for $s \in (s_0, +\infty)$ then there is no bifurcation point of (2.20) greater than s_0 by Proposition 2.7.. If, moreover, d intersects C transverally at $d(s_0)$ then s_0 really is a bifurcation point of (1.1), (1.3) under reasonable assumptions (see e.g. Nishiura, 1982).

Remark 2.9. Up to now, we considered only the space and the operators from our simple model situation from Remark 2.1.. We can consider also a general real Hilbert space $\mathbf{V}$, a linear, positive, symmetric, completely continuous operator $A : \mathbf{V} \to \mathbf{V}$ and nonlinear completely continuous operators $N_j : \mathbf{V} \times \mathbf{V} \to \mathbf{V}$ satisfying (2.3). In this case, we must speak about characteristic values κ_j of A instead of eigenvalues of $-\Delta$ in Proposition 2.7. – see Drábek and Kučera (1988). But all assertions (particularly Theorem 2.10. below) remain valid.

Theorem 2.10. *Let* (1.4) *hold, let* d *be a differentiable curve satisfying* (2.18) *and intersecting* C *transversally at a critical point* $d^0 = d(s_0)$ *such that* $E_B(d^0) = \text{Lin}\{U_0\}$, $U_0 \in K^0$, $U_0^* \in K^0$. *Consider a multivalued mapping* M *such that there exist a multivalued mapping* M_0 *and operators* $P^\tau(\tau \in [0, +\infty))$ *satisfying the assumptions* (2.6)–(2.15). *Then there exists a bifurcation point* $s_I > s_0$ *of* (2.19). *In more detail, there is* $\delta_0 > 0$ *such that for any* $\delta \in (0, \delta_0)$ *there exist* $s(\delta)$, $U(\delta)$ *satisfying* (2.19), $\|U(\delta)\|^2 = \delta$, $s(\delta) > s_0$ *and such that if* $\delta_n \to 0_+$, $s(\delta_n) \to s_I$ *then* $s_0 < s_I < +\infty$; $s(\delta)$, $U(\delta)$ *do not satisfy* (2.20).

In the proof of Theorem 2.10. (see Section 4), we shall use the system of equations

$$D(d(s))U - BAU + \frac{\tau}{1+\tau}N(U) + P^\tau(U) = 0 \qquad (2.21)$$

supplemented by the norm condition

$$\|U\|^2 = \frac{\delta\tau}{1+\tau} \qquad (2.22)$$

where $\delta > 0$ is fixed, $\tau \in R_+$ is an additional parameter. Notice that (2.21) reads as the linearized equation $D(d(s))U - BAU = 0$ for $\tau = 0$ and gives the inclusion (2.19) for $\tau \to +\infty$ (precisely see Lemma 3.2.). Hence, (2.21) can be understood in a certain sense as a homotopy joining the inclusion with the corresponding equation.

3. SOME AUXILIARY ASSERTIONS

In this section, the assumptions of Theorem 2.10. will be automatically considered.

Remark 3.1. If $[d_n, U_n, \tau_n] \in R_+^2 \times \mathbf{V} \times \mathbf{V} \times R_+$, $d_n \to d$, $U_n \rightharpoonup U$, $\tau_n \to \tau \in [0, +\infty]$,

$$D(d_n)U_n - BAU_n + \frac{\tau_n}{1+\tau_n}N(U_n) + P^{\tau_n}(U_n) = 0 \qquad (3.1)$$

then $U_n \to U$. If, moreover, $\|U\| = 0$ and $W_n = \frac{U_n}{\|U_n\|} \rightharpoonup W$ then $W_n \to W$. Indeed, multiplying our equation by U_n and U, we obtain

$$\langle U_n, U_n \rangle = \left\langle D^{-1}(d_n)\left[BAU_n - \frac{\tau_n}{1+\tau_n}N(U_n) - P^{\tau_n}(U_n)\right], U_n \right\rangle,$$

$$\langle U_n, U \rangle = \left\langle D^{-1}(d_n)\left[BAU_n - \frac{\tau_n}{1+\tau_n}N(U_n) - P^{\tau_n}(U_n)\right], U \right\rangle.$$

Hence, the compactness argument together with the assumption (2.11) imply

$$\|U\|^2 - \limsup \|U_n\|^2 = \liminf \langle P^{\tau_n}(U_n), U_n - U \rangle \geq 0$$

and the first assertion follows. The second one can be shown analogously if we divide our equation by $\|U_n\|$ and multiply it by W_n, W.

Lemma 3.2. *Let* $[d_n, U_n, \tau_n] \in R_+^2 \times \mathbf{V} \times \mathbf{V} \times R_+$, $d_n \to d, U_n \rightharpoonup U$, $\tau_n \to +\infty$ *and* (3.1) *holds. Then* $U_n \to U$ *and* d, U *satisfy*

$$D(d)U - BAU + N(U) \in -M(U).$$

Proof: The assertion follows directly from the compactness argument (see Remark 3.1.) and the assumption (2.12).

Lemma 3.3. *If $E_{B^*}(d^0) \cap K^0 \neq \emptyset$ then $E_I(d^0) = E_B(d^0) \cap K$.*

Proof: Clearly, always $E_B(d^0) \cap K \subset E_I(d^0)$. Suppose that $U_0^* \in E_{B^*}(d^0) \cap K^0$ and $U \in \mathbf{V} \times \mathbf{V}$ are such that

$$D(d^0)U - BAU \in -M_0(U), \quad D(d^0)U - BAU \neq 0. \tag{3.2}$$

Then there is Z such that $\langle D(d^0)U - BAU, Z \rangle < 0$. We have $U_0^* + tZ \in K$ for $t > 0$ small enough and it follows from (2.9) that

$$\left\langle D(d^0)U - BAU, U_0^* + tZ \right\rangle \geq 0.$$

But the left hand side should be negative by the choice of Z because of $\langle D(d^0)U - BAU, U_0^* \rangle = \langle D(d^0)U_0^* - B^*AU_0^*, U \rangle = 0$. This contradiction shows that if $U \in E_I(d^0)$ then $U \in E_B(d^0)$. In this case $0 \in M_0(U)$, i.e. $U \in K$ by the definition.

Lemma 3.4. *There exist δ_0, $C > 0$ such that if $\|U\| < \delta_0$ and (2.21) holds then $s \leq C$.*

Proof: Suppose by contradiction that there exist $U_n = [u_n, v_n]$, s_n such that $\|U_n\| \to 0$, $W_n = [w_n, z_n] = \frac{U_n}{\|U_n\|} \to W = [w, z]$, $s_n \to +\infty$ and (3.1) holds. It follows from Remark 3.1. that $W_n \to W$. Multiplying (3.1) by $\frac{U_n}{\|U_n\|^2}$ we obtain

$$d_1(s_n)\|w_n\|^2 + d_2(s_n)\|z_n\|^2 - b_{11} \langle Aw_n, w_n \rangle - b_{12} \langle Az_n, w_n \rangle -$$

$$-b_{21} \langle Aw_n, z_n \rangle - b_{22} \langle Az_n, z_n \rangle + \frac{\tau_n}{1 + \tau_n} \left\langle \frac{N(U_n)}{\|U_n\|}, W_n \right\rangle + \frac{\langle P^{\tau_n}(U_n), U_n \rangle}{\|U_n\|^2} = 0. \tag{3.3}$$

If $\|w_n\| \to 0$, then $\|z_n\| \to 1$. In this case, all expressions in the last equation tend to zero with the exception of $d_1(s_n)\|w_n\|^2 + d_2(s_n)\|z_n\|^2 - b_{22} \langle Az_n, z_n \rangle + \frac{\langle P^{\tau_n}(U_n), U_n \rangle}{\|U_n\|^2}$ which is positive according the positiveness of A and the assumptions (1.4), (2.10) and (2.18). This is the contradiction and therefore $\|w_n\| \geq \varepsilon > 0$. Hence, $d_1(s_n)\|w_n\|^2 \to +\infty$ by (2.18) and this leads to the contradiction again.

Remark 3.5. It follows from the equations describing C_j that if the curve d satisfies (2.18) and intersects C_j transversally at $d^0 = d(s_0)$ then $\frac{(\kappa_j d_2 - b_{22})^2}{b_{12} b_{21}} d_1'(s_0) + d_2'(s_0) < 0$.

Lemma 3.6. *If $[s_n, U_n, \tau_n] \to [s_0, 0, 0]$, $W_n = \frac{U_n}{\|U_n\|} \to -U_0$ and (3.1) holds then*

$$\liminf \frac{s_n - s_0}{\tau_n} > 0.$$

Proof: Set $U_0 = [u_0, v_0]$, $U_0^* = [u_0^*, v_0^*] = [b_{21} b_{12}^{-1} u_0, v_0]$. We have

$$D(d(s_0))U_0^* - B^*AU_0^* = 0 \tag{3.4}$$

by the assumption and the last assertion of Proposition 2.7.. Multiplying (3.1) by $\|U_n\|^{-1}U_0^*$, (3.4) by $\|U_n\|^{-1}U_n$ and subtracting the resulting equations we obtain

$$[d_1(s_n) - d_1(s_0)]\langle w_n, u_0^* \rangle + [d_2(s_n) - d_2(s_0)]\langle z_n, v_0^* \rangle =$$

$$-\frac{\tau_n}{1 + \tau_n}\frac{\langle N(U_n), U_0^* \rangle}{\|U_n\|} - \frac{\langle P^{\tau_n}(U_n), U_0^* \rangle}{\|U_n\|}.$$

It can be written as

$$\frac{s_n - s_0}{\tau_n}R_n = -\frac{\langle N(U_n), U_0^* \rangle}{(1 + \tau_n)\|U_n\|} - \frac{\langle P^{\tau_n}(U_n), U_0^* \rangle}{\tau_n\|U_n\|}. \tag{3.5}$$

where $R_n = d_1'(\bar{s}_n)\langle w_n, u_0^* \rangle + d_2'(\tilde{s}_n)\langle z_n, v_0^* \rangle$ with some $\bar{s}_n$, $\tilde{s}_n$ lying between s_n, s_0. We have $\langle w_n, u_0^* \rangle \to -\langle u_0, u_0^* \rangle = -\frac{(d_2\kappa_j - b_{22})^2}{b_{12}b_{21}}\|v_0\|^2$, $\langle z_n, v_0^* \rangle \to -\langle v_0, v_0^* \rangle = -\|v_0\|^2$ by Proposition 2.7.. It follows from Remark 3.5. that

$$\lim R_n = -\left[\frac{(d_2\kappa_j - b_{22})^2}{b_{12}b_{21}}d_1'(s_0) + d_2'(s_0)\right] \cdot \|v_0\|^2 > 0.$$

The first term in the right-hand side of (3.5) tends to zero by (2.3). We have $U_0^* \in K^0$, $-U_0 \notin K$ because $U_0 \in K^0$ and our assertion follows from (3.5) and (2.15).

Lemma 3.7. *There exists $\delta_0 > 0$ such that if $\delta \in (0, \delta_0)$, s_n, U_n, τ_n satisfy* (2.21), (2.22), *$U_n \notin K$, $\frac{U_n}{\|U_n\|} \to W$, $[s_n, U_n, \tau_n] \to [s_0, U, \tau]$, $\tau \in [0, +\infty)$ then $W \notin K$.*

Proof: Suppose that there are $\delta_m \to 0_+$ such that for any m there exist s_n^m, U_n^m, τ_n^m $(n = 1, 2, \ldots)$ satisfying

$$D(d(s_n^m))U_n^m - BAU_n^m + \frac{\tau_n^m}{1 + \tau_n^m}N(U_n^m) + P^{\tau_n^m}(U_n^m) = 0, \tag{3.6}$$

$$\|U_n^m\|^2 = \frac{\delta_m\tau_n^m}{1 + \tau_n^m}, \tag{3.7}$$

$U_n^m \notin K$ for all n, m, $\frac{U_n^m}{\|U_n^m\|} \to W^m \in K$, $[s_n^m, U_n^m, \tau_n^m] \to [s_0, U^m, \tau^m]$ if $n \to +\infty$ for any m fixed. First, let $\tau^m = 0$ for some m. Dividing (3.6) by $\|U_n^m\|$, the limiting process by using (2.3), (2.13) gives

$$D(d^0)W^m - BAW^m = 0.$$

That means $W^m = \pm U_0$, $U_0 \in K^0$, $-U_0 \notin K$ and therefore $W^m = U_0 \in K^0$ which contradicts the assumption $U_n^m \notin K$. Hence, $\tau^m > 0$ for all m. It follows $U_n^m \to U^m = \sqrt{\frac{\delta_m\tau^m}{1 + \tau^m}}W^m \in \partial K$ and the limiting process in (3.6) gives (by using (2.10))

$$D(d^0)U^m - BAU^m + \frac{\tau^m}{1 + \tau^m}N(U^m) = 0.$$

We can suppose $\frac{U^m}{\|U^m\|} \to W$ if $m \to +\infty$ and Remark 3.1. implies $\frac{U^m}{\|U^m\|} \to W \in \partial K$. Dividing the last equation by $\|U^m\|$ we obtain

$$D(d^0)W - BAW = 0$$

and therefore $W = \pm U_0$, $U_0 \in K^0$ which is the contradiction again.

317

Remark 3.8. The assumption (2.10) implies that if $[\tau_n, U_n] \in R_+ \times \mathbf{V} \times \mathbf{V}$, $\frac{P^{\tau_n}(U_n)}{\|U_n\|} \rightharpoonup F$ then

$$\langle F, W \rangle = \lim \frac{\langle P^{\tau_n}(U_n), W \rangle}{\|U_n\|} \leq 0 \quad \text{for all } W \in K.$$

Further, if $F \neq 0$ then $\langle F, W \rangle < 0$ for all $W \in K^0$. Indeed, otherwise there are $V \in \mathbf{V} \times \mathbf{V}$, $W \in K^0$ such that $\langle F, V \rangle > 0$, $\langle F, W \rangle = 0$. It follows that $\langle F, W + tV \rangle > 0$ for all $t > 0$ but this contradicts the first assertion because $W + tV \in K$ for t small.

Lemma 3.9. *There exists $\delta_0 > 0$ such that if $[s, U, \tau]$ satisfies (2.21), $U \notin K$, $\|U\| < \delta_0$ then $s \neq s_0$.*

Proof: Suppose by contradiction that there are U_n, τ_n satisfying

$$D(d^0)U_n - BAU_n + \frac{\tau_n}{1 + \tau_n}N(U_n) + P^{\tau_n}(U_n) = 0, \tag{3.8}$$

$\|U_n\| \to 0$, $U_n \notin K$, $W_n = \frac{U_n}{\|U_n\|} \rightharpoonup W$. Multiplying (3.8) by $\|U_n\|^{-1}U_0^*$, (3.4) by $\|U_n\|^{-1}U_n$ and subtracting the resulting equations we obtain

$$\frac{\tau_n}{1 + \tau_n}\frac{\langle N(U_n), U_0^* \rangle}{\|U_n\|} + \frac{\langle P^{\tau_n}(U_n), U_0^* \rangle}{\|U_n\|} = 0.$$

Suppose that $\tau_n \to 0$. Dividing (3.8) by $\|U_n\|$, we obtain by using (2.3), (2.13)

$$D(d^0)W - BAW = 0. \tag{3.9}$$

Hence, $W = \pm U_0$. But $W = U_0 \in K^0$ is impossible because $U_n \notin K$ and $W = -U_0$ would contradict Lemma 3.6.. This contradiction shows that $\tau_n \geq \varepsilon > 0$. It follows from (3.8) that $\frac{P^{\tau_n}(U_n)}{\|U_n\|}$ are bounded and therefore we can suppose $\frac{P^{\tau_n}(U_n)}{\|U_n\|} \rightharpoonup F$,

$$D(d^0)W - BAW + F = 0.$$

Multiplying this equation by U_0^*, (3.4) by W and subtracting, we obtain $\langle F, U_0^* \rangle = 0$. Remark 3.8. implies $F = 0$, i.e. we have (3.9) again. It follows $W = -U_0 \notin K$ and therefore it should be $\langle F, U_0^* \rangle < 0$ by (2.14) because $U_0^* \in K^0$ which is the contradiction.

Lemma 3.10. *There exists $\delta_0 > 0$ such that if $[s, U, \tau]$ satisfies (2.21), $s > s_0$, $\|U\| < \delta_0$ then $U \notin \partial K$.*

Proof: Suppose the contrary. Then there are $\delta_n \to 0$, $[s_n, U_n, \tau_n]$ satisfying (3.1), $s_n > s_0$, $\|U_n\| < \delta_n$, $U_n \in \partial K$. We have $P^{\tau_n}(U_n) = 0$ by (2.10) and therefore

$$D(d(s_n))U_n - BAU_n + \frac{\tau_n}{1 + \tau_n}N(U_n) = 0.$$

We can suppose $\frac{U_n}{\|U_n\|} \rightharpoonup U$. Dividing the last equation by $\|U_n\|$ we obtain by using (2.3) and the compactness argument $\frac{U_n}{\|U_n\|} \to U \in \partial K$, $D(d(s))U - BAU = 0$. That means $s = s_0$ (see Proposition 2.7. and (2.18)) and therefore $U = \pm U_0$. But $U_0 \in K^0$, $-U_0 \notin K$ and this is the contradiction with $U \in \partial K$.

4. PROOF OF MAIN RESULT

Consider a real Banach space Y with the norm $\| \cdot \|$ and an equation of the type

$$U - T(s)U + H_\tau(s, U) = 0 \tag{4.1}$$

where s, τ are real parameters and the following assumptions are fulfilled:

for any $s \in R$, $T(s)$ is a linear completely continuous operator in Y;
the mapping $s \to T(s)$ of R into the space of linear continuous
mappings in Y (with the operator norm) is continuous;
the mapping $Q : R \times Y \times R \to Y$ defined by
$Q(s, U, \tau) = T(s)U - H_\tau(s, U)$ is completely continuous; $\tag{4.2}$

$$\lim_{\|U\| + |\tau| \to 0} \frac{\|H_\tau(s, U)\|}{\|U\| + |\tau|} = 0 \text{ uniformly on bounded } s\text{-intervals.} \tag{4.3}$$

Supplement (4.1) by the norm condition

$$\|U\|^2 = \frac{\delta \tau}{1 + \tau} \tag{4.4}$$

with a fixed $\delta > 0$. The equation (2.21) can be written in the form (4.1) with $Y = \mathbf{V} \times \mathbf{V}$,

$$
\begin{aligned}
T(s)U &= D^{-1}(d(s))BAU, \\
H_\tau(s, U) &= D^{-1}(d(s)) \left[\frac{\tau}{1 + \tau} N(U) + P^\tau(U) \right]
\end{aligned}
\tag{4.5}
$$

and if we set $P^\tau(U) = P^{-\tau}(U)$ for $\tau < 0$ then the assumptions (4.2), (4.3) are satisfied.

Remark 4.1. Denote by $\theta(s)$ the sum of algebraic multiplicities of all positive eigenvalues of the operator $T(s) - I$. By a critical point of T we shall mean a parameter s such that $\mathrm{Ker}(T(s) - I) \neq \{0\}$. It is well-known that $\deg(I - T(s), 0, B_r) = (-1)^{\theta(s)}$. This holds for a general linear completely continuous operator in a Banach space – see e.g. Nirenberg (1974). Hence, the condition (4.6) below implies that

$$\deg(I - T(s_0 + \zeta), 0, B_r) \neq \deg(I - T(s_0 - \zeta), 0, B_r) \quad \text{for any } \zeta \in (0, \zeta_0), \ r > 0$$

where B_r denotes the ball with the radius r centered at the origin.

The proof of Theorem 2.10. will be based on the following continuation theorem.

Theorem 4.2. *Let K be a closed convex cone in Y with its vertex at the origin, $K \neq Y$, let the mappings T, H satisfy (4.2), (4.3). Suppose that s_0 is the greatest critical point of T, $\mathrm{Ker}(I - T(s_0)) = \mathrm{Lin}\{U_0\}$, $U_0 \in K^0$,*

$$\theta_T(s_0 + \xi) - \theta_T(s_0 - \xi) \text{ is odd for any } \xi \in (0, \xi_0) \tag{4.6}$$

with some $\xi_0 > 0$. Suppose that the following implications hold for any $[s, U, \tau]$ and $[s_n, U_n, \tau_n]$ satisfying (4.1), (4.4), $\tau \in [0, +\infty)$:

$$\text{if } U \notin K \text{ then } s \leq C \text{ (with some } C > 0 \text{ fixed)} ; \tag{4.7}$$

$$\text{if } U_n \notin K, \ \tau_n > 0, [s_n, U_n, \tau_n] \to [s_0, 0, 0], \ \frac{U_n}{\|U_n\|} \to -U_0$$

$$\text{then } s_n > s_0 \text{ for all } n \geq n_0; \tag{4.8}$$

$$\text{if } U_n \notin K, \ \tau_n > 0, [s_n, U_n, \tau_n] \to [s_0, U, \tau], \ \frac{U_n}{\|U_n\|} \to W$$

$$\text{with some } W \in K \text{ then } s_n < s_0 \text{ for all } n \geq n_0; \tag{4.9}$$

$$\text{if } U \notin K \text{ then } s \neq s_0; \tag{4.10}$$

$$\text{if } s > s_0, \ \|U\| \neq 0 \text{ then } U \notin \partial K. \tag{4.11}$$

Then for any $\delta \in (0, \delta_0)$ ($\delta_0 > 0$ small) there exists a closed connected set C_δ^- in $R \times Y \times R$ containing $[s_0, 0, 0]$ and such that
if $[s_0, 0, 0] \neq [s, U, \tau] \in C_\delta^-$ then (4.1), (4.4) are fulfilled and $s > s_0$, $U \notin K$;
for any $\tau > 0$ there is at least one couple $[s, U]$ such that $[s, U, \tau] \in C_\delta^-$.

Proof: This theorem is a special version of a general continuation theorem proved by Kučera (1988) (Theorem 1.2, Remark 1.6). The detailed proof is rather complicated and we shall recall here only main ideas. Set $X = Y \times R$. For a given $\delta > 0$ define

$$S(s)x = S(s)(U, \tau) = [T(s)U, 0],$$

$$R(s, x) = R(s, U, \tau) = [H_\tau(s, U), -\tfrac{1+\tau}{\delta}\|U\|^2] \text{ for all } x = [U, \tau] \in X.$$

Then (4.1), (4.4) are equivalent to the equation

$$x - S(s)x + R(s, x) = 0 \tag{4.12}$$

where $S(s)$ for s fixed is a linear completely continuous operator, R is a small compact perturbation. The assumptions of Theorem 4.2. imply $\mathrm{Ker}\,(I - S(s_0)) = \mathrm{Lin}\{x_0\}$, $x_0 = [U_0, 0]$,

$$\deg(I - S(s_0 + \xi), 0, B_r) \neq \deg(I - S(s_0 - \xi), 0, B_r) \text{ for all } \xi \in (0, \xi_0), r > 0.$$

This enables us to use considerations from the proof of well-known global bifurcation theorem given by Dancer (1974), where a special type of equations (4.12) with $S(s) = sS$ is considered. These considerations imply that there exist two closed connected subsets C_0^+ and C_0^- of the closure of the set of nontrivial solutions of (4.12) (i.e. also of (2.21), (2.22)) starting at $[s_0, 0]$ in the directions x_0 and $-x_0$, respectively. Moreover, these branches either meet each other in a point different from $[s_0, 0]$ or they are both unbounded. Our main idea is to show that the first possibility is excluded under the assumptions (4.8)–(4.11). The principle of this proof is the following. It follows from the fact that C_0^- starts at $[s_0, 0]$ in the direction $[-U_0, 0]$, $-U_0 \notin K$ and from the assumption (4.8) that it lies outside of K (in U) and above s_0 (in s) at its beginning. The assumption (4.11) ensures that C_0^- remains outside of K as long as $s > s_0$. (C_0^- cannot intersect ∂K in $[s, 0, 0]$, $s > s_0$ because we suppose that s_0 is the greatest critical point.) Further, (4.10) implies that C_0^- remains above s_0 as long as $U \notin K$ and (4.9) means that C_0^- cannot come to s_0 in s from above and from outside of K with $\tau \neq 0$. It follows that $s > s_0$, $U \notin K$ for all $[s, U, \tau] \in C_0^-$, $[s, U, \tau] \neq [s_0, 0, 0]$. On the other hand, C_0^+ starts into K^0 because $U_0 \in K^0$ and this leads to the conclusion that C_0^+, C_0^- cannot intersect in a point different from $[s_0, 0, 0]$. Hence, C_0^- must be unbounded.

It is bounded in U by (4.4) and bounded in s by (4.7) and the above considerations. Hence, it is unbounded in τ. Simultaneously $\tau > 0$ along this branch according (4.4) and the connectedness. The assertion follows.

Proof of Theorem 2.10.: We shall show that the assumptions of Theorem 4.2. are fulfilled with the operators (4.5) and with K from (2.6). It follows from Proposition 2.7. and (2.18) that s_0 from the assumptions of Theorem 2.10. is the greatest critical point of T and $\mathrm{Ker}(I - T(s_0)) = E_B(d^0) = \mathrm{Lin}\{U_0\}$, with $U_0 \in K^0$. The assumption (4.6) follows from the end of Proposition 2.7., the assumptions (4.7), (4.8), (4.9), (4.10) and (4.11) follow from Lemmas 3.4., 3.6., 3.7., 3.9. and 3.10., respectively. Hence, it follows from Theorem 4.2. that for any $\delta \in (0, \delta_0)$ fixed there are $[s_n, U_n, \tau_n]$ satisfying (4.1), (4.4) (i.e. (2.21), (2.22)), $U_n \notin K$, $s_n \to s(\delta) \geq s_0$, $\tau_n \to +\infty$. We can suppose $\frac{U_n}{\|U_n\|} \to U(\delta)$ and Lemma 3.2. implies that $\frac{U_n}{\|U_n\|} \to U(\delta)$ and $s(\delta)$, $U(\delta)$ satisfy (2.5), $U(\delta) \notin K^0$.

It remains to show that $s(\delta) > s_0 + \varepsilon$ with some fixed $\varepsilon > 0$ for all δ small enough. Suppose that there are $\delta_n \to 0$ such that $s(\delta_n) \to s_0$. Then

$$D(d(s_n))U(\delta_n) - BAU(\delta_n) + N(U(\delta_n)) \in -M(U(\delta_n)).$$

We can suppose $\frac{U(\delta_n)}{\|U(\delta_n)\|} \to W$. Dividing the last inclusion by $\|U(\delta_n)\|$ and using the compactness of A, (2.3), (2.8), we obtain $\frac{U(\delta_n)}{\|U(\delta_n)\|} \to W$, $W \notin K^0$,

$$D(d^0)W - BAW \in -M_0(W).$$

Lemma 3.3. implies $W \in E_B(d^0) \cap K$ and that means $W = U_0$ because $E_B(d^0) = \mathrm{Lin}\{U_0\}$, $U_0 \in K^0$, $-U_0 \notin K$. This is the contradiction because $W \notin K^0$.

REFERENCES

DANCER, E.N. (1974), On the structure of solutions of non-linear eigenvalue problem, *Indiana Univ. Math. J.* **23**, p. 1069–1076.

DRÁBEK, P., KUČERA, M. and M. MÍKOVÁ (1985), Bifurcation points of reaction-diffusion systems with unilateral conditions, *Czechoslovak Math. J.* **35**, p. 639–660.

DRÁBEK, P. and M. KUČERA (1988), Reaction-diffusion systems: Destabilizing effect of unilateral conditions, *Nonlinear Analysis, T. M. A.* **12**, p. 1173-1192.

DUVAUT, G. and J.L. LIONS (1972), *Les Inéquations en Mechanique et en Physique*, (Dunod, Paris).

EISNER, J. and M. KUČERA (1995), Hopf bifurcation and ordinary differential inequalities, *Czechoslovak Math. J.* (to appear).

FUČÍK, S. and A. KUFNER (1980), *Nonlinear Differential Equations*, (Elsevier, Amsterdam).

GIERER, A. and H. MEINHARDT (1972), A theory of biological pattern formation, *Kybernetik* **12**, p. 30–39.

KIELHÖFER, H. (1974), Stability and semilinear evolution equations in Hilbert space, *Arch. Rat. Mech. Analysis* **57**, p. 150–165.

KUČERA, M. (1982), Bifurcation points of variational inequalities, *Czechoslovak Math. J.* **32**, p. 208–226.

KUČERA, M. (1988), A global continuation theorem for obtaining eigenvalue and bifurcation points, *Czechoslovak Math. J.* **38**, p. 120–137.

KUČERA, M. (1991), A bifurcation of periodic solutions to ordinary differential inequalities, *Colloq. Math. Soc. J. Bolyai* **62**, *Differential Equations* (Budapest), p. 227-255.

NEČAS, J. (1967), *Les Méthodes Directes en Théorie des Equations Elliptiques*, (Academia, Praha).

NISHIURA, Y. (1982), Global structure of bifurcating solutions of some reaction-diffusion systems, *SIAM J. Math. Analysis* **13**, p. 555–593.

NIRENBERG, L. (1974), *Topics in Nonlinear Functional Analysis*, (Academic Press, New York).

QUITTNER, P. (1986), Spectral analysis of variational inequalities, *Commentat. Math. Univ. Carol.* **27**, p. 605–629.

QUITTNER, P. (1987), Bifurcation points and eigenvalues of inequalities of reaction-dif- fusion type, *J. reine angew. Math.* **380**, p. 1–13.

STABILITY THEORY FOR FLOWS GOVERNED BY IMPLICIT EVOLUTION EQUATIONS

Christiaan Le Roux and Niko Sauer

Faculty of Science
University of Pretoria
Pretoria 0002, South Africa

1. INTRODUCTION

The use of energy methods for the study of stability of the rest state (null solution) of evolution equations of the form

$$Bu_t = Lu + Nu \tag{1}$$

with B a diagonal matrix with positive entries, L a linear operator and N a homogeneous nonlinear operator has recently been explored in an extensive way in Galdi and Padula (1990). The approach was to view the equation (1) in a Hilbert space setting and to derive conditions for stability and instability from the spectral properties of the operator L. The strength of the method was illustrated by a number of examples from fluid dynamics and related fields.

The stability of the motion of incompressible fluids of second grade was studied in Dunn and Fosdick (1974) and Galdi, Padula and Rajagopal (1990) also using energy methods. These equations may be written as implicit evolution equations of the form

$$\frac{d}{dt}[Bu(t)] = Lu(t) + Nu(t) \tag{2}$$

where B, L and N are operators defined on some space X with values in a different space Y; B and L being unbounded linear operators and N a homogeneous nonlinear operator (see Galdi, Grobbelaar-van Dalsen and Sauer, 1993).

Equations such as (2) also occur in so-called dynamical boundary conditions such as in problems where a rigid body rotates in a fluid, taking the dynamics of the rigid body into account (see Grobbelaar-van Dalsen and Sauer, 1989, 1993).

Navier-Stokes Equations and Related Nonlinear Problems
Edited by A. Sequeira, Plenum Press, New York, 1995

In this paper we put forth an energy method for the study of stability of the solutions of equations such as (2) based on the ideas in Galdi and Padula (1990) and Sauer (1989).

2. THE GENERAL SETTING

Let X be a real vector space and $\langle Y, (\cdot, \cdot), \| \cdot \| \rangle$ a real Hilbert space. Let $B, L : \mathcal{D} \subset X \to Y$ be linear operators and $N : \mathcal{D} \to Y$ a nonlinear operator such that $N(0) = 0$. Let A be a symmetric, nonnegative linear operator with $\mathcal{D}(A) \supset B[\mathcal{D}] \cup L[\mathcal{D}]$ in Y and put $C = AB$. We define the bilinear forms R and S on $\mathcal{D}$ as

$$R(u, v) = (Lu, Cv); \quad S(u, v) = (Bu, Cv); \quad u, v \in \mathcal{D}.$$

The following general assumptions will be made throughout:

(G1): R is symmetric.

(G2): If $u(t)$ is a solution of (2), then the mappings $t \to Cu(t)$ and $t \to ALu(t)$ are weakly continuous in Y.

Remark 2.1. When we refer to a solution of (2) it will be taken to mean that $Bu(t)$ is differentiable in the norm topology of Y and that (2) holds for every $t > 0$.

Remark 2.2. From the symmetry of A we conclude that S is symmetric. Furthermore, the nonnegativity of A implies that $S(u, u) \geq 0$ for every $u \in \mathcal{D}$.

Remark 2.3. It is the introduction of the (somewhat) arbitrary operator A which makes the energy method possible. In applications, A has to be suitably chosen. In many instances $C = B$ in which case A becomes the identity.

Lemma 2.4. If $u(t)$ is a solution of (2), $R(u(t), u(t))$ and $S(u(t), u(t))$ are differentiable and

$$\frac{d}{dt} S(u, u) = 2[(ALu, Bu) + (ANu, Bu)]$$

$$\frac{d}{dt} R(u, u) = 2[(ALu, Lu) + (ALu, Nu)]. \tag{3}$$

Proof. For any given function $g : t \to g(t); \quad t > 0$ we define the difference $g(t; h) = h^{-1}[g(t + h) - g(t)]$. From the symmetry of S we obtain

$$(Bu, Cu)(t; h) = (Bu(t; h), Cu(t + h)) + (Bu(t), Cu(t; h))$$
$$= (Bu(t; h), Cu(t + h) + Cu(t)).$$

Similarly,

$$(ALu, Bu)(t; h) = (Bu(t; h), ALu(t + h) + ALu(t)).$$

The differentiability of $S(u, u)$ now follows from (G2). Indeed,

$$\frac{d}{dt} S(u, u) = 2([Bu]_t, Cu) = 2(Lu + Nu, Cu).$$

The expression for the derivative of $R(u, u)$ is obtained similarly.

3. A STABILITY METHOD

For the purposes of this section two additional assumptions (S1) and (S2) are made:

(S1): $-R$ is positive definite; i.e $-R(u, u) \geq 0$ for every $u \in \mathcal{D}$ with equality if and only if $u = 0$.

Following Galdi and Padula (1990) we introduce the norms $\| \cdot \|_1$ and $\| \cdot \|_2$ on $\mathcal{D}$ by

$$\|u\|_1^2 = -R(u, u)$$
$$\|u\|_2^2 = \|u\|_1^2 + \tfrac{1}{2}(ALu, Lu). \tag{4}$$

By virtue of (S1) and the nonnegativity of A the definitions are appropriate. We may now formulate the other assumption:

(S2): There exist constants $c_1 > 0$ and $\alpha \geq 1$ such that for every $u \in \mathcal{D}$

$$(ANu, Nu) \leq [c_1 \|u\|_1^\alpha \|u\|_2]^2. \tag{5}$$

It is evident that the norms introduced here involve the operators B, L and C. In applications the operator C will be determined by B and L so that the expressions will involve only the 'physics' of the problem. The *energy* associated with $u \in \mathcal{D}$ is defined as

$$E_u = \tfrac{1}{2}[S(u, u) - R(u, u)] = \tfrac{1}{2}(Bu - Lu, Cu). \tag{6}$$

If $u(t)$ is a solution of (2) we shall write $E(t)$ instead of $E_{u(t)}$. From Lemma 2.4 it is clear that $E(t)$ is differentiable and

$$E'(t) = (Lu, Cu) + (Nu, Cu) - (ALu, Lu) - (ALu, Nu). \tag{7}$$

Theorem 3.1. *If*

$$E(0) < \tfrac{1}{2} \left[\frac{\sqrt{3} - 1}{c_1} \right]^{\frac{2}{\alpha}} \tag{8}$$

the null-solution of (2) is monotonically and asymptotically stable with respect to $\| \cdot \|_1$. *Moreover*

$$\|u(t)\|_1^2 \leq \frac{[\|u(0)\|_1^2 + E_0/\gamma]e^{\kappa E_0/\gamma}}{1 + t}$$

where $E_0 = E(0)$, $\kappa = \tfrac{1}{2}c_1^2(2E_0)^{\alpha-1}$ *and* $\gamma = 1 - 2^{\alpha/2}c_1 E_0^{\alpha/2} - 2^{\alpha-1}c_1^2 E_0^\alpha$.

Proof. From (4) and (8), using the Cauchy-Schwarz inequality in the form $|(Au, v)| \leq (Au, u)^{1/2}(Av, v)^{1/2}$ which holds because of the nonnegativity and symmetry of A, (5) and (6), we obtain

$$
\begin{aligned}
E'(t) &= -\|u\|_1^2 + (Nu, ABu) - (ALu, Lu) + (ALu, Nu) \\
&\leq -\|u\|_1^2 + (ABu, Bu)^{1/2}(ANu, Nu)^{1/2} - (ALu, Lu) \\
&\qquad + (ALu, Lu)^{1/2}(ANu, Nu)^{1/2} \\
&\leq -\|u\|_1^2 + [S(u, u)]^{1/2}(ANu, Nu)^{1/2} - (ALu, Lu) \\
&\qquad + \tfrac{1}{2}(ALu, Lu) + \tfrac{1}{2}(ANu, Nu) \\
&\leq -\|u\|_1^2 - \tfrac{1}{2}(ALu, Lu) + (2E)^{1/2}c_1\|u\|_1^\alpha\|u\|_2 + \tfrac{1}{2}c_1^2\|u\|_1^{2\alpha}\|u\|_2^2 \\
&\leq -\|u\|_2^2 + c_1(2E)^{1/2}\|u\|_1^{\alpha-1}\|u\|_2^2 + \tfrac{1}{2}c_1^2\|u\|_1^{2\alpha}\|u\|_2^2 \\
&\leq -\|u\|_2^2[1 - c_1(2E)^{\alpha/2} - \tfrac{1}{2}c_1^2(2E)^\alpha].
\end{aligned}
\tag{9}
$$

It is now clear that if (8) is satisfied, $E(t)$ will be decreasing and the monotonic stability is proved.

It is evident that (8) is the same as stating that $\gamma > 0$, in which case we obtain from (9) that $E'(t) \leq -\gamma\|u(t)\|_2^2$. This, in turn, implies that

$$
\int_0^t \|u(s)\|_2^2 \, ds \leq \frac{E_0 - E(t)}{\gamma} \leq \frac{E_0}{\gamma}.
\tag{10}
$$

By Lemma 2.4 we have

$$
\begin{aligned}
\tfrac{1}{2}\frac{d}{dt}\|u(t)\|_1^2 &= -\tfrac{1}{2}\frac{d}{dt}R(u, u) \\
&= -(ALu, Lu) - (ALu, Nu) \\
&\leq -(ALu, Lu) + (ALu, Lu)^{1/2}(ANu, Nu)^{1/2} \\
&\leq -(ALu, Lu) + \frac{\varepsilon}{2}(ALu, Lu) + \frac{1}{2\varepsilon}(ANu, Nu).
\end{aligned}
$$

Choosing $\varepsilon = 2$, we obtain

$$
\frac{d}{dt}\|u(t)\|_1^2 \leq \tfrac{1}{2}(ANu, Nu) \leq \tfrac{1}{2}c_1^2\|u\|_1^{2\alpha}\|u\|_2^2.
\tag{11}
$$

From here we proceed exactly as in Galdi and Padula (1990), using (10), to solve the differential inequality (11) and complete the proof of the theorem. $\qquad \square$

Remark 3.2. The case where Y is a finite Cartesian product of Hilbert spaces and B a positive diagonal matrix, was studied in Galdi and Padula (1990) and relates to the present study. In this case $X = Y$. If we take $C = I$, the energy becomes

$$
E_u = \tfrac{1}{2}[(Bu, u) - (Lu, u)]
$$

which is the expression used in Galdi and Padula (1990). The associated norms are derived from

$$
\|u\|_1^2 = -(Lu, u); \quad \|u\|_2^2 = \|u\|_1^2 + \tfrac{1}{2}\|B^{-1}Lu\|^2.
$$

The first norm is the same as and the second is equivalent to that defined in Galdi and Padula (1990). Also, the condition (5) is equivalent to the corresponding condition in Galdi and Padula (1990), (1.4), p.197.

Remark 3.3. It is sometimes more appropriate to define the energy simply as $E = \frac{1}{2}S(u,u)$ in which case

$$E'(t) = -\|u\|_1^2 + (ANu, Bu).$$

If, in addition, it can be proved that S is bounded and positive definite with respect to $\|\cdot\|_1$ in the sense that there exist positive constants c_1, c_2 and c_3 such that for some $\alpha > 1$

$$c_3\|u\|_1^2 \leq S(u,u) \leq c_2\|u\|_1^2$$
$$(ANu, Bu) \leq c_1\|u\|_1^{2\alpha},$$

then the steps in the proof of Theorem 3.1 can be modified to find

$$E'(t) \leq -\|u\|_1^2[1 - \frac{c_1}{c_3^{\alpha-1}}(2E)^{\alpha-1}]$$

and it follows that $E'(t) \leq 0$ provided that

$$E_0 \leq \gamma := \frac{c_3}{2c_1^{1/(\alpha-1)}}.$$

From here it is quite easy to show that

$$E'(t) \leq -\gamma\|u\|_1^2 \leq -\frac{2\gamma}{c_2}E(t)$$

and it follows that E and hence $\|u\|_1$, decay exponentially.

4. AN INSTABILITY METHOD

The stability method disussed in the previous section is largely based on the negativity of R. Combined with the nonnegativity of S, this property implies that for every $\omega > 0$ and for every $u \in \mathcal{D}$, $R(u,u) - \omega S(u,u) = (Lu - \omega Bu, Cu)$ is negative definite. For the study of instability we shall make, in addition to (G1) and (G2) the assumptions (I1, ... I4), the first of which is the negation of the negative definiteness.

(I1): *There exists $a > 0$ and $\varphi_0 \in \mathcal{D}$ such that $R(\varphi_0, \varphi_0) - aS(\varphi_0, \varphi_0) > 0$.*

If we let $M := L - aB$, this condition may be re-stated in the following form:

$$(AM\varphi_0, B\varphi_0) = (M\varphi_0, C\varphi_0) > 0. \tag{12}$$

(I2): *There exists a constant $k_a > 0$ such that for every $u \in \mathcal{D}$*

$$(AMu, Mu) \geq k_a^2 S(u,u). \tag{13}$$

(I3): *There is a norm $\|\cdot\|_Z$ on $\mathcal{D}$ with respect to which R is bounded, i.e. there exists a constant $c > 0$ such that for all $u, v \in \mathcal{D}$*

$$|R(u,v)| \leq c\|u\|_Z\|v\|_Z. \tag{14}$$

(I4): *There exist positive constants c_1 and α such that for every $u \in \mathcal{D}$*

$$|(ANu, Nu)| \leq c_1^2\|u\|_Z^{2\alpha}[(S(u,u))^{1/2} + (ALu, Lu)^{1/2}]^2. \tag{15}$$

Theorem 4.1. *Under the general assumptions* (G1), (G3) *and the instability assumptions* (I1), ... (I4), *the null-solution of* (2) *is unstable with respect to* $\| \cdot \|_Z$.

Proof. Let $u(t)$ be a solution of (2) and set $w(t) = e^{-at}u(t)$. Then it is seen that $[Bw]_t = Mw + e^{-at}Nu$. We observe also that $(AMw, Bw) = (ALw, Bw) - a(ABw, Bw) = R(w, w) - aS(w, w)$. Thus, after a slight adaptation of Lemma 2.4, we obtain

$$
\begin{aligned}
\tfrac{1}{2}\frac{d}{dt}(AMw, Bw) &= \tfrac{1}{2}\frac{d}{dt}[R(w, w) - aS(w, w)] \\
&= ([Bw]_t, ALw) - a([Bw]_t, ABw) \\
&= (Mw + e^{-at}Nu, AMw) \\
&\geq (AMw, Mw) - e^{-at}(AMw, Mw)^{1/2}(ANu, Nu)^{1/2} \\
&\geq (AMw, Mw)^{1/2}[(AMw, Mw)^{1/2} - (ANu, Nu)^{1/2}] \quad (16)
\end{aligned}
$$

From the inverse triangle inequality, applied to the semi-norm (Ax, x), we obtain

$$
(AMw, Mw)^{1/2} \geq (ALw, Lw)^{1/2} - a(ABw, Bw)^{1/2}.
$$

From (13) we obtain for $q > 0$,

$$
q(AMw, Mw)^{1/2} \geq qk_a[S(w, w)]^{1/2} = qk_a(ABw, Bw)^{1/2}.
$$

Therefore,

$$
(1 + q)(AMw, Mw)^{1/2} \geq (ALw, Lw)^{1/2} + (qk_a - a)(ABw, Bw)^{1/2}.
$$

Choose $q > a/k_a$ and let

$$
c_a = \min\left[\frac{1}{1+q}, \frac{qk_a - a}{1 + q}\right].
$$

Then

$$
(AMw, Mw)^{1/2} \geq c_a\{[S(w, w)]^{1/2} + (ALw, Lw)^{1/2}\}. \quad (17)
$$

Combination of (15), (16) and (17) yields

$$
\tfrac{1}{2}\frac{d}{dt}(AMw, Bw) \geq (AMw, Mw)^{1/2}[c_a - c_1\|u\|_Z^\alpha]\{[S(w, w)]^{1/2} + (ALw, Lw)^{1/2}\} \quad (18)
$$

Suppose that the null-solution of (2) is stable with respect to $\| \cdot \|_Z$, i.e. for given $\eta > 0$ a $\delta > 0$ can be found such that $\|u_0\|_Z < \delta$ implies that $\|u(t)\|_Z < \eta$ for all $t > 0$, where $u_0 = u(0)$. Choose η so that $0 < \eta < (c_a/c_1)^{1/\alpha}$ and $\delta > 0$ accordingly. Then the right hand side of (18) is nonnegative which means that $(AMw, Bw) = (Mw, w)$ is nondecreasing. This implies that

$$
(Lw(t), Cw(t)) \geq a(Bw(t), Cw(t)) + (AMu_0, Bu_0) \geq (AMu_0, Bu_0).
$$

From (14) we may now deduce that

$$
\|w(t)\|_Z^2 \geq c^{-1}(Lw(t), Cw(t)) \geq c^{-1}(AMu_0, Bu_0).
$$

In terms of $u(t)$ this amounts to

$$
\|u(t)\|_Z^2 \geq c^{-1}(AMu_0, Bu_0)e^{2at}.
$$

Rescaling of φ_0 so that $\|\varphi_0\| < \delta$ and taking $u_0 = \varphi_0$, shows that a situation exists in which $\|u(t)\|$ becomes larger than η after a finite time and the proof is concluded by *reductio ad absurdum*.

5. A PROBLEM WITH DYNAMICAL BOUNDARY CONDITIONS

In this section we illustrate the methods described in abstract terms by studying the stability of a reaction-diffusion type of equation in which material is created according to a logistic law, under dynamical boundary conditions. Let Ω be a bounded domain of R^n ($n \leq 4$) with smooth boundary Γ. The equations are

$$
\begin{aligned}
v_t &= \Delta v + \lambda v(p - v) \text{ on } \Omega \times (0, \infty) \\
[\gamma_0 v]_t &= -\beta \gamma_1 v \text{ on } \Gamma \times (0, \infty)
\end{aligned}
\tag{19}
$$

where Δ denotes the Laplacian, $\gamma_0 v$ the value of v on Γ and $\gamma_1 v$ the normal derivative of v. The second equation states that the rate of increase in the boundary is explained by the flux at the boundary. The constants β and p are taken to be positive. If we make the substitution $u = p - v$ the equations (19) become

$$
\begin{aligned}
u_t &= \Delta u - \lambda u(p - u) \\
[\gamma_0 u]_t &= -\beta \gamma_1 u.
\end{aligned}
\tag{20}
$$

We shall consider the stability and instability of the rest state $u = 0$ ($v = p$) of (20). For the space X we choose $L^2(\Omega)$ and we take $\mathcal{D} = H^2(\Omega)$. Let $Y = L^2(\Omega) \times L^2(\Gamma)$ with inner product

$$
(\langle u, f \rangle, \langle v, g \rangle) = (u, v) + (f, g)_\Gamma = \int_\Omega u(x)v(x)\,dx + \int_\Gamma fg\,ds.
$$

The various operators are chosen as follows: $Bu = \langle u, \gamma_0 u \rangle$, $Lu = \langle \Delta u - \lambda pu, -\beta \gamma_1 u \rangle$ and $A\langle f, g \rangle = \langle \beta f, g \rangle$. A is evidently in accordance with Section 2. Also,

$$
\begin{aligned}
S(u, v) &= \beta(u, v) + (\gamma_0 u, \gamma_0 v)_\Gamma \\
R(u, v) &= -\beta[(\nabla u, \nabla v) + \lambda p(u, v)] \\
(ANu, Nu) &= \beta \lambda^2 \int_\Omega u^4\,dx.
\end{aligned}
$$

The last integral exists by virtue of the Sobolev embedding theorem (Adams, 1975).

Let us assume that λ is positive. The the stability assumption (S1) is satisfied. Indeed,

$$
\begin{aligned}
\|u\|_1^2 &= \beta[\lambda p\|u\|^2 + \|\nabla u\|^2] \\
\|u\|_2^2 &= \|u\|_1^2 + \frac{\beta}{2}[\|\Delta u - \lambda pu\|^2 + \beta\|\gamma_1 u\|_\Gamma^2] \\
E_u &= \tfrac{1}{2}[\beta(1 + \lambda p)\|u\|^2 + \beta\|\nabla u\|^2 + \|\gamma_0 u\|_\Gamma^2].
\end{aligned}
$$

From the Sobolev embedding theorem it is also clear that

$$
(ANu, Nu) \leq \beta \lambda^2 K \|u\|_1^4 \leq \beta \lambda^2 K \|u\|_1^2 \|u\|_2^2
$$

so that (S2) holds with $\alpha = 1$. Hence the rest state is stable if E_0 is sufficiently small. In fact, the trace theorem implies that $E(t) \to 0$ as $t \to \infty$.

Suppose that $\lambda < 0$. Then
$$(AMu, Bu) = \beta[(|\lambda|p - a)\|u\|^2 - \|\nabla u\|^2] - a\|\gamma_0 u\|_\Gamma^2.$$

Let μ be the smallest eigenvalue of $-\Delta$ under homogeneous Dirichlet boundary conditions, and φ_0 the normalized eigenfunction. Then
$$(AM\varphi_0, B\varphi_0) = \beta(|\lambda|p - \mu - a) \tag{21}$$

If $|\lambda|p > \mu$, the expression (21) will be positive for $a \in (0, |\lambda|p - \mu)$ and (I1) will be satisfied.

Next, let
$$\nu = \inf\{\|\nabla u\|^2 + \|\gamma_0 u\|_\Gamma^2 : \|u\|^2 = 1; u \in H^1(\Omega)\}.$$

It is evident that $\mu > \nu > 0$. In addition, suppose that
$$|\lambda|p \leq \nu \text{ and } 2\beta < |\lambda|p - \mu. \tag{22}$$

If we choose a in the open interval $(2\beta, |\lambda|p - \mu)$, we have
$$\begin{aligned}(AMu, Mu) &= (ALu, Lu) + 2\beta a(\|\nabla u\|^2 + \|\gamma_0 u\|_\Gamma^2) + \beta a^2 \|u\|^2 - 2\beta a|\lambda|p\|u\|^2 \\ &\quad + a(a - 2\beta)\|\gamma_0 u\|_\Gamma^2 \\ &\geq 2\beta a(\nu - |\lambda|p)\|u\|^2 + a^2[\beta\|u\|^2 + \frac{a - 2\beta}{a}\|\gamma_0 u\|_\Gamma^2] \\ &\geq a^2[\beta\|u\|^2 + \frac{a - 2\beta}{a}\|\gamma_0 u\|_\Gamma^2].\end{aligned}$$

For $k_a^2 = a^2 \min\{1, 1 - 2\beta/a\} = a(a - 2\beta)$ we therefore have
$$(AMu, Mu) \geq k_a^2[\beta\|u\|^2 + \|\gamma_0 u\|_\Gamma^2] = k_a^2 S(u, u)$$

and (I2) is satisfied.

If we let $\|\cdot\|_Z$ be the norm of the Sobolev space $H^1(\Omega)$, then (I3) is satisfied. Finally,
$$(ANu, Nu) = \beta\lambda^2 \int_\Omega u^4\, dx \leq \beta\lambda^2 K\|u\|_Z^4.$$

Suppose that $|\lambda|p$ is not an eigenvalue of $-\Delta$ under homogeneous Neumann boundary conditions. Then it can be proved that there is a positive constant K_1 such that
$$(ALu, Lu) \geq K_1\|u\|_Z^2$$

and hence
$$\begin{aligned}(ANu, Nu) &\leq \beta\lambda^2 \frac{K}{K_1}\|u\|_Z^2[(ALu, Lu)^{\frac{1}{2}}]^2 \\ &\leq \beta\lambda^2 \frac{K}{K_1}\|u\|_Z^2[[S(u, u)]^{\frac{1}{2}} + (ALu, Lu)^{\frac{1}{2}}]^2\end{aligned}$$

and (I4) is proved. Thus we have

Theorem 5.1. *Suppose $\lambda < 0$. If $\mu > |\lambda|p \geq \nu$, $|\lambda|p$ is not an eigenvalue of $-\Delta$ under homogeneous Neumann boundary conditions, and $2\beta < |\lambda|p - \mu$, then the null solution of (20) is unstable with respect to $\|u\|_Z$. If $\lambda > 0$, the null solution is stable under small perturbations.*

Remark 5.2. The results of this section in terms of (19) means that the constant solution $v = p$ is stable and the null-solution can be unstable.

6. FLUIDS OF SECOND GRADE

As a further illustration we consider the equations of motion of an incompressible fluid of second grade under homogeneous Dirichlet boundary conditions in a bounded domain Ω. Stability of the null solution has been proved in Dunn and Fosdick (1974) under more general boundary conditions but under restrictions imposed by thermodynamical considerations.

Let $\boldsymbol{u}(x,t)$ denote the velocity field ($x \in \Omega$ with $\Omega \subset R^3$ bounded). The equations in question may be written in the form

$$[\rho \boldsymbol{u} - \alpha_1 \nabla \cdot \boldsymbol{A}(\boldsymbol{u})]_t = \mu \nabla \cdot \boldsymbol{A}(\boldsymbol{u}) + \nabla \cdot \boldsymbol{S}(\boldsymbol{u})$$
$$\nabla \cdot \boldsymbol{u} = 0,$$

with

$$\boldsymbol{A} = \nabla \boldsymbol{u} + (\nabla \boldsymbol{u})^T$$
$$\boldsymbol{S} = -p\boldsymbol{I} + \alpha_1 (\boldsymbol{u} \cdot \nabla)\boldsymbol{A} + \alpha_1 [\boldsymbol{A}\nabla \boldsymbol{u} + (\nabla \boldsymbol{u})^T \boldsymbol{A}] + \alpha_2 \boldsymbol{A}^2 - \rho \boldsymbol{u} \otimes \boldsymbol{u}.$$

Here $\otimes$ denotes the tensor product. The constants ρ, μ and α_1 are taken to be positive. For X we take the space of square integrable solenoidal vector fields and we set $\mathcal{D} = X \cap H_0^1(\Omega) \cap H^3(\Omega)$. The spae Y is taken as $L^2(\Omega)$. On $\mathcal{D}$ we define the operators B, L and N by $B\boldsymbol{u} = P[\rho \boldsymbol{u} - \alpha_1 \nabla \cdot \boldsymbol{A}]$, $L\boldsymbol{u} = P[\mu \nabla \cdot \boldsymbol{A}]$ and $N\boldsymbol{u} = P[\nabla \cdot \boldsymbol{S}]$ where P is the projection onto the subspace of solenoidal vectors which are tangential at the boundary. The operator $A = B^{-1}$ is well-defined and is easily seen to be symmetric and nonnegative. We also find, because the members of $\mathcal{D}$ vanish on the boundary, that

$$S(\boldsymbol{u}, \boldsymbol{v}) = (B\boldsymbol{u}, \boldsymbol{v}) = \rho(\boldsymbol{u}, \boldsymbol{v}) + \alpha_1(\boldsymbol{A}(\boldsymbol{u}), \boldsymbol{A}(\boldsymbol{v}))$$
$$R(\boldsymbol{u}, \boldsymbol{v}) = (L\boldsymbol{u}, \boldsymbol{v}) = -\mu(\boldsymbol{A}(\boldsymbol{u}), \boldsymbol{A}(\boldsymbol{v})).$$

S and $-R$ are positive definite and by Poincaré's inequality there exist positive constants c_2 and c_3 such that

$$c_3 \|\boldsymbol{u}\|_1^2 \leq S(\boldsymbol{u}, \boldsymbol{u}) \leq c_2 \|\boldsymbol{u}\|_1^2.$$

Furthermore, a standard calculation shows that

$$(AN\boldsymbol{u}, B\boldsymbol{u}) = (N\boldsymbol{u}, \boldsymbol{u}) = -\tfrac{1}{2}(\alpha_1 + \alpha_2)(\boldsymbol{A}^2(\boldsymbol{u}), \boldsymbol{A}(\boldsymbol{u})) \leq \tfrac{1}{2}|\alpha_1 + \alpha_2|.\|\boldsymbol{u}\|_1^3.$$

If, in this case, we define the energy as $E = \tfrac{1}{2}S(\boldsymbol{u}, \boldsymbol{u})$, Remark 3.3 applies with $\alpha = 3/2$, and the stability of the null-solution follows if $\alpha_1 + \alpha_2 \neq 0$. In this case $\gamma = c_3/2c_1^2$, and the convergence factor in the exponent is $c_3/c_1^2 c_2$ with $c_1 = |\alpha_1 + \alpha_2|/2$. The case $\alpha_1 + \alpha_2 = 0$, which was treated in Dunn and Fosdick (1974), is simpler and does not require the initial energy to be sufficiently small. The case where $\alpha_1 = 0$ in the presence of body forces has recently been treated in Amann (1994).

REFERENCES

ADAMS, R.A. (1975), *Sobolev Spaces*, (Academic Press, New York, San Francisco, London).

AMANN, H. (1994), Stability of the rest state of a viscous incompressible fluid. *Arch. Rational Mech. Anal.*, **126**, p. 231–242

DUNN, J.E. and R.L. FOSDICK (1974), Thermodynamics, stability, and boundedness of fluids of complexity 2 and fluids of second grade. *Arch. Rational Mech. Anal.*, **56**, p. 191–252.

GALDI, G.P. and M. PADULA (1990), A new approach to energy theory in the stability of fluid motion. *Arch. Rational Mech. Anal.*, **110**, p. 187–286.

GALDI, G.P., PADULA, M. and K.R. RAJAGOPAL (1990), On the conditional stability of the rest state of a fluid of second grade in unbounded domains. *Arch. Rational Mech. Anal.*, **109**, p. 173–182.

GALDI, G.P., GROBBELAAR-VAN DALSEN, M. and N. SAUER (1993), Existence and uniqueness of classical solutions of the equations of motion for second-grade fluids. *Arch. Rational Mech. Anal.*, **124**, p. 221–237.

GROBBELAAR-VAN DALSEN, M. and N. SAUER (1989), Dynamic boundary conditions for the Navier-Stokes equations. *Proc. Royal Soc. Edinburgh*, **113A**, p. 1–11.

GROBBELAAR-VAN DALSEN, M. and N. SAUER (1993), The solutions in Lebesque spaces of the Navier-Stokes equations with dynamic boundary conditions. *Proc. Royal Soc. Edinburgh*, **123A**, p. 745–761.

SAUER, N. (1989), The Friedrichs extension of a pair of operators. *Quaestiones Mathematicae*, **12**, p. 239–249.

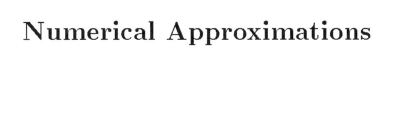

Numerical Approximations

A VORTICITY CREATION ALGORITHM FOR THE NAVIER-STOKES EQUATIONS IN ARBITRARY DOMAIN

G.-H. Cottet

LMC-IMAG, Université de Grenoble, BP 53x,
38041 Grenoble Cédex, France

ABSTRACT

We propose an algorithm of vorticity creation type for the Navier-Stokes equation in arbitrary two dimensional domains. This algorithm can be seen as a rephrasing of Chorin's popular algorithm and one can prove its convergence in the linear case under the sole assumption that the initial enstrophy is finite. The convergence proof follows from delicate energy estimates.

1. INTRODUCTION

The problem of vorticity boundary conditions for the incompressible two dimensional Navier-Stokes equations is a challenging mathematical and numerical question.

On the one hand the value of the vorticity at the boundary of obstacles is a physically relevant quantity. The vorticity, the generation of which is very often restricted at boundaries, is crucial in the understanding of most of the features of the flow (e.g. boundary layers, wakes, small scales or large eddies). This is one of the motivation for the use of numerical algorithms based on the vorticity formulation of the equations, in particular vortex methods (Chorin, 1973). For these algorithms the determination of correct -exact or at least approximate- vorticity boundary condtions is clearly essential.

On the other hand, the value of the vorticity at boundaries is not directly accessible form the mathematical model where, most often, boundary conditions are given in terms of the velocity, for instance the so-called no-slip condition on which this paper will focus.

From the mathematical point of view, the need to prescribe vorticity values at the boundary stems form the fact that, in absence of such boundary condition the convection diffusion equation on the vorticity is underdetermined, while, in the system allowing to compute the divergence free velocity in terms of the vorticity, enforcing values of both normal and tangential components of the velocity would lead to an overdetermined problem. The issue is therefore to determine the right boundary condition, either in the Dirichlet or Neumann form, to prescribe on the vorticity, such that, the vorticity solution to the convection-diffusion equation is at all time admissible in the sense that the related velocity with zero normal component has also zero tangential component. The Neumann form of this boundary condition is sometimes prefered because it is directly interpretable in terms of local vorticity production at the boundary.

There are several ways to construct exact vorticity boundary conditions. One, proposed by Anderson (1989), is to compute the time derivative of the tangential derivative of the velocity in terms of the time derivative fo the vorticity in the domain. Using the advection diffusion equation satisfied by the vorticity and integrating by parts lead to an integral-differential equation for the normal derivative of the vorticity at the boundary. Another way, proposed in Cottet (1988), is to differentiate the system giving the velocity in terms of the vorticity by writing $-\Delta u = \text{curl}\,\omega$. This system is now naturally completed by the values of both components of the velocity at the boundary. The role of the vorticity boundary condition is then to guarantee that the velocity is divergence free, or equivalently that $\omega = \text{curl}\,u$, by enforcing this last identity at the boundary.

All these formulations involve a delicate coupling of the velocity and vorticity through boundary conditions and equations inside the computational domain. It is worth noticing that, although they are equivalent to the Navier-Stokes equations in the original velocity-pressure formulation and that numerical methods based on these formulations have been succesfully implemented, it is not clearly understood under which conditions natural discretizations of the equations will lead to stable numerical algorithms. The reason is that, to our knowledge, no energy estimates are available for these models. Energy estimates, which will play a central role in this paper, are crucial both to check the well-posedness of the continous models, and to ensure that numerical or round-off errors in the computation of the velocity will not dramatically affect the computation of the vorticity at the boundary and ultimately, through the convection diffusion equations in the domain, the whole vorticity field.

As a matter of fact, we only know one complete numerical algorithm based on the vorticity formulation of the Navier-Stokes equation, for which convergence has been proved (Hou and Wetton, 1992). This is a finite-difference algorithm using a second order approximate Dirichlet vorticity boundary condition. The stability of the algorithm in the energy norm for the velocity comes from a fortunate sign in the definition of this algorithm.

Our primary goal here was to understand the convergence properties of Chorin's vorticity creation. Rather than trying to find explicitly a correct vorticity boundary condition, the idea in this algorithm is based on the physical intuition of the mechanisms which are underlying vorticity creation. It involves a splitting of the system into 3 substeps. In the 2 first, the Navier-Stokes equations are solved without any vorticity production at the boundary, which means homogeneous Neumann boundary conditions. In the third step, a vortex sheet is introduced at the boundary in order to cancel the slip resulting from the 2 previous steps. This vortex sheet can be interpreted as the single layer potential necessary for the computation of the stream function associated to the vorticity. The difficulty in the convergence analysis of this scheme is that it does not clearly result from the implementation of a given vorticity boundary condition -either on Dirichlet or Neumann form- and, as a result, it is difficult to define it in an intrinsic way, independent of the geometry of the domain. As a matter of fact, the only convergence proof so far for this algorithm (Benfatto and Pulvirenti, 1986) is restricted to the half plane case and is heavily based on extensions of vorticity over the entire plane and the use of integral representations in the plane. Note that, to our knowledge, even for the Stokes problem there is no convergence proof valid for more general geometries. For further discussions on splitting and vorticity creation formulas we refer to Beale and Chorin, Hugues, Mac-Cracken and Marsden (1978).

In Cottet (1988), we suggested an alternative formulation of Chorin's algorithm based on some duality relations between normal derivatives of the vorticity and time derivatives of the tangential component of the velocity. In this paper, we explore further this direction and propose a splitting based vorticity creation algorithm which can be formulated in any geometry. In the half-plane case, this algorithm leads to Chorin's algorithm through one additional natural time-discretization.

We first define, in section 2, our algorithm for the Stokes case. We prove that this algorithm is enstrophy decreasing, and use ensergy estimates to prove its convergence. We then turn in section 3 to the Navier-Stokes equation. In this case we were not able to prove energy estimates similar to the ones valid for the Stokes case. One can actually not expect the algorithm to be enstrophy decreasing, since this property is not true for the continuous Navier-Stokes equations. Energy estimates on the vorticity are rather subtle even in the continuous case, and it seems difficult to reproduce them in the discrete case. However we show that, if we assume the same stability properties for the algorithm as in the linear case, convergence towards the exact solution follows essentially from the same lines.

2. THE STOKES PROBLEM

As already stressed above our analysis will be possible due to the reformulation of this algorithm in an intrinsic way, that is a way which does not make use of any mapping of the domain or extension of the vorticity outside the domain.
In their vorticity formulation the Stokes equations read

$$\frac{\partial \omega}{\partial t} - \Delta \omega = 0 \text{ in } \Omega \tag{1}$$

$$\omega(\cdot, 0) = \omega_0 \text{ for } x \in \Omega \text{ and } t \in [0, T] \tag{2}$$

$$\frac{\partial \omega}{\partial \nu} = g \text{ on } \partial \Omega \tag{3}$$

where the (unknown) boundary condition g is such that for every time t the solution to the elliptic system

$$-\Delta \psi = \omega \text{ in } \Omega \tag{4}$$

$$\psi = 0 \text{ on } \partial \Omega \tag{5}$$

also satisfies

$$u \cdot \tau = -\frac{\partial \psi}{\partial \nu} = 0 \text{ on } \partial \Omega \tag{6}$$

In the above equations ν denotes the normal to $\partial \Omega$ oriented outward, and τ is the tangent to $\partial \Omega$ (such that (ν, τ) defines the positive orientation of the plane). The velocity field u will be said to be associated to ω, in the sense that $u = \text{curl } \psi$, where ψ statisfies (4)-(5). For simplicity we will assume that the domain Ω is bounded and with smooth boundary.

Let h be a positive number and define $t_n = nh$ for $0 \leq n \leq N$, with $Nh = T$. In our formulation, the algorithm consists in recursively defining two vorticity fields ω_1^h, ω_2^h

and a sequence ω_n in the following way: first we initialize the sequence ω_n by the exact initial vorticity filed, then, to obtain ω_{n+1} from ω_n we first solve, for $t \in [t_n, t_{n+1}]$

$$\frac{\partial \omega_1^h}{\partial t} - \Delta \omega_1^h = 0 \text{ in } \Omega \tag{7}$$

$$\omega_1^h(\cdot, t_n) = \omega_n \text{ in } \Omega \tag{8}$$

$$\frac{\partial \omega_1^h}{\partial \nu} = 0 \text{ on } \partial\Omega \tag{9}$$

We then define u_1^h as the velocity field associated to ω_1^h and we solve, still in the time interval $[t_n, t_{n+1}]$

$$\frac{\partial \omega_2^h}{\partial t} - \Delta \omega_2^h = 0 \text{ in } \Omega \tag{10}$$

$$\omega_2^h(\cdot, 0) = \omega_n \text{ in } \Omega \tag{11}$$

$$\frac{\partial \omega_2^h}{\partial \nu} = g^h \text{ on } \partial\Omega \tag{12}$$

with

$$g^h = -\frac{\partial}{\partial t}(u_1^h \cdot \tau) \tag{13}$$

Finally we set

$$\omega_{n+1} = \omega_2^h(\cdot, t_{n+1}) \tag{14}$$

Observe that this defines a continous (with respect to time) vorticity field ω_2^h, and that the (discontinuous) vorticity ω_1^h is only used to "feed" the problem to be solved for ω_2^h with the correct boundary condition. Also notice that the value of g^h at time t_n, that is at times when u_1^h is not continuous, is not necessary to well define the algorithm (the normal derivative of the vorticity needs only be defined for almost all times).

A first important remark about this algorithm is that is produces the right net amount of vorticity: for the exact solution the mean vorticity, which is equal to the circulation of the velocity along the boundary, is zero in the no-slip case. In the algorithm, the net amount of vorticity produced by (10)-(12) is equal to the time integral of the integral along the boundary of the normal derivative of the vorticity:

$$\int_\Omega (\omega_{n+1} - \omega_n)\, dx = \int_{t_n}^{t_{n+1}} \int_{\partial\Omega} g^h\, dx\, dt.$$

But, due to (9), we have, for $t \in [t_n, t_{n+1}]$,

$$\int_\Omega \omega_1^h(x, t)\, dx = \int_\Omega \omega_n(x)\, dx$$

Thus, if the vorticity has mean value 0 at time t_n, the circulation of u_1^h remains zero for all time, and there is no net production of vorticity at the next time step.

2.1. Consistency of the algorithm with Chorin's vorticity creation algorithm in the half-space case

Let us assume in this section that

$$\Omega = \{x = (x_1, x_2) \in \mathbf{R}^2 \; ; \; x_2 > 0\}$$

Let us focus on one step $t_n \to t_{n+1}$ of the above algorithm. We first make the following classical observation: by extending ω_2^h throughout $\partial\Omega$ in an even fashion, problem (10)-(12) is equivalent to the heat equation in the whole space with a forcing term localized on the axis $x_2 = 0$. More precisely, if one defines, for $x = (x_1, x_2) \in \mathbf{R}^2$, $\bar{\omega}_2^h(x_1, x_2)$ to be $\omega_2^h(x_1, x_2)$ if $x_2 > 0$ and $\omega_2^h(x_1, -x_2)$ if not, then (10)-(12) is equivalent to

$$\frac{\partial \bar{\omega}_2^h}{\partial t} - \Delta \bar{\omega}_2^h = 2g^h(x_1) \otimes \delta(x_2) \text{ in } \mathbf{R}^2 \tag{15}$$

A consistant time-splitting of the above equation leads to considering two substeps: one consisting of the heat equation with a zero forcing term, and the next one as a simple incrementation of the vorticity with the forcing term. The first substep is now equivalent to the Stokes equation in Ω with homogeneous Neumann boundary condition, that is (7)-(9), the result of which we denote by $\tilde{\omega}_{n+1}$. The final result of this algorithm is now

$$\omega'_{n+1} = \tilde{\omega}_{n+1} - 2 \int_{t_n}^{t_{n+1}} \frac{\partial}{\partial t}(u_1^h \cdot \tau)(\cdot, t) \otimes \delta \, dt \tag{16}$$

Recall that here u_1^h is the velocity field corresponding to ω_1^h, the vorticity field solution of the heat equation with homogeneous Neumann boundary condition and inital condition ω'_n. To prove that this algorithm is actually equivalent to Chorin's algorithm, we need to show the following result.

Proposition 2.1. Let ω_0 a smooth vorticity with zero mean value in the half plane, u_0 its velocity, with zero normal component at the boundary, and $\bar{\omega}_0$ its even extension to the whole plane. Let $\omega'_0 = \bar{\omega}_0 - 2u_0 \otimes \delta_{\partial\Omega}$, $\bar{\omega}(\cdot, t)$ the solution of the heat equation in the plane with initial condition ω'_0 and $u(\cdot, t)$ the associated velocity. Then

$$u(\cdot, t) \cdot \tau \to 0 \text{ in } H^{-1/2}(\partial\Omega), \text{ as } t \to 0^+ \tag{17}$$

We postpone the proof of this result to the end of the paper. As a consequence, a straightforward induction argument shows that (16) can be rewritten as

$$\omega'_{n+1} = \tilde{\omega}_{n+1} - 2(u_1^h \cdot \tau)(\cdot, t_{n+1}) \tag{18}$$

This is precisely Chorin's product formula: one step of homogeneous heat equation followed by the creation of a vortex sheet at the boundary to cancel the slip at the boundary resulting from this step.

In other words we have demonstrated that, compared to our algorithm, Chorin's algorithm involves one additional, natural, time discretization.

2.2. Stability in the enstrophy norm of the algorithm

From now on we will assume that Ω is a bounded, smooth, simply connected domain. Our only assumption regarding the smoothness of the continuous solution of the Stokes equation is that the initial vorticity ω_0 is in $L^2(\Omega)$ and that the associated initial velocity vanishes at the boundary. We prove in this section that the algorithm is enstrophy decreasing. We recall that this poperty is indeed true for the continuous equations. To check it, one has to multiply the equations in the primitive variables (u, p) by the projection of Δu onto the space of divergence free vector fields with normal components vanishing at the boundary. The property follows then from routine integration by parts.

We will denote by $|\cdot|$ the L^2 norm in Ω. For simplicity in the notations we will drop everywhere the superscripts h, so that the vorticity fields produced by the algorithm are denoted by ω_1 and ω_2.

We focus here on a given time step $[t_n, t_{n+1}]$. To obtain energy estimates in this time interval, we first multiply (10) by ω_2 and integrate over Ω. We get

$$\frac{1}{2}\frac{d}{dt}|\omega_2|^2 + |\nabla\omega_2|^2 = \int_{\partial\Omega} \omega_2 \frac{\partial\omega_2}{\partial\nu}\,dx \tag{19}$$

Notice that a straightforward use of (12), combined with elliptic regularity and trace results to control the boundary term, would introduce multiplicative constants in the energy which would prevent any uniform control of the enstrophy. We thus need to estimate carefully the right hand side of (19). To this end we first observe that we can rewrite (13) as follows

$$\frac{\partial\omega_2}{\partial\nu} = \frac{\partial}{\partial t}\left(\frac{\partial\psi_1}{\partial\nu}\right) = \frac{\partial}{\partial\nu}\left(\frac{\partial\psi_1}{\partial t}\right) \text{ on } \partial\Omega$$

Moreover, by the definition of ψ_1 from ω_1 and (7):

$$-\Delta\left(\frac{\partial\psi_1}{\partial t}\right) = \frac{\partial\omega_1}{\partial t} = \Delta\omega_1 \text{ in } \Omega \tag{20}$$

$$\frac{\partial\psi_1}{\partial t} = 0 \text{ on } \partial\Omega \tag{21}$$

If we set

$$\phi = \frac{\partial\psi_1}{\partial t}$$

we thus have, upon integrating by parts

$$\int_{\partial\Omega} \omega_2 \frac{\partial\omega_2}{\partial\nu}\,dx = \int_{\partial\Omega} \omega_2 \frac{\partial\phi}{\partial\nu}\,dx = \int_\Omega \Delta\phi\,\omega_2\,dx + \int_\Omega \nabla\phi\cdot\nabla\omega_2\,dx$$

$$= -\int_\Omega \Delta\omega_1\,\omega_2\,dx + \int_\Omega \nabla\phi\cdot\nabla\omega_2\,dx$$

$$= \int_\Omega \nabla\omega_1\cdot\nabla\omega_2\,dx + \int_\Omega \nabla\phi\cdot\nabla\omega_2\,dx \tag{22}$$

Note that in the integration by parts leading to the last equality we have used the fact that $\partial\omega_1/\partial\nu = 0$ on $\partial\Omega$. Combining (19) and (22) yields

$$\frac{1}{2}\frac{d}{dt}|\omega_2|^2 = \int_\Omega \nabla\omega_2\cdot(\nabla\omega_1 - \nabla\omega_2)\,dx + \int_\Omega \nabla\phi\cdot\nabla\omega_2\,dx$$

$$= \int_\Omega \nabla\omega_2\cdot\nabla e\,dx + \int_\Omega \nabla\phi\cdot\nabla\omega_1\,dx - \int_\Omega \nabla\phi\cdot\nabla e\,dx \tag{23}$$

where we have set $e = \omega_1 - \omega_2$. Next we write

$$\int_\Omega \nabla\phi\cdot\nabla\omega_1\,dx = -\int_\Omega \phi\Delta\omega_1\,dx = \int_\Omega \phi\Delta\phi\,dx = -|\nabla\phi|^2 \tag{24}$$

To establish this we used first that $\partial\omega_1/\partial\nu = 0$ on $\partial\Omega$, then that $\Delta\phi = -\Delta\omega_1$ and finally that $\phi = 0$ on $\partial\Omega$. From (23) and (24) we thus have

$$\frac{1}{2}\frac{d}{dt}|\omega_2|^2 = \int_\Omega \nabla\omega_2\cdot\nabla e\,dx - |\nabla\phi|^2 - \int_\Omega \nabla\phi\cdot\nabla e\,dx \tag{25}$$

To recover informations on e, let us now substract (10) from (7), multiply it by e and integrate over Ω. We obtain:

$$\frac{1}{2}\frac{d}{dt}|e|^2 + |\nabla e|^2 = -\int_{\partial\Omega} e\frac{\partial\phi}{\partial\nu}\,dx$$

and arguing as for (22):

$$\frac{1}{2}\frac{d}{dt}|e|^2 + |\nabla e|^2 = -\int_\Omega \nabla\omega_1 \cdot \nabla e\, dx - \int_\Omega \nabla\phi \cdot \nabla e\, dx \qquad (26)$$

Adding (25) to (26) leads to

$$\frac{1}{2}\frac{d}{dt}|\omega_2|^2 + \frac{1}{2}\frac{d}{dt}|e|^2 + |\nabla e|^2 = -|\nabla e|^2 - |\nabla\phi|^2 - 2\int_\Omega \nabla e\nabla\phi\, dx = -|\nabla e + \nabla\phi|^2 \le 0. \quad (27)$$

Integrating this inequality over the time step $[t_n, t_{n+1}]$ and using the crucial fact that $e(\cdot, t_n + 0) = 0$, since ω_1 and ω_2 start from the same initial condition, we obtain:

$$|\omega_{n+1}|^2 - |\omega_n|^2 + |\tilde\omega_{n+1} - \omega_{n+1}|^2 + \int_{t_n}^{t_{n+1}} |\nabla e(\cdot, s)|^2\, ds \le 0 \qquad (28)$$

We thus have proved, in particular, the following stability result

Theorem 2.1. *The algorithm defined by (7)-(13) is enstrophy decreasing:*

$$|\omega_n|^2 \le |\omega_0|^2, \forall n > 0$$

2.3. Convergence towards the exact solution of the Stokes equation

Let us first sketch the main lines of the convergence arguments which will follow. First it is clear that theorem 2.2 shows that, for a subsequence, the approximate vorticity will converge weakly, in a space to be specified, to some vorticity ω. That ω is a distribution solution of the Stokes equation clearly results from the linearity of this equation. The key point is thus to prove that the velocity u associated to this vorticity ω vanishes at the boundary.

To establish this kind of property it seems at first glance that this would require the control of ω in a space where one can talk of traces, for instance $H^1(\Omega)$. Unfortunately the energy estimate (28), does not give any information of that kind. However we do have informations on the gradient of $e = \omega_1 - \omega_2$. On the other hand, (28) tells us that the series $\sum_n |e(t_n - 0)|^2$ is convergent, something which, as we will prove next, implies that the weak limit of e must be zero. We will then prove that this in turn tells us that the boundary conditions on ω_1 and ω_2 cannot differ significantly, and thus that the slip of u_1 tends to zero when $h \to 0$.

The following result summarizes the estimates available for e^h (to avoid confusion we now restore the superscript h).

Proposition 2.2. *The distributions e^h are uniformly bounded (with respect to h) in the space $L^\infty(0, T; L^2(\Omega)) \cap L^2(0, T; H^1(\Omega))$. Furthermore we can write*

$$\frac{\partial e^h}{\partial t} = v^h + \sum_{0 < nh < T} \delta(t - t_n)\alpha_n^h \qquad (29)$$

where the functions v^h and the sequences (α_n^h) are respectively uniformly bounded in $L^\infty(0, T; H^{-2})$ and in $l_1 \cap l_\infty(\mathbf{N}; H^{-2})$, and there exists a constant C such that, for all positive n

$$\|\alpha_n^h\|_{H^{-2}(\Omega)} \le Ch \qquad (30)$$

Proof: First notice that the Dirac masses in the right hand side of (29) translate the jumps that ω_1^h (and thus e^h) undergo at every time-step.

Our first claim concerning the uniform control of e^h in $L^\infty(0,T;L^2(\Omega)) \cap L^2(0,T;H^1(\Omega))$ readily follows from (28). As for (29), it results from the fact that, if $t \in]t_n, t_{n+1}[$

$$\frac{\partial e^h}{\partial t} = \Delta e^h$$

Thus (29) is satisfied with $v^h = \Delta e^h$ and $\alpha_n^h = [e^h]_{|t=t_n}$.
e^h is bounded in $L^\infty(0,T;L^2(\Omega))$ and thus v^h is bounded in $L^\infty(0,T;H^{-2}(\Omega))$. Since $e^h(t) \to 0$ when $t \to t_n, t > t_n$ we can write

$$\alpha_n^h = -\int_{t_n}^{t_{n+1}} \frac{\partial e^h}{\partial t}(\cdot,s)\,ds = -\int_{t_n}^{t_{n+1}} \Delta e^h(\cdot,s)\,ds \leq h\|e^h\|_{L^\infty(0,T;H^{-2}(\Omega))}$$

from which (30) follows. $\qquad\square$

We are now in a situation where one can think of applying classical results to get a uniform control with respect to time of the family e^h. However in these results the time derivatives are in general supposed to be controled in L^p spaces rather than in measures, so we need to adapt their proof to our specific situation

Proposition 2.3. *There exists a distribution λ in $L^\infty(0,T;L^2(\Omega)) \cap L^2(0,T;H^1(\Omega))$ such that, for a subsequence,*

$$e^h \quad \to \quad \lambda \text{ strongly in } L^\infty(0,T;L^2(\Omega)) \tag{31}$$
$$e^h \quad \rightharpoonup \quad \lambda \text{ weakly in } L^2(0,T;H^1(\Omega)) \tag{32}$$

Proof: We already know that there exists λ such that, upon extracting a subsequence, e^h tends to λ weakly in $L^2(0,T;H^1(\Omega))$. To simplify we will assume that $\lambda = 0$.

Let us focus on the case $t = 0$ (the other times would be dealt with through obvious modifications in the proof below).

We set $f^h(\cdot,t) = e^h(\cdot,\beta t)$, where β is a positive parameter to be specified later. If ϕ is a smooth function of time satisfying $\phi(0) = -1, \phi(T) = 0$, we can write

$$e^h(\cdot,0) = f^h(\cdot,0) = \int_0^T (\phi f^h)'(\cdot,s)\,ds$$

where of course the time derivative in the right hand side above is to be taken in a distribution sense. Using the definition of f^h and (29) we find:

$$\int_0^T (\phi f^h)'(\cdot,s)\,ds = \int_0^T (\phi'(\cdot,s)e^h(\cdot,\beta s) + \beta\phi(\cdot,s)v^h(\cdot,\beta s)\,ds + \sum_{nh \leq \beta T} \alpha_n^h \phi(\cdot,t_n/\beta)$$
$$= I_1^h + I_2^h + I_3^h$$

Next we write

$$\|I_2^h\|_{H^{-2}} \leq C\beta^{1/2}\|v^h\|_{L^2(0,T;H^{-2})}$$

and, due to (30)

$$\|I_3^h\|_{H^{-2}} \leq \|\phi\|_{L^\infty}\beta T$$

So, given $\varepsilon > 0$, we can find β small enough so that I_2^h and I_3^h are smaller than ε in the H^{-2} norm for all h. We fix this value of β. Then we observe that I_1^h is uniformly bounded in H^1 and tends to 0 weakly in H^{-2}; by the compactness of H^1 in H^{-2} this implies that, for a subsequence, I_1^h tends to 0 strongly in H^{-2}. We have thus proved that, for h small enough in a subsequence, we have

$$\|e^h(\cdot, 0)\|_{H^{-2}} \leq \varepsilon$$

which ends our proof. □

So far we have used the informations available for e^h and its derivative. We are now going to use the remaining informations contained in the energy estimate (29) to prove that the limit λ of e^h is necessarily zero:

Proposition 2.4.

$$e^h \to 0 \ \ weakly \ in \ L^2(0, T; H^1(\Omega))$$

Proof: By (31) we have

$$h \sum_n \|e^h(\cdot, t_n)\|_{H^{-2}} - h \sum_n \|\lambda(\cdot, t_n)\|_{H^{-2}} \to 0 \tag{33}$$

On the other hand we know by (28) that the sum

$$\sum_{t_n \in [0,T]} \|e^h(\cdot, t_n)\|_{L^2}$$

is bounded independantly of h. Combined with (33) this implies that

$$h \sum_{t_n \in [0,T]} \|\lambda(\cdot, t_n)\|_{H^{-2}}^2 \to 0 \text{ as } h \to 0.$$

Finally we notice that, by (29), $\frac{\partial \lambda}{\partial t}$ is necessarily in the space of time measures with values in $H^{-2}(\Omega)$ and thus is continuous with respect to time with values in H^{-2}. Therefore

$$h \sum_{t_n \in [0,T]} \|\lambda(\cdot, t_n)\|_{H^{-2}}^2 \to \|\lambda\|_{L^2(0,T;H^{-2})}^2$$

and $\lambda = 0$. □

We can now prove our convergence result

Theorem 2.2. *The sequence ω_2^h converges weakly in $L^2(0, T; L^2)$ to the exact solution of the Stokes equation as h tends to 0.*

Proof: Because of the energy estimate (28), we know that there exists a vorticity field ω such that, for a subsequence, w_2^h converges weakly to ω in $L^2(0, T; L^2)$. Since ω_2^h satisfies (10) for almost every time, it is clear that (1) is satisfied in the sense of distribution. It only remains to prove that the velocity field u related to ω (still in the sense that curl $u = \omega$ in Ω and $u \cdot \nu = 0$ on $\partial\Omega$) vanishes at the boundary. By (7) and (10) and the fact that we assumed Ω simply connected, we can find 2 pressure fields p_1^h and p_2^h such that

$$\frac{\partial u_1^h}{\partial t} - \Delta u_1^h = \nabla p_1^h \ ; \ \frac{\partial u_2^h}{\partial t} - \Delta u_2^h = \nabla p_2^h$$

Using obvious vector identities together with (9),(12) we get

$$\frac{\partial}{\partial \tau}(p_1^h - p_2^h) = -\frac{\partial}{\partial t}(u_2^h \cdot \tau) \ ; \ \frac{\partial}{\partial \nu}(p_1^h - p_2^h) = -\frac{\partial}{\partial \tau}(\omega_1^h - \omega_2^h) \tag{34}$$

In passing, let us emphasize that these identities, which somehow summarize all the informations contained in the algorithm, are essential in understanding how normal derivatives of the vorticity are naturally linked to time derivatives of tangential velocities.

Let now ϕ denote a smooth test function on $\partial\Omega$ and consider the (smooth) function ψ defined on Ω as the solution to the following problem

$$\Delta\psi = 0 \text{ in } \Omega \ ; \ \frac{\partial\psi}{\partial\nu} = \frac{\partial\phi}{\partial\tau} \text{ on } \partial\Omega$$

Green's formula along with (34) enables us to write

$$
\begin{aligned}
\int_{\partial\Omega} \frac{\partial}{\partial t}(u_2^h \cdot \tau)\phi \, dx &= \int_{\partial\Omega}(p_1^h - p_2^h)\frac{\partial\phi}{\partial\tau} \, dx = -\int_{\partial\Omega}\frac{\partial}{\partial\nu}(p_1^h - p_2^h)\psi \, dx \\
&= -\int_{\partial\Omega}(\omega_1^h - \omega_2^h)\frac{\partial\psi}{\partial\tau} \, dx
\end{aligned}
\tag{35}
$$

But we have proved in propositions 2.3 and 2.4 that $\omega_1^h - \omega_2^h$ tends to 0 weakly in $L^2(0,T;H^1(\Omega))$. Therefore its trace on $\partial\Omega$ tends to 0 weakly in $L^2(0,T;H^{1/2}(\partial\Omega))$. Due to (35) we deduce that $\frac{\partial}{\partial t}u_2^h \cdot \tau$ tends to 0 in $L^2(0,T;\mathcal{D}'(\partial\Omega))$. Since $u_2^h \cdot \tau(\cdot,0) = 0$ (u_2^h is at time 0 the velocity associated to the exact initial vorticity field), this implies that

$$u_2^h \cdot \tau \to 0 \text{ in } L^2(0,T;\mathcal{D}'(\partial\Omega))$$

The limit u of u_2^h has thus to vanish at the boundary. Finally the uniqueness of all possible limits of subsequences proves that the whole sequence ω_2^h tends to the exact solution of the Stokes equation. □

3. THE NAVIER-STOKES SYSTEM

To define our vorticity creation algorithm we set, as for the linear case, $t_n = nh$. We define two vorticity fields ω_1^h and ω_2^h and a sequence ω_n by solving in $]t_n, t_{n+1}[$, first

$$\frac{\partial\omega_1^h}{\partial t} + \text{div}(u_1^h\omega_1^h) - \Delta\omega_1^h = 0 \text{ in } \Omega \tag{36}$$

$$\omega_1^h(\cdot, t_n) = \omega_n \text{ in } \Omega \tag{37}$$

$$u_1^h = \text{curl}\left(-\Delta_d^{-1}(\omega_1^h)\right) \text{ in } \Omega \tag{38}$$

$$\frac{\partial\omega_1^h}{\partial\nu} = 0 \text{ on } \partial\Omega \tag{39}$$

In (38), Δ_d^{-1} denotes the solution operator for the Laplace equation in Ω with homogeneous Dirichlet boundary condition on $\partial\Omega$. We next define ω_2^h as the solution of

$$\frac{\partial\omega_2^h}{\partial t} + \text{div}(u_1^h\omega_2^h) - \Delta\omega_2^h = 0 \text{ in } \Omega \tag{40}$$

$$\omega_2^h(\cdot, t_n) = \omega_n \text{ in } \Omega \tag{41}$$

$$\frac{\partial\omega_2^h}{\partial\nu} = g^h \text{ on } \partial\Omega \tag{42}$$

with

$$g^h = \left[\frac{\partial u_1^h}{\partial t} + (u_1^h \cdot \nabla)u_1^h\right] \cdot \tau \tag{43}$$

Observe that we have chosen to use the same velocity field u_1^h in the convection-diffusion systems giving ω_1^h and ω_2^h, but this is probably not crucial.

The verification that the algorithm creates the right net amount of vorticity follows from the same argument as for the Stokes case (because of the conservative form of the vorticity equations, the non linear terms in (36),(38) do not contribute to this).

3.1. Recovering Chorin's algorithm

Assume that Ω is the half plane. If we argue as in section 2.1 and split into a free convection-diffusion equation (which itself can be split into a convection step with no through flow condition, then a diffusion step with homogeneous Neumann boundary condition) followed by a forcing step at the boundary, we are led to the following natural scheme

$$\omega'_{n+1} = \tilde{\omega}_{n+1} - 2\left[\int_{t_n}^{t_{n+1}} (\frac{\partial u_1^h}{\partial t} + (u_1^h \cdot \nabla)u_1^h)\, ds\right] \cdot \tau \otimes \delta \qquad (44)$$

where we have denoted by $\tilde{\omega}_{n+1}$ the value at time t_{n+1} of ω_1^h. Let us now set $U_1^h = \frac{\partial u_1^h}{\partial t} + (u_1^h \cdot \nabla)u_1^h$ and denote by $X_1^h(t; x, s)$ the value at time t of the characteristics associated to the velocity u_1^h and passing at time s through x. We make the following observation: if $t \in [t_n, t_{n+1}]$ then $X_1^h(t; x, t_{n+1}) - x \to 0$ as $h \to 0$ as soon as u_1^h is smooth enough. Thus, under suitable smoothness assumptions:

$$U_1^h(X_1^h(t; x, t_{n+1}), t) - U_1^h(x, t) = O(\Delta t)$$

if $t \in [t_n, t_{n+1}]$. On the other hand we can write

$$U_1^h(X_1^h(t; x, t_{n+1}), t) = \frac{\partial}{\partial t}u_1^h(X_1^h(t; x, t_{n+1}), t)$$

If $x \in \partial\Omega$, $X_1^h(t; x, t_{n+1})$ remains on $\partial\Omega$ for all time since $u_1^h \cdot \nu = 0$ on $\partial\Omega$. It thus follows from what precedes that (43) is consistent with the following scheme:

$$\omega'_{n+1} = \tilde{\omega}_{n+1} - 2\left[\int_{t_n}^{t_{n+1}} (\frac{\partial}{\partial t}(u_1^h(X_1^h(t; x, t_{n+1}), t)\, ds\right] \cdot \tau \otimes \delta \qquad (45)$$

Finally it can be proven by arguments similar to those used for the Stokes equation that the effect of a free convection diffusion step is to immediately cancel the slip at the boundary, so (45) can be rewritten as

$$\omega'_{n+1} = \tilde{\omega}_{n+1} - 2u_1^h(\cdot, t_{n+1}) \cdot \tau \otimes \delta$$

The conclusion is that, as in the Stokes case, although in a more involved way, Chorin's vorticity creation algorithm can be interpreted as an additional time discretization over our algorithm.

3.2. Consistency of the algorithm

As we already said, stability estimates seem hard to establish in the non linear case. However, in order to validate the algorithm, it is worthwhile to check that, if we assume that a priori estimates similar to (28) are valid, convergence of the algorithm follows.

We will thus assume that there exists a constant M such that for all $n > 0$

$$|\omega_n|^2 + \sum_{k=0}^{n} |\omega_k - \tilde{\omega}_k|^2 + \int_0^{t_n} |\nabla e(\cdot, s)|^2\, ds \le M \qquad (46)$$

As for the linear case, one important consequence of the energy estimate (46) is the fact that e^h tends to 0. This follows from Proposition 2.3 which extends without modification to the non linear case. To check it we write, for $t \in]t_n, t_{n+1}[$,

$$\frac{\partial e^h}{\partial t} = \Delta e^h - \operatorname{div}(u_1^h e^h)$$

It is readily seen from (36)-(38) that, for $t \in [t_n, t_{n+1}]$, $\|\omega_1^h(\cdot, t)\|_{L^2} \leq \|\omega_n\|_{L^2} \leq \mathbf{M}$. By ellpitic regularity u_1^h is therefore bounded in $L^\infty(0, T; L^p(\Omega))$ for all finite p. So $u_1^h e^h$ is bounded in, say, $L^\infty(0, T; L^{3/2}(\Omega)) \subset L^\infty(0, T; H^{-1}(\Omega))$, and $\operatorname{div}(u_1^h e^h)$ is bounded in $L^\infty(0, T; H^{-2}(\Omega))$. The decomposition (29) and the estimate (30) follows then from the same arguments as in the linear case. Propositions 2.4 and 2.5 in turn remain true, as they are only based on these 2 results and on the fact that the series $\sum_n \|e^h(\cdot, t_n - 0)\|_{L^2}$ are uniformly bounded .

One can then prove

Theorem 3.3. *Assume (46). The sequence ω_2^h converges weakly in $L^2(0, T; L^2(\Omega))$ towards the solution ω of the Navier-Stokes equation. Morevoer the velocities u_2^h converge strongly to the exact velocity in $L^2(0, T; L^p(\Omega))$ for all finite p.*

Proof: Let us first summarize the estimates available for (u_2^h, ω_2^h). By (46)

$$u_2^h \text{ is bounded in } L^\infty(0, T; H^1(\Omega))$$

By (29) and the fact already used that u_1^h is bounded in $L^\infty(0, T; L^p(\Omega))$ for all finite p, we also have that

$$\frac{\partial u_2^h}{\partial t} \text{ is bounded in } L^\infty(0, T; H^{-2}(\Omega))$$

For all finite p, $H^1(\Omega)$ is compactly imbedded in $L^p(\Omega)$, which is imbedded in $H^{-2}(\Omega)$. By classical compactness arguments we deduce that there exists a velocity u, such that, for a subsequence,

$$u_2^h \to u \text{ strongly in } L^\infty(0, T; L^p(\Omega))$$

for all finite p. But weak compactness also shows that, still for a subsequence, ω_2^h tends to $\omega = \operatorname{curl} u$ weakly in $L^\infty(0, T; L^2(\Omega))$ and thus

$$u_2^h \omega_2^h \rightharpoonup u\omega \text{ weakly in } L^\infty(0, T; L^{3/2}(\Omega)).$$

This first shows that (u, ω) is solution of the Navier-Stokes equation in Ω. As for the linear case, the key point is now to check that u vanishes at the boundary. We proceed as for the Stokes problem, with however 2 slight modifications. The first one is motivated by the fact that we used, for the non linear term in the definition of ω_2^h, u_1^h and not u_2^h. This prevents us from writing $(u_1^h \cdot \nabla)\omega_2^h$ on a rotational form.

Let

$$v_1^h = \operatorname{curl}\left[(u_1^h \cdot \nabla)u_2^h\right] - (u_1^h \cdot \nabla)\omega_2^h$$

and define ϕ_1^h as the solution to the elliptic problem

$$\Delta\phi_1^h = v_1^h \text{ in } \Omega \;;\; \frac{\partial\phi_1^h}{\partial\nu} = 0 \text{ on } \partial\Omega$$

We claim that

$$\phi_1^h \to 0 \text{ in } L^2(0, T; W^{1,4/3}(\Omega)) \tag{47}$$

The reason is that, since curl $\left[(u_2^h \cdot \nabla)u_2^h\right] = (u_2^h \cdot \nabla)\omega_2^h$, we can rewrite

$$v_1^h = \text{curl}\left[((u_1^h - u_2^h) \cdot \nabla)u_2^h\right] - ((u_1^h - u_2^h) \cdot \nabla)\omega_2^h$$

and thus

$$\|v_1^h\|_{W^{-1,4}} \le C\|u_1^h - u_2^h\|_{L^4}\|\omega_2^h\|_{L^2}$$

where $W^{-1,4}$ denotes the topological dual of $W^{1,4/3}(\Omega)$. In what precedes, we have proved that $e^h \rightharpoonup 0$ weakly in $L^2(0,T;H^1(\Omega))$. By compactness of $H^1(\Omega)$ into $L^4(\Omega)$ this implies that

$$u_1^h - u_2^h \to 0 \text{ in } L^2(0,T;L^4(\Omega))$$

Therefore $v_1^h \to 0$ strongly in $L^2(0,T;W^{-1,4}(\Omega))$ which proves (47).

We now come back to our velocities u_1^h and u_2^h. We have

$$\text{curl}\left[\frac{\partial u_1^h}{\partial t} + (u_1^h \cdot \nabla)u_1^h - \Delta u_1^h\right] = 0$$

$$\text{curl}\left[\frac{\partial u_2^h}{\partial t} + (u_1^h \cdot \nabla)u_2^h - \Delta u_2^h - \text{curl}\,\phi_1^h\right] = 0$$

This yields 2 pressure fields p_1^h and p_2^h such that

$$\frac{\partial u_1^h}{\partial t} + (u_1^h \cdot \nabla)u_1^h - \Delta u_1^h = \nabla p_1^h$$

$$\frac{\partial u_2^h}{\partial t} + (u_1^h \cdot \nabla)u_2^h - \Delta u_2^h - \text{curl}\,\phi_1^h = \nabla p_2^h$$

We have

$$\Delta(p_1^h - p_2^h) = \text{div}\left[(u_1^h \cdot \nabla)(u_1^h - u_2^h)\right]$$

Using (43) we get

$$\frac{\partial}{\partial \tau}(p_1^h - p_2^h) = -\frac{\partial}{\partial t}(u_2^h \cdot \tau) + (u_1^h \cdot \nabla)u_2^h \cdot \tau$$

$$\frac{\partial}{\partial \nu}(p_1^h - p_2^h) = -\frac{\partial}{\partial \tau}(\omega_1^h - \omega_2^h) + \frac{\partial \phi_1^h}{\partial \tau} + ((u_1^h \cdot \nabla)(u_1^h - u_2^h)) \cdot \nu$$

from which we deduce that, for a given smooth test function ϕ and if ψ denotes the solution of $\Delta\psi = 0$ in Ω, $\partial\psi/\partial\nu = \partial\phi/\partial\tau$ on $\partial\Omega$

$$\int_{\partial\Omega}\left[\frac{\partial u_2^h}{\partial t} - (u_1^h \cdot \nabla)u_2^h\right] \cdot \tau\phi\,dx = \int_{\partial\Omega}(p_1^h - p_2^h)\frac{\partial\phi}{\partial\tau}\,dx = \int_{\partial\Omega}\frac{\partial\psi}{\partial\nu}(p_1^h - p_2^h)\,dx$$

Integrations by parts yield

$$\int_{\partial\Omega}\frac{\partial\psi}{\partial\nu}(p_1^h - p_2^h)\,dx =$$

$$= \int_{\Omega}\Delta\psi(p_1^h - p_2^h)\,dx + \int_{\Omega}\nabla\psi\nabla(p_1^h - p_2^h)\,dx$$

$$= -\int_{\Omega}\psi\Delta(p_1^h - p_2^h)\,dx + \int_{\partial\Omega}\psi\frac{\partial}{\partial\nu}(p_1^h - p_2^h)\,dx$$

$$= -\int_{\Omega}\psi\,\text{div}\left[(u_1^h \cdot \nabla)(u_1^h - u_2^h)\right]\,dx + \int_{\partial\Omega}\psi\frac{\partial}{\partial\nu}(p_1^h - p_2^h)\,dx$$

$$= \int_{\Omega}\nabla\psi \cdot \left[(u_1^h \cdot \nabla)(u_1^h - u_2^h)\right]\,dx + \int_{\partial\Omega}\psi\left[\frac{\partial}{\partial\nu}(p_1^h - p_2^h) - (u_1^h \cdot \nabla)(u_1^h - u_2^h) \cdot \nu\right]\,dx$$

$$= \int_{\Omega}\nabla\psi \cdot \left[(u_1^h \cdot \nabla)(u_1^h - u_2^h)\right]\,dx + \int_{\partial\Omega}(-\phi_1^h + \omega_1^h - \omega_2^h)\frac{\partial\psi}{\partial\tau}\,dx \qquad (48)$$

But, on the one hand, we know that, as in the linear case, $\omega_1^h - \omega_2^h$ tends to 0 weakly in $L^2(0, T; H^{1/2})$, and, on the other hand, we deduce from (47) that ϕ_1^h tends to 0 in $L^2(0, T; W^{7/12,4/3})$ ($W^{7/12,4/3}$ is the trace space associated to $W^{1,4/3}$). Whence we obtain that the boundary integral in the right hand side above tends to 0. We also have

$$\|(u_1^h \cdot \nabla)(u_1^h - u_2^h)\|_{L^1} \le C|u_1^h|_2 |e^h|_2$$

By (31), this implies that the first integral tends to zero. Passsing to the limit in (48) thus shows that the scalar function $\gamma = u \cdot \tau$ is solution to the following equation:

$$\frac{\partial \gamma}{\partial t} - \frac{1}{2}\frac{\partial(\gamma^2)}{\partial \tau} = 0$$

Since the initial velocity, corresponding to the exact vorticity field ω_0, vanishes at the boundary, this is enough to ensure that γ remains zero for all time, and our proof is completed. □

4. PROOF OF PROPOSITION 2.1

Let us consider the function $\psi'(\cdot, t)$ solution to

$$-\Delta\psi' = \bar\omega \text{ in } \Omega \ ; \ \frac{\partial\psi'}{\partial\nu} = 0 \text{ on } \partial\Omega \tag{49}$$

and let us prove

Lemma 4.1.

$$\psi'(\cdot, t) \to 0 \text{ in } H^{1/2}(\partial\Omega) \text{ as } t \to 0^+ \tag{50}$$

We claim that this is actually equivalent to the result we are looking for in Proposition 2.1: if we write $u = \text{curl}\,\psi$ with

$$-\Delta\psi = \bar\omega \text{ in } \Omega \ ; \ \psi = 0 \text{ on } \partial\Omega$$

and, if we assume that (50) is true, then $\psi'' = \psi - \psi'$ satisfies

$$-\Delta\psi'' = 0 \text{ in } \Omega \ ; \ \psi''(\cdot, t) \to 0 \text{ in } H^{1/2}(\partial\Omega)$$

This immediately yields

$$\frac{\partial\psi''}{\partial\nu} \to 0 \text{ in } H^{-1/2}(\partial\Omega)$$

But we precisely have $u \cdot \tau = -\frac{\partial\psi''}{\partial\nu}$ and our result follows. □

To prove (50), let us first show that, when $t \to 0^+$

$$\bar\omega(\cdot, t) \to \bar\omega_0 - 2u_0 \otimes \delta_{\partial\Omega} \text{ in } H^{-1}(\mathbf{R}^2). \tag{51}$$

Observe that, since ω_0 has zero mean value in $\mathbf{R}^2$, u_0 is in $H^1(\mathbf{R}^2)$ and its trace on $\partial\Omega$ is therefore in particular in $L^2(\partial\Omega)$. It is then readily seen that $\bar\omega_0 - 2u_0 \otimes \delta_{\partial\Omega}$ is in $H^{-1}(\mathbf{R}^2)$ so that (52) is actually nothing but the continuity with respect to time of the solution of the heat equation in $H^{-1}(\mathbf{R}^2)$. The proof of this result can be easily derived using the Fourier transform. If λ denotes the Fourier transform of $\bar\omega_0 - 2u_0 \otimes \delta_{\partial\Omega}$, then $\hat\omega(\cdot, t)$, Fourier transform of $\bar\omega(\cdot, t)$ is given by

$$\hat\omega(\xi, t) = e^{-|\xi|^2 t}\lambda(\xi)$$

and, for $t \to 0^+$,

$$\|(1 + |\xi|^2)^{-1/2}[\hat{\omega}(\xi, t) - \lambda(\xi)]\|_{L^2_\xi} \to 0$$

by Lebesgue's theorem, which yields (51).

Let us finally prove (50). We first observe that the solution to (49) can be extended in an even way across $\partial\Omega$, and thus can be written as

$$\psi' = G * \bar{\omega}$$

where $G(x) = -(2\pi)^{-1}\log|x|$ is the elementary solution for the Laplace equation in 2-D. By (51), this implies that

$$\psi'(\cdot, t) \to G * [\bar{\omega}_0 - 2u_0 \otimes \delta_{\partial\Omega}] \text{ in } H^{1/2}(\mathbf{R}^2)$$

But $u_0 \cdot \tau = -\partial\psi_0/\partial\nu$, where ψ_0 is the stream function associated to u_0, so that, upon developping the integrals in the above convolution, it is readily seen that it yields the integral representation of ψ_0. Since $\psi_0 = 0$ on $\partial\Omega$, we in particular get (50).

REFERENCES

ANDERSON, C. (1989), Vorticity boundary conditions and boundary vorticity generation for two-dimensional viscous incompressible flows, *J. Comp. Phys.* **80**, p. 72-97.

BEALE, J.T. and C. GREENGARD (1993), Convergence of Euler-Stokes splitting of the Navier-Stokes equations, *Comm. Pure Appl. Math.* **43**.

BENFATTO, G. and M. PULVIRENTI (1986), Convergence of Chorin-Marsden product formula in the half-plane, *Comm. Math. Phys.* **106**, p. 427-458.

CHORIN, A.J. (1973), Numerical study of slightly viscous flow, *J. Fluid Mech.* **57**, p. 785-796.

CHORIN, A.J., HUGUES, T.J.R. and J.E. MARSDEN (1978), Product formula and numerical algorithms, *Comm. Pure Appl. Math.* **31**, p. 205-256.

COTTET, G.-H. (1988), Vorticity boundary conditions and the deterministic vortex method for the Navier-Stokes equations in exterior domains, in: Caflish, R. (ed.), *Mathematical aspects of vortex dynamics*, (SIAM, Philadelhia).

HOU, T. and B. WETTON (1992), Convergence of a finite-difference method for the Navier-Stokes equations using vorticity boundary conditions, *SIAM J. Num. Anal.* **29**, p. 615-639.

NEGATIVE ISOTROPIC EDDY VISCOSITY: A COMMON PHENOMENON IN TWO DIMENSIONS

S. Gama[1,2], M. Vergassola[1] and U. Frisch[1]

[1] CNRS, Observatoire de Nice,
B.P. 229, 06304 Nice Cedex 4, France.

[2] FEUP, Universidade do Porto,
R. Bragas, 4099 Porto Codex, Portugal.

ABSTRACT

We show the existence of two-dimensional flows having an isotropic and negative eddy viscosity. Such flows, when subject to a very weak large-scale perturbation of wavenumber k will amplify it isotropically with a rate proportional to k^2.

1. INTRODUCTION

Using the multiscale techniques, as discussed by Dubrulle and Frisch (1991), we calculate the eddy viscosities. The result is given in terms of auxiliary problems (Dubrulle and Frisch, 1991), which possess analytical solutions only in special cases (for instance, the basic flow depends only on a single space-coordinate like the Kolmogorov flow). This is why the only flows known for sure to possess a negative eddy viscosity were highly anisotropic (Meshalkin and Sinai 1961). So, it was an open problem if two-dimensional flows exist having an *isotropic* eddy viscosity which is negative.

2. THE DECORATED HEXAGONAL FLOW

In Vergassola, Gama and Frisch (1993) it is demonstrated the existence of at least one deterministic time-independent, space-periodic flow, which has an isotropic negative eddy viscosity above some critical Reynolds number. The flow is:

$$\Psi(x_1, x_2) = -\frac{1}{2}[\cos 2x_1 + \cos(x_1 + \sqrt{3}\,x_2) + \cos(x_1 - \sqrt{3}\,x_2)] \tag{1}$$
$$+\frac{1}{2}[\cos(4x_1 + 2\sqrt{3}\,x_2) + \cos(5x_1 - \sqrt{3}\,x_2) + \cos(x_1 - 3\sqrt{3}\,x_2)]$$

$$-\frac{1}{2}\left[\cos\left(4x_1\right) + \cos\left(2x_1 + 2\sqrt{3}\,x_2\right) + \cos\left(2x_1 - 2\sqrt{3}\,x_2\right)\right]$$
$$+\frac{1}{2}\left[\cos\left(4x_1 - 2\sqrt{3}\,x_2\right) + \cos\left(5x_1 + \sqrt{3}\,x_2\right) + \cos\left(x_1 + 3\sqrt{3}\,x_2\right)\right]\,.$$

The stream-lines are shown in Figure 1. We have called it the 'decorated hexagonal flow'. This flow possesses a center of symmetry, so that the AKA-effect is absent (Frisch, She and Sulem 1987b), and is invariant under rotations of $\pi/3$, ensuring the isotropy of fourth order tensors.

Figure 1. Stream-lines of the 'decorated hexagonal flow', given by (1), which has a negative eddy viscosity.

2.1 Numerical Methods

We use two different methods to solve the auxiliary problems emerging from the multiscale techniques. First, there is an expansion in powers of the Reynolds number which can be carried out to large orders, and then extended analytically (thanks to a meromorphy property) beyond the disk of convergence. For meromorphic functions, a very robust method of analytic continuation is by Padé approximants.

Second, there is a pseudo-spectral method on a square $(N \times N)$ regular grid with dealiasing by truncation beyond wavenumber $N/3$. The general idea is to evaluate multiplications in the physical space (x-space) and derivatives in Fourier space (k-space), moving back and forth by fast Fourier transforms. The spectral calculations were done on the Connection Machine CM-200 with 16k processors of INRIA- Sophia-Antipolis (France). For technical details see Gama, Vergassola and Frisch (1994a).

These two methods agree within a fraction of 1%. The decorated hexagonal flow achieves negative eddy-viscosities. Table 1 shows the results for the [10/10]-diagonal Padé approximant and the spectral calculations at two different resolutions.

3. HOW COMMON IS NEGATIVE EDDY VISCOSITY?

We have then shown that two-dimensional flows having an isotropic and negative eddy viscosity indeed exist. The next question is to ask if this property is exceptional in the class of parity-invariant flow with six-fold rotational symmetry (Gama, Vergassola and Frisch 1994a). We have therefore analyzed a large number of Gaussian time-independent flows with random Fourier components. The energy spectrum characterizing statistically the flow is of the form k^{-n} with a cutoff at $K=7$, beyond which the energy spectrum is zero. The molecular viscosity is initially $\nu=2$ and it is halved until one of the following situations is realized: (i) the eddy viscosity becomes negative, (ii) the eddy viscosity begins to increase, (iii) a small-scale instability appears. In the third event, a refined search in-between the last two values is made.

Table 1. Eddy-viscosity ν_E for the decorated hexagonal flow.

ν	Spectral 64^2 ν_E	Spectral 256^2 ν_E	Padé Prediction ν_E
1	1.348	1.348	1.348
0.8	0.961	0.961	0.961
0.7	0.642	0.642	0.642
0.6	0.134	0.134	0.134
0.59	0.066	0.066	0.066
0.58	−0.007	−0.007	−0.007
0.57	−0.085	−0.085	−0.085
0.56	−0.169	−0.169	−0.169
0.55	−0.259	−0.259	−0.260

We found that about 30% of the flows eventually developed a negative eddy viscosity, when lowering the molecular viscosity. During our investigation, we have also found some rather amusing flows with negative eddy viscosities, such as the one shown in Figure 2.

Figure 2. One of the many Gaussian flows with negative eddy viscosity which has six-fold symmetry and a k^{-1} energy spectrum.

We have also tried to find a simple rule for guessing if a flow can have negative eddy viscosity. Visual inspection of a number of such flows shown that they usually possess regions of rather closely packed stream-lines, such as the roughly circular structure encircling the 'gear' in Figure 2, suggesting that these flows locally resemble the Kolmogorov flow which is known to have a negative (albeit highly anisotropic) eddy viscosity.

4. NONLINEAR DYNAMICS

For basic flow driven by a prescribed external force, the multiscale analysis is extended to the nonlinear régime. It is crucial to distinguish the cases of mirror-symmetric (non-chiral) and chiral forcing.

In the former case, the large-scale dynamics is governed by an equation of the 'form' Navier-Stokes or Navier-Stokes-Kuramoto-Sivashinsky (NSKS) for positive eddy viscosities or marginally negative eddy viscosities, respectively. Here, 'form' means that the nonlinearities emerging from the multiscales techniques appear *renormalized* by a coefficient a depending on the basic flow. The constant a is, in general, not equal to one because Galilean invariance is broken by the force maintaining the basic flow (Gama and Vergassola 1994b). The latter symmetry can be restored by a rescaling of time and viscosity, as is done in the theory of lattice gases (Frisch *et al.* 1986, 1987a), except when the coefficient a vanishes (Gama and Vergassola 1994b).

The NSKS equation:

$$\partial_t \mathbf{u} + \mathbf{u} \cdot \nabla \mathbf{u} = -\nabla p - \nabla^2 \mathbf{u} - \mu \nabla^4 \mathbf{u}, \tag{1}$$

$$\nabla \cdot \mathbf{u} = 0, \tag{2}$$

where $\mathbf{u} = \mathbf{u}(t, x, y)$ is a two-dimensional incompressible velocity field and $\mu > 0$ is a control parameter, has been investigated numerically in detail by Gama, Frisch and Scholl (1991) at a time when the existence of negative and isotropic eddy viscosities was only conjectured. Up to millions of time steps at the resolution 256^2 and tens of thousands at the resolution 1024^2 were performed. A linear growth phase, a disorganized inverse cascade phase and a structured vortical phase were successively observed. In the vortical phase, monopolar and multipolar structures were found to proliferate and displayed strongly depleted nonlinearities. By this we understand that the evolution from quite arbitrary initial data leads eventually to strongly depleted nonlinearities.

When the driving force is not mirror-symmetric, a new 'chiral' nonlinearity appears (Vergassola 1993). In special cases, the large-scale equation reduces to the Burgers equation. In general, chiral forcing produces a strong enhancement or depletion of the large-scale vortices, depending on their cyclonicity.

ACKNOWLEDGMENTS

We have benefited from extensive discussions with R. Benzi, P. Collet, B. Dubrulle, J. Sommeria and A. Vespignani. This work was supported by grants from DRET (91/112), NATO Scientific Studies Program under grant 11/A/89/PO and from the European Community (Human Capital and Mobility ERBCHRXCT920001). The numerical calculations were done on the CM-2/CM-200 of the 'Centre Régional de Calcul PACA, antenne INRIA-Sophia-Antipolis' through the R3T2 network.

REFERENCES

DUBRULLE, B. AND U. FRISCH (1991), The eddy-viscosity of parity-invariant flow. *Phys. Rev. A* **43**, 5355-5364.

GAMA, S., FRISCH, U. and H. SCHOLL (1991), The two-dimensional Navier-Stokes equations with a large-scale instability of the Kuramoto-Sivashinsky type: numerical exploration on the Connection Machine. *J. Sci. Comp.* **6**, 425-452.

GAMA, S., VERGASSOLA, M. and U. FRISCH (1994a), Negative eddy viscosity in isotropically forced two-dimensional flow: linear and nonlinear dynamics. *J. Fluid Mech.* **260**, 95-126.

GAMA, S. and M. VERGASSOLA (1994b), Slow-down of nonlinearityin 2-D Navier-Stokes flow, *Physica* D **76**, 291-296.

FRISCH, U., HASSLACHER, B. and Y. POMEAU (1986), Lattice gas automata for the Navier-Stokes equation. *Phys. Rev. Lett.* **56**(14), 1505-1508.

FRISCH, U., HUMIÈRES, D., HASSLACHER, B., LALLEMAND, P., POMEAU, Y. and J.P. RIVET (1987a), Lattice gas hydrodynamics in two and three dimensions. *Complex Systems* **1**, 649-707; also reproduced in *Lattice Gas Methods For Partial Differential Equations*. Ed. G.D, Doolen, Addison-Wesley, 77-135, 1990.

FRISCH, U., SHE, Z.S. and P.L. SULEM (1987b), Large-scale flow driven by the anisotropic kinetic alpha effect. *Physica* D **28**, 382-392.

MESHALKIN, L.D. and IA.G. SINAI (1961), Investigation of the stability of a stationary solution of a system of equations for the plane movement of an incompressible viscous liquid. *Appl. Math. Mech.* **25**, 1700-1705.

VERGASSOLA, M. (1993), Chiral nonlinearities in forced 2-D Navier-Stokes flows. *Europhys. Lett.* **24**, 41-45.

VERGASSOLA, M., GAMA, S. and U. FRISCH (1993), Proving the existence of negative isotropic eddy-viscosity. *Theory of Solar and Planetary Dynamos*, M.R.E. Proctor, P.C. Matthews and A.M. Rucklidge eds., Cambridge University Press, 321-327.

A PARTICLE IN CELL METHOD FOR THE ISENTROPIC GAS DYNAMIC SYSTEM

S. Mas-Gallic

Centre de Mathématiques Appliquées, École Polytechnique
91128 Palaiseau Cedex, France

SUMMARY

The aim of this paper is to present a numerical method and some numerical results, relative to the flow of a barotropic gas. The equations of conservation of mass and momentum are written (compressible Euler system) and the Helmholtz decomposition of the velocity is used to transform the system. The resulting Euler system, written in vorticity, density and potential variables is numerically solved by an extention of the *classical* P.I.C. method for incompressible fluids.

1. INTRODUCTION

A lot of results are known about the flows of incompressible viscous fluids (see e.g. R. Temam, 1977) or inviscid fluids (see Marchioro and Pulvirenti, 1993, and also Cottet, 1982, Raviart, 1985, for applications to particle methods) in several dimensions. In the case of compressible fluids, in dimension greater than one, only few and partial results concerning the existence and regularity of flows of barotropic gas exist. Recently, global existence of solution have been proved by Lions (1993a) (see also Lions, 1993b, 1993c for related results) in the viscous case. In the stationnary case still viscous, we also mention the work of Novotný (1991), Novotný and Padula (1991). In the present paper, we shall only be concerned with the presentation of a numerical method of resolution of the Euler system obtained from the mass and momentum conservation equations in the study of the flow of a barotropic gas. The method presented here is deeply inspired by the particle in cell method (P.I.C.) first introduced to compute flows of incompressible inviscid fluids and is now *classically* used (see Harlow, 1964; Harlow and Amsden, 1964; Christiansen, 1973 and more recently Cottet, 1987). Let us briefly recall that the P.I.C. method couples a lagrangian method to transport the physical quantities convected by the flow and an eulerian method for the remaining quantities.

Navier-Stokes Equations and Related Nonlinear Problems
Edited by A. Sequeira, Plenum Press, New York, 1995

Written in velocity-vorticity variables $(\mathbf{u}, \omega)$ the Euler system reads

$$\begin{cases} \dfrac{\partial \omega}{\partial t} + \mathbf{u}.\nabla \omega = 0 \\ \nabla.\mathbf{u} = 0 \\ \nabla \times \mathbf{u} = \omega \end{cases} \tag{1.1}$$

and we notice first that the vorticity by exactly convected by the flow and second that the pressure has disappeared. Once the formulation is obtained, the vorticity is discretized into vorticity elements which follow the characteristic curves of the velocity field. Then, we rewrite the relation between $(\mathbf{u}$ and $\omega)$ as

$$- \Delta \mathbf{u} = \nabla \times \omega \tag{1.2}$$

and this equation is solved by application of a standard finite difference scheme.

Another formulation of the system expressed in current function - vorticity variables (ψ, ω), where the current function ψ is defined by $\mathbf{u} = \nabla \times \psi$ may seem more convenient especially from the boundary condition point of view (see the work of Huberson, Jollès and Shen, 1992).

In the present work, we consider the flow of a two dimensional barotropic gas governed by the Euler system which results from the mass and momentum conservation. Denoting by ρ, $\mathbf{u}$ and p respectively the density, the velocity and the pressure of the gas and by γ its specific heat ratio, the equations of conservation of mass and of momentum as well as the barotropic gas law read

$$\begin{cases} \dfrac{\partial \rho}{\partial t} + \nabla.(\rho\mathbf{u}) = 0, \\ \dfrac{\partial \mathbf{u}}{\partial t} + (\mathbf{u}.\nabla)\mathbf{u} + \dfrac{1}{\rho}\nabla p = 0, \\ p = k\rho^{\gamma}. \end{cases} \tag{1.3}$$

The system is completed with initial conditions and boundary conditions that will be made precise later.

Since we are interested in the extension of the P.I.C. method to the calculation of the flow of such a fluid, we shall need first to regognize which quantities are to be transported by the flow with no other effect and to attain this goal we shall introduce a new formulation of the system.

Using the Helmholtz decomposition of the velocity, a little algebra will yield a system essentially written in density, vorticity and potential variables, plus the rotational velocity and a *pseudo*-pressure. This system will be extendely constituted of two purely convective equations (one for the density ρ and one for the vorticity ω), a Hamilton-Jacobi equation (for the potential ϕ) and two elliptic equations (one for the velocity $\mathbf{v}$ and one for the pseudo-pressure q). Let us mention that this formulation was first introduced in Louaked, Mas-Gallic and Pironneau, 1993 (see also Louaked, 1990, in the case of a potential flow).

From the formulation, it is clear at first glance that the density is a convected quantity. A slightly different formulation essentially based on the Helmholtz decomposition of the velocity will give us the expected system in which the second convected quantity, the vorticity, will clearly appear. We consider from now on that the problem is periodic in the two directions. Since the method couples a particle method and a finite difference method, we shall need to define the two schemes used as well as the coupling operators which allow the exchange of information between the particles and the grid. The first operator which is the assignment operator of lagrangian values to the grid (vorticity and density, for example) is due to Brackbill and Ruppel (1986) (it has the important advantage of being conservative). The second operator is a simple interpolation of Eulerian quantities (velocity for example) on the particles.

In the next Section, the formulation is presented precisely and the appearance of the variables is made clear. Section 3 is devoted to the presentation of the numerical scheme and some test results.

2. PRESENTATION OF THE FORMULATION

Let us now present the system to be solved in its actual formulation. We start from the system (1.3) and consider the two dimensional case. In order to mimic the different steps of the PIC method when applied to the case of a compressible fluid, we introduce the Helmholtz decomposition of the velocity field $\mathbf{u}$

$$\mathbf{u} = \nabla\phi + \mathbf{v}, \qquad \nabla.\mathbf{v} = 0 \tag{2.1}$$

where $\mathbf{v}$ is the rotationnal part of the velocity. This decomposition separates the potential part, $\nabla\phi$, and the incompressible part, $\mathbf{v}$ of the velocity and we easily obtain the following system

$$\begin{cases} \dfrac{\partial\rho}{\partial t} + \nabla.(\rho(\mathbf{v}+\nabla\phi)) = 0 \\[2mm] \dfrac{\partial\mathbf{v}}{\partial t} - (\mathbf{v}+\nabla\phi)\times\nabla\times\mathbf{v} + \nabla q = 0 \\[2mm] \nabla.\mathbf{v} = 0 \\[2mm] \dfrac{\partial\phi}{\partial t} + \dfrac{1}{2}\mid \mathbf{v}+\nabla\phi\mid^2 + k\dfrac{\gamma}{\gamma-1}\rho^{\gamma-1} = q \end{cases} \tag{2.2}$$

Thus, we take definitely advantage of this natural separation by introducing the vorticity $\omega = \nabla\times\mathbf{u} = \nabla\times\mathbf{v}$ and we get the following system

$$\begin{cases} \dfrac{\partial\rho}{\partial t} + \nabla.(\rho(\nabla\phi+\mathbf{v})) = 0 \\[2mm] \dfrac{\partial\omega}{\partial t} + \nabla.(\omega(\nabla\phi+\mathbf{v})) = 0 \\[2mm] \dfrac{\partial\phi}{\partial t} + \dfrac{1}{2}|\mathbf{v}+\nabla\phi|^2 + k\dfrac{\gamma}{\gamma-1}\rho^{\gamma-1} = q \\[2mm] \nabla.\mathbf{v} = 0 \\[2mm] \nabla\times\mathbf{v} = \omega \\[2mm] \Delta q = \omega^2 + \nabla\omega\times(\nabla\phi+\mathbf{v}) \end{cases} \tag{2.3}$$

supplemented with the initial conditions

$$\rho(.,0) = \rho_0, \quad \omega(.,0) = \omega_0, \quad \phi(.,0) = \phi_0. \tag{2.4}$$

After some algebra, essentially based on the following relation

$$(\mathbf{u}.\nabla)\mathbf{u} = -\mathbf{u}\times(\nabla\times\mathbf{u}) + \dfrac{1}{2}\nabla(|\mathbf{u}|^2) \tag{2.5}$$

359

we end up with the following formulation of Euler equations

$$\begin{cases} \dfrac{\partial \rho}{\partial t} + \nabla.(\rho(\nabla \phi + \mathbf{v})) = 0 \qquad (i) \\[2mm] \dfrac{\partial \omega}{\partial t} + \nabla.(\omega(\nabla \phi + \mathbf{v})) = 0 \qquad (ii) \\[2mm] \dfrac{\partial \phi}{\partial t} + \dfrac{1}{2}|\mathbf{v} + \nabla \phi|^2 + k\dfrac{\gamma}{\gamma - 1}\rho^{\gamma-1} = q \qquad (iii) \\[2mm] -\Delta \mathbf{v} = \nabla \times \omega \qquad (iv) \\[2mm] \Delta q = \omega^2 + \nabla \omega \times (\nabla \phi + \mathbf{v}) \qquad (v) \end{cases} \qquad (2.6)$$

Under this form we notice that both the density ρ and the vorticity ω are convected by the flow with the velocity $\mathbf{u} = \nabla \phi + \mathbf{v}$. In the method that shall be presented later, the convection equations $(2.6)(i)$ and $(2.6)(ii)$ are solved by a particle method whereas the Hamilton-Jacobi equation $(2.6)(iii)$ and the elliptic equations $(2.6)(iv)$ and $(2.6)(v)$ are solved by finite difference schemes. Now, rewriting equation $(2.6)(iii)$ under the form

$$\frac{\partial \phi}{\partial t} + \frac{1}{2}|\nabla \phi|^2 + \mathbf{v}.\nabla \phi = f, \qquad (2.7)$$

where $f = q - 1/2|\mathbf{v}|^2$, we recognize a Hamilton-Jacobi equation with a convection term (see Crandall and Lions, 1988 for a detailed analysis). This equation will be numerically solved by a E.N.O. type scheme introduced by S. Osher and J. Sethian (1988).

Notice that, if at time zero the vorticity is equal to zero, it remains so for all time t and the flow is potential. The sytem then reads

$$\begin{cases} \dfrac{\partial \rho}{\partial t} + \nabla.(\rho \nabla \phi) = 0 \\[2mm] \dfrac{\partial \phi}{\partial t} + \dfrac{1}{2}|\nabla \phi|^2 + k\dfrac{\gamma}{\gamma - 1}\rho^{\gamma-1} = 0 \end{cases} \qquad (2.8)$$

and results of numerical results for this model can be found in M. Louaked (1990).

In order to couple the difference and the particle method, the lagrangian quantities (density and vorticity) have to be distributed on the grid and the eulerian quantities (velocity) need to be given a value on the particles. We thus define two operators. The assignment operator of lagrangian values to the grid is the one introduced by J.U. Brackbill and H.M. Ruppel (1986) (it has the important advantage of being conservative) and the eulerian quantities are interpolated.

3. NUMERICAL RESOLUTION

The proposed method is semi-lagrangian, the convection equations $(2.6)(i)-(ii)$ are solved by a particle method, the density and the vorticity are discretized into elements which follow the integral curves of the velocity field and equations $(2.6)(iii)-(v)$ are solved by finite difference schemes.

We fix a grid of width ε and denote by S_{ij} its vertices. The function ψ will denote the basis function (either piecewise constant or piecewise bilinear) of the finite difference approximation. At time 0 the discretized potential ϕ_h, $\mathbf{v}_h$ and q_h are defined by

$$\phi_h^0(\mathbf{x}) = \sum_{ij} \phi_{ij}^0 \psi(\mathbf{x} - S_{ij}), \qquad q_h^0(\mathbf{x}) = \sum_{ij} q_{ij}^0 \psi(\mathbf{x} - S_{ij})$$

$$\mathbf{v}_h^0(\mathbf{x}) = \sum_{ij} \mathbf{v}_{ij}^0 \psi(\mathbf{x} - S_{ij})$$

The particle quantities such as the vorticity ω and the density ρ are approximated by linear combinations of regularized Dirac measures as follows at time 0 ; we choose a set of weighted points $(W_p^0, \mathbf{x}_p^0)$ and define

$$\omega_h^0(\mathbf{x}) = \sum_p W_p^0 \omega_p^0 \psi(\mathbf{x} - \mathbf{x}_p^0), \qquad \rho_h^0(\mathbf{x}) = \sum_p W_p^0 \rho_p^0 \psi(\mathbf{x} - \mathbf{x}_p^0).$$

The coefficients ω_p^0 and ρ_p^0 respectively stand for approximations of $\omega_0(\mathbf{x}_p^0)$ and $\rho_0(\mathbf{x}_p^0)$. Then, we introduce a time step $\Delta t > 0$ and define the time discretization of the equations.

Let us now define the n-th step. The finite difference schemes used to solve the potential equation $(2.6)(iii)$ were, in the potential case, the classical leap frog scheme

$$\phi_{ij}^{n+1} = \phi_{ij}^{n-1} - \Delta t[(\frac{\phi_{i+1,j}^n - \phi_{i-1,j}^n}{2\Delta x})^2 + (\frac{\phi_{i,j+1}^n - \phi_{i,j-1}^n}{2\Delta y})^2 - \qquad (3.1)$$

$$-2k\frac{\gamma}{\gamma - 1}|\rho_{ij}^n|^{\gamma - 1}]$$

This scheme (as well as the ENO scheme to be presented in next remark) is stable under a CFL condition.

Remark 3.1. In the case of a general flow the following extension of the ENO scheme was used (see Osher and Sethian, 1988) :

$$\frac{\phi_{ij}^{n+1} - \phi_{ij}^n}{\Delta t} = H_{H-J}(\bar{\phi})_{ij} + \qquad (3.2)$$

$$+v_1^+ D_-^x \phi_{ij}^n + v_1^- D_+^x \phi_{ij}^n + v_2^+ D_-^y \phi_{ij}^n + v_2^- D_+^y \phi_{ij}^n + f_{ij}^n$$

where

$$D_+^x \phi_{ij}^n = \frac{\phi_{i+1,j}^n - \phi_{ij}^n}{\Delta x}, \qquad D_-^x \phi_{ij}^n = \frac{\phi_{ij}^n - \phi_{i-1,j}^n}{\Delta x}.$$

The discrete Hamiltonian H_{H-J} is defined by

$$H_{H-J}(\bar{\phi})_{ij}^n = [|D_+^x \phi_{ij}^n|_+^2 + |D_-^x \phi_{ij}^n|_-^2] + [|D_+^y \phi_{ij}^n|_+^2 + |D_-^y \phi_{ij}^n|_-^2] \qquad (3.3)$$

where $\bar{\phi}^n = (\phi_{ij}^n)$. We set

$$f_{ij}^n = q_{ij}^n - k\frac{\gamma}{\gamma - 1}|\rho_{ij}^n|^{\gamma - 1} - \frac{1}{2}|v_{ij}^n|^2 \qquad (3.4)$$

This scheme is no more centered but upwind which is somehow natural since an advection term appears in the Hamilton-Jacobi equation. □

The Laplace operators for $\mathbf{v}$ and q are discretized either by the standard s centered scheme or by a finite difference solver introduced by R. W. Hockney (1971) (specifically well adapted to the case of periodic boundary conditions).

Let us now present the particle-grid operator. The density ρ_{ij}^n on the grid is defined by the assignment method due to Brackbill and Ruppel (1986), for each grid point S_{ij}, we set

$$\rho_{ij}^n = \frac{\sum_p W_p^0 \rho_p^0 \psi(S_{ij} - \mathbf{x}_p^n)}{\sum_p W_p^n \psi(S_{ij} - \mathbf{x}_p^n)} \simeq \rho(S_{ij}, n\Delta t). \qquad (3.5)$$

Remark 3.2. The method of assignment previously defined is conservative (i.e. if $\rho_p \equiv 1$, then $\rho_{ij} \equiv 1$). The simplest assignment method defines

$$\rho_{ij}^n = \sum_p W_p^0 \rho_p^0 \psi(S_{ij} - \mathbf{x}_p^n)$$

but leads to a non conservative method, property which is of rather high importance from the physical point of view (conservation of mass for example)

Remark 3.3. Notice that, since the fluid is compressible, the coefficient W_p (which is the weight of the particle in the quadrature rule and stands for a volume of the particle) and the coefficient ρ_p are separately not constant in time. More precisely, their time evolution is given by the divergence of the velocity as follows

$$\frac{dW_p}{dt}(t) = \nabla.\mathbf{u}_h(\mathbf{x}_p(t), t)W_p(t), \tag{3.6}$$

$$\frac{d\rho_p}{dt}(t) + \nabla.\mathbf{u}_h(\mathbf{x}_p(t), t)\rho_p(t) = 0 \tag{3.7}$$

and it turns out that their product $W_p\rho_p$ is constant in time which somehow explains the expression (3.5). □

The positions of the particles are solutions of the following system of ordinary differential equations

$$\frac{d\mathbf{x}_p}{dt}(t) = \mathbf{u}_h(\mathbf{x}_p(t), t) \tag{3.8}$$

which is solved by a second order Runge-Kutta scheme

$$\mathbf{x}_p^{n+1} = \mathbf{x}_p^n + \Delta t(\frac{3}{2}\mathbf{u}_p^n - \frac{1}{2}\mathbf{u}_p^{n-1})$$

where $\mathbf{u}_p^n = \mathbf{v}_p^n + (\nabla_h\phi_h)_p^n$ and $\nabla_h\phi_h$ is the finite difference approximation of the gradient on the grid. Similarly, the volumes are obtained by a Runge-Kutta scheme applied to equation (3.6)

$$W_p^{n+1} = W_p^n + \Delta t(\frac{3}{2}(\nabla_h.\mathbf{u}_h)_p^n W_p^n - \frac{1}{2}(\nabla_h.\mathbf{u}_h)_p^{n-1}W_p^{n-1}).$$

Remark 3.4. Notice that the computation of $\nabla_h.\mathbf{u}_h$ needs the introduction of a scheme for $\Delta_h\phi_h$ and again the 5-points scheme

$$(\Delta_h\phi_h)_{ij}^n = \frac{\phi_{i+1,j}^n + \phi_{i-1,j}^n - 2\phi_{i,j}^n}{\Delta x^2} + \frac{\phi_{i,j+1}^n + \phi_{i,j-1}^n - 2\phi_{i,j}^n}{\Delta y^2}$$

was used. □

Finally, the second exchange operator is a simple interpolation. We used either the N.G.P. or the bilinear interpolation and thus the particle quantities were distributed either on the nearest grid point or on the four grid points surrounding the particle.

We present now some numerical results and since the two dimensional computations have not been compared yet to any others, we shall only present some results concerning a one dimensional problem. Namely, we consider an infinite tube along the x-axis with two different sets of initial data. In the first test, we start from a gas at rest with constant density ρ_0 which fills half of the tube and a piston which is pulled on the left. The results of this test are illustrated in the first two figures (left picture for the velocity and right picture for the density). On these figures, the velocity and density computed with the PIC method presented here and the results obtained by Glaister's solver (Glaister, 1991) are plotted.

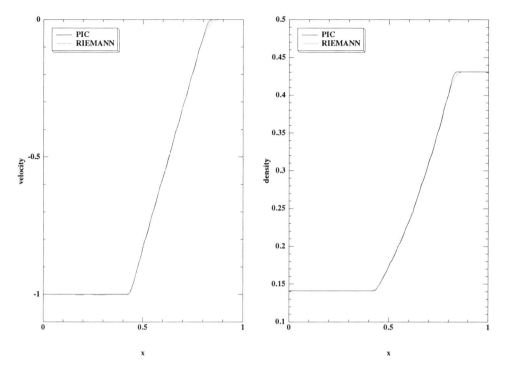

Figure 1. Density and velocity profiles for PIC and Glaister's method at $t = 1$ for $\Delta x = .01$ and $CFL = .1$

In the second test, we start from a non uniform density which presents a shock and we refer to Engquist and Osher (1981) for more details (see also Louaked, Mas-Gallic and Pironneau, 1993, for some comments on a slightly different method). The results of this test are illustrated by the second figures, where again the velocities (left figure) and densities (right figure) computed by the PIC method and Glaister's method are compared.

The results of these two tests for the PIC as illustrated by the figures show a good agreement with Glaister's solver although the PIC method is slightly more diffusive for the computation of both the density and the velocity in the first test and for the computation of the velocity in the second one. The reason is that our PIC method is based on a Lax-Friedrichs scheme for the computation of the velocity and this drawback can actually be corrected. Let us also point out that by contrast, the density-shock is better captured by the PIC method, illustrating the fact that particle methods are well suited for the resolution of transport equations.

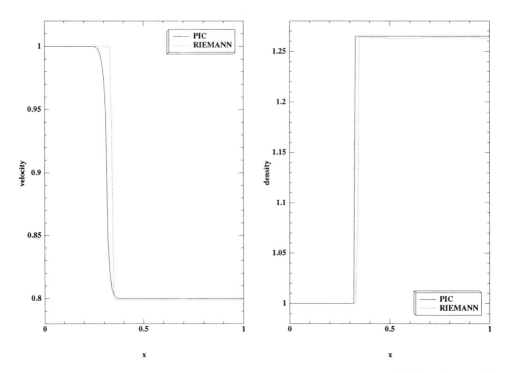

Figure 2. Density and velocity profiles for PIC and Glaister's method at $t = 7.93$ for $\Delta x = .01$ and $CFL = .1$

REFERENCES

BRACKBILL, J.U. and H.M. RUPPEL (1986), Flip: A method for adaptively zoned, Particle-in-cell calculations of fluid flows in two dimensions, *J. Comp. Phys.* **65**, p. 314-343.

CHRISTIANSEN, J.P. (1973), Numerical simulation of hydrodynamics by the method of point vortices, *J. Comp. Phys.* **13**, p. 363-379.

COTTET, G.-H. (1987), Convergence of a Vortex In Cell method for the two-dimensional Euler equations, *Math. Comp.* **49**, p. 407-425.

COTTET, G.-H. (1982), Dynamique cochléaire en dimension un - Méthodes particulaires pour l'équation d'Euler dans le plan, *Thèse de 3 ème cycle*, Paris.

CRANDALL, M.G. and P.L. LIONS (1988), *Math. Comp.* **43**, p. 1.

ENGQUIST, B. and S. OSHER (1981), Upwind difference schemes for systems of conservation laws-potential flow equations, *M.R.C. technical report* **2186**.

GLAISTER, P. (1991), A Riemann solver for barotropic flow, *J. Comp. Phys.*, **93**, p. 477-480.

HARLOW, F. and A. AMSDEN (1964), Slip instability, *Phys. of Fluids* **7** , p. 327-334.

HARLOW, F. (1964), The particle-in-cell computing method for fluid dynamics, *Methods in Comp. Phys.* **3**.

HOCKNEY, R.W. (1971), *Comp. Phys. Comm.* **2**.

HUBERSON, S. and A. JOLLÈS (1989), *C. R. Acad. Sci.* **309**, Série II, p. 445-448, Paris.

HUBERSON, S., JOLLÈS, A. and W. SHEN, W (1992), Numerical Simulation of Incompressible Viscous Flows by Means of Particle Methods, in ANDERSON, C. and GREENGARD, C. (eds.), Vortex Dynamics and Vortex Methods, *A.M.S. Lectures in Applied Mathematics*, vol **28**, p. 369-384.

LIONS, P.L. (1993a), Existence globale de solutions pour les équations de Navier-Stokes compressibles isentropiques, *Note C. R. Acad. Sc. Paris.*

LIONS, P.L (1993b), Compacité des solutions pour les équations de Navier-Stokes compressibles isentropiques, *Note C. R. Acad. Sc. Paris.*

LIONS, P.L. (1993c), Limites incompressible et acoustique pour des fluides visqueux compressibles isentropiques, *Note C. R. Acad. Sc. Paris.*

LOUAKED, M (1990), Une méthode P.I.C (particle-in-cell) pour les équations d'Euler compressibles, *Rapport interne* **90016**, Lab. Anal. Num.

LOUAKED, M., MAS-GALLIC, S. and O. PIRONNEAU (1993), A particle in cell method for the 2-D compressible Euler equations, in BEALE, J.T., COTTET, G.-H. and S. HUBERSON (eds.), *Vortex Flows and Related Numerical Methods*, NATO ASI Series C, Mathematical and Physical Sciences **395**, (Kluwer, Dordrecht).

MARCHIORO, C. and M. PULVIRENTI (1993), *Mathematical Theory of Incompressible Non-Viscous Fluids*, Applied Math. Sciences **96**, (Springer, New York).

NOVOTNÝ, A. (1991), *Existence and uniqueness of stationary solutions for viscous compressible heat conductive fluid with large potential and small nonpotential forces*, in PERELLO, C. (ed.), proc. EQUADIFF 91, (Barcelone).

NOVOTNÝ, A. and M. PADULA (1991), Existence and uniqueness of stationary solutions for viscous compressible heat conductive fluid with large potential and small nonpotential external forces, *preprint* **164**, Univ. Ferrara.

OSHER, S. and F. SALOMON (1982), Upwind difference schemes for hyperbolic systems of conservation laws, *Math. Comp.* **38**, p. 339-374.

OSHER, S. and J. SETHIAN (1988), Fronts propagating with curvature-dependent speed : Algorithms based on Hamilton-Jacobi formulations, *J. Comp. Phys.* **79**, p. 12-49.

RAVIART, P.-A. (1985), An analysis of particle methods, in BREZZI, F. (ed.), Numerical Methods in Fluid Dynamics *Lecture Notes in Mathematics*, vol. **1127** (Springer Verlag, Berlin).

TEMAM, R. (1977), *Navier-Stokes Equations* (North Holland, Amsterdam New-York).

DOMAIN DECOMPOSITION METHODS FOR FLUID DYNAMICS

F. Nataf[1], F. Rogier[2] and E. de Sturler[3]

[1]CMAP, CNRS URA756, Ecole Polytechnique
91128 Palaiseau Cedex, France

[2]Division Calcul parallèle, ONERA, 29 Av. de la Division Leclerc
92322 Châtillon, France

[3]CSCS, ETH, Via Cantonale
6928 Manno, Switzerland

1. INTRODUCTION

In order to solve the Navier-Stokes equations, it is essential to have at one's disposal efficient numerical algorithms of solution of the linear convection-diffusion equations. Indeed, despite its nonlinear nature, the solving of the Navier-Stokes equations can be reduced to the solving of linear equations. Consider the incompressible Navier-Stokes equations in vorticity-stream function (ω, ψ) formulation in 2-D:

$$\frac{\partial \omega}{\partial t} + u\frac{\partial \omega}{\partial x} + v\frac{\partial \omega}{\partial y} - \nu\Delta\omega = 0$$
$$\Delta\psi = -\omega$$
$$\begin{pmatrix} u \\ v \end{pmatrix} = \text{rot } \psi$$

In order to solve this system, a classical and widely used algorithm is the following. Let $(\omega^n, \psi^n, u^n, v^n)$ denote the approximation of (ω, ψ, u, v) at time $n\Delta t$, $(\omega^{n+1}, \psi^{n+1}, u^{n+1}, v^{n+1})$ is obtained by the solution of two linear boundary value problems and by the computation of a curl:

$$\frac{\omega^{n+1} - \omega^n}{\Delta t} + u\frac{\partial \omega^{n+1}}{\partial x} + v\frac{\partial \omega^{n+1}}{\partial y} - \nu\Delta\omega^{n+1} = 0$$

then,

$$\Delta\psi^{n+1} = -\omega^{n+1}$$

Navier-Stokes Equations and Related Nonlinear Problems
Edited by A. Sequeira, Plenum Press, New York, 1995

and

$$\begin{pmatrix} u^{n+1} \\ v^{n+1} \end{pmatrix} = \text{rot } \psi^{n+1}$$

The solution of the Navier-Stokes equations is thus reduced to the successive solutions of symmetric (Δ) and nonsymmetric (convection-diffusion) linear systems. More generally, the convection-diffusion equation

$$\mathcal{L}(C) = \frac{\partial C}{\partial t} + a\frac{\partial C}{\partial x} + b\frac{\partial C}{\partial y} - \nu\Delta C = f$$

plays an important role in numerous physical phenomena. It models the transport of a quantity C along a vector field (a, b) and its diffusion, proportional to a viscosity coefficient ν. In fluid dynamics, for instance, C may be the concentration of a dye or of a pollutant in water (river, sea, estuary, ...) or air (plume of smoke, ...).

We consider here the numerical solution on parallel computers (MIMD, Multiple Instruction Multiple Data) of the linear systems arising from the discretization of the convection-diffusion equation. These computers are new and present many interesting features. They are made up of many processors called nodes linked together by a communication network. Each node is a computer by itself in the sense where it has its own direct access memory. Each node can host its own program possibly distinct from the ones of the other nodes. The nodes can work concurrently. They communicate with each other thanks to messages they can send and receive. These messages can be used to synchronize the nodes and to send and receive data. The programming language is Fortran 77 or C except for the orders corresponding to the sending and receiving of messages (eg. csend, crecv). These machines are potentially very powerful. The benefits in terms of memory is obvious. In order to have an actual speed of computation close to the optimum (speed of a processor × nbr of processors) it is necessary to have algorithms fitted to parallel machines.

Domain decomposition methods seem a natural and promising approach. Many works have been devoted to symmetric systems and have led to efficient methods. The discretization of the convection-diffusion equation leads to non symmetric linear systems of equations. For a large viscosity, the algorithms designed for symmetric systems can be used and perform well. Nevertheless, for convection dominated flows (small viscosity), these methods perform poorly and there are few theoretical results. In this work, we present a method based on the use of artificial boundary conditions. Indeed, the rate of convergence of domain decomposition methods is very sensitive to the choice of the interface conditions. The original Schwarz method is based on the use of Dirichlet boundary conditions. In order to increase the efficiency of the algorithm, it has been proposed to replace the Dirichlet boundary conditions with more general boundary conditions, see Lions (1989). In the usual Schur method, Dirichlet and Neumann boundary conditions are used. In Hagstrom, Tewarson and Jazcilevich (1988), they are replaced with artificial boundary conditions. More generally, it has been remarked that absorbing (or artificial) boundary conditions are a good choice (see, Hagstrom, Tewarson and Jazcilevich, 1988; Despres, 1991; Nataf, 1992; Givois, 1992, where such boundary conditions are used). In this report, we partially clarify this question.

In § 2.1, we set the problem to be solved and reformulate it as a problem whose unknowns are functions from the boundaries of the subdomains to **R**. In § 2.2, we take as interface conditions exact artificial boundary conditions (ABC). We show that GMRES and BICGSTAB algorithms converge in a number of steps equal to the number of subdomains minus one. These exact ABC are difficult to use and we outline in § 2.3

how to approximate them by partial differential operators. In § 3, we show numerical results.

2. DOMAIN DECOMPOSITION METHOD FOR THE CONVECTION-DIFFUSION EQUATION

2.1. Reformulation of the problem

For sake of simplicity, we consider the continuous convection-diffusion equation set on the whole plane $\mathbf{R}^2$:

$$\mathcal{L}(u) = f \text{ in } \mathbf{R}^2 \tag{2.1}$$

$$\tag{2.2}$$

where f is a given function. The plane is decomposed into N vertical strips with or without overlap. We have $\mathbf{R}^2 = \cup_{i=1}^{N} \bar{\Omega}_i$ where $\Omega_i =]l_i, L_i[\times \mathbf{R}$. In this §, we write an equivalent form of problem (2.1) where the unknowns are functions defined on the boundaries of the subdomains. Let $\mathcal{B}_{i,l}$, $\mathcal{B}_{i,r}$ be operators so that the following BVP are well-posed:

$$\begin{align*}
\mathcal{L}(v) &= f_i \text{ in } \Omega_i \\
\mathcal{B}_{i,l}(v) &= g_{i,l} \text{ on } \{l_i\} \times \mathbf{R} \\
\mathcal{B}_{i,r}(v) &= g_{i,r} \text{ on } \{L_i\} \times \mathbf{R}
\end{align*} \tag{2.3}$$

for any functions f_i, $g_{i,l}$ and $g_{i,r}$. We define the operator S_i by

$$v = S_i(g_{i,l}, g_{i,r}, f_i).$$

It is clear that the knowledge of $\mathcal{B}_{i,l}(u)$, $\mathcal{B}_{i,r}(u)$, $1 \leq i \leq N$ enables to recover the value of u by solving the BVPs (2.3). Let us denote H the $2(N-1)$-tuple $(\mathcal{B}_{2,l}(u), \ldots, \mathcal{B}_{N,l}(u),$ $\mathcal{B}_{1,r}(u), \ldots, \mathcal{B}_{N-1,r}(u))$. It is possible to write a linear system for H. Indeed, we have for $\mathcal{B}_{i,l}(u)$:

$$\begin{align*}
\mathcal{B}_{i,l}(u) &= \mathcal{B}_{i,l}(S_{i-1}(\mathcal{B}_{i-1,l}(u), \mathcal{B}_{i-1,r}(u), f)) \\
&= \mathcal{B}_{i,l}(S_{i-1}(\mathcal{B}_{i-1,l}(u), \mathcal{B}_{i-1,r}(u), 0)) + \mathcal{B}_{i,l}(S_{i-1}(0, 0, f))
\end{align*}$$

and for $\mathcal{B}_{i,r}(u)$ a similar relation. It is thus possible to define a linear operator $\mathcal{T}$ so that the previous linear system writes in a compact form:

$$(Id - \mathcal{T})(H) = G \tag{2.4}$$

where G can be easily computed and depends on f. When there is no overlap, system (2.4) is well-posed and when there is no overlap, one needs the extra condition that $\mathcal{B}_{i,r} - \mathcal{B}_{i+1,l}$ is invertible.

2.2. Choice of the interface conditions

The principle of substructuring methods (also called Schur methods) is to solve (2.4) with conjugate gradient like methods as GMRES or BICGSTAB. The speed of convergence will of course depend on $\mathcal{T}$ and thus on the choice of the interface conditions $\mathcal{B}_{i,l}$ and $\mathcal{B}_{i,r}$. We consider here the case where we take for $\mathcal{B}_{i,l}$ and $\mathcal{B}_{i,r}$ (resp. $\mathcal{B}_{i,l}$) exact artificial boundary conditions which are defined by $\mathcal{B}_{i,r} = \partial_x - \Lambda_{i,r}$ (resp. $\mathcal{B}_{i,l} = -\partial_x - \Lambda$) where $\Lambda_{i,l}$ is the Dirichlet to Neumann operator of the right (resp. left) half-plane $]L_i, \infty[\times\mathbf{R}$ (resp. $]-\infty, l_i[\times\mathbf{R})$. Then it can be seen that $\mathcal{T}$ is nilpotent operator of order $N-1$, $\mathcal{T}^{N-1} = 0$ (see Nataf, Rogier and de Sturler, 1994). We deduce from this that GMRES and BICGSTAB methods applied to (2.4) will converge in $N-1$ steps while (2.4) is an infinite dimensional system. This shows that exact arificial boundary conditions are a very good choice as interface conditions.

Indeed, let H^0 be the initial approximation to the solution to (2.4). Let $r_0 = G - (Id - \mathcal{T})(H^0)$ be the initial residual. We seek for $\tilde{H}$ such that $H = H^0 + \tilde{H}$ i.e. $\tilde{H}$ satisfies:

$$(Id - \mathcal{T})(\tilde{H}) = r_0$$

The GMRES method minimizes the residual norm over the Krylov space $K^n(Id - \mathcal{T}), r_0) \equiv span\{r_0, (Id - \mathcal{T})r_0, \ldots, (Id - \mathcal{T})^{n-1}r_0\}$. Clearly, $\tilde{H} \in K^{N-1}(Id - \mathcal{T}), r_0)$ so that $N-1$ iterations are necessary for the solution of (2.4). Thus, we have just proved

Proposition 2.1. *The GMRES algorithm applied to (2.4) converges in at most $N-1$ steps.*

Let us now consider the convergence of Bi-CGSTAB Van der Vorst (1992) for the solution of the linear system (2.4). We shall see that

Proposition 2.2. *If there is no breakdown of Bi-CGSTAB, we have convergence of Bi-CGSTAB applied to (2.4) in at most $N-1$ steps.*

Because Bi-CGSTAB is based on BiCG (Fletcher, 1976) we will first discuss the convergence of BiCG. We choose some $\tilde{r}_0 \neq 0$, (for example $\tilde{r}_0 = r_0$). Now the BiCG algorithm generates two sequences of polynomials, the residuals $r_i = P_i(Id - \mathcal{T})r_0$:

$$r_0, r_1, r_2, \ldots$$

and $\tilde{r}_i = P_i((Id - \mathcal{T})^T)\tilde{r}_0$:

$$\tilde{r}_0, \tilde{r}_1, \tilde{r}_2, \ldots$$

where P_i indicates a polynomial of degree i. These sequences satisfy the following relations (Fletcher, 1976):

$$r_i^T \tilde{r}_j = 0 \qquad i \neq j \qquad (2.5)$$
$$r_i^T \tilde{r}_i \neq 0 \qquad\qquad (2.6)$$

If $r_i^T \tilde{r}_i = 0$ then BiCG would break down, but we will not discuss this problem here. For the residuals we have $r_i = P_i(Id - \mathcal{T})r_0 \in span\{r_0, (Id - \mathcal{T})r_0, (Id - \mathcal{T})^2 r_0, \ldots, (Id - \mathcal{T})^i r_0\} = K^{i+1}(Id - \mathcal{T}, r_0)$, and furthermore we have $K^{i+1}((Id - \mathcal{T}), r_0) = K^{i+1}(\mathcal{T}, r_0)$. Together this gives

$$r_i \in K^{i+1}(\mathcal{T}, r_0) \qquad (2.7)$$

Proposition 2.3. *Let $\{r_0, r_1, \ldots, r_{k-1}\}$ be independent and $r_k \in span\{r_0, r_1, \ldots, r_{k-1}\}$, then $r_k = 0$ and BiCG converges in k steps.*

Although being similar, this property differs from the finite termination properties for BiCG (Fletcher, 1976), for CG if the operator is a low rank perturbation of the identity, which leads, as in this case, to convergence in a number of steps equal to the rank of the perturbation (Golub and Van Loan, 1989), and for GMRES Saad and

Schultz (1986). In these other cases, the residual is necessarily zero because it is both an element of and orthogonal to the same space, whereas the present property is derived from the residual being an element of one space and orthogonal to another, in principle completely different space.

Proof: For r_k we have the following two relations:

$$r_k \in span\{r_0, r_1, \ldots, r_{k-1}\} \tag{2.8}$$

$$r_k \perp span\{\tilde{r}_0, \tilde{r}_1, \ldots, \tilde{r}_{k-1}\} \tag{2.9}$$

So (2.8) implies $r_k = \sum_{i=0}^{k-1} \alpha_i r_i$, and then (2.9) gives

$$\forall j : 0 \le j \le k-1 : \tilde{r}_j^T \left(\sum_{i=0}^{k-1} \alpha_i r_i \right) = 0 \Leftrightarrow \sum_{i=0}^{k-1} \alpha_i \tilde{r}_j^T r_i = 0$$

Together with (2.5) this leads to

$$\forall j : 0 \le j \le k-1 : \alpha_j \tilde{r}_j^T r_j = 0,$$

which means, using (2.6), that $\alpha_j = 0, \ 0 \le j \le k-1$. Therefore we have

$$r_k = 0,$$

and hence BiCG has converged.

Proposition 2.4. *For the linear system defined in (2.4) BiCG will converge in at most $N-1$ iterations if there is no breakdown.*

Proof: From $\mathcal{T}^{N-1} = 0$, we can derive that $K^N(\mathcal{T}, r_0) = K^{N-1}(\mathcal{T}, r_0)$. Together with (2.7) this leads to $r_{N-1} \in K^{N-1}(\mathcal{T}, r_0)$, so that $r_{N-1} \in span\{r_0, r_1, \ldots, r_{N-2}\}$. Proposition 2.3. then proves that $r_{N-1} = 0$, and therefore BiCG has converged. $\square$

Note that if the set $\{r_0, r_1, \ldots, r_k\}$ becomes dependent before $k = N-1$ BiCG will have converged as well.

It is not difficult to see that if the BiCG-residual $r_{N-1} = 0$, then also the Bi-CGSTAB-residual $r_{N-1}^{stab} = 0$. Bi-CGSTAB constructs its residual r_i^{stab} such as to be a polynomial of the form $r_i^{stab} = Q_i(Id - M)P_i(Id - M)r_0$, where $P_i(Id - M)r_0$ is still the BiCG-residual (Van der Vorst, 1992). So that, if the BiCG-residual $r_i = P_i(Id - \mathcal{T})r_0 = 0$, then also $r_i^{stab} = 0$, and Bi-CGSTAB will have converged as well.

Assuming that the norm of $\mathcal{T}^{N-2}G$ is sufficiently large, the equality $H = \sum_{i=0}^{N-2} \mathcal{T}^i G$ also indicates that GMRES cannot solve the set of equations (2.4) in less iterations than BiCG (however with half the number of matrix vector products).

2.3. Approximate artificial boundary conditions and DDM

We have seen that exact artificial boundary conditions lead to very interesting convergence properties. Unfortunately, they are difficult to use in a code. Indeed, operators $\Lambda_{i,r}$ or l are not partial differential operators. Moreover, in general, we do not have an explicit form of these operators. Nevertheless, it is usually possible to approximate them by partial differential operators as it is done for approximating exact artificial boundary conditions (see e.g. Engquist and Majda, 1977, 1979). In this section, we explain how these exact artificial boundary conditions are approximated by local operators (i.e. partial differential operators). This enables us to write a Schur type formulation for an arbitrary decomposition of the domain and to remove the restriction of a decomposition into vertical strips. In § 3, this strategy is applied to the convection-diffusion operator and numerical results are shown.

Our goal is to approximate at some point x_0 of the boundary of a subdomain the operators $\Lambda_{i,r \text{ or } l}$ by partial differential operators. In order to be able to follow the strategy developed in Engquist and Majda (1977), we assume that the coefficients of the operator $\mathcal{L}$ vary slowly so that they can be approximated by their values at x_0. By making use of the Fourier transform with respect to the tangential variable, we obtain an approximation of $\Lambda_{i,r \text{ or } l}$ in the form of a convolution operator. This operator is itself approximated by a partial differential operator by approximating its symbol by a polynomial (for more details, see Nataf and Rogier, 1995; Nataf, 1993). In some cases, it is possible to make less restrictive assumptions (see e.g. in the context of absorbing boundary conditions or of paraxial approximations Bamberger, Engquist, Halpern and Joly, 1988a, 1988b; Lohéac, 1991; Nataf, Rogier and de Sturler, 1994; Halpern and Rauch, 1993).

We want to write a system analogous to system (2.4) but based on the approximate ABC. Since these operators are local, we are not restricted any more to decompositions into vertical strips. We will thus obtain a substructuring formulation which can be solved by conjugate gradient like methods. The resulting algorithm is what we call a Schur type algorithm (or substructuring methods).

Let Ω be a bounded open set of $\mathbf{R}^2$. Let $\Omega_{i,1 \leq i \leq N}$ be a finite sequence of sets embedded in Ω such that $\bar{\Omega} = \cup_{i=1}^{N} \bar{\Omega}_i$. Let $\Gamma = \partial\Omega$, $\Gamma_i = \partial\Omega_i - \Gamma$. The outward normal from Ω_i is $\vec{n}_i$ and $\vec{\tau}_i$ is a tangential unit vector. Let us denote by $\mathcal{B}_{i,1 \leq i \leq N}$ the approximations to the exact ABC. Since the operators $\mathcal{B}_i$ are local, the subscript r or l is meaningless and will not be used here. We assume the operators $\mathcal{B}_{i,1 \leq i < N}$ to lead to well posed boundary value problems (see below BVP (2.10)). We assign to each subdomain i an operator S_i: Let f be a function from Ω_i to $\mathbf{R}$ and h a function from Γ_i to $\mathbf{R}$, $S_i(h, f, g)$ is the solution v of the following boundary value problem:

$$
\begin{aligned}
\mathcal{L}(v) &= f(x), & x \in \Omega_i \\
\mathcal{B}_i(v) &= h(x), & x \in \Gamma_i \\
\mathcal{C}(v) &= g(x), & x \in \partial\Omega_i \cap \Gamma
\end{aligned}
\tag{2.10}
$$

In order to take multiple overlaps into account, we introduce a sequence (η_i^j), $1 \leq i \leq N$, $1 \leq j \leq N, i \neq j$ of functions defined on the boundaries of the subdomains which satisfy:

$$
\begin{aligned}
&i) \quad \eta_i^j : \partial\Omega_i \longrightarrow [0,1] \\
&ii) \quad \eta_i^j = 0 \text{ on } \partial\Omega_i - \bar{\Omega}_j \\
&iii) \quad \textstyle\sum_{j,j \neq i} \eta_i^j(x) = 1, \quad x \in \partial\Omega_i
\end{aligned}
$$

Remark 2.5. η_i^j is zero if $\partial\Omega_i \cap \bar{\Omega}_j = \emptyset$.

It is now possible to write a substructuring formulation. Let u be the solution to (2.1) and $u_i = u_{|\Omega_i}$. We write a system for $\mathcal{B}_i(u_i)$:

$$
\begin{aligned}
\mathcal{B}_i(u_i) &= \sum_{j,j \neq i} \eta_i^j \mathcal{B}_i(u_i) = \sum_{j,j \neq i} \eta_i^j \mathcal{B}_i(u_j) \\
&= \sum_{j,j \neq i} \eta_i^j \mathcal{B}_i(S_j(\mathcal{B}_j(u_j), f_{|\Omega_j}, g)) \\
&= \sum_{j,j \neq i} \eta_i^j \mathcal{B}_i(S_j(0, f_{|\Omega_j}, g)) + \sum_{j,j \neq i} \eta_i^j \mathcal{B}_i(S_j(\mathcal{B}_j(u_j), 0, 0))
\end{aligned}
$$

Thus, $(\mathcal{B}_i(u_i))_{1 \leq i \leq N}$ solves the following linear system:

$$\mathcal{B}_i(u_i) - \sum_{j,j \neq i} \eta_i^j \mathcal{B}_i(S_j(\mathcal{B}_j(u_j), 0, 0)) = \sum_{j,j \neq i} \eta_i^j \mathcal{B}_i(S_j(0, f_{|\Omega_j}, g)), \ 1 \leq i \leq N \qquad (2.11)$$

Let $H = (H_i)_{1 \leq i \leq N}$ and $G = (G_i)_{1 \leq i \leq N}$ be the vectors

$$H = \begin{bmatrix} \mathcal{B}_1(u_1) \\ \vdots \\ \mathcal{B}_N(u_N) \end{bmatrix} \text{ and } G = \begin{bmatrix} \sum_{j,j \neq 1} \eta_1^j \mathcal{B}_1(S_j(0, f_{|\Omega_j}, g)) \\ \vdots \\ \sum_{j,j \neq N} \eta_N^j \mathcal{B}_N(S_j(0, f_{|\Omega_j}, g)) \end{bmatrix}$$

and $\mathcal{T}$ be the linear operator defined by

$$\mathcal{T}(H) = \begin{bmatrix} \sum_{j,j \neq 1} \eta_1^j \mathcal{B}_1(S_j(\mathcal{B}_j(u_j), 0, 0)) \\ \vdots \\ \sum_{j,j \neq N} \eta_N^j \mathcal{B}_N(S_j(\mathcal{B}_j(u_j), 0, 0)) \end{bmatrix}$$

System (2.11) may now be written in the following compact form:

$$(Id - \mathcal{T})(H) = G \qquad (2.12)$$

We consider three algorithms for the solution of (2.12), GMRES, BiCGSTAB and Jacobi:

$$H^{n+1} = \mathcal{T}(H^n) + G$$

The last algorithm corresponds to the additive Schwarz method. Since the operator $\mathcal{T}$ is no longer nilpotent, the Schwarz method should not converge in a finite number of steps. GMRES and BiCGSTAB (except if breakdown occurs) always converge in a finite number of steps (ignoring round-off errors) for a finite dimensional problem.

3. NUMERICAL RESULTS FOR THE CONVECTION-DIFFUSION EQUATION

We apply the strategy explained above to the convection-diffusion equation. Let

$$\mathcal{L} = \frac{1}{\Delta t} + a(x,y)\frac{\partial}{\partial x} + b(x,y)\frac{\partial}{\partial y} - \nu \Delta \qquad (3.1)$$

where $\vec{a} = (a, b)$ is the velocity field, ν is the viscosity. Δt is a constant which could correspond for instance to a time step for a backward-Euler scheme for the time dependent convection-diffusion equation.

For a subdomain Ω_i, the approximations to the exact ABC obtained using the method outlined in § 2.3 read as follows ($\vec{a}$ is the velocity field (a, b), $\vec{n}_i$ is the outward normal from Ω_i and $\vec{\tau}_i$ is a tangential unit vector on $\partial \Omega_i$):

$$\mathcal{B}_i^0 = \frac{\partial}{\partial \vec{n}_i} - \frac{\vec{a}.\vec{n}_i - \sqrt{(\vec{a}.\vec{n}_i)^2 + \frac{4\nu}{\Delta t}}}{2\nu} \qquad (3.2)$$

or

$$\mathcal{B}_i^2 = \frac{\partial}{\partial \vec{n}_i} - \frac{\vec{a}.\vec{n}_i - \sqrt{(\vec{a}.\vec{n}_i)^2 + \frac{4\nu}{\Delta t}}}{2\nu} + \frac{\vec{a}.\vec{\tau}_i}{\sqrt{(\vec{a}.\vec{n}_i)^2 + \frac{4\nu}{\Delta t}}} \frac{\partial}{\partial \vec{\tau}_i}$$

$$-\frac{\nu}{\sqrt{(\vec{a}.\vec{n_i})^2 + \frac{4\nu}{\Delta t}}}(1 + \frac{(\vec{a}.\vec{\tau_i})^2}{(\vec{a}.\vec{n_i})^2 + \frac{4\nu}{\Delta t}})\frac{\partial^2}{\partial \vec{\tau_i}^2}$$

where the superscript denotes the order of the approximation, for more details see Nataf and Rogier (1995), Nataf (1993). The boundary conditions $\mathcal{B}_i^0$ or ^{2} are far field boundary conditions (also called Outflow B.C., Absorbing B.C., Artificial B.C., Radiation B.C.,..., see Engquist and Majda, 1977; Halpern, 1986).

We use a two-dimensional test problem to illustrate the validity of the method. We solve the following problem:

$$\begin{cases} \mathcal{L}(u) = \frac{u}{\Delta t} + a(x,y)\frac{\partial u}{\partial x} + b(x,y)\frac{\partial u}{\partial y} - \nu \Delta u = 0, & 0 \le x \le 1, \ 0 \le y \le 1 \\ u(0,y) = 1, & 0 < y < 1 \\ \frac{\partial u}{\partial y}(x,1) = 0, & 0 < x < 1 \\ \frac{\partial u}{\partial x}(1,y) = 0, & 0 < y < 1 \\ u(x,0) = 0, & 0 < x < 1 \end{cases}$$

The operator $\mathcal{L}$ is discretized by a standard upwind finite difference scheme of order 1 (see Fletcher (1991)) and $\mathcal{B}_{i,1 \le i \le N}$ by a finite difference approximation. We used a rectangular finite difference grid. The mesh size is denoted by h. The unit square is decomposed into overlapping rectangles. The resulting discretization of system (2.12) is denoted by:

$$(Id - T_h)(H_h) = G_h \tag{3.3}$$

The test problem has been implemented at ONERA on an IPSC860.

Remark 3.6. Any other discretization could be used as well.

From the definition of T_h, we see that the computation of T_h applied to some vector H_h amounts to the solution of N independent boundary value subproblems (one subproblem in each subdomain) which can be solved in parallel. We have considered three algorithms in order to solve (3.3): GMRES(∞), Bi-CGSTAB and a Jacobi algorithm:

$$H_h^{n+1} = T_h(H_h^n) + G_h \tag{3.4}$$

which corresponds to an additive Schwarz method (ASM) whose convergence in the continuous case has been studied in Nataf and Rogier (1995) for outflow boundary conditions.

In tables 1 and 2, we give the number of subproblems solved so that the maximum of the error is smaller than 10^{-6}. One iteration of GMRES(∞) or of ASM counts for computing the solution for each subdomain once and one iteration of BiCGSTAB counts for computing the solution for each subdomain twice. In the tables, Id corresponds to the use of Id as interface condition (Dirichlet problems). The tests include the case $\mathcal{B}_i = Id$ since it corresponds to the classical Schwarz method when the Jacobi algorithm is used.

The results in Table 1 were obtained using the following parameters: 8×1 subdomains, 21×120 points in each subdomain, overlap $= 2h$, $\nu = 0.1$, $\Delta t = 10^{40}$, $a = y$, $b = 0$.

Table 1: Computational cost vs. interface conditions and solvers

Boundary Cond.	ASM	Bi-CGSTAB	GMRES
Id	844	88	61
$\mathcal{B}_0$	86	38	33
$\mathcal{B}_2$	46	28	24

The results in Table 2 were obtained using the following parameters:
4×4 subdomains, 35×35 points in each subdomain, overlap $= 2h$, $\nu = 0.1$, $\Delta t = 1$, $a = y$, $b = 0$.

Table 2: Computational cost vs. interface conditions and solvers

Boundary Cond.	ASM	Bi-CGSTAB	GMRES
Id	479	64	50
$\mathcal{B}_0$	27	22	19
$\mathcal{B}_2$	18	16	16

The use of outflow boundary conditions leads to a significant improvement whatever iterative solver is used. Bi-CGSTAB and GMRES give similar results with an advantage to GMRES in terms of computational cost and to BiCGSTAB in terms of storage requirements, since only two directions have to be stored.

REFERENCES

DESPRES, B. (1991), *Domain Decomposition Method and the Helmholtz Problem*, Mathematical and Numerical aspects of wave propagation phenomena (SIAM), p. 44-52.

BAMBERGER, A., ENGQUIST, B., HALPERN, L. and P. JOLY (1988a), Parabolic wave equation approximations in heterogenous media, *SIAM J. Appl. Math.* **48** No 1, p. 99-128.

BAMBERGER, A., ENGQUIST, B., HALPERN, L. and P. JOLY (1988b), Higher order paraxial wave approximations in heterogeneous media, *SIAM J. Appl. Math.* **48** No 1, p. 129-154.

ENGQUIST, B. and A. MAJDA (1977), Absorbing Boundary Conditions for the Numerical Simulation of Waves, *Math. Comp.* **31** (139), p. 629-651.

ENGQUIST, B. and A. MAJDA (1979), Radiation Boundary Conditions for Acoustic and Elastic Wave Calculations, *Comm. on Pure and Appl. Math.* vol. **XXXII**, p. 313-357.

FLETCHER, C.A.J. (1991), *Computational Techniques for Fluid Dynamics* (Springer Series in Computational Physics).

FLETCHER, R. (1976), *Conjugate gradient methods for indefinite systems* in: WATSON, G.A. (ed.), Numerical Analysis Dundee 1975, *Lecture Notes in Mathematics* **506**, p. 73–89, (Berlin, Heidelberg, New York: Springer-Verlag).

GIVOIS, E. (1992), *Etude et implémentation de deux méthodes de décomposition de domaines. Une approche monodimensionnelle pour l'initiation de la détonique à l'échelle moléculaire.* (Thèse de l'Université Paris Dauphine).

GOLUB, G.H. and C.F. VAN LOAN (1989), *Matrix Computations* (2nd edition, John Hopkins University Press).

HAGSTROM, T., TEWARSON, R.P. and A. JAZCILEVICH (1988), Numerical Experiments on a Domain Decomposition Algorithm for Nonlinear Elliptic Boundary Value Problems, *Appl. Math. Lett.* **1**, No 3, p. 299-302.

HALPERN, L. (1986), Artificial Boundary Conditions for the Advection-Diffusion Equations, *Math. Comp.* **174**, p. 425-438.

HALPERN, L. and J. RAUCH (1993), Absorbing Boundary Conditions for Diffusion Equations, *Prépublications Mathématiques de l'Université de Paris Nord* **93-05**.

LIONS, P.L. (1989), *On the Schwarz Alternating Method III: A Variant for Nonoverlapping Subdomains*, Third International Symposium on Domain Decomposition Methods for Partial Differential Equations, *SIAM*, p. 202-223.

LOHÉAC, J.P. (1991), An Artificial Boundary Condition for an Advection-Diffusion Equation, *Math. Meth. in the Appl. Sci.* **14**, p. 155-175.

LOHÉAC, J.P., NATAF, F. and M. SCHATZMAN (1993), Parabolic Approximations of the Convection-Diffusion Equation, *Math. of Comp.* **60** (202), p. 515-530.

NATAF, F. (1992), Méthodes de Schur généralisées pour l'équation d'advection-diffusion (Generalized Schur methods for the advection-diffusion equation), *C.R. Acad. Sci. Paris* t. **314**, Série I, p. 419-422.

NATAF, F. (1993), On the Use of Open Boundary Conditions in Block Gauss-Seidel Methods for the Convection-Diffusion Equations, *Rapport interne CMAP* **284**.

NATAF, F. and F. ROGIER (1995), Factorization of the Convection-Diffusion Operator and the Schwarz Algorithm, M^3AS (to appear).

NATAF, F., ROGIER, F. and E. DE STURLER (1994), Optimal interface conditions for domain decomposition methods, *Rapport interne* n° **301**, CMAP (Ecole Polytechnique).

SAAD, Y. and M. SCHULTZ (1986), GMRES: A generalized minimal residual algorithm for solving nonsymmetric linear systems, *SIAM J. Sci. Statist. Comput.* **7**, p. 856–869.

VAN DER VORST, H.A. (1992), A fast and smoothly converging variant of BI-CG for the solution of nonsymmetric linear systems, *SIAM J. Sci. Statist. Comput.* **13**, p. 631–644.

A REGULARIZING PROPERTY OF ROTHE'S METHOD TO THE NAVIER-STOKES EQUATIONS

Reimund Rautmann

Universität Paderborn, Fachbereich Mathematik-Informatik,
Warburger Str. 100, D-33098 Paderborn, Germany

INTRODUCTION

The first order semi-discrete scheme introduced by Rothe (1930) approximates evolution problems by a sequence of boundary value problems for some resolvent equation, which have to be solved step by step. In case of the Navier-Stokes evolution equation the resolvent equation of some suitable linearization has the regularization properties of second order elliptic systems (of course, in dependence on the step length). Therefore on a bounded domain $\Omega \subset \mathbb{R}^3$, the convergence of Rothe's scheme in $L^2(\Omega)$ and its boundedness in $H^2(\Omega)$ implies its convergence even in $H^{2,r}(\Omega)$ for any $r \in [2, \infty)$, as we will see below. Since we have established L^2-convergence and H^2-boundedness (or even H^2-convergence) in Rautmann (1993b, 1993c) under suitable assumptions, our result in this note is boundedness and convergence of Rothe's scheme in $H^{2,r}(\Omega)$ for all $r \in [2, \infty)$.

We will show elsewhere that our result also applies in the study of more refined schemes of the product formula type. Namely, convergence proofs for such schemes depend on $H^{1,\infty}$- bounds for the flow velocity which follow from $H^{2,r}$-bounds if $r > 3$ by Sobolev's imbedding theorem. Product formula approximations have firstly been proposed in Chorin (1973), Chorin, Hughes, Mc Cracken and Marsden (1978), Pironneau (1982), in order to overcome the destabilizing effect of the nonlinear transport term in the Navier-Stokes equation by the stabilizing influence of its elliptic part. Since then similar schemes have been studied by several authors, see the references in the recent publications Beale and Greengard (1992) and Rautmann and Masuda (1994), where explicit convergence rates for different splitting schemes of this type have been proved.

Having explained the notations in Section 1 we formulate our result in Theorem 1.2. For a linearized Rothe scheme introduced in Section 2, in Section 3 we establish L^q-bounds for its right hand side by means of a former result on H^2-boundedness (which we have recalled in Theorem 1.1 of Section 1). From this in Section 4 we find the existence of Rothe approximations in $H^{2,q}$. Then due to Solonnikov, Miyakawa, Giga and v.Wahl's Stokes resolvent estimates a recent result of Ashyralyev and Sobolevskii applies which, by a bootstrap argument, leads to uniform norm-and Hölder semi-norm bounds in $H^{s,r}(\Omega)$ for all $r \in [2, \infty)$ and some $s \in (2, 2 + \frac{1}{r})$. From this in Section 5 $H^{2,r}$- convergence (and even convergence in a little stronger norms) results from a compactness and uniqueness argument. In the Appendix (Section 6) we list basic estimates in a form which we will use frequently.

1. NOTATIONS AND MAIN RESULT

Let Ω denote a bounded open and pathwise connected set in the three-dimensional Euclidean space $\mathbb{R}^3$. In order to avoid technical difficulties we assume that the boundary $\partial\Omega$ is smooth of class C^∞. The velocity $u(t, x) = (u_1, u_2, u_3)$ and the kinematic pressure $p(t, x) \geq 0$ of a viscous incompressible flow in Ω at time $t \geq 0$ obey the Navier-Stokes equations

$$\frac{\partial}{\partial t}u - \Delta u + u \cdot \nabla u + \nabla p = f, \qquad \nabla \cdot u = 0 \qquad \text{in } \Omega \text{ for } t \geq 0, \qquad (1.1)$$

$$u|_{\partial\Omega} = 0 \quad \text{on} \quad \partial\Omega \quad \text{for} \quad t \geq 0,$$
$$u|_{t=0} = u_0.$$

Here we set the viscosity constant and the mass density equal 1. The function f denotes the prescribed density of the outer forces.

We will consider (1.1) in the framework of the Lebesgue spaces $L^q = L^q(\Omega)$, $q \geq 2$. By the well known results of H. Weyl (1940) and Fujiwara and Morimoto (1977) the space

$$L^q = X^q \oplus G^q$$

is a direct sum of the spaces

$$X^q = \text{ closure of } C^\infty_{c,0} \text{ in } L^q$$

and

$$G^q = \text{ closure of } \{\nabla\varphi | \varphi \in C^1(\bar{\Omega})\} \text{ in } L^q,$$

where $C^\infty_{c,o}$ denotes the linear space of C^∞ - vectorfunctions on Ω which are divergence - free and have compact support in Ω. Let

$$P_q : L^q \to X^q$$

denote the projection along the space G^q of generalized gradients.

Since Ω is bounded, for the Sobolev spaces $H^{m,q}(\Omega)$ with norms

$$\|g\|_{H^{m,q}} = \Big\{ \sum_{|\alpha| \leq m} \int_\Omega |\partial^\alpha g(x)|^q dx \Big\}^{\frac{1}{q}},$$

where $\alpha = (\alpha_1, \alpha_2, \alpha_3), \alpha_j = 0, ..., m, |\alpha| = \sum_{j=1}^3 \alpha_j, m \in \mathbb{N}, 1 \leq q < \infty$ the imbedding $H^{m,r} \hookrightarrow H^{m,q}$ holds if $q \leq r$. We will write $H^{m,2} = H^2, H^{0,q} = L^q$. Therefore the construction of

$$P_r u = u - \nabla\varphi$$

in Fujiwara and Morimoto (1977), p.694, shows that we have

$$P_r = P_{q|L^r} = P_{2|L^r}$$

if $2 \leq q \leq r$, $P_{2|L^r}$ denoting the restriction of P_2 to L^r. In the following we will restrict us to spaces L^q with $q \geq 2$. Therefore we will write

$$P = P_2.$$

The projection P is bounded on each Sobolev space $H^{m,q}, m \in \mathbb{N}, q \in [2, \infty)$, (von Wahl, 1985, p.XXIII). - In this place I would like to thank Professor Tom Beale for a useful discussion concerning the projections P_r.

As usual for $s \in (0, m)$ we denote by $H^{s,q}$ the complex interpolation space between $H^{m,q}$ and L^q. $H^{s,q}$ is independent of $m > s$, for details see Triebel (1978). In addition $\overset{\circ}{H}{}^{s,q}$ stands for the closure in $H^{s,q}$ of the subspace C_c^∞ of C^∞ -vector functions which have compact support in Ω.

A formal application of the projection P on (1.1) leads us to the Navier-Stokes evolution equation

$$\begin{aligned}
\partial_t u + A_q u + P u \cdot \nabla u &= Pf, & t > 0, \\
u(0) &= u_0
\end{aligned} \tag{1.2}$$

for a strong solution $u(t) \in D_{A_q}$, where the Stokes operator A_q is the closure of $-P\Delta$ in the space X^q. Its domain is

$$D_{A_q} = H^{2,q} \cap \overset{\circ}{H}{}^{1,q} \cap X^q, \tag{1.3}$$

(Cattabriga, 1961; Solonnikov, 1964). The operator A_q fulfills the apriori estimate

$$||u||_{H^{2,q}} \leq c||A_q u||_{L^q} \tag{1.4}$$

and the resolvent estimate

$$||(A_q + \lambda)^{-1}||_{L^q} \leq \frac{c_{\varepsilon,q}}{|\lambda|} \tag{1.5}$$

with some constant $c_{\varepsilon,q} > 0$ for all complex numbers

$$\lambda \in \Sigma_\varepsilon = \{z \in \mathbb{C} | |z| > 0, |arg z| < \pi - \varepsilon\},$$

where $\varepsilon > 0$. Because of (1.4), A_q is a closed operator, and due to (1.5), $-A_q$ generates the bounded holomorphic Stokes semigroup $e^{-tA_q}, t \geq 0$, acting on X^q, (Solonnikov, 1973; Miyakawa, 1981; Giga, 1981a; Giga, 1981b; Giga, 1985; von Wahl, 1985). Moreover on the right halfplane $Re\lambda \geq 0$ (and even on a neighborhood of $\lambda = 0$) the estimate

$$||(A_q + \lambda)^{-1}||_{L^q} \leq \frac{c_q}{1 + |\lambda|} \tag{1.6}$$

holds with some constant $c_q > 0$, (von Wahl, 1985, p.79). Thus the fractional powers A_q^α with dense domain $D_{A_q^\alpha} \subset X^q$ are defined for all real $\alpha > 0$. In the case $\alpha < \beta$ the domain $D_{A_q^\beta}$ is dense in $D_{A_q^\alpha}$, the imbedding $D_{A_q^\beta} \hookrightarrow D_{A_q^\alpha}$ being compact (Friedman, 1969, p.158; Pazy, 1983, p.69-74).

Let J be a real interval, X a Banach space. Then by $C^0(J, X)$ we denote the set of all uniformly bounded and continuous functions f defined on J with values in X. By $c, c_o, ...,$ we will denote positive constants which may have different values in different places below. In the following we will always assume $Pf = 0$. This implies that the density of the outer forces in (1.1) is a gradient field. We use this unessential restriction only in order to simplify the notations below.

For sufficiently small $T > 0$, the existence of a unique solution $u \in C^o([0,T], D_{A_q})$ of (1.2) for given $u_0 \in D_{A_q}$ is well known. If the initial value u_o has some more regularity, i.e. if we assume

$$u_0 \in D_{A_q^{1+\varsigma}}$$

for some $\varsigma \in (0, 1/2q)$, then the unique local solution u of (1.2) fulfills $u \in C^o([0,T], D_{A_q^{1+\varsigma}})$ for some $T > 0$, (von Wahl, 1985, p.127-128; Rautmann, 1983c), without any non-realistic compatibility condition concerning u_o. In the following we will assume that a solution $u \in C^o([0,T], D_{A_q^{1+\varsigma}})$ of (1.2) is given for some $q \in [2, \infty)$, $\varsigma \in (0, 1/2q)$.

Starting from some given initial value $v_0^h = u_o \in D_{A_q^{1+\varsigma}}, \varsigma \in (0, 1/2q)$, we can try to approximate a solution $u \in C^o([0,T], D_{A_q})$ of (1.2) by means of Rothe's scheme

$$\frac{v_k^h - v_{k-1}^h}{h} + A_q v_k^h = -P v_{k-1}^h \cdot \nabla v_k^h, \tag{1.7}$$

$k = 1, ..., K$, where v_k^h stands for some approximate value of $u(t)$ in the grid point $t = t_k$ of a suitable time grid

$$t_k = k \cdot h, \ h = \frac{T}{K}, \ k = 0, .., K, \ K = 1, 2, ...$$

We will prove that the scheme (1.7) converges strongly and uniformly in $H^{2,q}$ for each $q \in [2, \infty)$, if the approximations are uniformly bounded in H^2 and uniformly strongly convergent in $L^2(\Omega)$. In Rautmann (1993b, 1993c), we have proved

Theorem 1.1. *Let $u \in C^o([0,T], D_{A_2})$ denote a solution of the Navier-Stokes initial value problem (1.2) with right hand side $Pf = 0$, v_k^h the approximations from (1.7). Assume the initial values $v_o^h \in D_{A_2}$ are uniformly bounded in H^2 and satisfy*

$$||v_0^h - u_0|| \leq ch \tag{1.8}$$

with a constant $c \geq 0$. Then we have $v_k^h \in D_{A_2}$, and for any $h \in (0, h_0]$, h_0 being sufficiently small, the error estimates

$$||v_k^h - u(t_k)|| \leq b_o h, \tag{1.9}$$

$$||A_2^\beta (v_k^h - v(t_k))|| \leq b_\beta h^{1-\beta} \tag{1.10}$$

for $\beta \in [0,1]$ hold uniformly in $k = 1, .., K \leq \frac{T}{h_0}$. The constants b_0, b_β depend on $\sup_{t \in [0,T]} ||A_2 v(t)||, c, T$, the constant b_β additionally on β, too.

For Navier-Stokes solutions $u \in C([0,T], D_{A_2^{1+\varsigma}}), \varsigma \in (0, \frac{1}{4})$, even convergence rates in H^2 have been established in Rautmann (1993b), Theorem II.

Thus the main result of this note will be

Theorem 1.2. *Let $u \in C^o([0,T], D_{A_q^{1+\varsigma}})$ for some $q \in [2, \infty)$ and $\varsigma \in (0, 1/2q)$ denote a solution of (1.2) with initial value $u_0 \in D_{A_q^{1+\varsigma}}$. Then the approximations v_k^h in Rothe's scheme (1.7) with initial value $v_0^h = u_0$ converge in $H^{2,q}$ (and even in $D_{A_q^{1+\eta}}$ for all $\eta \in [0, \varsigma)$) to $u(k \cdot h)$ with $K \to \infty$ uniformly in $k = 1, ..., K$ and $h = T/K$.*

2. A LINEARIZED ROTHE SCHEME IN $H^{2,q}$, $q > 2$.

For some $q \in (2, \infty)$ and $\zeta \in (0, \frac{1}{2q})$ let $u \in C^0([0,T], D_{A_q^{1+\zeta}})$ denote a solution of (1.2) with initial value $u_0 \in D_{A_q^{1+\zeta}}$. In order to prove Theorem 1.2 we note that with $v_0^h = u_0$ the assumptions of Theorem 1.1 hold because of the imbeddings $L^q \hookrightarrow L^2$ and

$$D_{A_q^{1+\zeta}} \hookrightarrow D_{A_q} = H^{2,q} \cap \overset{\circ}{H}{}^{1,q} \cap X^q \hookrightarrow H^2 \cap \overset{\circ}{H}{}^1 \cap X^2 = D_{A_2}.$$

Let (v_k^h), $h = \frac{T}{K}$, $k = 0, .., K$, $K = 1, 2, ...$, be the sequence of Rothe approximations $v_k^h \in D_{A_2}$, $||v_k^h||_{H^2} \leq M_2$, from (1.7) which exist by Theorem 1.1 and which are of first order convergent to $u(kh)$ in L^2, uniformly with respect to k and h. The constant M_2 is the sum of the constant b_1 in Theorem 1.1 and the bound $\sup_{t \in [0,T]} ||A_2 u(t)||_{L^2} < \infty$.

Starting with (v_k^h) we will find a sequence (u_k^h) of approximations to u in $H^{2,q}$ from the linearized scheme

$$\frac{u_k^h - u_{k-1}^h}{h} + A_q u_k^h = -P v_{k-1}^h \cdot \nabla v_k^h = f_k, \quad k = 1, ..., K, \quad u_0^h = v_0^h. \tag{2.1}$$

Namely in the next section after having established suitable bounds for f_k in X^q inductively we will see the existence of $u_k^h \in D_{A^q} \subset D_{A^2}$ and conclude $u_k^h = v_k^h$ from the uniqueness of the linear Stokes resolvent boundary value problem in D_{A_2}. Then in Section 4 using Ashyralyev and Sobolevkskii's coercivity inequality (Ashyralyev and Sobolevskii, 1994, p. 93, (2.11)) we will find uniform $H^{2,q}$ bounds for u_k^h.

3. BOUNDS IN L^q FOR THE CONVECTIVE TERM

First we mention the useful characterization of the domains of fractional powers of A_q by $H^{s,q}$-spaces:

Proposition 3.1. *Assume* $0 \leq s = 2\zeta < \frac{1}{q}$, $1 < q < \infty$. *Then* $X^q \cap \overset{\circ}{H}{}^{s,q} = X^q \cap H^{s,q} = D_{A_q^\zeta}$.

For the proof see von Wahl (1985), p. 92, 96, Giga (1981b).

Remark 3.1. In order to prove that $f_k \in D_{A_q^\zeta}$ holds for values of ζ as above and to estimate $||A_q^\zeta f_k|||_{L^q}$, by Proposition 3.1 it will be enough to find bounds for $||f_k||_{H^{s,q}}$.

Proposition 3.2. *The Projection*

$$P : H^{s,q} \to H^{s,q} \cap X^q$$

is bounded for all $q \in [2, \infty)$.

The proof is immediate, because $P : H^{m,q} \to H^{m,q} \cap X^q$ is bounded for any $m \in \mathbb{N}$, see von Wahl (1985), p. XXIII, and $H^{s,q}$ is interpolation space between $H^{0,q}$ and $H^{m,q}$, $m > s$, if $0 < s \notin \mathbb{N}$.

Proposition 3.3. *Assume* $s \in [0, 1)$. *Then in case* $3 < q$ *the inequalities*

$$a) \qquad ||f \cdot g||_{H^{s,q}} \leq c||f||_{H^{1,q}} \cdot ||g||_{H^{s,q}}$$

and

$$b) \qquad ||f \cdot \nabla g||_{H^{s,q}} \leq c||f||_{H^{1,q}} \cdot ||g||_{H^{1+s,q}}$$

hold for all $f \in H^{1,q}, g \in H^{s,q}$ *in* (a), $g \in H^{1+s,q}$ *in* (b).

A more general inequality has been stated in Kato (1990), p. 60.

Proof: Because of the imbedding $H^{1,q} \hookrightarrow L^{\infty}$, the first statement
a) follows from the two evident estimates

$$||f \cdot g||_{L^q} \le ||f||_{L^{\infty}}||g||_{L^q} \le c_o||f||_{H^{1,q}} \cdot ||g||_{L^q} \quad \text{if } g \in L^q,$$

and

$$\begin{aligned} ||\nabla(f \cdot g)||_{L^q} &\le ||(\nabla f) \cdot g||_{L^q} + ||f \cdot (\nabla g)|_{L^q} \\ &\le ||\nabla f||_{L^q}||g||_{L^{\infty}} + ||f||_{L^{\infty}}||\nabla g||_{L^q} \\ &\le c_1||f||_{H^{1,q}} \cdot ||g||_{H^{1,q}} \quad \text{if } g \in H^{1,q}. \end{aligned}$$

Both series of estimates together show that for any given $f \in H^{1,q}$ the map $\phi : g \to f \cdot g$ is a bounded linear one of L^q into L^q and of $H^{1,q}$ into $H^{1,q}$. Therefore $\phi : H^{s,q} \to H^{s,q}$ is bounded, too, $H^{s,q}$ being the interpolation space between L^q and $H^{1,q}$. By the interpolation inequality (Triebel, 1978, pp. 59,185,317) which holds for $||\phi||_{H^{s,q}}$ we get (a).

In the special case $g = \nabla G$ by definition of the norm $|| \cdot ||_{H^{1+s,q}}$ we see $||\nabla G||_{H^{s,q}} \le ||G||_{H^{1+s,q}}$, thus (b) results from (a).

Proposition 3.4. *a) The imbedding* $H^2 \hookrightarrow H^{1+s,q}$ *and the inequality*

$$||g||_{H^{1+s,q}} \le c_o||g||_{H^2}^{\lambda}||g||_{L^2}^{1-\lambda} \le c_1||g||_{H^2}$$

hold for all $q \in [2,4], s \in [0, \frac{1}{q}), g \in H^2$, *with* $\lambda = \frac{3}{2}\{\frac{1}{2} - \frac{1}{q} + \frac{1+s}{3}\}$ *and*
b) the imbedding $H^{2,q} \hookrightarrow H^{1+s,r}$ *and the estimate*

$$||g||_{H^{1+s,r}} \le c_o||g||_{H^{2,q}}^{\mu}||g||_{L^q}^{1-\mu} \le c_1||g||_{H^{2,q}}$$

hold for all $q \ge 3, r \in [q, \infty), s \in [0, \frac{1}{r}), g \in H^{2,q}$, *and*

$$\mu = \frac{3}{2}\{\frac{1}{q} - \frac{1}{r} + \frac{1+s}{3}\}.$$

Proof: Proposition 3.4 is a special case of Lemma 0.2.1 in von Wahl (1985), p. XVIII, p. XX]. For details see Triebel (1978), pp. 181-186, 317, 327 - 328.

In the following for suitably given $v_k, k = 0, ..., K$ we denote by $f_k = -Pv_{k-1} \cdot \nabla v_k$ the right hand side in (2.1), and set $D_{A_q^\varrho} = X^q$.

Proposition 3.5. *a) We consider numbers* q, s, ζ *with* $3 < q \le 4, 0 \le 2\zeta = s < \frac{1}{q}$.
Assume the vector functions $v_k \in H^2$ *fulfill*
(a1) $||v_k||_{H^2} \le M_2, k = 0, .., K$, *and*
(a2) $||v_k - v_{k-1}||_{L^2} \le M_1 h, k = 1, .., K$, *with constants* $h > 0, M_j > 0$.
Then for all $q \in (3,4]$ *and* $\zeta \in [0, 1/2q]$ *we have*
(a3) $f_k \in D_{A_q^\zeta}, ||A_q^\zeta f_k||_{L^q} \le N_q$, *for all* $k = 1, ..., K$, *and*
(a4) $||A_q^\zeta(f_{k+m} - f_k)||_{L^q}(\frac{k}{m})^\alpha \le N_\alpha$,
for all $k, m \in \text{IN}, 1 \le k < k + m \le K$, *with some suitable constants* $\alpha \in (0,1]$ *and* $N_q, N_\alpha > 0$, *which are independent of* $m, k \le K$, K *and* $h = \frac{T}{K}$.

b) Assume for some $q > 3$ *the vector functions* $v_k \in H^{2,q}$ *fulfill*
(b1) $||v_k||_{H^{2,q}} \le M_q$, $k = 0, ..., K$, *and*
(b2) $||v_k - v_{k-1}||_{L^q} \le M_1 h$, $k = 1, ..., K$, *with constants* $h > 0, M_j > 0$.

Then for all $r \in [q, \infty)$ and all $\zeta \in [0, 1/2r)$ we have
(b3) $f_k \in D_{A_r^\zeta}, \|A_r^\zeta f_k\|_{L^r} \leq N_r$ for all $k = 1, .., K$, and
(b4) $\|A_r^\zeta(f_{k+m} - f_k)\|_{L^r}(\frac{k}{m})^\beta \leq N_\beta$
for all $k, m \in \mathbb{N}, 1 \leq k < k+m \leq K$, with some suitable constants $\beta \in (0, 1], N_r, N_\beta > 0$.

Proof: Because of our assumption on v_k and q, s, ζ in (a), Propositions 3.1 - 3.4 show that (a_3) holds.
In order to prove (a_4), first from (a_1) we see

$$\|v_{k+m} - v_k\|_{H^2} \leq 2M_2,$$

and estimating

$$v_{k+m} - v_k = \sum_{j=1}^{m}(v_{k+j} - v_{k+j-1})$$

by (a_2) gives

$$\|v_{k+m} - v_k\|_{L^2} \leq mM_1 h,$$

thus

$$\|v_{k+m} - v_k\|_{H^{1,q}} \leq c_0\|v_{k+m} - v_k\|_{H^{1+s,q}} \leq N_q'(mh)^\alpha \qquad (3.1)$$

follows from the imbedding $H^{1+s,q} \hookrightarrow H^{1,q}$ and Proposition 3.4 (a). The constants N_q' and

$$\alpha = 1 - \frac{3}{2}\{\frac{1}{2} - \frac{1}{q} + \frac{1+s}{3}\} \in (0, 1] \qquad (3.2)$$

are independent of $m, k \leq K, K$ and $h = \frac{T}{K}$.
Writing for the difference

$$A_q^\zeta(f_{k+m} - f_k) = A_q^\zeta P(v_{k+m-1} - v_{k-1}) \cdot \nabla v_k + A_q^\zeta P v_{k-1} \cdot \nabla(v_{k+m} - v_k),$$

from Propositions 3.1 - 3.3 we find

$$d_q = \|A_q^\zeta(f_{k+m} - f_k)\|_{L^q}(\frac{k}{m})^\alpha \leq c\|v_{k+m-1} - v_{k-1}\|_{H^{1,q}}(\frac{k}{m})^\alpha \cdot \|v_{k+m}\|_{H^{1+s,q}} +$$

$$+ c\|v_{k-1}\|_{H^{1,q}}\|v_{k+m} - v_k\|_{H^{1+s,q}} \cdot (\frac{k}{m})^\alpha.$$

From this and (3.1) by multiplication with $(kh)^\alpha \leq T^\alpha$ using Proposition 3.4 (a) we get

$$d_q \leq c_1 N_q' T^\alpha,$$

which gives (a_4) with $N_\alpha = c_1 N_q' T^\alpha$.
Similarly because of our assumptions on v_k and q in (b), Propositions 3.1 - 3.4 show that (b_3) holds. Interpolating as in the proof of (a_4) above from the imbedding $H^{1+s,r} \hookrightarrow H^{1,r}$ and Proposition 3.4 (b) we get

$$\|v_{k+m} - v_k\|_{H^{1,r}} \leq c_0\|v_{k,m} - v_k\|_{H^{1+s,r}} \leq N_r'(mh)^\beta, \qquad (3.3)$$

where the constants N_r' and

$$\beta = 1 - \frac{3}{2}\{\frac{1}{q} - \frac{1}{r} + \frac{1+s}{3}\} \in (0, 1] \qquad (3.4)$$

are independent of $m, k \leq K, K$ and $h = \frac{T}{K}$.

From Propositions 3.1 - 3.3 we find

$$d_r = ||A_r^\zeta(f_{k+m} - f_k)||_{L^r} \cdot (\frac{k}{m})^\beta \leq$$

$$\leq c||v_{k+m-1} - v_{k-1}||_{H^{1,r}} \cdot (\frac{k}{m})^\beta ||v_{k+m}||_{H^{1+s,r}} + c||v_{k-1}||_{H^{1,q}}||v_{k+m} - v_k||_{H^{1+s,r}} \cdot (\frac{k}{m})^\beta.$$

From this and (3.3) by multiplication with $(kh)^\beta \leq T^\beta$, using Proposition 3.4 (b), we get

$$d_r \leq c_1 N_r' T^\beta$$

which gives (b_4) with $N_\beta = c_1 N_r' T^\beta$.

4. EXISTENCE OF ROTHE APPROXIMATIONS IN $H^{2,r}$. APPLICATION OF ASHYRALYEV AND SOBOLEVSKII'S COERCIVITY INEQUALITY.

We denote by $v_0^h = u_0 \in D_{A_q^{1+\varsigma}} \hookrightarrow D_{A_2}$ for arbitrary $q \in (3,4], \zeta \in [0, \frac{1}{2q})$ the prescribed initial value in (1.2) and (1.7), and assume that the solution $u \in C^0([0,T], D_{A_q}) \hookrightarrow C^0([0,T], D_{A_2})$ of (1.2) exists. In virtue of Theorem 1.1, inequality (a_1) in Proposition 3.5 holds (with $M_2 = b_1 + \sup_{t\in[0,T]} ||A_2 u(t)||$).

Next we verify inequality (a_2) : P being an orthogonal projection of L^2, Hölder's inequality (6.3), Sobolev's inequality (6.4), the multiplicative inequality (6.6) and the momentum inequality (6.7) give

$$\begin{aligned}
||Pv_{k-1}^h \cdot \nabla v_k^h||_{L^2} &\leq ||v_{k-1}^h||_{L^6}||\nabla v_k^h||_{L^3} \\
&\leq c_0||\nabla v_{k-1}^h||_{L^2}||\nabla v_k^h||_{L^2}^{1/2}||A_2 v_k^h|||_{L^2}^{1/2} \\
&\leq c_1 M_0^{3/4} M_2^{5/4},
\end{aligned} \tag{4.1}$$

where M_0 denotes a uniform bound for $||v_k^h||_{L^2} (e.g. M_0 = ||u_0||_{L^2} + b_0 h_0$ by Theorem 1.1). Thus from equation (1.7) we get (a_2) with $N_2 = c_2 M_2 \cdot \{1 + M_0^{3/4} M_2^{1/4}\}$.

From Proposition 3.5 (a_3) firstly with $\zeta = 0$ we find $f_k \in X^q, ||f_k||_{L^q} \leq N_q$ for all $q \in (3,4], k = 1,..,K$.

Writing the linearized scheme (2.1) in the equivalent form

$$(1 + hA_q)u_k^h = u_{k-1}^h - hf_k, \tag{4.2}$$

$k = 1,...,K$, from the L^q - theory of the Stokes operator (Giga, 1981b; Miyakawa, 1981; Solonnikov, 1973) we see by induction on k that the solutions $u_k^h \in D_{A_q} \subset D_{A_2}$ exist, $k = 1,..,K$, since the real numbers $\frac{1}{h} > 0$ belong to the resolvent set of $-A_q$. Recalling (1.7) and the uniqueness of the linear Stokes resolvent boundary value problem in D_{A_2} we conclude

$$u_k^h = v_k^h, \quad k = 1,...,K, \quad h = \frac{T}{K}. \tag{4.3}$$

In addition from Proposition 3.5 (a 3) we conclude $f_k \in D_{A_q^\varsigma}$ and $||A_q^\varsigma f_k||_{L^q} \leq N_q$ for the values $q \in (3,4]$ and $\zeta \in [0, 1/2q)$ given above. By induction with respect to k we find that we can apply A_q^ς on both sides of the equation

$$u_k^h = (1 + hA_q)^{-1}\{u_{k-1}^h - hf_k\}$$

which is equivalent to (4.2). Therefore we have $u_k^h \in D_{A_q^{1+\varsigma}}$, $k = 0, .., K$, and again $A_q^\varsigma u_k^h = A_q^\varsigma v_k^h$ follows by the same reasoning which led us to (4.3). As a consequence we can apply the fractional power A_q^ς on both sides of (2.1) getting the scheme

$$\frac{A_q^\varsigma u_k^h - A_q^\varsigma u_{k-1}^h}{h} + A_q A_q^\varsigma u_k^h = A_q^\varsigma f_k, \quad k = 1, ..., K, \ A_q^\varsigma u_0^h = A_q^\varsigma v_0^h. \tag{4.4}$$

Noting that (4.4) with its given right hand sides $A_q^\varsigma f_k \in X^q$ is a linear scheme, we can apply Ashyralyev and Sobolevskii's coercivity inequality (Ashyralyev and Sobolevskii, 1994).

In order to formulate it, to any Banach space X with norm $|| \cdot ||_X$ and point grid

$$J_h = \{t_k | t_k = k \cdot h, k = 1, ..., K\} \subset [0, T], \qquad h = \frac{T}{K},$$

we consider the Banach space X_h of X-valued functions

$$\varphi^h : J_h \to X, \qquad \varphi^h(t_k) = \varphi_k^h$$

equipped with the norm

$$||\varphi^h||_{C(X_h)} = \max_{1 \le k \le K} ||\varphi_k^h||_X. \tag{4.5}$$

In addition for any $\alpha \in (0, 1]$ we introduce on X_h the weighted Hölder norm

$$||\varphi^h||_{C_0^\alpha(X_h)} = ||\varphi^h||_{C(X_h)} + \max_{1 \le k < k+m \le K}\{||\varphi_{k+m}^h - \varphi_k^h||_X \cdot (\frac{k}{m})^\alpha\}. \tag{4.6}$$

We will define the function $D_h \varphi^h \in X_h$ by its components

$$(D_h \varphi^h)_k = \frac{\varphi_k^h - \varphi_{k-1}^h}{h}, \ k = 1, ... K,$$

if in addition $\varphi_0^h \in X$ is prescribed.

Theorem 4.1. (Ashyralyev and Sobolevskii) *Assume $-A$ generates a holomorphic semigroup working on the Banach space X with Norm $||\cdot||_X$ and the $w_k^h \in D_A \subset X$ are calculated by means of the linear scheme*

$$\frac{w_k^h - w_{k-1}^h}{h} + A w_k^h = f_k, \tag{4.7}$$

$k = 1, .., K$, *where $w_0^h \in D_A$ and the f_k are given in X, $h = \frac{T}{K}$.*
Then the coercivity inequality

$$||D_h w^h||_{C_0^\alpha(X_h)} + ||A w^h||_{C_0^\alpha(X_h)} \le c_0 ||A w_o^h||_X + c_1 ||f^h||_{C_0^\alpha(X_h)} \tag{4.8}$$

holds, where the constants $c_0, c_1 > 0$ are independent of w, k, h and K.

For the proof see Ashyralyev-Sobolevskii (1994), p. 93-96, 76-79. As an immediate consequence of Theorem 4.1 and the bounds $(a_3), (a_4)$ in Proposition 3.5 for f_k we find

Corollary 4.1. *The functions $u_k^h = v_k^h$ in (4.4) with $\zeta = 0$ fulfill the estimates* (b_1) *and* (b_2) *in Proposition 3.5 uniformly in $k = 1, .., K, h = \frac{T}{K}$ and K for arbitrarily prescribed $q \in (3, 4]$.*

Therefore beginning with any initial value $u_0^h = v_0^h \in D_{A_q}$ for arbitrary $q \in (3, 4]$ we get the uniform norm bound M_q in (b_1) of $u_k^h = v_k^h$ in $D_{A_q} \hookrightarrow H^{2,q}$ and the bound (b_2) in L^q for the difference $v_k - v_{k-1}$. Reading Proposition 3.5 (b_3) firstly with $\zeta = 0$ now we find

$$f_k \in X^r, \quad \|f_k\|_{L^r} \le N_r \tag{4.9}$$

for all $r \ge q$.

In order to get Rothe approximations $u_k^h \in H^{2,r}$ in (2.1) for $r > 4$ we start with some initial value $v_0^h = u_0 \in D_{A_r^{1+\zeta}}$ for arbitrary $r \in (4, \infty)$, $\zeta \in [0, 1/2r)$ and calculate u_k^h by means of

$$\frac{u_k^h - u_{k-1}^h}{h} + A_r u_k^h = f_k, \qquad k = 1, ..., K. \tag{4.10}$$

Repeating the conclusions above which led us to $A_q^\zeta v_k^h = A_q^\zeta u_k^h \in D_{A_q}$, from (4.10) with $f_k \in X^r$ now using Proposition 3.5 b) we find

$$A_r^\zeta v_k^h = A_r^\zeta u_k^h \in D_{A_r}. \tag{4.11}$$

Consequently we can apply the fractional power A_r^ζ on both sides of (4.10) getting the scheme

$$\frac{A_r^\zeta u_k^h - A_r^\zeta u_{k-1}^h}{h} + A_r A_r^\zeta u_k^h = A_r^\zeta f_k, \qquad k = 1, ..., K,$$

$$A_r^\zeta u_0^h = A_r^\zeta v_0^h, \tag{4.12}$$

which is linear in $A_r^\zeta u_k^h$ for any given $A_r^\zeta f_k \in X^r$. An immediate consequence of Theorem 4.1 and the bounds $(b_3), (b_4)$ in Proposition 3.5 for $A_r^\zeta f_k$ is

Corollary 4.2. *For all exponents $r \in (4, \infty)$ and all $\zeta \in [0, 1/2r)$ the solutions $A_r^\zeta u_k^h = A_r^\zeta v_k^h$ in (4.12) fulfill the estimates*

$$\|A_r A_r^\zeta u_k^h\|_{L^r} \le M_{r,\zeta} \tag{4.13}$$

and

$$\|A_r^\zeta u_k^h - A_r^\zeta u_{k-1}^h\|_{L^r} \le N_{r,\zeta} h. \tag{4.14}$$

The constants $M_{r,\zeta}$ and $N_{r,\zeta}$ are independent of u_k^h and $k, K, h = \frac{T}{K}$, where $1 \le k \le K$.

Remark 4.1. Because of $D_{A_r} \subset D_{A_q}$ for $q \le r$, after having fixed some arbitrary initial value $u_0^h = v_0^h \in D_{A_r^{1+\zeta}}$, in virtue of Theorem 1.1 we can always fulfill the assumptions $(a_1), (a_2)$ of Proposition 3.5. Thus Corollary 4.1 and 4.2 together show that for arbitrary $r \in (2, \infty)$ beginning with any initial value $u_0^h = v_0^h \in D_{A_r^{1+\zeta}}$, from (2.1) and (4.4), (4.12) we get Rothe approximations $u_k^h = v_k^h$ which are uniformly bounded in $D_{A_r^{1+\zeta}} \hookrightarrow H^{2,r}$ and which have uniformly bounded difference quotients in $D_{A_r^\zeta} \hookrightarrow L^r$.

5. CONVERGENCE OF THE SEQUENCE OF ROTHE APPROXIMATIONS IN $H^{2,r}$.

By means of the approximations u_k^h calculated from (4.10) or (4.12) we define the sequence of Rothe functions

$$\tilde{u}^h(t) = u_{k-1}^h + \frac{t - t_{k-1}}{h}\{u_k^h - u_{k-1}^h\} \tag{5.1}$$

for all $t \in [t_{k-1}, t_k)$. Evidently each function $\tilde{u}^h$ on $[t_{k-1}, t_k]$ is linearly interpolating its values $u_{k-1}^h, u_k^h \in D_{A_r^{1+\varsigma}}$ and is therefore strongly $D_{A_r^{1+\varsigma}}$-continuous on [0,T]. In virtue of (4.14) a straightforward calculation shows that $\tilde{u}^h$ fulfills the Lipschitz condition

$$||A_r^\varsigma(\tilde{u}^h(t_2) - \tilde{u}^h(t_1))||_{L^r} \leq N_{r,\varsigma}|t_2 - t_1| \tag{5.2}$$

uniformly in $K, h = \frac{T}{K}$ and $t_1, t_2 \in [0,T]$. Interpolating (5.2) and the immediate consequence

$$||A_r A_r^\varsigma \tilde{u}(t)|| \leq M_{r,\varsigma} \text{ or } ||A_r A_r^\varsigma(\tilde{u}^h(t_2) - \tilde{u}^h(t_1))||_{L^r} \leq 2M_{r,\varsigma} \tag{5.3}$$

of (4.13), from (5.2) and (5.3) we see

$$||A_r^{1+\varsigma'}(\tilde{u}^h(t_2) - \tilde{u}^h(t_1))||_{L^r} \leq (2M_{r,\varsigma})^{1+\varsigma'} N_{r,\varsigma}^\eta|t_2 - t_1|^\eta, \tag{5.4}$$

where $\eta = \varsigma - \varsigma' > 0$. Thus the sequence $(\tilde{u}^h)_{h=\frac{T}{K}}$ is on [0,T] strongly equicontinuous in $D_{A^{1+\varsigma'}}$ for all $0 \leq \varsigma' < \varsigma$.

The imbedding $D_{A_r^{1+\varsigma}} \hookrightarrow D_{A_r^{1+\varsigma'}}$ being compact, from (5.3) and (5.4) by Ascoli-Arzelà's theorem we conclude that the sequence $(\tilde{u}^h)$ is relatively sequentially compact in $D_{A^{1+\varsigma'}}$. Therefore there exists a subsequence $(\tilde{u}^{h'})$ of $(\tilde{u}^h)$ which is on [0,T] uniformly strongly convergent to some v in $A_{A_r^{1+\varsigma'}}$. Since the topology of the latter space is stronger than the topology of the space D_{A_2}, and the whole sequence $(\tilde{u}^h)$ converges in D_{A_2} to the unique Navier-Stokes solution $u \in C^0([0,T], D_{A_2})$, we have $v = u$. Because of the relative sequential compactness of the sequence $(\tilde{u}^h)$ in $D_{A^{1+\varsigma'}}$, the latter conclusion also shows that the whole sequence $(\tilde{u}^h)$ converges strongly in $D_A^{1+\varsigma'}$ to u uniformly on [0,T]. Thus in virtue of the imbeddings $D_{A_r^{1+\varsigma'}} \hookrightarrow D_{A_r} \hookrightarrow H^{2,r}$, Theorem 1.2 has been proved. In addition, interpolation methods lead us easily to explicit convergence rates in $H^{2,r}$. This will be shown elsewhere.

6. APPENDIX

The following basic inequalities apply in the proofs above several times. They are quoted below in the special form in which we use them.

Hölder's inequalities

$$\int_\Omega |f||g|dx \leq ||f||_{L^q} \cdot ||g||_{L^{q'}}, \qquad \frac{1}{q} + \frac{1}{q'} = 1, q \geq 1, \tag{6.1}$$

$$\int_\Omega |f||g||h|dx \leq ||f||_{L^2} \cdot ||g||_{L^3} \cdot ||h||_{L^6}, \tag{6.2}$$

$$||f \cdot g||_{L^p} \leq ||f||_{L^p} \cdot ||g||_{L^{q*}}, \qquad \frac{1}{q} + \frac{1}{q*} = \frac{1}{p}, \qquad q \geq p \geq 1. \tag{6.3}$$

Poincaré-Sobolev's inequality

$$||f||_{L^q} \le c_q ||\nabla f||, \qquad f \in X^q \cap \overset{\circ}{H}^{1,2}, \qquad q \in [2,6]. \tag{6.4}$$

Multiplicative inequalities

$$||u||_{L^q} \le ||\nabla u||_{L^p}^{\alpha} \cdot ||u||_{L^r}^{1-\alpha} \quad \text{if} \quad u \in \overset{\circ}{H}^{1,p} \tag{6.5}$$

(or $u \in H^{1,p}$ and $\int_{\Omega} u(x)dx = 0$),

$$||\nabla u||_{L^q} \le c||\nabla^2 u||^{\alpha}_{L^p} \cdot ||\nabla u||_{L^r}^{1-\alpha} \tag{6.6}$$

with

$$\alpha = (\frac{1}{r} - \frac{1}{q})(\frac{1}{3} - \frac{1}{p} + \frac{1}{r})^{-1} \in [0,1],$$

if $\nabla u \in H^{1,p}$ and $\int_{\Omega} \nabla u dx = 0, 1 < p, r, \Omega \subset \mathrm{I\!R}^3$ open bounded, (Ladyžhenskaya, Solonnikov and Ural'ceva, 1968, p.62-63, Theorem 2.2; Friedman (1969) p. 27, Theorem 10.1).

Moment inequality for any m-sectorial operator A:

$$||A^{\beta} f|| \le c||A^{\gamma} f||^{\frac{\beta-\alpha}{\gamma-\alpha}} \cdot ||A^{\alpha} f||^{\frac{\gamma-\beta}{\gamma-\alpha}}, \alpha \le \beta \le \gamma, \alpha < \gamma, f \in D_{A^{\gamma}}, \tag{6.7}$$

c.p. Friedman (1969), p.159, Theorem 14.1.

REFERENCES

ADAMS, R.A. (1975), *Sobolev Spaces*, (Academic Press, New York).

ASHYRALYEV, A. and P.E. SOBOLEVSKII P.E. (1994), *Well-posedness of parabolic difference equations*, Operator Theory Advances and Applications **69**, (Birkhäuser Basel, Boston, Berlin).

BEALE, J.T. and C. GREENGARD,C. (1992), Convergence of Euler-Stokes splitting of the Navier-Stokes equations, *IBM Research Report* **RC 18072** (79337) 6/11/92 Mathematics 30..

CATTABRIGA, L. (1961), Su un problema al contorno relativo al sistema di equazioni di Stokes, *Rend. Mat. Sem. Univ. Padova* **31**, p. 308-340.

CHORIN, A.J. (1973), Numerical study of slightly viscous flow, *J. Fluid Mechanics* **57**, p. 785-796.

CHORIN, A.J., HUGHES, T.J.R., MC CRACKEN and J.E. MARSDEN (1978), Product formulas and numerical algorithms, *Comm. Pure Appl. Math.* **XXXI**, p. 205-256.

FRIEDMAN, A. (1969), *Partial differential equations*, (Holt, Rinehart and Winston, New York).

FUJITA, H. (1977), On the semidiscrete finite element approximation for the evolution equation $u_t + A(t)u = 0$ of parabolic type, in: MILLER, J.J. (ed.), *Topics in numerical analysis III*, (Academic Press, New York), p. 143-157.

FUJITA, H. and H. MORIMOTO (1970), On fractional powers of the Stokes operator, *Proc. Japan. Acad.* **46**, p. 1141-1143.

FUJIWARA, D. (1969), On the asymptotic behaviour of the Green operators for elliptic boundary problems and the pure imaginary powers of some second order operators, *J. Math. Soc. Japan* **21**, p. 481-521.

FUJIWARA, D. and H. MORIMOTO (1977), An L_r-theorem of the Helmholtz decomposition of vector fields, *J. Fac. Sci. Univ. Tokyo Sect. IA Math* **24**, p. 685-700.

GIGA, Y. (1981a), The Stokes operator in L_r spaces, *Proc. Japan Acad.* **57** Ser. A, p. 85-89.

GIGA, Y. (1981b), Analyticity of the semigroup generated by the Stokes Operator in L_r spaces, *Math. Z.* **178**, p. 297-329.

GIGA, Y. (1985), Domains of fractional powers of the Stokes operators in L_p spaces, *Arch. Rat. Mech. Anal.* **89**, p. 251-265.

GIGA, Y. and T. MIYAKAWA (1985), Solutions in L_r of the Navier-Stokes initial value problem, *Arch. Rational Mech. Anal.* **89**, p. 267-281.

GRUBB, G. (1991), Initial value problems for the Navier-Stokes equations with Neumann conditions, in: HEYWOOD, J.G., MASUDA, K., RAUTMANN, R. AND V.A. SOLONNIKOV (eds.), The Navier-Stokes Equations II, Theory and Numerical Methods, *Lecture Notes in Mathematics* **1530**, (Springer, Berlin), p. 262-283.

KACUR, J. (1985), *Method of Rothe in evolution equations*, (Teubner, Leipzig).

KATO, T. (1990), Liapunov functions and monotonicity in the Navier-Stokes equations, in: FUJITA, H., IKEBE, T. AND S.T. KURODA (eds.), Functional-Analytic Methods for Partial Differential Equations, *Lecture Notes in Mathematics* **1450**, (Springer Berlin).

KOBAYASHI, T. and T. MURAMATU (1991), Abstract Besov space approach to the nonstationary Navier-Stokes equations, *Inst. Math. Univ. of Tsukuba, Preprint*.

KREIN, S.G. (1971), Linear differential equations in Banach spaces, *AMS-Translation of Math. Monographs* Vol. **29**, Providence Rhode Island.

LADYŽHENSKAYA, O.A. (1969), *The mathematical theory of viscous incompressible flow*, Second ed., (Gordon and Breach, New York).

LADYŽHENSKAYA, O.A., SOLONNIKOV, V.A. and N.N. URAL'CEVA (1968), *Linear and quasilinear equations of parabolic type*, (AMS Providence, Rhode Island).

MASUDA, K. (1987), Remarks on compatibility conditions for solutions of Navier-Stokes equations, *J.Fac. Sci. Univ. Tokyo Sect. IA Math.* **34**, p. 155-164.

MIYAKAWA, T. (1981), On the initial value problem for the Navier-Stokes equations in L^p spaces, *Hiroshima Math. J.* **11**, p. 9-20.

PAZY, A. (1983), *Semigroups of linear operators and applications to partial differential equations*, (Springer, Berlin).

PIRONNEAU, O. (1982), On the transport diffusion algorithm and its applications to the Navier-Stokes equations, *Num.Math.* **38**, p. 309-332.

RAUTMANN, R. (1980), On the convergence rate of nonstationary Navier-Stokes approximations, in: Proc. IUTAM Symp. Paderborn 1979, *Springer Lecture Notes in Math.* **771**, p. 425-449.

RAUTMANN, R. (1983a), A semigroup approach to error estimates for nonstationary Navier-Stokes approximations, Proc. Conference Oberwolfach 1982, *Methoden Verfahren Math. Physik* **27**, p. 63-77.

RAUTMANN, R. (1983b), Zur Konvergenz des Rothe-Verfahrens für instationäre Stokes-Probleme in dreidimensionalen Gebieten, *Z. Angew. Math. Mech.* **64**, T 387-388.

RAUTMANN, R. (1983c), On optimum regularity of Navier-Stokes solutions at time $t = 0$, *Math. Z.* **184**, p. 141-149.

RAUTMANN, R. (1988), Ein Vektorpotentialmodell für die Wirbelbildung am Rand umströmter Körper, *Z. Angew. Math. Mech.* **68**, p. 383-387.

RAUTMANN, R. (1989a), Eine konvergente Produktformel für linearisierte Navier-Stokes-Probleme, *Z. Angew. Math. Mech.* **69**, p. 181-183.

RAUTMANN, R. (1989b), A convergent product formula approach to three dimensional flow computations, *Finite Approximations in Fluid Mechanics* **II**, p. 322-325.

RAUTMANN, R. (1992), H^2-convergent linearizations to the Navier-Stokes initial value problem, in: BUTAZZO, G., GALDI, G.P. and L. ZANGHIRATI (eds.), Proc. Intern. Conf. on *New developments in partial differential equations and applications to mathematical physics*, Ferrara 14-18 October 1991, (Plenum Press, New York), p. 135-156.

RAUTMANN, R. (1993a), Optimum regularity of Navier-Stokes solutions at time $t = 0$ and applications, *Acta Mech. supp.* **4**, p. 1-11.

RAUTMANN, R. (1993b), H^2-convergence of Rothe's scheme to the Navier-Stokes equations, *Journal of Nonlinear Analysis* (to appear).

RAUTMANN, R. (1993c), A remark on the convergence of Rothe's scheme to the Navier-Stokes equations (accepted for publication).

RAUTMANN, R. (1994), A direct construction of very smooth local Navier-Stokes solutions, *Acta Appl. Math.* **37**, p. 153-168.

RAUTMANN, R. and K. MASUDA (1994), H^2-convergent approximation schemes to the Navier-Stokes equations, *Comm. Math. Univ. Sancti Pauli* **43**, p. 55-108.

RECTORYS, K. (1982), *The method of discretization in time and partial differential equations*, (Reidel, Dordrecht).

ROTHE, E. (1930), Zweidimensionale parabolische Randwertaufgaben als Grenzfall eindimensionaler Randwertaufgaben, *Math. Ann.* **102**, p. 650-670.

SHINBROT, M. (1973), *Lectures on fluid mechanics*, (Gordon and Breach, New York).

SOLONNIKOV, V.A. (1964), On differential properties of the solutions of the first boundary-value problem for nonstationary of Navier-Stokes equations, *Trudy Mat. Inst. Steklov* **73**, p. 221-291, English transl.: British Library Lending Div., RTS 5211.

SOLONNIKOV, V.A. (1973), Estimates for the solutions of nonstationary Navier-Stokes equations, *Zap. Nauchn. Sem. Leningrad Mat.Steklova* **38**, p. 153-231, English transl: *J. Sov. Math.* **8** (1977), p. 467-529.

STRANG, G. (1969), Approximating semigroups and the consistency of difference schemes, *Proc. Amer. Math. Soc.* **20**, p. 1-7.

TANABE, H. (1979), *Equations of evolution*, (Pitman, London).

TEMAM, R. (1977), *Navier-Stokes equations*, (North-Holland, Amsterdam), rev.ed. 1979, 3rd. rev.ed. 1984.

TRIEBEL, H. (1978), *Interpolation theory, function spaces, differential operators*, (North-Holland, Amsterdam).

VARNHORN, W. (1991), Time stepping procedures for the nonstationary Stokes equations, preprint 1353, Technische Hochschule Darmstadt.

VON WAHL, W. (1985), *The equations of Navier-Stokes and abstract parabolic equations*, (Vieweg Braunschweig).

WEYL, H. (1940), The method of orthogonal projection in potential theory, *Duke Math. J.* **7**, p. 411-444.

Supported by Deutsche Forschungsgemeinschaft.
Paderborn, September 1994

SIMULATION OF MULTIPLE FREE SURFACE TRANSIENT NEWTONIAN AND NON-NEWTONIAN FLOWS

Murilo F. Tome,[1] Brian Duffy,[2] Sean McKee[2]

[1]ICMSC - USP de São Carlos - Depto. de Ciência da Computação e Estat.
Av. Dr. Carlos Botelho, 1465 - Caixa Postal 668
13560-970 - São Carlos - SP - Brazil

[2]University of Strathclyde
Department of Mathematics
Livingstone Tower - 26 Richmond street
Glasgow G1 1XH - Scotland - U. K.

1. INTRODUCTION

The Marker-and-Cell (MAC) method is a finite difference solution technique based on a staggered grid for investigating the dynamics of an incompressible viscous fluid. It employs the primitive variables of pressure and velocity and has particular aplication to the modelling of fluid flows with free surfaces. One of the key features is the use of virtual particles whose coordinates are stored and which move from one cell to the next according to the latest computed velocity. If a cell contains a particle it is considered to contain fluid, thus providing flow visualisation of the free surface. The original ideas date back to 1965 and Harlow and Welch (1965) who wrote the first computer code. There have been a number of developments in the intervening years. The principal authors include Amsden and Harlow (1971), Pracht (1971), Golafshani (1988), Viecelli (1971), McQueen and Ruther (1983) and Markham and Proctor (1983).

More recently, the authors have developed FREEFLOW which is embodied in the computer code GENSMAC. Although based on the ideas of the Marker-and-Cell technique, it has a number of novel features. A user supplied data file of coordinates prescribes the fluid domain which can be quite general and need only be connected. With a view to parallelisation the momentum equations are solved explicitly, but a automatic step-changing routine optimises the stability restriction. A conjugate gradient solver is used to invert the discrete Poisson equation which, due to the concept of a virtual boundary, has always a symmetric and positive definite matrix. An accurate approximation to the stress conditions on the free surface(s) leads to improved visualisation. Finally, the code is written in structured FORTRAN with features from FORTRAN 90. Full details can be found in Tome and McKee, (1994) and Tome, Duffy and McKee, (1994). The

Navier-Stokes Equations and Related Nonlinear Problems
Edited by A. Sequeira, Plenum Press, New York, 1995

aim of this paper is simply to illustrate the effectiveness of this code by applying it to several industrial problems: these include sloshing, injection moulding and dye swelling.

2. A BRIEF DESCRIPTION OF THE CODE

The GENSMAC code solves the two-dimensional time dependent mass and momentum conservation equations for an incompressible viscous fluid which may be Newtonian or non-Newtonian (generalized fluid). For a Newtonian these reduce to the familiar Navier-Stokes equations.

2.1 Basic Equations

The basic equations are the two-dimensional mass and conservation equations which in non-dimensional form can be written as

$$\nabla.\mathbf{u} = 0 \tag{1}$$

$$\frac{\partial \mathbf{u}}{\partial t} = -\nabla p + N(\mathbf{u}) \tag{2}$$

where $N(\mathbf{u})$ has components

$$
\begin{aligned}
N_1 \;=\; & -\frac{\partial(u^2)}{\partial x} - \frac{\partial(uv)}{\partial y} + \frac{1}{Re}\nu(q)\frac{\partial}{\partial y}\left(\frac{\partial u}{\partial y} - \frac{\partial v}{\partial x}\right) \\
& +\frac{1}{Re}\left[2\frac{\partial u}{\partial x}\frac{\partial v}{\partial x} + \left(\frac{\partial u}{\partial y} + \frac{\partial v}{\partial x}\right)\frac{\partial v}{\partial y}\right] + \frac{1}{F_r^2}g_x \;,
\end{aligned}
$$

$$
\begin{aligned}
N_2 \;=\; & -\frac{\partial(uv)}{\partial x} - \frac{\partial(v^2)}{\partial x} - \frac{1}{Re}\nu(q)\frac{\partial}{\partial x}\left(\frac{\partial u}{\partial y} - \frac{\partial v}{\partial x}\right) \\
& +\frac{1}{Re}\left[\left(\frac{\partial u}{\partial y} + \frac{\partial v}{\partial x}\right)\frac{\partial v}{\partial x} + 2\frac{\partial v}{\partial y}\frac{\partial v}{\partial y}\right] + \frac{1}{F_r^2}g_y \;,
\end{aligned}
$$

with $\mathbf{g} = g_x\mathbf{i} + g_y\mathbf{j}$ and

$$
q = \left[2(\frac{\partial u}{\partial x})^2 + 2(\frac{\partial v}{\partial y})^2 + \left(\frac{\partial u}{\partial y} + \frac{\partial v}{\partial x}\right)^2\right]^{1/2} \;,
$$

$Re = UL/\nu_0$ and $F_r = U/\sqrt{Lg}$ denoting the Reynolds number and the Froude number respectively. Note that L and U are the length and velocity scales respectively, ν_0 is a reference viscosity and $g = |\mathbf{g}|$. Further, $\mathbf{u} = (u,v)^T$ are the non-dimensional components of velocity while p is the non-dimensional pressure per unit density. Note that when $\nu(q) = \nu_0$ (a constant) equations (2) reduce to the Navier-Stokes equations.

2.2 Solution procedure

For simplicity of exposition only the solution procedure for Newtonian flow will be described; non-Newtonian flow, although involving essentially the same ideas, is considerably more complicated. A full description can be found in Tome, Duffy and McKee, (1994).

It is supposed that at a given time t_0, the velocity field $\mathbf{u}(\mathbf{x}, t_0)$ is known and boundary conditions for the velocity and pressure are given. The updated velocity field $\mathbf{u}(\mathbf{x}, t)$ at $t = t_0 + \delta t$ is calculated as follows:

1. Let $\tilde{p}(\mathbf{x}, t)$ be a pressure field which satisfies the correct pressure condition on the free surface.

2. Calculate the intermediate velocity field $\tilde{\mathbf{u}}(\mathbf{x}, t)$ from the explicit discretised form of

$$\frac{\partial \tilde{u}}{\partial x} = \left[-\frac{\partial(u^2)}{\partial x} - \frac{\partial(uv)}{\partial y} - \frac{\partial \tilde{p}}{\partial x} + \frac{1}{Re}\frac{\partial}{\partial y}\left(\frac{\partial u}{\partial y} - \frac{\partial v}{\partial x}\right) + \frac{1}{F_r^2}g_x \right]_{t=t_0}$$

$$\frac{\partial \tilde{v}}{\partial y} = \left[-\frac{\partial(uv)}{\partial x} - \frac{\partial(v^2)}{\partial y} - \frac{\partial \tilde{p}}{\partial y} - \frac{1}{Re}\frac{\partial}{\partial x}\left(\frac{\partial u}{\partial y} - \frac{\partial v}{\partial x}\right) + \frac{1}{F_r^2}g_y \right]_{t=t_0}$$

where $\tilde{\mathbf{u}}(\mathbf{x}, t_0) = (u, v)^T$ using the correct boundary conditions for $\mathbf{u}(\mathbf{x}, t_0)$. The intermediate velocity field is then correct by

$$\mathbf{u}(\mathbf{x}, t) = \tilde{\mathbf{u}}(\mathbf{x}, t) - \psi(\mathbf{x}, t)$$

with

$$\nabla^2 \psi(\mathbf{x}, t) = \nabla.\tilde{\mathbf{u}}(\mathbf{x}, t) .$$

Thus, $\mathbf{u}(\mathbf{x}, t)$ now satisfies

$$\nabla.\mathbf{u}(\mathbf{x}, t) = 0 .$$

3. Solve the Poisson equation

$$\nabla^2 \psi(\mathbf{x}, t) = \nabla.\tilde{\mathbf{u}}(\mathbf{x}, t) .$$

4. Compute the velocity

$$\mathbf{u}(\mathbf{x}, t) = \tilde{\mathbf{u}}(\mathbf{x}, t) - \psi(\mathbf{x}, t)$$

5. Compute the pressure

$$p(\mathbf{x}, t) = \tilde{p}(\mathbf{x}, t) + \psi(\mathbf{x}, t)/\delta t$$

2.3 Boundary conditions

Boundary condition must be imposed both on fixed boundaries and on free surfaces. On the fixed boundaries we can impose no-slip, free-slip, prescribed inflow, prescribed outflow and continuative outflow (for details see Tome, 1993).
The appropriate free-surface boundary conditions are the vanishing of the normal and the tangential stresses:

$$p - 2/Re\left[n_x^2 \frac{\partial u}{\partial x} + n_x n_y \left(\frac{\partial u}{\partial y} + \frac{\partial v}{\partial x}\right) + n_y^2 \frac{\partial v}{\partial y}\right] = 0$$

$$\frac{1}{Re}\left[2n_x m_y \frac{\partial u}{\partial x} + (n_x m_y + n_y m_x)\left(\frac{\partial u}{\partial y} + \frac{\partial v}{\partial x}\right) + 2n_y m_x \frac{\partial v}{\partial y}\right] = 0$$

where $\mathbf{n} = (n_x, n_y)^T$ and $\mathbf{m} = (m_x, m_y)^T$ are the normal and tangential directions cosines to the free surface respectively. These conditions are applied by making accurate local finite difference approximations on the free surface (see Tome and McKee, 1994).

The appropriate boundary condition for the Poisson equation (see Amsden and Harlow, 1971) is

$$\frac{\partial \psi}{\partial n} = 0 \quad \text{on fixed boundaries and} \quad \psi = 0 \quad \text{on the free surfaces.}$$

3. ILLUSTRATIVE INDUSTRIAL APPLICATIONS

In this section several test problems with industrial application are solved to illustrate the effectiveness and versatility of the methodology.

3.1 Sloshing

The Petro-chemical industry has considerable interest in the transport of petroleum products. A phenomenon common to both lories and ships is sloshing: this can have both a destabilising effect and can lead to greater evaporation.
This example shows a square block of Newtonian fluid falling under the force of gravity into a circular cavity. No-slip conditions were applied at the cavity wall while free-slip conditions were applied at the other walls. Figure 1 displays the particle configuration at different time intervals and shows the dynamics of the fluid motion as it sloshes back and forth under the influence of gravity. The final picture displays the approaching stationary configuration of the free surface. In this example the Reynolds number is $Re = 6.67$ and the Froude number is $F_r = 1.0$.

a   b

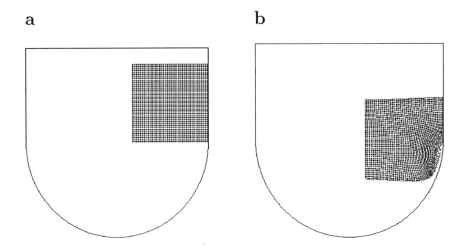

Figure 1. Simulation of sloshing at times $t = 0.00, 1.57, 3.13, 4.70, 7.83, 9.39, 12.53$ and 18.79, respectively.

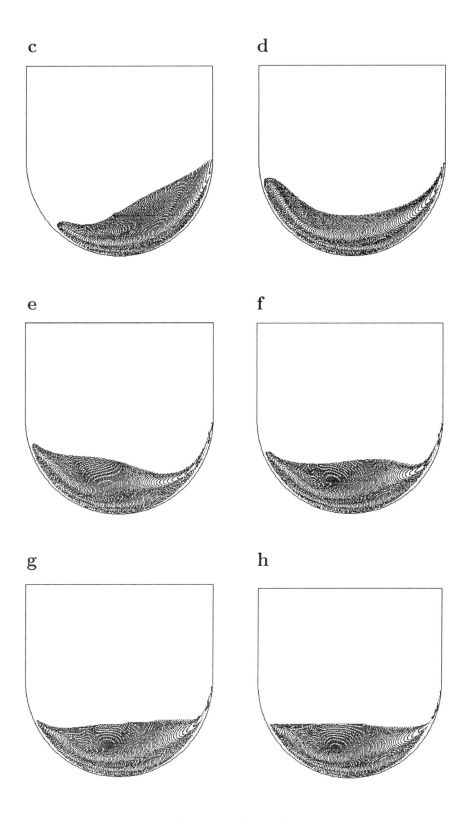

Figure 1. Continued.

3.2 Tub filling

The food industry is concerned with the automatic and rapid filling of containers with a variety of food stuffs with differing rheological properties (eg. yogurt, margerine, etc). This example illustrates the filling process for a Newtonian fluid; the simulation of non-Newtonian fluid, such as the above, does not create any essential difficulties.

Two runs were performed. In the first the inlet velocity is 5.0 m/s and the dynamic viscosity $\mu = 1$ Pa sec. In figure 2 it is observed that this filling speed is too rapid, leading to the possible spillage of the fluid (although it is necessary to be cautious about such statements as the model is only two-dimensional). In the second run the inlet velocity was chosen to be 1.0 m/s with the viscosity unchanged. It is clear from figure 3 that this will lead to even filling.

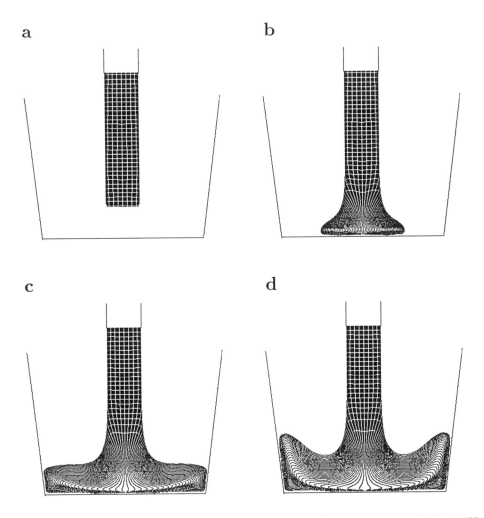

Figure 2. Simulation of tub filling - $\mu = 5$ Pa sec - $U = 5$ m/s, at times $t = 3.7461, 5.4128, 7.9044, 9.5711, 14.4120$ and 19.5505, respectively.

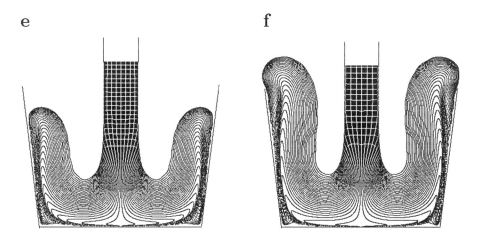

Figure 2. Continued.

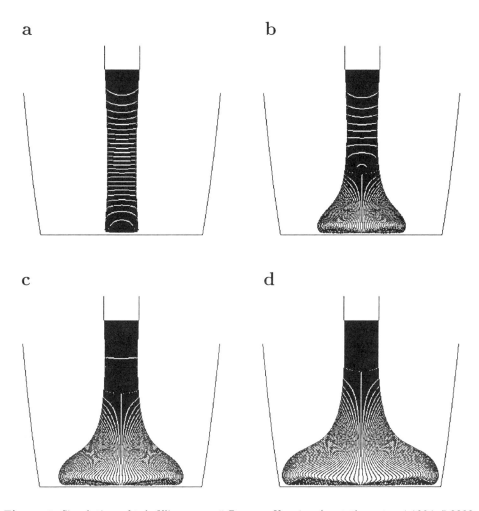

Figure 3. Simulation of tub filling - $\mu = 5$ Pa sec - $U = 1$ m/s, at times $t = 4.1334$, 5.8000, 7.4667, 9.13340, 14.9813 and 19.9785, respectively.

e   f

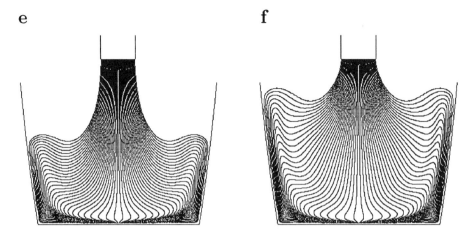

Figure 3. Continued.

3.3 Dye swell

Fibre spinning, in which the extrudate is drawn down under tension from a capillary, is one example of an industrial process that exhibits dye swell. Since the dimensions of the final product must take account of the final thickness it is important for industry to have a thorough understanding of this process.

With this in mind two runs, one for a Newtonian fluid and one for a non- Newtonian fluid, were made. The computation of $\nu(q)$ uses the Cross model (see, for example, Barnes, Hutton and Walters, 1989)

$$\frac{\nu(q) - \nu_\infty}{\nu_0 - \nu_\infty} = \frac{1}{(1 + (Kq)^m)}$$

where the constants were chosen to be:

$$\nu_\infty = 0.001 \ m^2/s$$
$$\nu_0 = 0.050 \ m^2/s$$
$$m = 1.0$$

and the values of K in figure 4 are:

$K = 0.0$ (Figure 4(a))

$K = 0.15$ (Figure 4(b))

$K = 0.75$ (Figure 4(c))

$K = 1.50$ (Figure 4(d))

The flow was started at the left hand side of the capilliary and "Poisseuille" flow was allowed to develop before the extrusion process takes place. Figure 4 displays at single snapshot at (non-dimensional) time $t = 6.250$. The case $K = 0.0$, of course, corresponds to Newtonian flow.

a

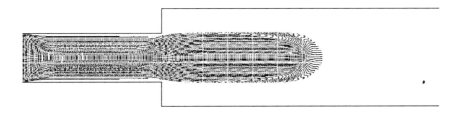

b

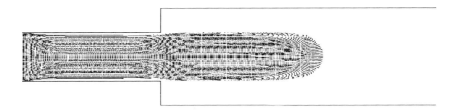

c

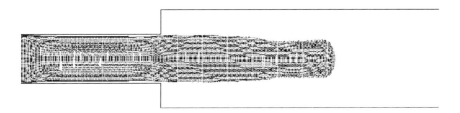

d

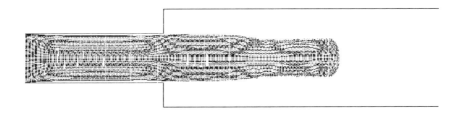

Figure 4. Die-swell.

ACKNOWLEDGMENTS

The authors are grateful to Drs. Crilly and Barrat for invaluable discussions. The first author would like to acknowledge the financial support provided by FAPESP - Fundação de Amparo a Pesquisa do Estado de São Paulo.

REFERENCES

AMSDEN, A. A. and F. H. HARLOW (1971), The SMAC Method: A Numerical Technique for Calculating Incompressible Fluid Flow, Los Alamos Scientific Lab., Report LA-4370, Los Alamos, New Mexico.

BARNES, H. A., J. F. HUTTON and K. WALTERS (1989), An Introduction to Rheology, Elsevier, Amsterdam.

GOLAFSHANI, M. (1988), A Simple Numerical Technique for Transient Creeping Flows with Free Surfaces, *Int. J. Numer. Meth. Fluids*, **8**, 897-912.

HARLOW. H. and A. A. AMSDEN (1971), A Numerical Fluid Dynamics Calculation for All Flow Speeds, *J. Comp. Phys.*, **8**, 197-213.

HARLOW, F. and J. E. WELSH (1965), Numerical Calculation of Time-Dependent Viscous Incompressible Flow of Fluid with a Free Surface, *Phys. Fluids*, **8**, 2182-2189.

MARKHAM, G. and M. V. PROCTOR (1983), Modifications to the two-dimensional incompressible fluid flow code ZUNI to provide enhanced performance, C.E.G.B. Report TPRD/L/0063/M82, Leatherhead, England.

MCQUEEN, J. F. and P. RUTTER (1983), Outline description of a recently implemented fluid flow code ZUNI, C.E.G.B. Report LM/PHYS/258, Leatherhead, England.

MIYATA, H. and S. NISHIMURA (1985), Finite-Difference Simulation of non-Linear Waves Generated by Ships of Arbitrary Three-dimensional Configuration, *J. Comp. Phys.*, **60**, 391-436.

PRACHT, W. E. (1971), Calculating Three-Dimensional Fluid Flows at All Speeds with an Eulerian-Lagrangian Computing Mesh, *J. Comp. Phys.*, **17**, 132-159.

TOME, M. F. (1993), GENSMAC: A Multiple Free Surface Fluid Flow Solver, Ph.D. Thesis, Department of Mathematics, University of Strathclyde, Glasgow.

TOME, M. F., B. R. DUFFY and S. MCKEE (1994), A Numerical Method for Solving Unsteady non-Newtonian Free Surface Flows, submitted for publication.

TOME, M. F. and S. MCKEE (1994), GENSMAC: A Computational Marker and Cell Method for Free Surface Flows in General Domains, *J. Comp. Phys.*, **110**, 171-186

VIECELLI, J. A. (1971), A Computing Method for Incompressible Flows Bounded by Moving Walls, *J. Comp. Phys.*, **8**, 119-143.

INDEX